T0231928

SYNTHESIS OF
ARITHMETIC CIRCUITS

SYNTHESIS OF ARITHMETIC CIRCUITS

FPGA, ASIC, and Embedded Systems

JEAN-PIERRE DESCHAMPS
University Rovira i Virgili

GÉRY JEAN ANTOINE BIOUL
National University of the Center of the Province of Buenos Aires

GUSTAVO D. SUTTER
University Autonoma of Madrid

WILEY-INTERSCIENCE

A JOHN WILEY & SONS, INC., PUBLICATION

Library of Congress Cataloging-in-Publication Data:

Deschamps, Jean-Pierre, 1945-
    Synthesis of arithmetic circuits: FPGA, ASIC and embedded systems/Jean-Pierre Deschamps, Gery Jean Antoine Bioul, Gustavo D. Sutter.
       p. cm.
    ISBN-13 978-0471-68783-2 (cloth)
    ISBN-10 0-471-68783-9 (cloth)
       1. Computer arithmetic and logic units. 2. Digital electronics. 3. Embedded computer systems.
I. Bioul, Gery Jean Antoine. II. Sutter, Gustavo D. III. Title.

    TK7895.A65D47 2006
    621.39'5 - - dc22   2005003237

10   9   8   7   6   5   4   3

CONTENTS

14 Other Arithmetic Operators

15 Circuits for Finite Field Operations

PREFACE

From the beginnings of digital electronic science, the synthesis of circuits carrying out arithmetic operations has been a central topic. As a matter of fact, it is an activity directly related to computer development. From then on, a well-known technical discipline was born: computer arithmetic. Traditionally, the study of arithmetic circuits has been oriented toward applications to general-purpose computers, which provide the most important applications of digital circuits. However, the electronic market share corresponding to specific systems (embedded systems) is significant. It is important to point out that the huge business volume that corresponds to general-purpose computers (personal computers, servers, main frames) is distributed among a relatively reduced number of different models. Therefore the number of designers involved in general-purpose computer development is not as big as it might seem and is much less than the number of engineers dedicated to production and sales. The case of embedded systems is different. Embedded systems are circuits designed for specific applications (special-purpose devices), so a great diversity of products exist in the market, and the design effort per fabricated unit can be a lot bigger than in the case of general-purpose computers. In consequence, the design of specific computers is an activity in which numerous engineers are involved, in all type of companies—even small ones—within numerous countries.

In this book methods and examples for synthesis of arithmetic circuits are described with an emphasis somewhat different from the classic texts on computer arithmetic.

- It is not limited to the description of the arithmetic units of computers.
- Descriptions of computation algorithms are presented in a section apart from the one dedicated to their materialization or implementation by digital circuits. The development of an embedded system is an operation of hardware–software codesign for which it is not known beforehand what tasks will be executed by a microprocessor and what other tasks by specific coprocessors. For this reason, it

appeared useful to describe the algorithms in an independent manner, without any assumption on subsequent executions by an existent processor (software) or by a new customized circuit (hardware).

- A special, although not exclusive, importance has been given to user programmable devices (field programmable devices such as FPGAs), especially to the families Spartan II and Virtex. Those devices are very commonly used for the realization of specific systems, mainly in the case of small series and prototypes. The particular architecture of those components leads the designer to use synthesis techniques somewhat different from the ones applied for ASICs (application-specific integrated circuits) for which standard cell libraries exist.
- In what concern circuits description, logic schemes are presented, sometimes with some VHDL models, in such a way that the corresponding circuits can easily be simulated and synthesized.

After an introductory chapter, the book is divided in two parts. The first one is dedicated to mathematical aspects and algorithms: mathematical background (Chapter 2), number representation (Chapter 3), addition and subtraction (Chapter 4), multiplication (Chapter 5), division (Chapter 6), other arithmetic operations (Chapter 7), and operations in finite fields (Chapter 8). The second part is dedicated to the central topic—the synthesis of arithmetic circuits: hardware platforms (Chapter 9), general principles of synthesis (Chapter 10), adders and subtractors (Chapter 11), multipliers (Chapter 12), dividers (Chapter 13), other arithmetic primitives (Chapter 14), operators for finite fields (Chapter 15), and floating-point unit.

Numerous VHDL models, and other source files, can be downloaded from http:// www.ii.uam.es/~gsutter/arithmetic/. This will be indicated in the text (e.g., *complete VHDL source code available*). As regards the VHDL models, they are of two types: some of them have been developed for simulation purposes only, so the working of the corresponding circuit can be observed; others are synthesizable models that have been implemented within commercial programmable components (FPGA's).

The authors thank the people who have helped them in developing this book, especially Dr. Tim Bratten, for correcting the text, and Paula Mirón, for the cover design. They are grateful to the following universities for providing them the means for carrying this work through to a successful conclusion: University Rovira i Virgili (Tarragona, Spain), University Rey Juan Carlos (Madrid, Spain), State University UNCPBA (Tandil, Argentina), University FASTA (Mar del Plata, Argentina), and Autonomous University of Madrid (Spain).

JEAN-PIERRE DESCHAMPS
University Rovira i Virgili

GÉRY JEAN ANTOINE BIOUL
National University of the Center of the Province of Buenos Aires

GUSTAVO D. SUTTER
University Autonoma of Madrid

ABOUT THE AUTHORS

Jean-Pierre Deschamps received a MS degree in electrical engineering from the University of Louvain, Belgium, in 1967, a PhD in computer science from the Autonomous University of Barcelona, Spain, in 1983, and a PhD degree in electrical engineering from the Polytechnic School of Lausanne, Switzerland, in 1984. He has worked in several companies and universities. He is currently a professor at the University Rovira i Virgili, Tarragona, Spain. His research interests include ASIC and FPGA design, digital arithmetic, and cryptography. He is the author of six books and about a hundred international papers.

Géry Jean Antoine Bioul received a MS degree in physical aerospace engineering from the University of Liège, Belgium. He worked in digital systems design with PHILIPS Belgium and in computer-aided industrial logistics with several Fortune-100 U.S. companies in the United States, and Africa. He has been a professor of computer architecture in several universities mainly in Africa and South America. He is currently a professor at the State University UNCPBA of Tandil (Buenos Aires), Argentina, and a professor consultant at the Saint Thomas University FASTA of Mar del Plata (Buenos Aires), Argentina. His research interests include logic design and computer arithmetic algorithms and implementations. He is the author of about 50 international papers and patents on fast arithmetic units.

Gustavo D. Sutter received a MS degree in Computer Science from the State University UNCPBA of Tandil (Buenos Aires), Argentina, and a PhD degree from the Autonomous University of Madrid, Spain. He has been a professor at the UNCPBA, Argentina and is currently a professor at the University Autonoma of Madrid, Spain. His research interests include ASIC and FPGA design, digital arithmetic, and development of embedded systems. He is the author of about 30 international papers and communications.

1

INTRODUCTION

The design of embedded systems, that is, circuits designed for specific applications, is based on a series of decisions as well as on the use of several types of development techniques. For example:

- Selection of the data representation
- Generation or selection of algorithms
- Selection of hardware platforms
- Hardware–software partitioning
- Program generation
- New hardware synthesis
- Cosimulation, coemulation, and prototyping

Some of these activities have a close relationship with the study of arithmetic algorithms and circuits, especially in the case of systems including a great amount of data processing (e.g., ciphering and deciphering, image processing, digital signature, biometry).

1.1 NUMBER REPRESENTATION

When using general-purpose equipment, the designer has few possible choices concerning the internal representation of data. He must conform to some fixed

Synthesis of Arithmetic Circuits: FPGA, ASIC, and Embedded Systems
By Jean-Pierre Deschamps, Géry J. A. Bioul, and Gustavo D. Sutter
Copyright © 2006 John Wiley & Sons, Inc.

and predefined data types such as *integer*, *floating-point*, *double precision*, and *character*. On the contrary, if a specific system is under development, the designer can choose, for each data, the most convenient type of representation. It is no longer necessary to choose some standard fixed-point or floating-point numeration system. Nonstandard specific formats can be used. In Chapter 3 the main number representation methods will be defined.

1.2   ALGORITHMS

Every complex data processing operation must be decomposed into simpler operations — the computation primitives — executable either by the main processor or by some specific coprocessor. The way the computation primitives are used in order to perform the complex operation is what is meant by *algorithm*. Obviously, knowledge of algorithms is of fundamental importance for developing arithmetic procedures (software) and circuits (hardware). It is the topic of Chapters 4–8.

1.3   HARDWARE PLATFORMS

The selection of a hardware platform is based on the answer to the following question. How do we get the desired behavior at the lowest cost, while fulfilling some additional constraints? As a matter of fact, the concept of cost must be carefully defined in each particular case. It can cover several aspects: for example, the unit production cost, the nonrecurring engineering costs, and the implicit cost for a late introduction of the product to the market. Some examples of additional technical constraints are the size of the system, its power consumption, and its reliability and maintainability.

For systems requiring little data processing capability, *microcontrollers* and low-range *microprocessors* can be the best choice. If the computation needs are greater, more powerful microprocessors, or even *digital signal processors* (DSPs), should be considered. This type of solution (microprocessors and DSPs) is very flexible as the development work mainly consists in generating programs.

For getting higher performances, it may be necessary to develop specific circuits. A first option is to use a programmable device, for example, a *field-programmable gate array* (FPGA). It could be an interesting option for prototypes and small series. For greater series, an *application-specific integrated circuit* (ASIC) should be developed. ASIC vendors offer several types of products: for example, *gate arrays*, with relatively small prototyping costs, or *standard cell libraries*, integrating a complete *system-on-chip* (SOC) including processors, program memories, data memories, logic, macrocells, and analog interfaces.

A brief presentation of the most common hardware platforms is given in Chapter 9.

1.4 HARDWARE–SOFTWARE PARTITIONING

The hardware–software partitioning consists of deciding which operations will be executed by the central processing unit (the software) and which ones by specific coprocessors (the hardware). As a matter of fact, the platform selection and the hardware–software partitioning are tightly related operations. For systems requiring little data processing capability, the whole system is implemented in software. If higher performances are necessary, the noncritical operations, as well as control of the operation sequence, are executed by the central processing unit, while the critical ones are implemented within specific coprocessors.

1.5 SOFTWARE GENERATION

The operations belonging to the software block of the chosen partition must be programmed. In Chapters 4–8 the algorithms are presented in an Ada-like language that can easily be translated to C or even to the assembly language of the chosen microprocessor.

1.6 SYNTHESIS

Once the hardware–software partition has been defined, all the tasks assigned to the specific hardware (FPGA, ASIC) must be translated into circuit descriptions. Some important synthesis principles and methods are described in Chapter 10. The synthesis of arithmetic circuits, based on the algorithms of Chapters 4–8, is the topic of Chapters 11–15, and an additional chapter (16) is dedicated to the implementation of floating-point arithmetic.

1.7 A FIRST EXAMPLE

Common examples of application fields resorting to embedded solutions are cryptography, access control, smart cards, automotive, avionics, space, entertainment, and electronic sales outlets. In order to illustrate the main steps of the design process, a small digital signature system will now be developed (complete assembly language and VHDL code available).

1.7.1 Specification

The system under development (Figure 1.1) has three inputs,

- `character` is an 8-bit vector.
- `new_character` is a signal used for synchronizing the input of successive characters.
- `sign` is a control signal ordering the computation of a *digital signature*.

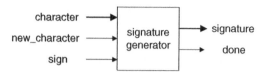

Figure 1.1   System under development.

and two outputs,

- done is a status variable indicating that the signature computation has been completed,
- signature is a 32-bit vector, namely, the signature of the message.

The working of the system is shown in Figure 1.2: a sequence $c_1, c_2, \ldots, c_n$ of any number n of characters (the message), synchronized by the signal new_character, is inputted. When the sign control signal goes high, the done flag is lowered and the signature of the message is computed. The done flag will be raised as soon as the signature s is available.

In order to sign the message two functions must be defined:

- a hash function associating a 32-bit vector (the summary) to every message, whatever its length;
- an encode function computing the signature corresponding to the summary.

The following (naive) hash function is used:

Algorithm 1.1   Hash Function

```
summary:=0;
while not(end_of_message) loop
   get(character);
   a:=(summary(7 downto 0)+character) mod 256;
   summary(23 downto 16):=summary(31 downto 24);
   summary(15 downto 8):=summary(23 downto 16);
```

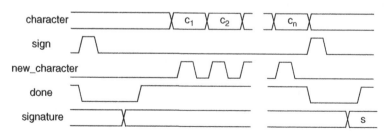

Figure 1.2   Input and output signals.

```
    summary(7 downto 0):=summary(15 downto 8);
    summary(31 downto 24):=a;
end loop;
```

As an example, assume that the message is the following (every character can be equivalently considered as an 8-bit vector or a natural number smaller than 256, i.e. a base-256 digit; see Chapter 3):

$$12, 45, 216, 1, 107, 55, 10, 9, 34, 72, 215, 114, 13, 13, 229, 18.$$

The summary is computed as follows:

$$summary = (0, 0, 0, 0),$$
$$summary = (12, 0, 0, 0),$$
$$summary = (45, 12, 0, 0),$$
$$summary = (216, 45, 12, 0),$$
$$summary = (1, 216, 45, 12),$$
$$summary = (119, 1, 216, 45),$$
$$summary = (100, 119, 1, 216),$$
$$summary = (226, 100, 119, 1),$$
$$summary = (10, 226, 100, 119),$$
$$summary = (153, 10, 226, 100),$$
$$summary = (172, 153, 10, 226),$$
$$summary = (185, 172, 153, 10),$$
$$summary = (124, 185, 172, 153),$$
$$summary = (166, 124, 185, 172),$$
$$summary = (185, 166, 124, 185),$$
$$summary = (158, 185, 166, 124),$$
$$summary = (142, 158, 185, 166).$$

The final result, translated from the base-256 to the decimal representation, is

$$summary = 142 \times 256^3 + 158 \times 256^2 + 185 \times 256 + 166 = 2392766886.$$

The encode function computes

$$encode(y) = y^x \bmod m$$

x being some *private key*, and m a 32-bit number. Assume that

$$x = 1937757177 \quad \text{and} \quad m = 2^{32} - 1 = 4294967295.$$

Then the signature of the previous message is

$$s = (2392766886)^{1937757177} \bmod 4294967295 = 37998786.$$

1.7.2 Number Representation

In this example all the data are either 8-bit vectors (the characters) or 32-bit vectors (the summary, the key, and the module m). So instead of representing them in the decimal numeration system, they should be represented in the binary or, equivalently, the hexadecimal system. The message is

0C, 2D, D8, 01, 6B, 37, 0A, 09, 22 48, D7, 72, 0D, 0D, E5, 12.

The summary, the key, the module, and the signature are

$$summary = 8E9EB9A6,$$
$$private\ key = 737FD3F9,$$
$$m = \text{FFFFFFFF},$$
$$s = 0243D0C2.$$

1.7.3 Algorithms

The hash function amounts to a mod-256 addition, that is, a simple 8-bit addition without output carry. The only complex operation is the mod m exponentiation.

Assume that x, y, and m are n-bit numbers. Then

$$x = x(0) + 2.x(1) + \cdots + 2^{n-1}.x(n-1),$$

and e can be written in the form

$$e = ((\cdots((1^2.y^{x(n-1)})^2.y^{x(n-2)})^2 \cdots)^2.y^{x(1)})^2.y^{x(0)} \bmod m.$$

The corresponding algorithm is the following (Chapter 8, Algorithm 8.14).

Algorithm 1.2   Exponentiation

```
e:=1;
for i in 1..n loop
  e:=(e*e) mod m;
  if x(n-i)=1 then e:=(e*y) mod m; end if;
end loop;
```

The only computation primitive is the modulo m product, which, in turn, is equivalent to a natural multiplication followed by a modulo m reduction, that is, an integer division by m. The following algorithm (Chapter 8, Algorithm 8.5) computes $r = x.y \bmod m$. It uses two procedures: *multiply*, which computes the product z of two natural numbers x and y, and *divide*, which generates q (the quotient) and r (the remainder) such that $z = q.m + r$ with $r < m$.

Algorithm 1.3   Modulo m Multiplication

```
multiply (x, y, z);
divide (z, m, q, r);
```

A classical method for computing the product z of two natural numbers x and y is the *shift and add* algorithm (Chapter 5, Algorithm 5.3). In base 2:

Algorithm 1.4   Natural Multiplication

```
p(0):=0;
for i in 0..n-1 loop
    p(i+1):=(p(i)+x(i)*y)/2;
end loop;
z:=p(n)*(2**n);
```

For computing q and r such that $z = q.m + r$ with $r < m$, the classical restoring division algorithm can be used (Chapter 6, Algorithms 6.1 and 6.2). Given x and y (the operands) such that $x < y$, and p (the desired precision), the restoring division algorithm computes q and r such that

$$x.2^p = q.y + r. \tag{1.1}$$

Within the exponentiation algorithm 1.2, the operands e and y are n-bit numbers. Furthermore, e is always smaller than m, so that both products $z = e * e$ or $z = e * y$ are $2.n$-bit numbers satisfying the relation

$$z < m.2^n.$$

Thus by substituting x by z, p by n, and y by $m.2^n$ in (1.1), the restoring division algorithm computes q and r' such that

$$z.2^n = q.(m.2^n) + r' \quad \text{with} \quad r' < m.2^n,$$

that is,

$$z = q.m + r \quad \text{with} \quad r = r'.2^{-n} < m.$$

The restoring algorithm is similar to the pencil and paper method. At every step the latest obtained remainder, say, $r(i-1)$, is multiplied by 2 and compared with the divider y. If $2.r(i-1)$ is greater than or equal to y, then the new remainder is

$r(i) = 2.r(i-1) - y$ and the corresponding quotient bit is equal to 1. In the contrary case, the new remainder is $r(i) = 2.r(i-1)$ and the corresponding quotient bit equal to 0. The initial remainder $r(0)$ is the dividend.

Algorithm 1.5   Restoring Division

```
r(0):=z; y:=m*(2**n);
for i in 1..n loop
  if 2*r(i-1)-y<0 then q(i):=0; r(i):=2*r(i-1); else
  q(i):=1; r(i):=2*r(i-1)-y; end if;
end loop;
r:=r(n)/(2**n);
```

By merging Algorithms 1.4 and 1.5, the following modular product algorithm is obtained.

Algorithm 1.6   Modular Product

```
p(0):=0;
for i in 0..n-1 loop
  p(i+1):=(p(i)+x(i)*y)/2;
end loop;
r(0):=p(n)*(2**n); y:=m*(2**n);
for i in 1..n loop
  if 2*r(i-1)-y<0 then q(i):=0; r(i):=2*r(i-1); else
  q(i):=1; r(i):=2*r(i-1)-y; end if;
end loop;
r:=r(n)/(2**n);
```

Observe that the multiplication of $p(n)$ and m by 2^n, as well as the division of $r(n)$ by 2^n can be deleted. Then $r(0) = p(n)$ is a $2.n$-bit fixed-point number (Chapter 3) smaller than 2^n and the divider is equal to m. The quotient q and the remainder $r(n)$ satisfy the relation $p(n).2^n = q.m + r(n)$ so that $r = r(n)$.

1.7.4   Hardware Platform

For implementing this illustrative example, a prototyping board will be used, namely, an XSA-100 board from XESS Corporation. It includes an XC2S100 FPGA (Spartan-II family of Xilinx) integrating the complete digital signature system. The design environment includes virtual components (synthesizable VHDL models, Chapter 9), among others PicoBlaze, an 8-bit microprocessor, and its program memory ([XIL2002]).

1.7.5   Hardware–Software Partitioning

As mentioned above, the only complex operation is the computation of y^x modulo m. All the other operations can be carried out by the processor. The corresponding system architecture is shown in Figure 1.3. It works as follows:

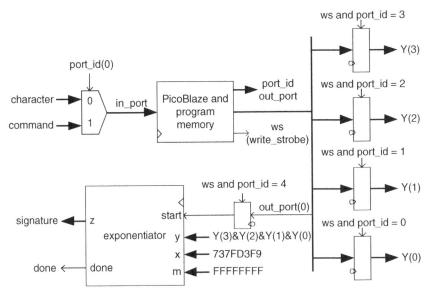

Figure 1.3 System architecture.

- PicoBlaze reads the `character` input at address 0 and the `command` input at address 1, where

  `command = 0 0 0 0 0 0 sign new_character`.

- It computes the 32-bit summary and writes it, under the form of four separate bytes,

  `summary = Y(3) Y(2) Y(1) Y(0)`,

  into four registers whose addresses are 3, 2, 1 and 0, respectively.

- A specific coprocessor receives the `start` signal from PicoBlaze at address 4, computes

  $s = (summary)^{737FD3F9} \bmod FFFFFFFF$,

  and generates the `done` flag.

1.7.6 Program Generation

The program executed by PicoBlaze is made up of three parts (assembly language code available):

- reading of the `new_character` and `sign` input signals,
- reading of the `character` input and updating of the `summary`,
- writing of the `summary` and of the `start` command within the interface registers:

```
summary:=(0, 0, 0, 0);
start:=0;
loop
  --wait for command=0
  while command>0 loop null; end loop;
  --wait for command=1 (new_character) or 2 (sign)
  while command=0 loop null; end loop;
  if command=1 then
    a:=(summary(0)+character) mod 256;
    summary(0):=summary(1);
    summary(1):=summary(2);
    summary(2):=summary(3);
    summary(3):=a;
  elsif command=2 then
    Y(3):=summary(3);
    Y(2):=summary(2);
    Y(1):=summary(1);
    Y(0):=summary(0);
    start:=1;
    summary:=(0, 0, 0, 0);
    start:=0;
  end if;
end loop;
```

1.7.7 Synthesis

The synthesis (complete VHDL code available) of the *exponentiator* block of Figure 1.3 is based on the algorithms of Section 1.7.3. A summary of the main principles for translating an algorithm to a circuit is given in Chapter 10. The data path of Figure 1.4 allows executing Algorithm 1.2. It includes:

- two 32-bit registers: a parallel register storing e, and a loadable shift register, initially storing x and allowing to successively read the value of $x(n-1)$, $x(n-2), \ldots , x(0)$;
- a mod m multiplier with a start input signal and a done output flag;
- a 32-bit 2-to-1 multiplexer selecting either e or y as the second multiplier operand.

The complete circuit is described by the following VHDL model (including the control unit):

```
entity exponentiator is
port (
  x, y, m: in std_logic_vector(n-1 downto 0);
  z: inout std_logic_vector(n-1 downto 0);
  clk, reset, start: in std_logic;
  done: out std_logic
);
end exponentiator;
```

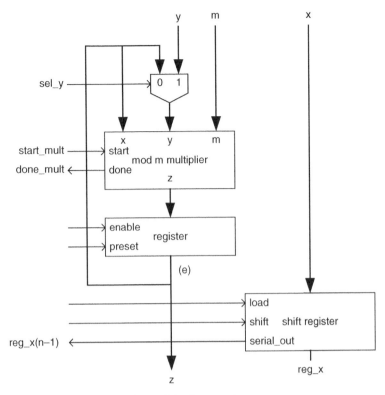

Figure 1.4 Exponentiator.

```
architecture circuit of exponentiator is
  component sequential_mod_mult..end component;
  signal start_mult, sel_y, done_mult: std_logic;
  signal reg_x, input_y, output_z: std_logic_vector(n-1 downto
  0);
  subtype step_number is natural range 0 to n;
  signal count: step_number;
  subtype internal_states is natural range 0 to 14;
  signal state: internal_states;
begin
  label_1: sequential_mod_mult port map(z, input_y, m,
  output_z, clk, reset, start_mult, done_mult);
  with sel_y select input_y<=z when '0', y when others;
  process (clk, reset)
  begin
    if reset='1' then
      state<=0; done<='0'; start_mult<='0'; count<=0;
    elsif clk'event and clk='1' then
      case state is
        when 0=>if start='0' then state<=state+1; end if;
```

```
              when 1=>if start='1' then state<=state+1; end if;
              when 2=>z<=conv_std_logic_vector(1, n);
                reg_x<=x; count<=0; done<='0'; state<=state+1;
              when 3=>
                sel_y<='0'; start_mult<='1'; state<=state+1;
              when 4=>state<=state+1;
              when 5=>start_mult<='0'; state<=state+1;
              when 6=>
                if done_mult='1' then state<=state+1; end if;
              when 7=>z<=output_z;
                if reg_x(n-1)='1' then state<=state+1;
                else state<=13; end if;
              when 8=>
                sel_y<='1'; start_mult<='1'; state<=state+1;
              when 9=>state<=state+1;
              when 10=>start_mult<='0'; state<=state+1;
              when 11=>
                if done_mult='1' then state<=state+1; end if;
              when 12=>z<=output_z; state<=state+1;
              when 13=>reg_x(0)<=reg_x(n-1);
                for i in 1 to n-1 loop reg_x(i)<=reg_x(i-1);
                end loop;
                count<=count+1; state<=state+1;
              when 14=>
                if count>=n then done<='1'; state<=0;
                else state<=3; end if;
          end case;
        end if;
    end process;
end circuit;
```

1.7.8 Prototype

All the files (complete source files available) necessary for programming an XSA-100 board are included in the file section1_7.zip:

- exponentiator.vhd is the complete description of the exponentiation circuit (including the modular multiplier model);
- signatu.psm is the assembly language program;
- kpcsm.vhd is the PicoBlaze model;
- signatu.vhd is the program memory model generated from the assembly language program with kcpsm.exe (the PicoBlaze assembler released by Xilinx [XIL2002]).

In order to test the complete system, the circuit of Figure 1.5 has been synthesized. It is made up of:

- the circuit of Figure 1.3 including PicoBlaze, its program memory, the interface registers, and the *exponentiator*;

- a finite state machine generating the commands and characters corresponding to the example of Section 1.7.1;
- a circuit that interfaces the board with signals d(7..0) controllable from the host computer ([XSA2002]):

  d(7) cannot be used,

  d(3..0) are used for selecting one of the outputs (out_0 to out_15) or inputs (in_0 to in_15),

  d(6..4) are control signals,

d(6.4)	Command
000	nop
001	write
010	read
011	reset
100	address strobe

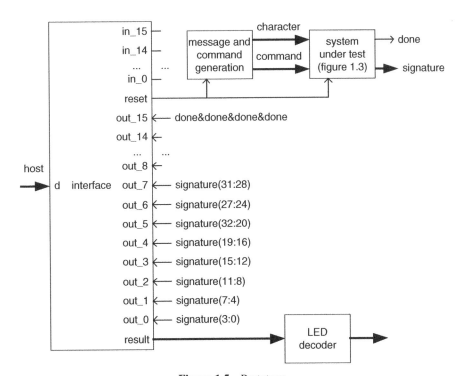

Figure 1.5 Prototype.

in this application the `write` and `address strobe` commands are not used; when the `read` command is active, the hexadecimal representation of the 4-bit vector selected with `d(3..0)` is displayed on the LED of the board;

- the 7-segment LED decoder.

The VHDL model of the circuit of Figure 1.5 (`firma.vhd`) is also included in *section*1_*7.zip* as well as the file describing the pin assignment (`pines.ucf`). The whole system (Figure 1.5) can be synthesized with ISE, the synthesis program of Xilinx, and downloaded to the XSA-100 board.

1.8 BIBLIOGRAPHY

[DEM2002] G. De Micheli, R. Ernst, and W. Wolf (eds.), *Readings in Hardware/Software Co-Design*, Morgan Kaufmann Publishers, San Francisco, CA, 2002.

[GAJ1994] D. D. Gajski, F. Vahid, S. Narayan, and J. Gong, *Specification and Design of Embedded Systems*, Prentice Hall, Englewood Cliffs, NJ, 1994.

[XIL2002] Xilinx Inc, *PicoBlaze 8-bit Microcontroller for Virtex-E and Spartan-II/IIE Devices*, application note XAPP213 (v2.0), Dec. 2002; `http://www.xilinx.com`

[XSA2002] *XSA Board V1.1, V1.2 User Manual*, June 2002; `http://www.xess.com`.

2

MATHEMATICAL BACKGROUND

This chapter presents some topics in mathematics; it is intended to make this book self-contained. For further details the reader is referred to textbooks on algebra ([COH1993], [GIL2003], [HER1975], [HUN1974]), mathematical analysis ([APO1974], [RUD1976]), number theory ([KOB1994], [ROS1992]), finite fields ([McC1987]), and cryptography ([MEN1996]).

2.1 NUMBER THEORY

2.1.1 Basic Definitions

Definitions 2.1

1. The set of natural numbers[1] $N = \{0, 1, 2, 3, \ldots\}$.
2. The set of integers $Z = \{ \ldots, -3, -2, -1, 0, 1, 2, 3, \ldots \}$.

Definition 2.2 Given two integers x and y, y *divides* x (y is a *divisor* of x) if there exists an integer z such that $x = z.y$.

---

[1]For convenience, the element *zero* has been included in N.

Synthesis of Arithmetic Circuits: FPGA, ASIC, and Embedded Systems
By Jean-Pierre Deschamps, Géry J. A. Bioul, and Gustavo D. Sutter
Copyright © 2006 John Wiley & Sons, Inc.

Definition 2.3   Given two integers x and y, with $y > 0$, there exist two integers q (the *quotient*) and r (the *remainder*) such that

$$x = q.y + r, \quad \text{where} \quad 0 \le r < y.$$

It can be proved that q and r are unique. Then (notation)

$$r = x \bmod y, \quad q = x \text{ div } y.$$

An alternative definition is the following.

Definition 2.4 (Integer Division)   Given two integers x and y, with $y > 0$, there exist two integers q (the *quotient*) and r (the *remainder*) such that

$$x = q.y + r, \quad \text{where} \quad 0 \le r < y \text{ if } x \ge 0 \quad \text{and} \quad -y < r \le 0 \text{ if } x < 0.$$

It can be proved that q and r are unique. Then (notation)

$$r = x \text{ rem } y, \quad q = x/y.$$

Examples 2.1

1. $x = -16, y = 3$:

$$-16 \bmod 3 = 2, \ -16 \text{ div } 3 = -6, \ -16 = -6.3 + 2,$$
$$-16 \text{ rem } 3 = -1, \ -16/3 = -5, \ -16 = -5.3 + (-1)$$

2. $x = -15, y = 3$:

$$-15 \bmod 3 = 0, \ -15 \text{ div } 3 = -5, \ -15 = -5.3 + 0,$$
$$-15 \text{ rem } 3 = 0, \ -15/3 = -5, \ -15 = -5.3 + 0.$$

Definitions 2.5

1. Given two integers x and y, z is the *greatest common divisor* of x and y if

   z is a natural number (nonnegative integer),
   z divides both x and y,
   any other common divider of x and y is also a divider of z.

   Notation: $z = gcd(x, y)$.
2. Given two integers x and y, they are said to be *relatively prime* if $gcd(x, y) = 1$.
3. An integer $p > 1$ is said to be *prime* if its only positive divisors are 1 and p.

2.1.2 Euclidean Algorithms

Given two natural numbers x and y, the *Euclidean algorithm* for natural numbers computes $gcd(x, y)$. It is based on a series of integer divisions:

$$r(i - 1) = q(i).r(i) + r(i + 1), \quad \text{where} \quad 0 \leq r(i + 1) < r(i).$$

Observe that any divider of $r(i - 1)$ and $r(i)$ is also a divider of $r(i)$ and $r(i + 1)$ so that

$$gcd(r(i - 1), r(i)) = gcd(r(i), r(i + 1)).$$

Initially,

$$r(0) = x \quad \text{and} \quad r(1) = y.$$

Then compute

$$r(0) = q(1).r(1) + r(2),$$
$$r(1) = q(2).r(2) + r(3),$$
$$r(2) = q(3).r(3) + r(4),$$
$$\cdots$$
$$r(n - 3) = q(n - 2).r(n - 2) + r(n - 1),$$
$$r(n - 2) = q(n - 1).r(n - 1) + r(n),$$

where $r(1) > r(2) \cdots > r(n) = 0$ and $gcd(r(i - 1), r(i)) = gcd(r(i), r(i + 1))$, so that

$$gcd(x, y) = gcd(r(0), r(1)) = \cdots = gcd(r(n - 1), r(n)) = gcd(r(n - 1), 0)$$

$$= r(n - 1).$$

Example 2.2   Let $r(0) = x = 8580$; $r(1) = y = 4070$;

$$8580 = 2.4070 + 440$$
$$4070 = 9.440 + 110$$
$$440 = 4.110 + 0$$

Then $gcd(8580, 4070) = 110$.

In the *extended Euclidean algorithm* a series of coefficients $b(i)$ and $c(i)$ are calculated in parallel with the computation of $r(0)$, $r(1)$, $r(2)$, ..., $r(n)$:

$$b(0) = 1, \qquad\qquad\qquad c(0) = 0,$$
$$b(1) = 0, \qquad\qquad\qquad c(1) = 1,$$
$$b(2) = b(0) - b(1).q(1), \qquad c(2) = c(0) - c(1).q(1),$$
$$\cdots \qquad\qquad\qquad\qquad \cdots$$
$$b(n-1) = b(n-3) - b(n-2).q(n-2), \quad c(n-1) = c(n-3) - c(n-2).q(n-2)$$

It can be demonstrated by induction that

$$r(i) = b(i).x + c(i).y, \quad \forall\, i = 0, 1, 2, \ldots, n-1.$$

In particular,

$$gcd(x,y) = r(n-1) = b(n-1).x + c(n-1).y.$$

In conclusion, the extended Euclidean algorithm expresses the greatest common divisor z of two natural numbers x and y as a linear combination of x and y, that is,

$$z = b.x + c.y. \tag{2.1}$$

Algorithm 2.1   Extended Euclidean Algorithm

```
if x=0 then z:=y; b:=0; c:=1;
elsif y=0 then z:=x; b:=1; c:=0;
else
   r_i:=x; r_iplus1:=y; b_i:=1; c_i:=0; b_iplus1:=0;
     c_iplus1:=1;
   while r_iplus1>0 loop
      q:=r_i/r_iplus1; r_iplus2:=r_i mod r_iplus1;
      b_iplus2:=b_i-b_iplus1*q; c_iplus2:=c_i-c_iplus1*q;
      r_i:=r_iplus1; r_iplus1:=r_iplus2;
      b_i:=b_iplus1; b_iplus1:=b_iplus2;
      c_i:=c_iplus1; c_iplus1:=c_iplus2;
   end loop;
   z:=r_i; b:=b_i; c:=c_i;
end if;
```

Example 2.3   Let $r_i = x = 230490$; $r_{i+1} = y = 43290$; $b_i = c_{i+1} = 1$; $b_{i+1} = c_i = 0$.

Step 1:

$$q = 230490/43290 = 5; \ r_{i+2} = 230490 \bmod 43290 = 14040;$$
$$b_{i+2} = 1 - 0 * 5 = 1; \ c_{i+2} = 0 - 1 * 5 = -5;$$
$$r_i = 43290; \quad r_{i+1} = 14040;$$
$$b_i = 0; \quad b_{i+1} = 1;$$
$$c_i = 1; \quad c_{i+1} = -5;$$

Step 2:

$$q = 43290/14040 = 3; \ r_{i+2} = 43290 \bmod 14040 = 1170;$$
$$b_{i+2} = 0 - 1 * 3 = -3; \ c_{i+2} = 1 + 5 * 3 = 16;$$
$$r_i = 14040; \ r_{i+1} = 1170;$$
$$b_i = 1; \quad b_{i+1} = -3;$$
$$c_i = -5; \quad c_{i+1} = 16;$$

Step 3:

$$q = 14040/1170 = 12; \quad r_{i+2} = 14040 \bmod 1170 = 0;$$
$$b_{i+2} = 1 + 3 * 12 = 37; \quad c_{i+2} = -5 - 16 * 12 = -197;$$
$$r_i = 1170; \quad r_{i+1} = 0;$$
$$b_i = -3; \quad b_{i+1} = 37;$$
$$c_i = 16; \quad c_{i+1} = -197;$$

$$b = b_i = -3; \quad c = c_i = 16; \quad gcd(230490, 432900) = z = r_i = 1170$$
$$= -3 * 230490 + 16 * 43290$$

2.1.3 Congruences

Definition 2.6 Given two integers x and y, and a positive integer n, x is *congruent to y modulo n* if n divides the difference $(x - y)$.

Notation:

$$x \equiv y(\bmod \ n).$$

Property 2.1 (Basic Properties of Congruences)
1. $x \equiv y \ (\bmod \ n)$ if and only if $(x \bmod n) = (y \bmod n)$ (Definition 2.3).
2. The relation $x \equiv y \ (\bmod \ n)$ is an equivalence relation (reflexive, symmetric, and transitive).

3. If $x_1 \equiv y_1$ (mod n) and $x_2 \equiv y_2$ (mod n), then

$$(x_1 + x_2) \equiv (y_1 + y_2)(\text{mod } n), \quad (x_1 - x_2) \equiv (y_1 - y_2)(\text{mod } n),$$
$$(x_1.x_2) \equiv (y_1.y_2)(\text{mod } n). \tag{2.2}$$

From Properties 2.1(1 and 2), it can be seen that the mod n congruence relation partitions Z into n equivalence classes. Each equivalence class contains exactly one element of the set $\{0, 1, 2, \ldots, n - 1\}$, namely, the common value (x mod n) for all elements x of the class. Furthermore, according to Property 2.1(3), the addition, subtraction, and multiplication of congruence classes can be defined. As a matter of fact, the set of equivalence classes is isomorphic to

$$Z_n = \{0, 1, 2, \ldots, n - 1\}$$

where the addition, the subtraction, and the multiplication are defined by

$$(x + y) \text{ mod } n, (x - y) \text{ mod } n, (x.y) \text{ mod } n, \quad \forall x \text{ and } y \text{ in } Z_n.$$

Definition 2.7   Given two elements x and y of Z_n, such that $x.y = 1$, then y is said to be the *multiplicative inverse* of x. If such an inverse exists, it is unique.

Notation:

$$y = x^{-1}.$$

Property 2.2 x has a multiplicative inverse if and only if $gcd(x, n) = 1$.

Proof If $x.y = 1$ mod n, then $x.y = q.n+1$ so that any divisor of x and n is also a divisor of 1. Thus $gcd(x, n) = 1$.
  If $gcd(x, n) = 1$, then (relation (2.1)) there exist b and c such that $1 = b.x + c.n$, so that $x^{-1} = b$.

More generally, we have the following.

Properties 2.3

1. Let $g = gcd(a, n)$. Then the equation $a.x \equiv d$ (mod n) has a solution x if and only if g divides d.
2. The solutions of $a.x \equiv d$ (mod n) are the same as the solutions of $(a/g).x \equiv (d/g)$ (mod n/g).
3. There are g solutions, all of them congruent modulo n/g.

Proof

1. If $ax \equiv d$ (mod n), then $a.x - d = q.n$. As g divides both a and n, it also divides d. If g divides d, then $d = q.g$. According to (2.1), g is a linear

combination of a and n; that is, $g = b.a + c.n$. So $d = q(b.a + c.n)$ and $x = q.b$ is a solution.

2. If g divides d and $a.x \equiv d \pmod{n}$, that is, $a.x - d = q.n$, then $(a/g).x - (d/g) = q.(n/g)$ and $(a/g).x \equiv (d/g) \pmod{n/g}$. Inversely, if $(a/g).x \equiv (d/g) \pmod{n/g}$ then $a.x \equiv d \pmod{n}$.

3. As a/g and n/g are relatively prime, then there is a unique solution within $Z_{n/g}$, namely, $x = x_0 = (d/g).(a/g)^{-1} \bmod n/g$. The complete set of solutions within Z_n is

$$x_k = x_0 + k.(n/g), \quad \forall k = 0, 1, \ldots, g - 1.$$

Observe that if $k < g$ and $x_0 < (n/g)$, then $x_k \leq (n/g) - 1 + (g - 1).(n/g) = n - 1$.

Properties 2.4 (*Chinese Remainder Theorem*) Consider s pairwise relatively prime integers $m_1, m_2, \ldots, m_s$ whose product is equal to M. Then the system

$$N \equiv r_1 \pmod{m_1},$$
$$N \equiv r_2 \pmod{m_2},$$
$$\cdots$$
$$N \equiv r_s \pmod{m_s}, \tag{2.3}$$

has a unique solution N within Z_M ($|a|_m$ stands for a mod m):

$$N = \left| \Sigma_{1 \leq i \leq s} m_i^* . |r_i/m_i^*|_{m_i} \right|_M, \tag{2.4}$$

where

$$M = \Pi_{1 \leq i \leq s} m_i; \quad m_i^* = M/m_i. \tag{2.5}$$

The r_i are called *residues* modulo m_i.

Proof In order to compute a solution of system (2.3) observe that every m_i is relatively prime with every m_j ($j \neq i$) so that every m_j is relatively prime with $m_j^* = M/m_j$. Then m_j^* has a multiplicative inverse and

$$N = (m_1^*).(r_1/m_1^*) \bmod m_1 + (m_2^*).(r_2/m_2^*) \bmod m_2 + \cdots$$
$$+ (m_s^*).(r_s/m_s^*) \bmod m_s, \tag{2.6}$$

is obviously a solution. The uniqueness is deduced from the fact that different systems have different solutions, and that there are exactly as many different systems as elements in Z_M.

The computation of $(m_i^*)^{-1} \bmod m_i$ can be performed with the extended Euclidean algorithm: as m_i is relatively prime with M/m_j, the algorithm generates b and c such that

$$1 = b.m_i + c.(M/m_j),$$

and

$$(m_i^*)^{-1} = c \bmod m_i.$$

Garner's algorithm 2.2 ([GAR1959], [MEN1996]) computes N using a technique slightly different from the straight computation of (2.4). It computes first the mixed-radix digits within a preliminary step of a procedure step computing the base-B digits through a mixed-radix to base-B conversion (see mixed-radix system—Chapter 3).

A procedure inversion_step using the Euclidean algorithm to compute $(m_j)^{-1} \bmod m_i$ is first defined as

```
procedure inversion_step (m(j), m(i): in natural; invm(j): out
natural);
```

Algorithm 2.2   Garner's Algorithm

Assume N is given, according to (2.3), by its set of residues $r_i = N \bmod m_i$:

```
for i in 2..s loop
  c(i):=1;
  for j in 1..(i-1) loop
   inversion_step (m(j), m(i), invm(j));
   c(i):=invm(j)*c(i) mod m(i);
  end loop;
end loop;
u:=r(1); x:=u; b(1):=1;
for i in 2..s loop
  b(i):=b(i-1)*m(i-1);
  end loop;
  for i in 2..s loop
  u:=(r(i) − x)*c(i) mod m(i);
  x:=x+u*b(i);
  end loop;
```

Examples 2.4

1. Let $\{r_i\} = \{1, 2, 3, 4, 5\}$ be the set of remainders (residual expression) of a natural number N with respect to the respective set of moduli $\{m_i\} = \{2, 3, 5, 7, 11\}$. To compute the base-10 expression of N using (2.4), one first needs to compute $\{m_i^*\}$ and $\{1/m_i^* \bmod m_i\}$. A straightforward base-10

calculation leads to

$$M = \Pi_{1 \leq i \leq s} m_i = 2.3.5.7.11 = 2310,$$
$$\{m_i^*\} = \{M/m_i\} = \{1155, 770, 462, 330, 210\},$$

while the Euclidean algorithm is used to compute

$$\{1/m_i^* \bmod m_i\} = \{1, 2, 3, 1, 1\}.$$

Formula (2.4) yields

$$N = |1155.|1.1|_2 + 770.|2.2|_3 + 462.|3.3|_5 + 330.|4.1|_7 + 210.|5.1|_{11}|_{2310},$$
$$N = |6143|_{2310} = 1523.$$

2. Garner's algorithm is now used to solve the same problem. The Euclidean algorithm is used in the first loop of Algorithm 2.2. It computes:

$$i := 2, \; j := 1 \rightarrow 1/m_1 \bmod m_2 = 1/2 \bmod 3 = 2; \, c(2) := 2;$$
$$i := 3, \; j := 1 \rightarrow 1/m_1 \bmod m_3 = 1/2 \bmod 5 = 3; \, c(3) := 3;$$
$$i := 3, \; j := 2 \rightarrow 1/m_2 \bmod m_3 = 1/3 \bmod 5 = 2; \, c(3) := 1;$$
$$i := 4, \; j := 1 \rightarrow 1/m_1 \bmod m_4 = 1/2 \bmod 7 = 4; \, c(4) := 4;$$
$$i := 4, \; j := 2 \rightarrow 1/m_2 \bmod m_4 = 1/3 \bmod 7 = 5; \, c(4) := 6;$$
$$i := 4, \; j := 3 \rightarrow 1/m_3 \bmod m_4 = 1/5 \bmod 7 = 3; \, c(4) := 4;$$
$$i := 5, \; j := 1 \rightarrow 1/m_1 \bmod m_5 = 1/2 \bmod 11 = 6; \, c(5) := 6;$$
$$i := 5, \; j := 2 \rightarrow 1/m_2 \bmod m_5 = 1/3 \bmod 11 = 4; \, c(5) := 2;$$
$$i := 5, \; j := 3 \rightarrow 1/m_3 \bmod m_5 = 1/5 \bmod 11 = 9; \, c(5) := 7;$$
$$i := 5, \; j := 4 \rightarrow 1/m_4 \bmod m_5 = 1/7 \bmod 11 = 8; \, c(5) := 1.$$

The second loop computes the weights $b(j)$ as $\Pi_{1 \leq j \leq i-1} m_i$:

$$b(1) := 1; \; b(2) := b(1).m_1 = 2; \; b(3) := b(2).m_2 = 2.3 = 6;$$
$$b(4) := b(3).m_3 = 6.5 = 30; \; b(5) := b(4).m_4 = 30.7 = 210.$$

The third loop finally computes x as

$$u := r(1) = 1; x := u = 1$$
$$i := 2; u := (r(2) - x).c(2) \bmod 3 = (2 - 1).2 \bmod 3 = 2;$$
$$x := x + u.b(2) = 1 + 2.2 = 5;$$
$$i := 3; u := (r(3) - x).c(3) \bmod 5 = (3 - 5).1 \bmod 5 = 3;$$
$$x := x - u.b(3) = 5 + 3.6 = 23;$$
$$i := 4; u := (r(4) - x).c(4) \bmod 7 = (4 - 23).4 \bmod 7 = 1;$$
$$x := x + u.b(4) = 23 + 1.30 = 53;$$
$$i := 5; u := (r(5) - x).c(5) \bmod 11 = (5 - 53).1 \bmod 11 = 7;$$
$$x := x + u.b(5) = 53 + 7.210 = 1523.$$

Observe that the first two loops are independent and therefore may be computed in parallel. Moreover, if the modulus system is fixed, the $c(i)$ and $b(i)$ are computed once then stored for further use.

Definitions 2.8

1. The set of elements x of Z_n relatively prime with n is the *multiplicative group* Z_n^*:

$$Z_n^* = \{x \in Z_n | gcd(x, n) = 1\}.$$

2. The *Euler phi function* $\phi(n)$ is the number of elements in Z_n^*.

According to Property 2.2, Z_n^* is the set of invertible elements of Z_n. In particular, if p is a prime number then

$$Z_p^* = \{1, 2, \ldots, p - 1\} \quad \text{and} \quad \phi(p) = p - 1.$$

Properties 2.5 (*Fermat's Little Theorem*) Let p be a prime.. Any integer x satisfies $x^p \equiv x \pmod{p}$, and any integer x not divisible by p satisfies $x^{p-1} \equiv 1 \pmod{p}$.

Proof If x is not divisible by p and if $i..x \equiv j.x \pmod{p}$, that is, $(i - j).x = q.p$, then $i \equiv j \pmod{p}$. Thus

$$(1.x).(2.x). \ldots ((p - 1).x) \equiv 1.2. \ldots (p - 1)(\bmod p),$$

as the $p - 1$ above multiples of x are distinct and nonzero, they must be congruent to $1, 2, 3, \ldots, p - 1$ in some order.

So

$$(p - 1)!.x^{p-1} \equiv (p - 1)! \pmod{p},$$

or

$$(p - 1)!.(x^{p-1} - 1) \equiv 0 \pmod{p}.$$

As p does not divide $(p - 1)!$,

$$(x^{p-1} - 1) \equiv 0 \pmod{p},$$

that is,

$$x^{p-1} \equiv 1 \pmod{p} \quad \text{and} \quad x^p \equiv x \pmod{p}.$$

If x is divisible by p, then $x^p \equiv x \equiv 0 \pmod{p}$.

Corollary 2.1 Let p be a prime.. If x is not divisible by p and if $r \equiv s \pmod{p-1}$, then

$$x^r \equiv x^s \pmod{p}.$$

Proof Assume that $r > s$. Then $r = q.(p-1) + s$ and $1 \equiv 1^q \equiv (x^{p-1})^q \equiv x^{r-s}$ (mod p), so that $x^r \equiv x^s \pmod{p}$.

Definitions 2.9

1. The *order* of an element x of Z_n^* is the least positive integer t such that $x^t \equiv 1$ (mod n).
2. If the order of x is equal to the number $\phi(n)$ of elements in Z_n^*, then x is said to be a *generator* or *primitive element* of Z_n^*.
3. If Z_n^* has a generator, then Z_n^* is said to be *cyclic*.

Observe that if x is a generator then $Z_n^* = \{x^1, x^2, x^3, \ldots, x^{\phi(n)}\}$.

Example 2.5

$Z_7 = \{0, 1, 2, 3, 4, 5, 6\}$ and $Z_7^* = \{1, 2, 3, 4, 5, 6\}$;
7 is prime and $\phi(7) = 6$;
$1^1 \equiv 1 \pmod{7}$, $2^3 \equiv 1 \pmod{7}$, $3^6 \equiv 1 \pmod{7}$, $4^3 \equiv 1 \pmod{7}$, $5^6 \equiv 1 \pmod{7}$, $6^2 \equiv 1 \pmod{7}$.
There are two generators: 3 and 5. For example,
$3^1 \equiv 3 \pmod{7}$, $3^2 \equiv 2 \pmod{7}$, $3^3 \equiv 6 \pmod{7}$, $3^4 \equiv 4 \pmod{7}$, $3^5 \equiv 5 \pmod{7}$, $3^6 \equiv 1 \pmod{7}$.

2.2 ALGEBRA

2.2.1 Groups

Definition 2.10 A *group* $(G, *, 1)$ consists of a set G with a binary operation $*$ and an *identity element* 1 satisfying the following three axioms:

1. $x * (y * z) = (x * y) * z, \forall x, y, z \in G$ (*associativity*);
2. $x * 1 = 1 * x = x, \forall x \in G$ (*identity element*);

3. for each element x of G there exists an element x^{-1}, called the *inverse* of x, such that

$$x * x^{-1} = x^{-1} * x = 1.$$

If, furthermore,

4. $x^*y = y^*x, \forall x, y \in G$ (*commutativity*), the group is said to be *commutative* (or *Abelian*).

Axioms 1 and 2 define a *semigroup*.

Examples 2.6

$(Z, +, 0), (Z_n, +, 0), (Z_n^*, \cdot, 1)$

The following definitions generalize Definitions 2.9.

Definitions 2.11

1. The *order* of an element x of a finite group G is the least positive integer t such that

$$x^t = x * x * \cdots * x = 1.$$

2. If the order of x is equal to the number n of elements in G, then x is said to be a *generator* of G.

3. If G has a generator, then G is said to be *cyclic*.

Property 2.6   The order of an element x of a finite group G divides the number of elements in G..

Proof   First observe that if H is a subgroup of G, then an equivalence relation on G can be defined: $g_1 \equiv g_2$ if there exists an element h in H such that $g_1.h = g_2$. The number of elements in an equivalence class is equal to the number $|H|$ of elements in H. Thus the number $|G|$ of elements in G is equal to $|H|. |G/H|$, with G/H the set of classes and $|G/H|$ the number of classes. In other words the number of elements of a subgroup divides the number of elements of the group. It remains to observe that the set $\{x, x^2, \ldots, x^t = 1\}$, where t is the order of x, is a subgroup, so that the number t of elements of the subgroup divides the number of elements in G.

Example 2.7

$(Z_7^*, \cdot, 1)$;
3 and 5 are generators;
the subgroup generated by 2 is $\{2, 4, 1\}$; the corresponding classes are then $\{2, 4, 1\}$ and $\{6, 5, 3\}$; the number of elements (3) of the subgroup divides the number of elements (6) of Z_7^*.

2.2.2 Rings

Definition 2.12 A *ring* $(R, +, *, 0, 1)$ consists of a set R with two binary operations $+$ and $*$, an *additive identity element* 0, and a *multiplicative identity element* 1 satisfying the following axioms:

1. $(R, +, 0)$ is a commutative group;
2. $x * (y * z) = (x * y) * z, \forall x, y, z \in R$ (*associativity*);
3. $x * 1 = 1 * x = x, \forall x \in R$;
4. $x * (y + z) = (x * y) + (x * z)$ and $(x + y) * z = (x * z) + (y * z), \forall x, y, z \in R$ (*distributivity*).
   If, furthermore,
5. $x * y = y * x, \forall x, y \in R$ (*commutativity*), the ring is said to be *commutative*.

Examples 2.8

$(Z, +, \cdot, 0, 1), (Z_n, +, \cdot, 0, 1)$

2.2.3 Fields

Definition 2.13 A *field* $(F, +, *, 0, 1)$ consists of a set F with two binary operations $+$ and $*$, an *additive identity element* 0, and a *multiplicative identity element* 1 satisfying the following axioms:

1. $(F, +, *, 0, 1)$ is a commutative ring;
2. all nonzero elements of F have a multiplicative inverse.

Example 2.9

$(Z_p, +, \cdot, 0, 1)$, where p is a prime.

2.2.4 Polynomial Rings

Definitions 2.14

1. If F is a field, then a *polynomial* in the indeterminate x over F is an expression of the form

$$f(x) = a_n . x^n + a_{n-1} . x^{n-1} + \cdots + a_1 . x + a_0,$$

where $a_i \in F, \forall i \in \{0, 1, \ldots, n\}$.

2. The largest integer m (if any) such that $a_m \neq 0$ is the *degree* of $f(x)$. It is denoted $deg(f)$ and a_m is called the *leading coefficient*. If all the coefficients of $f(x)$ are equal to 0, then $f(x)$ is called the *zero polynomial* and its degree defined to be equal to $-\infty$. The 0-degree polynomials are also called *constant polynomials*.

3. A *monic polynomial* is a polynomial whose leading coefficient is equal to 1.

4. The polynomial ring $F[x]$ is the ring formed by the set of all polynomials in the indeterminate x with coefficients in F. The two operations are the standard polynomial addition and multiplication, with coefficient arithmetic performed in F. The additive identity element 0 is the *zero polynomial*. The multiplicative identity element 1 is the monic constant polynomial.

Definition 2.15 Thanks to the fact that F is a field, so that all the nonzero coefficients have an inverse, the standard polynomial division can also be performed. Thus, if $g(x)$ and $h(x) \neq 0$ are polynomials in $F[x]$, then there exist two polynomials $q(x)$ (the *quotient*) and $r(x)$ (the *remainder*) in $F[x]$ such that

$$g(x) = q(x).h(x) + r(x), \quad \text{where } deg(r) < deg(h). \tag{2.7}$$

Notation:

$$r(x) = g(x) \bmod h(x), \quad q(x) = g(x) \text{ div } h(x).$$

Definitions 2.16

1. Given two polynomials $g(x)$ and $h(x)$, $h(x)$ divides $g(x)$ (or $h(x)$ is a divisor of $g(x)$) if there exists a polynomial $q(x)$ such that $g(x) = q(x).h(x)$.

2. Given two polynomials $g(x)$ and $h(x)$, not both equal to 0, the *greatest common divisor* of $g(x)$ and $h(x)$ is the monic polynomial of greatest degree which divides both $g(x)$ and $h(x)$.

3. $gcd(0, 0) = 0$.

4. A polynomial $f(x)$ of degree at least 1 is said to be irreducible if it cannot be written as the product of two polynomials, each of positive degree.

A variant of the *Euclidean algorithm* for polynomials (VZG2003) expresses the greatest common divider of two polynomials $g(x)$ and $h(x)$ in the form

$$gcd(g, h) = b(x).g(x) + c(x).h(x).$$

The algorithm is based on the fact that if $u(x)$ and $v(x)$ are two polynomials such that

$$deg(u) = m, \quad deg(v) = t \quad \text{and } m > t,$$

that is,

$$u(x) = u_m.x^m + u_{m-1}.x^{m-1} + \cdots + u_1.x + u_0,$$
$$v(x) = v_t.x^t + v_{t-1}.x^{t-1} + \cdots + v_1.x + v_0,$$

then

$$v(x).u_m.(v_t)^{-1}.x^{m-t} = (v_t.x^t + v_{t-1}.x^{t-1} + \cdots + v_1.x + v_0).u_m.(v_t)^{-1}.x^{m-t}$$
$$= u_m.x^m + r'(x)$$

where $deg(r') < m$, so that

$$u(x) = (v(x).u_m.(v_t)^{-1}.x^{m-t} - r'(x)) + u_{m-1}.x^{m-1} + \cdots + u_1.x + u_0$$
$$= v(x).u_m.(v_t)^{-1}.x^{m-t} + r(x) \qquad (2.8)$$

where

$$r(x) = u_{m-1}.x^{m-1} + \cdots + u_1.x + u_0 - r'(x)$$

so that

$$deg(r) < m \quad \text{and} \quad max(deg(r), deg(v)) < deg(u).$$

Furthermore,

$$gcd(u,v) = gcd(v,r).$$

The sequence of operations is almost the same as for computing the greatest common divider of two integers. A series of polynomials $r(0)$, $r(1)$, $r(2)$, ... are generated. Initially, assume that $deg(g) > deg(h)$ and define

$$r(0) = g(x) \quad \text{and} \quad r(1) = h(x).$$

At each step the decomposition (2.8) is used:

$$u(x) = r(i-1), \quad v(x) = r(i), \quad m = deg(r(i-1)), \quad t = deg(r(i)),$$
$$deg(r(i-1)) > deg(r(i))$$

so that

$$r(i-1) = q(i).r(i) + r(i+1)$$

where

$$q(i) = u_m.(v_t)^{-1}.x^{m-t}, \quad r(i+1) = r(i-1) - q(i).r(i),$$
$$deg(r(i+1)) < m = deg(r(i-1)).$$

At the end of the step, $r(i)$ and $r(i+1)$ are interchanged if $deg(r(i)) < deg(r(i+1))$.

Operations:

$r(0) = g(x),$

$r(1) = h(x),$

$r(0) = r(1).q(1) + r(2),$ if $deg(r(1)) < deg(r(2))$ interchange $r(1)$ and $r(2),$

$r(1) = r(2).q(2) + r(3),$ if $deg(r(2)) < deg(r(3))$ interchange $r(2)$ and $r(3),$

$r(2) = r(3).q(3) + r(4),$ if $deg(r(3)) < deg(r(4))$ interchange $r(3)$ and $r(4),$

. . .

$r(n-3) = r(n-2).q(n-2) + r(n-1),$ if $deg(r(n-2)) < deg(r(n-1))$
$$\text{interchange } r(n-2) \text{ and } r(n-1),$$

$r(n-2) = r(n-1).q(n-1) + r(n),$

where

$$deg(r(0)) > deg(r(1)) > \cdots > deg(r(n)) = 0$$

and

$$gcd(r(i), r(i+1)) = gcd(r(i+1), r(i+2)),$$

so that

$$gcd(g, h) = gcd(r(0), r(1)) = \cdots = gcd(r(n-1), r(n)).$$

Let r_0 be the coefficient of x^0 in $r(n)$. If $r_0 = 0$, then

$$gcd(g, h) = gcd(r(n-1), 0) = r(n-1).$$

If $r_0 \neq 0$, then

$$gcd(g, h) = gcd(r(n-1), r_0) = 1.$$

In parallel with the computation of $r(0), r(1), r(2), \ldots, r(n)$, two series of polynomials $b(i)$ and $c(i)$ are generated:

$b(0) = 1,$

$b(1) = 0,$

$b(2) = b(0) - b(1).q(1),$ if $deg(r(1)) < deg(r(2))$ interchange $b(1)$ and $b(2),$

. . .

$b(n-1) = b(n-3) - b(n-2).q(n-2),$ if $deg(r(n-2)) < deg(r(n-1))$
$$\text{interchange } b(n-2) \text{ and } b(n-1),$$

$b(n) = b(n-2) - b(n-1).q(n-1).$

$c(0) = 0,$

$c(1) = 1,$

$c(2) = c(0) - c(1).q(1), \quad$ if $deg(r(1)) < deg(r(2))$ interchange $c(1)$ and $c(2)$,

...

$c(n-1) = c(n-3) - c(n-2).q(n-2), \qquad$ if $deg(r(n-2)) < deg(r(n-1))$

$\qquad\qquad\qquad\qquad\qquad\qquad\qquad\qquad$ interchange $c(n-2)$ and $c(n-1)$,

$\quad c(n) = c(n-2) - c(n-1).q(n-1).$

It can be demonstrated by induction that

$$r(i) = b(i).g(x) + c(i).h(x), \quad \forall\, i = 0, 1, 2, \ldots, n.$$

So, if $r_0 = 0$, then

$$gcd(g, h) = r(n-1) = b(n-1).g(x) + c(n-1).h(x),$$

and if $r_0 \neq 0$, then

$$gcd(g, h) = 1 = r_0^{-1}.r(n) = r_0^{-1}.b(n).g(x) + r_0^{-1}.c(n).h(x).$$

In the following algorithm u stands for $r(i-1)$, v for $r(i)$, r for $r(i+1)$, b for $b(i-1)$, d for $b(i)$, bb for $b(i+1)$, c for $c(i-1)$, e for $c(i)$, and cc for $c(i+1)$:

Algorithm 2.3   Variant of the Extended Euclidean Algorithm for Polynomials

```
u:=g; v:=h; b:=1; c:=0; d:=0; e:=1;
m:=degree(u); t:=degree(v);
if t=0 then
   if v(0)=0 then z=u; else z:=1; b:=0; c:=(v(0))⁻¹; end if;
elsif m=0 then
   if u(0)=0 then z=v; b:=0; c:=1; else z:=1; b:=(u(0))⁻¹;
   end if;
else
   while t>0 loop
     if m<t then swap(u, v); swap(b, d); swap(c, e); swap(m, t);
      end if;
      q:=u(m)*(v(t))⁻¹*xᵐ⁻ᵗ; r:=u-v*q; bb:=b-d*q; cc:=c-e*q;
      u:=v; v:=r; b:=d; c:=e; d:=bb; e:=cc;
      m:=t; t:=degree(v);
   end loop;
   if v(0)=0 then z:=u; else z:=1; b:=d*(v(0))⁻¹;
   c:=e*(v(0))⁻¹;
    end if;
end if;
```

2.2.5   Congruences of Polynomial

Definition 2.17   Given three polynomials $g(x)$, $h(x)$, and $f(x)$ in $F[x]$, $g(x)$ is congruent to $h(x)$ modulo $f(x)$ if $f(x)$ divides $g(x) - h(x)$.

Notation:

$$g(x) \equiv h(x)(\text{mod } f(x)).$$

Properties 2.7 (Properties of Congruences)

1. $g(x) \equiv h(x)$ (mod $f(x)$)) if and only if $(g(x) \text{ mod } f(x)) = (h(x) \text{ mod } f(x))$ (Definition 2.15);
2. the relation $g(x) \equiv h(x)$ (mod $f(x)$)) is an equivalence relation (reflexive, symmetric, and transitive);
3. if $g_1(x) \equiv h_1(x)$ (mod $f(x)$) and $g_2(x) \equiv h_2(x)$ (mod $f(x)$)), then

$$g_1(x) + h_1(x) \equiv g_2(x) + h_2(x)(\text{mod } f(x)), \, g_1(x) - h_1(x) \equiv g_2(x) - h_2(x)$$
$$(\text{mod } f(x)), \, g_1(x).h_1(x) \equiv g_2(x).h_2(x)(\text{mod } f(x)). \qquad (2.9)$$

From Properties 2.7(1 and 2) it can be seen that the congruence relation partitions $F[x]$ into equivalence classes. If n is the degree of $f(x)$, then each equivalence class contains exactly one polynomial of degree $d < n$. So, if F is a finite field, then the number of equivalence classes is equal to $|F|^n$, where $|F|$ is the number of elements in F. Furthermore, according to Property 2.7(3), the addition, subtraction, and multiplication of congruence classes can be defined. As a matter of fact, the set of equivalence classes is isomorphic to

$$\{g(x) \in F[x] | deg(g) < n\}$$

where the addition, the subtraction, and the multiplication are defined by

$$(g(x) + h(x)) \text{ mod } f(x), \quad (g(x) - h(x)) \text{ mod } f(x), \quad (g(x).h(x)) \text{ mod } f(x).$$

The set of equivalence classes is denoted $F[x]/f(x)$.

Properties 2.8

1. $F[x]/f(x)$ is a commutative ring.
2. If $f(x)$ is irreducible, then $F[x]/f(x)$ is a field.

Proof
1. Consequence of Property 2.7(3).
2. If $f(x)$ is irreducible, then the greatest common divisor of $f(x)$ and $g(x) \neq 0$ is 1. Using the Euclidean algorithm (Algorithm 2.2), $b(x)$ and $c(x)$ can be

computed such that

$$1 = b(x).f(x) + c(x).g(x)$$

and

$$c(x) = (g(x))^{-1} \bmod f(x).$$

Definition 2.18 Let p be a prime, $F = Z_p$, and $f(x)$ be an irreducible polynomial of degree n over Z_p. The corresponding field $F[x]/f(x)$ contains $q = p^n$ elements and is called either F_q or $GF(q)$ (Galois field).

As a matter of fact, it can be demonstrated that any finite field contains $q = p^n$ elements, for some prime p and some positive integer n, and is isomorphic to F_q (whatever the irreducible polynomial $f(x)$ of degree n over Z_p). In particular, if $n = 1$, then the corresponding field F_p is isomorphic to Z_p.

The set of 0-degree polynomials (the constants) is a subfield of F_q isomorphic to F_p. If $g(x)$ is a 0-degree polynomial (an element of F_p) then, according to the Fermat's little theorem, $(g(x))^p = g(x)$. Conversely, it can be demonstrated that if a polynomial $g(x)$ satisfies the condition $(g(x))^p = g(x)$, then $g(x)$ is a constant.

Another interesting property of F_q is that the set F_q^* of nonzero polynomials is a cyclic group. Let $g(x)$ be a nonzero polynomial, that is, an element of F_q^*, and assume that the order of $g(x)$ is t. According to the Property 2.6, t divides $q - 1$, so that $(g(x))^{q-1} = (g(x))^{t.k} = 1^k = 1$. Consider now a polynomial $g(x)$ and define $h(x) = (g(x))^r$, where $r = (q - 1)/(p - 1)$. According to the previous property, $(h(x))^{p-1} = (g(x))^{q-1} = 1$ and $(h(x))^p = h(x)$, so that $h(x)$ is a constant polynomial.

A last property, useful for performing arithmetic operations, is that $(g(x) + h(x))^p = (g(x))^p + (h(x))^p$. It is a straightforward consequence of the fact that all the binomial coefficients $(p!/(i!).(p - i)!)$ are multiples of p, except for $i = 0$ or p.

To summarize:

Properties 2.9 (Some Useful Properties of Finite Fields)

1. The set of 0-degree polynomials in F_q is a subfield of F_q isomorphic to F_p.
2. Given $g(x)$ in F_p, then $(g(x))^p = g(x)$ (Fermat's little theorem).
3. Given $g(x)$ in F_q such that $(g(x))^p = g(x)$; then $g(x) \in F_p$.
4. The set of nonzero polynomials of F_q is a cyclic group denoted F_q^*.
5. Given $g(x)$ in F_q, then $(g(x))^q = g(x)$.
6. Given $g(x)$ and $h(x)$ in F_q, then $(g(x) + h(x))^p = (g(x))^p + (h(x))^p$.
7. If $r = (p^n - 1)/(p - 1)$, that is, $r = 1 + p + p^2 + \cdots + p^{n-1}$, and $g(x)$ is an element of F_q, then $(g(x))^r$ is an element of F_p.

Example 2.10　$p = 2$, $n = 4$, $f(x) = 1 + x + x^4$ so that $x^4 \equiv 1 + x \bmod f(x)$; $\alpha = x$ is a generator of the cyclic group F_{16}^*:

$$\alpha^1 = x,$$

$$\alpha^2 = x^2,$$

$$\alpha^3 = x^3,$$

$$\alpha^4 = x^4 \equiv 1 + x,$$

$$\alpha^5 = x..(1 + x) = x + x^2,$$

$$\alpha^6 = x.(x + x^2) = x^2 + x^3,$$

$$\alpha^7 = x.(x^2 + x^3) = x^3 + x^4 \equiv 1 + x + x^3,$$

$$\alpha^8 = (\alpha^4)^2 = (1 + x)^2 = 1 + x^2,$$

$$\alpha^9 = x.(1 + x^2) = x + x^3,$$

$$\alpha^{10} = x.(x + x^3) = x^2 + x^4 \equiv 1 + x + x^2,$$

$$\alpha^{11} = x.(1 + x + x^2) = x + x^2 + x^3,$$

$$\alpha^{12} = x.(x + x^2 + x^3) = x^2 + x^3 + x^4 \equiv 1 + x + x^2 + x^3,$$

$$\alpha^{13} = x.(1 + x + x^2 + x^3) = x + x^2 + x^3 + x^4 \equiv 1 + x^2 + x^3,$$

$$\alpha^{14} = x.(1 + x^2 + x^3) = x + x^3 + x^4 \equiv 1 + x^3,$$

$$\alpha^{15} = x.(1 + x^3) = x + x^4 \equiv 1;$$

Given a polynomial $g(x) = g_0 + g_1.x + g_2.x^2 + g_3.x^3$, then

$$(g(x))^2 = g_0 + g_1.x^2 + g_2.x^4 + g_3.x^6 \equiv g_0 + g_1.x^2 + g_2.(1 + x) + g_3.x^2.(1 + x)$$
$$= (g_0 + g_2) + g_2.x + (g_1 + g_3).x^2 + g_3.x^3;$$

if $(g(x))^2 = g(x)$, then

$$g_0 + g_2 = g_0, \, g_2 = g_1, \, g_1 + g_3 = g_3;$$

thus

$$g_1 = g_2 = g_3 = 0,$$

and $g(x) = g_0$, that is, an element of F_p (Property 2.9(3)).

2.3 FUNCTION APPROXIMATION

Numerous techniques are used to evaluate functions. According to the type of the function at hand, some evaluation methods may be more appropriate than others. For instance, a method well suited for a polynomial may not be the best for an exponential function. Polynomial approximation is most often recommended for function evaluation as any continuous function can be approximated in this way, and the implementation only consists of additions, multiplications, and powers.

Taylor and MacLaurin series are the most classic approaches to approximate functions. The series lead to precise numerical techniques to compute a function very near to one point, but precision can be lost for a bigger range of values. Trigonometric, logarithmic, and exponential function computations are typical applications.

Definition 2.19

1. *Taylor series.* If a function $f(x)$ has continuous derivatives up to $(n+1)$th order, then this function can be expanded in the following fashion:

$$\Sigma_{0\leq i\leq n}((x-a)^i/i!).(d^if(x)/dx^i)_{x=a} + R_n, \qquad (2.10)$$

called a *Taylor expansion* at point a. R_n is called the remainder after $n+1$ terms. When this expansion converges over a certain range of x, that is, when

$$\lim_{n\to\infty} R_n = 0, \qquad (2.11)$$

the expansion is called a *Taylor series* of $f(x)$ at point a.

2. *MacLaurin series.* If (2.10) is expressed at point $a = 0$, the series is called a *MacLaurin series*:

$$\Sigma_{0\leq i\leq n}((x)^i/i!).(d^if(x)/dx^i)_{x=0} + R_n. \qquad (2.12)$$

Examples 2.11 Taylor–MacLaurin series expansions of exponential functions:

$$e^x = 1 + x + x^2/2! + x^3/3! + \cdots \qquad \infty < x < \infty$$
$$(2..13)$$

$$e^{-x.x} = 1 - x^2 + x^4/2! - x^6/3! + x^8/4! - \cdots \qquad -\infty < x < \infty$$
$$(2.14)$$

$$a^x = e^{x.\ln a} = 1 + x.\ln a + (x.\ln a)^2/2! + (x.\ln a)^3/3! + \cdots \qquad -\infty < x < \infty$$
$$(2.15)$$

Taylor–MacLaurin series expansions of logarithmic functions:

$$\ln x = (x-1) - (x-1)^2/2 + (x-1)^3/3 - \cdots \qquad\qquad 0 < x \leq 2 \quad (2.16)$$

$$\ln x = 2.[(x-1)/(x+1) + ((x-1)/(x+1))^3/3$$
$$+ ((x-1)/(x+1))^5/5 + \cdots] \qquad\qquad x > 0 \qquad (2.17)$$

$$\ln x = (x-1)/x + ((x-1)/x)^2/2 + ((x-1)/x)^3/3 + \cdots \quad x \geq 1/2 \quad (2.18)$$

$$\ln (1+x) = x - x^2/2 + x^3/3 - x^4/4 + x^5/5 - \cdots \qquad -1 < x \leq 1 \quad (2.19)$$

Taylor–MacLaurin series expansions of trigonometric functions:

$$\sin x = x - x^3/3! + x^5/5! - x^7/7! + \cdots \qquad\qquad -\infty < x < \infty \quad (2.20)$$

$$\cos x = x - x^2/2! + x^4/4! - x^6/6! + \cdots \qquad\qquad -\infty < x < \infty \quad (2.21)$$

$$\tan x = x + x^3/3 + 2.x^5/15 + 17.x^7/315$$
$$+ \cdots + 2^{2.n}.(2^{2.n} - 1).B_n.x^{2.n-1}/(2.n)! + \cdots \qquad |x| < \pi/2 \qquad (2.22)$$

$$\cot x = 1/x - x/3 - x^3/45 - 2.x^5/945 - \cdots - 2^{2.n}.B_n.x^{2.n-1}(2.n)! + \cdots$$
$$0 < |x| < \pi \qquad (2.23)$$

where B_n are the Bernoulli numbers ([ROS2000]).

2.4 BIBLIOGRAPHY

[APO1974] T. M. Apostol, *Mathematical Analysism*, 2nd ed., Addison Wesley, Reading, MA, 1974.

[COH1993] H. Cohen, *A Course in Computational Algebraic Number Theory*, Springer-Verlag, Berlin, 1993.

[GAR1959] H. Garner, The residue number system. *IRE Trans. Electron. Comput.*, **EC-8**: 140–147 (1959).

[GIL2003] W. J. Gilbert and W. K. Nicholson, *Modern Algebra with Applications*, John Wiley & Sons, Hoboken, NJ, 2003.

[HER1975] I. N. Herstein, *Topics in Algebra*, 2nd ed., Xerox College Pub., Lexington, MA, 1975.

[HUN1974] T. W. Hungerford, *Algebra*, Holt, Rinehart and Winston, New York, 1974.

[KOB1994] N. Koblitz, *A Course in Number Theory and Cryptography*, Springer-Verlag, New York, 1994.

[McC1987] R. J. McCeliece, *Finite Fields for Computer Scientists and Engineeers*, Kluwer Academic Publishers, Boston, 1987.

[MEN1996] A. J. Menezes, P.C. van Oorschot, and S. C. Vanstone, *Handbook of Applied Cryptography*, CRC Press, Boca Raton, FL, 1996.

[ROS1992] K. H. Rosen, *Elementary Number Theory and Its Applications*, Addison-Wesley, Reading, MA, 1992.

[ROS2000] K. H. Rosen (Editor-in-Chief), *Handbook of Discrete and Combinatorial Mathematics*, CRC Press, Boca Raton, FL, 2000.

[RUD1976] W. Rudin, *Principles of Mathematical Analysis* (International Series in Pure & Applied Mathematics), McGraw-Hill Science/Engineering/Math, New York, 1976.

[VZG2003] J. von zur Gathen and J. Gerhard, *Modern Computer Algebra*, Cambridge University Press, Cambridge, UK, 2003.

3

NUMBER REPRESENTATION

Arithmetic deals with operations on numbers: addition, subtraction, and so on. Thus number representation is a fundamental topic in arithmetic ([ERC2004], [PAR1999]). The choice of a *number representation system* has repercussions on the complexity of the algorithms executing the arithmetic operations, and thus on the costs and performances of the circuits that implement those algorithms. Apart from the cost and performance, another aspect to take into account, when choosing a number representation system, is the interface with other circuits or, simply, the human interface. Consider an example: the *residue number system* (RNS) allows the implementation of very fast and cost-effective arithmetic circuits. Nevertheless, the RNS needs some type of relatively expensive input and output interfaces since human beings don't use it, and the AD/DA converters don't understand this type of representation. Thus the use of a RNS is limited to cases in which the extra cost of the RNS encoding and decoding is negligible with respect to the total cost. In this chapter the most common number representation systems are described. The chapter is divided into three sections corresponding to natural numbers, integers, and real numbers.

3.1 NATURAL NUMBERS

3.1.1 Weighted Systems

Any natural number (nonnegative integer) can be represented, in a unique way, in the form of a sum of powers B^i of some natural number B greater than 1, each of

Synthesis of Arithmetic Circuits: FPGA, ASIC, and Embedded Systems
By Jean-Pierre Deschamps, Géry J. A. Bioul, and Gustavo D. Sutter
Copyright © 2006 John Wiley & Sons, Inc.

them multiplied by a natural number smaller than B. The following theorem defines the *base-B numeration system*.

Theorem 3.1   Given a natural number B greater than 1, any natural number x smaller than B^n can be expressed in the form

$$x = x_{n-1}.B^{n-1} + x_{n-2}.B^{n-2} + \cdots + x_0.B^0$$

where every coefficient x_i is a natural number smaller than B. Furthermore, there is only one possible vector $(x_{n-1} \ x_{n-2} \cdots x_0)$ representing x.

The following algorithm computes the coefficients x_i:

Algorithm 3.1

```
for i in 0..n - 1 loop x(i):= x mod B; x:= x/B; end loop;
```

Definitions 3.1

1. The most commonly used values of B are 10 (*decimal system*), 2 (*binary system*), 16 (*hexadecimal system*), and 8 (*octal system*). The coefficients x_i of the base-B representation of x are called the base-B digits of x. The binary digits are called *bits*. The hexadecimal digits 10, 11, 12, 13, 14, and 15 are usually replaced by letters: A, B, C, D, E, and F.
2. This type of representation is called *positional* as the weight B^i associated with the digit x_i depends on i, that is, on the position of the digit within the vector $(x_{n-1} \ x_{n-2} \cdots x_0)$.
3. The base-B digits could in turn be encoded in another base. As an example, if $B = 10$ and the decimal digits are represented in the form of 4-bit binary vectors, the so-obtained system is called *binary-coded decimal* (BCD).

Example 3.1   Compute the hexadecimal representation of 287645:

$$287645 = 17977.16 + 13$$
$$17977 = 1123.16 + 9$$
$$1123 = 70.16 + 3$$
$$70 = 4.16 + 6$$
$$4 = 0.16 + 4$$

$$287645 = 4.16^4 + 6.16^3 + 3.16^2 + 9.16^1 + 13.16^0 = [4639D]_{\text{base}16}.$$

It is possible to define *mixed numeration systems* or *mixed-radix systems*, that is with several bases. For instance, the time is expressed in days of 24 hours, hours of

60 minutes, minutes of 60 seconds, seconds of 1000 milliseconds, and so on. The generalization of Theorem 3.1 is the following.

Theorem 3.2 Given n natural numbers $b_{n-1}, b_{n-2} \ldots b_0$, greater than 1, any natural number x smaller than $B_n = b_{n-1}.b_{n-2}.\cdots.b_0$, can be expressed in the form

$$x = x_{n-1}.B_{n-1} + x_{n-2}.B_{n-2} + \cdots + x_0.B_0$$

where

$$B_0 = 1, B_1 = b_0, B_2 = b_1.b_0, \ldots, B_{n-1} = b_{n-2}.b_{n-3}.\cdots.b_0,$$

and every coefficient x_i is a natural number smaller than b_i. Furthermore, there is only one possible vector $(x_{n-1} \, x_{n-2} \cdots x_0)$ representing x.

Base-B and *mixed-radix* numeration systems are weighted systems. In base B the weights are B^i, that is the successive powers of B, while the weights in the *mixed-radix system* are given by

$$B_i = b_{i-1}.b_{i-2}.\cdots.b_0.$$

The following algorithm computes the coefficients x_i.

Algorithm 3.2

```
for i in 0..n-1 loop x(i):=x mod b(i); x:=x/b(i); end loop;
```

Example 3.2 Compute the representation of 287645 in the mixed base (13, 12, 15, 11, 12):

$$287645 = 23970.12 + 5$$
$$23970 = 2179.11 + 1$$
$$2179 = 145.15 + 4$$
$$145 = 12.12 + 1$$
$$12 = 0.13 + 12$$
$$287645 = 12.(12.15.11.12) + 1.(15.11.12) + 4.(11.12) + 1.12 + 5.$$

Comment 3.1 Given a natural number s, the conversion from the base-B representation of x (Theorem 3.1) to its base-B^s representation, and inversely, is straightforward. Suppose that $n = s.q$ (if n were not divisible by s, then $(\lceil n/s \rceil.s - n)$ initial 0's should be added). Then

$$x = X_{q-1}.(B^s)^{q-1} + X_{q-2}.(B^s)^{q-2} + \cdots + X_0.(B^s)^0$$

where

$$X_i = x_{i.s+s-1}.B^{s-1} + x_{i.s+s-2}.B^{s-2} + \cdots + x_{i.s}.B^0.$$

As an example, the binary representation of the decimal number 287645 is 01000110001110011101. The conversion to its hexadecimal representation is straightforward:

$$[0100\ 0110\ 0011\ 1001\ 1101]_{base\,2} = [4639D]_{base\,16}.$$

3.1.2   Residue Number System

A residue number system (RNS) is defined by a set of s moduli $\{m_i\}$. If the m_is are pairwise prime, the RNS is called *nonredundant*. The *RNS-representation* of a given natural number N is the vector $R(N)$, whose components r_i are the respective *residues modulo* m_i, that is, the successive remainders of the integer division N/m_i

$$r_i = N \bmod m_i.$$

The least common multiple (*lcm*) of $\{m_i\}$ is the *range* of the RNS, generally denoted M. The greatest natural number that can be represented in the RNS defined by $\{m_i\}$ is

$$M - 1 = (m_1 - 1, m_2 - 1, \ldots, m_s - 1).$$

If the m_is are pairwise prime then

$$M = \Pi_{1 \leq i \leq s}\, m_i$$

Garner's algorithm 2.2, restricted to the computation of the successive values of u, provides the mixed-radix components with respect to the weights

$$B(1) = 1, \quad B(i) = \Pi_{1 \leq j \leq i-1} m_j, \quad s \geq i \geq 2$$

(see Example 2.4).

3.2   INTEGERS

The most natural way of representing an integer is the *sign-magnitude representation system*. Nevertheless, it is not the most convenient for executing arithmetic operations. Several representation methods are now described.

3.2.1   Sign-Magnitude Representation

Any integer can be represented in the form $+x$ or $-x$, where x is a natural number. The natural number x can be represented in base B (Theorem 3.1), and instead of using the '+' and '−' symbols, an additional (sign) digit equal to 0 (nonnegative number) or 1 (negative number) is added:

Definition 3.2   The integer represented in the form $x_{n-1} \, x_{n-2} \cdots x_1 \, x_0$, where x_{n-1} is the sign bit, is

$$x_{n-2}.B^{n-2} + x_{n-3}.B^{n-3} + \cdots + x_0.B^0 \quad \text{if } x_{n-1} = 0,$$

$$-(x_{n-2}.B^{n-2} + x_{n-3}.B^{n-3} + \cdots + x_0.B^0) \quad \text{if } x_{n-1} = 1.$$

The range of represented numbers is $-B^{n-1} < x < B^{n-1}$.

Comment 3.2   The number of vectors $(x_{n-1} \, x_{n-2} \cdots x_1 \, x_0)$, where x_{n-1} is the sign bit, is equal to $2.B^{n-1}$, while the range $-B^{n-1} < x < B^{n-1}$ only includes $2.B^{n-1} - 1$ integers. The difference is due to the fact that the vector $(100 \cdots 0)$ does not represent any number (*zero* is a natural number so that its sign bit should always be equal to 0). Nevertheless, the integer *zero* could also be accepted with two representations, namely, $000 \cdots 0$ (plus zero) and $100 \cdots 0$ (minus zero).

3.2.2   Excess-*E* Representation

Another way of representing a negative number x consists in associating a natural number $R(x)$ to x, where R is a one-to-one function, and $R(x)$ is represented in base B.

Definition 3.3   In the *excess-E numeration system*, where E is a natural number,

$$R(x) = x + E,$$

so that the integer represented in the form $x_{n-1} \, x_{n-2} \cdots x_1 \, x_0$ is

$$x_{n-1}.B^{n-1} + x_{n-2}.B^{n-2} + \cdots + x_0.B^0 - E$$

and the range of represented numbers is

$$-E \le x \le B^n - E.$$

Comments 3.3

1. If B is even, an E is chosen equal to $B^n/2$; then the number represented in the form $x_{n-1} \, x_{n-2} \cdots x_1 \, x_0$ is

$$x_{n-1}.B^{n-1} + x_{n-2}.B^{n-2} + \cdots + x_0.B^0 - E$$

$$= (x_{n-1} - B/2).B^{n-1} + x_{n-2}.B^{n-2} + \cdots + x_0.B^0.$$

The sign definition rule is the following one: if x is negative then $x_{n-1} < B/2$; if x is nonnegative then $x_{n-1} \geq B/2$.

2. In some practical cases the value of E is different from $B^n/2$. As an example, in the ANSI/IEEE simple-precision floating-point system (Section 3.3), the exponent is an 8-bit number representing an integer x belonging to the range $-127 \leq x \leq 128$, according to the excess-E method with $E = 127$ and not 128.

3. If $B = 2$ and $E = 2^{n-1}$, then the number represented in the form x_{n-1} $x_{n-2} \cdots x_1\, x_0$ is

$$(x_{n-1} - 1).2^{n-1} + x_{n-2}.2^{n-2} + \cdots + x_0$$

$$= -x'_{n-1}.2^{n-1} + x_{n-2}.2^{n-2} + \cdots + x_0,$$

where x'_{n-1} stands for the complement of x_{n-1}.

4. The representation function R is unate, so that the magnitude comparison is easy.

Example 3.3 Represent $x = -287645$ with $n = 6$ digits in base $B = 10$ with $E = 10^6/2$.

$$B^6 = 1000000,$$

$$B^6/2 = 500000,$$

$$R(x) = x + E = 500000 - 287645 = 212355$$

Observe that

$$(2 - 10/2).10^5 + 12355 = -300000 + 12355 = -287645$$

3.2.3 B's Complement Representation

As in the preceding case, a one-to-one function $R(x)$, associating a natural number to x, is defined as follows.

Definition 3.4 In the *B's complement numeration system* every integer x belonging to the range $-B^n/2 \leq x < B^n/2$ is represented by

$$R(x) = x \bmod B^n,$$

so that the integer represented in the form $x_{n-1} x_{n-2} \cdots x_1 x_0$ is

$$x_{n-1}.B^{n-1} + x_{n-2}.B^{n-2} + \cdots + x_0$$

$$\text{if } x_{n-1}.B^{n-1} + x_{n-2}.B^{n-2} + \cdots + x_0 < B^n/2, \tag{3.1}$$

$$x_{n-1}.B^{n-1} + x_{n-2}.B^{n-2} + \cdots + x_0 - B^n$$

$$\text{if } x_{n-1}.B^{n-1} + x_{n-2}.B^{n-2} + \cdots + x_0 \geq B^n/2. \tag{3.2}$$

Conditions (3.1) and (3.2) can be written in the form

$$x_{n-1}.B^{n-1} + x_{n-2}.B^{n-2} + \cdots + x_0$$

$$\text{if } (x_{n-1} - B/2).B^{n-1} + x_{n-2}.B^{n-2} + \cdots + x_0 < 0, \tag{3.3}$$

$$x_{n-1}.B^{n-1} + x_{n-2}.B^{n-2} + \cdots + x_0 - B^n$$

$$\text{if } (x_{n-1} - B/2).B^{n-1} + x_{n-2}.B^{n-2} + \cdots + x_0 \geq 0, \tag{3.4}$$

and if B is even the latter conditions are equivalent to

$$x_{n-1}.B^{n-1} + x_{n-2}.B^{n-2} + \cdots + x_0 \qquad \text{if } x_{n-1} < B/2, \tag{3.5}$$

$$x_{n-1}.B^{n-1} + x_{n-2}.B^{n-2} + \cdots + x_0 - B^n \quad \text{if } x_{n-1} \geq B/2 \tag{3.6}$$

(take into account that $x_{n-2}.B^{n-2} + \cdots + x_0 < B^{n-1}$). Thus, if B is even, the integer represented by $x_{n-1} x_{n-2} \cdots x_1 x_0$ is

$$x = x'_{n-1}.B^{n-1} + x_{n-2}.B^{n-2} + \cdots + x_0, \quad \text{where}$$

$$x'_{n-1} = x_{n-1} - B \quad \text{if } x_{n-1} \geq B/2 \text{ and } x'_{n-1} = x_{n-1} \quad \text{if } x_{n-1} < B/2, \tag{3.7}$$

and the sign definition rule is the following one:

$$\text{if } x \text{ is negative then } x_{n-1} \geq B/2;$$
$$\text{if } x \text{ is nonnegative then } x_{n-1} < B/2.$$

In particular, if $B = 2$ the number represented in the form $x_{n-1} x_{n-2} \cdots x_1 x_0$ is

$$-x_{n-1}.2^{n-1} + x_{n-2}.2^{n-2} + \cdots + x_0, \tag{3.8}$$

and the most significant bit x_{n-1} is also the *sign bit*:

$$\text{if } x < 0 \text{ then } x_{n-1} = 1, \quad \text{and if } x \geq 0 \text{ then } x_{n-1} = 0. \tag{3.9}$$

Comments 3.4

1. The B's complement system is based on a congruence, namely, $R(x) = x \bmod B^n$, so that the arithmetic operations are easy (Chapters 4 and 5).

2. In order to represent an n-digit number with $n + 1$ digits (*digit extension*), the following rule must be used (B even):

$$\text{if } x_{n-1} \geq B/2, \text{ then } x_n = B - 1, \quad \text{and if } x_{n-1} < B/2, \text{ then } x_n = 0.$$

Actually, in the first case,

$$(B - 1 - B).B^n + x_{n-1}.B^{n-1} = -B^n + x_{n-1}.B^{n-1} = (x_{n-1} - B).B^{n-1},$$

and in the second case,

$$0.B^n + x_{n-1}.B^{n-1} = x_{n-1}.B^{n-1}.$$

3. If $B = 2$ (2's complement system) the $(n + 1)$-bit vector $x_{n-1} \, x_{n-1} \, x_{n-2} \cdots x_1 \, x_0$ represents the same number as the n-bit vector $x_{n-1} \, x_{n-2} \cdots x_1 \, x_0$ (*sign bit extension*).

The B's complement method is almost exclusively used with $B = 2$, in which case the most significant bit is also the sign bit (3.9). In the general case, the most significant digit must be compared with $B/2$ in (3.7) in order to deduce the sign of x. A *reduced B's complement* numeration system could also be defined in which the most significant digit x_{n-1} is either 0 or $B - 1$.

Definition 3.5 In the *reduced B's complement numeration system* (B even), every integer x belonging to the range $-B^{n-1} \leq x < B^{n-1}$ is represented by

$$R(x) = x \bmod B^n.$$

If $0 \leq x < B^{n-1}$, then

$$R(x) = x < B^{n-1} \text{ and } x_{n-1} = 0,$$

and if $-B^{n-1} \leq x < 0$, then

$$R(x) = B^n + x \geq B^n - B^{n-1} = (B - 1).B^{n-1} \text{ and } x_{n-1} = B - 1.$$

Thus the integer represented by $x_{n-1} \, x_{n-2} \cdots x_1 \, x_0$ is

$$x = -B^{n-1} + x_{n-2}.B^{n-2} + \cdots + x_0 \text{ if } x_{n-1} = B - 1 \text{ and}$$

$$x = x_{n-2}.B^{n-2} + \cdots + x_0 \text{ if } x_{n-1} = 0, \tag{3.10}$$

and the sign definition rule is the following one:

$$\text{if } x \text{ is negative then } x_{n-1} = B - 1; \quad \text{if } x \text{ is nonnegative then } x_{n-1} = 0.$$

In fact, the reduced B's complement representation is deduced from the nonreduced one by adding a digit (*digit extension*, Comment 3.4(2)) if the most significant digit is different from 0 or $B - 1$.

As in the binary case the $(n + 1)$-digit vector $x_{n-1} x_{n-1} x_{n-2} \cdots x_1 x_0$ represents the same number as the n-digit vector $x_{n-1} x_{n-2} \cdots x_1 x_0$.

Example 3.4   Represent $x = -287645$ with $n = 6$ digits in B's complement form with $B = 10$:

$$B^6 = 1000000,$$

$$B^6/2 = 500000,$$

$$R(x) = x + B^6 = 712355$$

Observe that

$$x_5' = 7 - 10 = -3,$$

$$-3.10^5 + 12355 = -287645.$$

In reduced B's complement form, $n = 7$ digits are necessary $(-287645 < - B^{n-1} = -50000)$:

$$R(x) = x + B^7 = 9712355.$$

Observe that $-10^6 + 712355 = -287645$ and that 9712355 is deduced from 712355 by adding one digit according to the digit extension rule.

3.2.4 Booth's Encoding

According to relation (3.8) the 2's complement representation $x_{n-1} x_{n-2} \cdots x_1 x_0$ of an integer x could also be seen as a *signed-digit* representation

$$x_{n-1}.2^{n-1} + x_{n-2}.2^{n-2} + \cdots + x_0$$

where $x_{n-1} \in \{-1, 0\}$ and all other digits $x_i \in \{0, 1\}$. The Booth's encoding ([BOO1951]) generates another signed-digit representation:

Definition 3.6   Consider an integer y whose 2's complement representation is $x_{n-1} x_{n-2} \cdots x_0$ and define

$$\begin{aligned}
y_0 &= -x_0, \\
y_1 &= -x_1 + x_0, \\
y_2 &= -x_2 + x_1, \\
&\vdots \quad \vdots \\
y_{n-1} &= -x_{n-1} + x_{n-2}.
\end{aligned} \tag{3.11}$$

Then by multiplying the first equation by 2^0, the second by 2^1, the third one by 2^2, and so on, and adding up the n equations, the following relation is obtained:

$$y_{n-1}.2^{n-1} + y_{n-2}.2^{n-2} + \cdots + y_0.2^0 = -x_{n-1}.2^{n-1}$$
$$+ x_{n-2}.2^{n-2} + \cdots + x_0.2^0.$$

The vector $(y_{n-1} \; y_{n-2} \cdots y_0)$ whose components y_i belong to $\{-1, 0, 1\}$ is the *Booth*-1 *representation* of x and

$$x = y_{n-1}.2^{n-1} + y_{n-2}.2^{n-2} + \cdots + y_0.2^0 \tag{3.12}$$

Observe that the Booth's representation of an integer is formally the same as the binary representation of a natural number. The Booth's encoding method can be generalized.

Definition 3.7   Consider an integer whose 2's complement representation is x_{n-1} $x_{n-2} \cdots x_0$, with $n = 2.m$ bits, and define

$$y_0 = -2.x_1 + x_0,$$
$$y_1 = -2.x_3 + x_2 + x_1,$$
$$y_2 = -2.x_5 + x_4 + x_3,$$
$$\vdots \qquad \vdots$$
$$y_{m-1} = -2.x_{2.m-1} + x_{2.m-2} + x_{2.m-3} \tag{3.13}$$

Then by multiplying the first equation by 4^0, the second by 4^1, the third by 4^2, and so on, and adding up the m equations, the following relation is obtained:

$$y_{m-1}.4^{m-1} + y_{m-2}.4^{m-2} + \cdots + y_0.4^0 = -x_{n-1}.2^{n-1}$$
$$+ x_{n-2}.2^{n-2} + \cdots + x_0.2^0.$$

The vector $(y_{m-1} \; y_{m-2} \cdots y_0)$ whose components y_i belong to $\{-2, -1, 0, 1, 2\}$ is the *Booth*-2 *representation* of x and

$$x = y_{m-1}.4^{m-1} + y_{m-2}.4^{m-2} + \cdots + y_0.4^0 \tag{3.14}$$

More generally, a *Booth-r representation* can be defined as follows:

Definition 3.8 Let x be an integer whose 2's complement representation is x_{n-1} $x_{n-2} \cdots x_0$, with $n = r.m$ bits, and define

$$y_0 = -x_{r-1}.2^{r-1} + x_{r-2}.2^{r-2} + \cdots + x_1.2 + x_0,$$

$$y_i = -x_{i.r+r-1}.2^{r-1} + x_{i.r+r-2}.2^{r-2} + \cdots + x_{i.r+1}.2 + x_{i.r}$$
$$+ x_{i.r-1}, \quad \forall i \in \{1, 2, \ldots, m-1\}. \tag{3.15}$$

The vector $(y_{m-1} \, y_{m-2} \cdots y_0)$ whose components y_i belong to

$$\left\{ -2^{r-1}, \; -(2^{r-1}-1), \ldots, -2, \; -1, 0, 1, 2, \ldots, 2^{r-1}-1, 2^{r-1} \right\} \tag{3.16}$$

is the *Booth-r representation* of x and

$$x = y_{m-1}.B^{m-1} + y_{m-2}.B^{m-2} + \cdots + y_0.B^0, \quad \text{where } B = 2^r. \tag{3.17}$$

Comments 3.5

1. Given an integer x whose 2's complement representation is $x_{n-1} \, x_{n-2} \ldots x_0$, with $n = r.m$ bits, the following signed digits could be defined (one for each r-bit slice):

$$yi = x_{i.r+r-1}.2^{r-1} + x_{i.r+r-2}.2^{r-2} + \cdots + x_{i.r+1}.2$$
$$+ x_{i.r}, \quad \forall i \in \{0, 1, 2, \ldots, m-2\}$$
$$y_{m-1} = -x_{m.r-1}.2^{r-1} + x_{m.r-2}.2^{r-2} + \cdots + x_{m.r+1}.2 + x_{m.r}, \tag{3.18}$$

so that

$$yi \in \{0, 1, 2, \ldots, 2^r - 1\} \forall i \in \{0, 1, 2, \ldots, m-2\}, \tag{3.19}$$
$$y_{m-1} \in \{ -2^{r-1}, \; -(2^{r-1}-1), \ldots, -2, \; -1, 0\},$$

and

$$x = y_{m-1}.B^{m-1} + y_{m-2}.B^{m-2} + \cdots + y_0.B_0, \quad \text{where } B = 2^r. \tag{3.20}$$

Nevertheless, for $r > 1$, the total range defined by (3.19), namely

$$\{ -2^{r-1}, \; -(2^{r-1}-1), \ldots, -2, \; -1, 0, 1, 2, \ldots, 2^r - 1\},$$

is larger then the range defined by (3.16).

2. The range (3.16) contains $B+1$ values, from $-B/2$ to $B/2$, where $B = 2^r$. This means that the total number of expressions (3.17) is equal to $(B+1)^m$. The

numbers x defined by (3.17) are included between $-(B/2).(B^m - 1)/(B - 1)$ and $(B/2).(B^m - 1)/(B - 1)$, so that the range of x contains $1 + B.(B^m - 1)/(B - 1)$ integers. Except when $m = 1$, the following inequality is satisfied:

$$1 + B.(B^{m-1})/(B - 1) < (B + 1)^m.$$

Thus the set of digits (3.16) is *redundant* as the number of different expressions is greater than the range of the represented numbers.

Example 3.5 Compute the Booth's encoding of -287645; the 2's complement representation of -287645 is;

$$1\ 0\ 1\ 1\ 1\ 0\ 0\ 1\ 1\ 1\ 0\ 0\ 0\ 1\ 1\ 0\ 0\ 0\ 1\ 1;$$

according to (3.11) its Booth-1 representation is

$$-1\ 1\ 0\ 0\ -1\ 0\ 1\ 0\ 0\ -1\ 0\ 0\ 1\ 0\ -1\ 0\ 0\ 1\ 0\ -1,$$

and according to (3.13) its Booth-2 representation is

$$-1\ 0\ -2\ 2\ -1\ 0\ 2\ -2\ 1\ -1.$$

By substituting two successive bits by a 4-valued digit (comments 3.5) the following representation is obtained:

$$-2\ 3\ 2\ 1\ 3\ 0\ 1\ 2\ 0\ 3.$$

Other expressions can be deduced from the previous one by applying simple rules such as

$$(\underline{-2}).4 + \underline{3}.1 = (\underline{-1}).4 + (\underline{-1}).1,\ \underline{1}.4 + \underline{3}.1 = \underline{2}.4 + (\underline{-1}).1,\ \underline{0}.4 + \underline{3}.1$$

$$= \underline{1}.4 + (\underline{-1}).1.$$

Thus, in the preceding expression $-2\ 3\ 2\ \underline{1\ 3}\ 0\ 1\ 2\ \underline{0\ 3}$, the underlined pairs can be substituted by $-1\ -1$, $2\ -1$ and $1\ -1$, respectively, yielding the following equivalent expression:

$$-1\ -1\ 2\ 2\ -1\ 0\ 1\ 2\ 1\ -1.$$

Observe that the latter is different from the Booth-2 representation, in spite of using the same digits (Comment 3.5(2)).

3.3   REAL NUMBERS

As regards the real numbers, there are two types of approximations: *fixed-point* and *floating-point* numeration systems. The fixed-point system is a simple extension of the integer representation system; it allows the representation of a relatively *reduced range* of numbers with some constant *absolute precision*. The floating point system allows the representation of a very large range of numbers, with some constant *relative precision*.

Definitions 3.9

1. In a *fixed-point numeration system*, the number represented in the form

$$x_{n-p-1} \; x_{n-p-2} \cdots x_1 \; x_0 . x_{-1} \; x_{-2} \cdots x_{-p} \tag{3.21}$$

   is x/B^p, where x is the integer represented by the same sequence of digits without point.

2. Let x_{min} and x_{max} be the minimum and maximum integers that can be represented with n digits, that is, $x_{min} = 1 - B^{n-1}$ and $x_{max} = B^{n-1} - 1$ in sign-magnitude representation, and $x_{min} = -B^n/2$ and $x_{max} = B^n/2 - 1$ in B's complement or excess-$B^n/2$ representation. Then, any real number x belonging to the interval

$$B^{-p}.x_{min} \le x \le B^{-p}.x_{max}$$

   can be represented in the form (3.21) with some *error* equal to the absolute value of the difference between x and its representation.

3. The *distance d* between exactly represented numbers is equal to the *unit in the least significant position (ulp)*, that is, B^{-p}, so that the *maximum error* is equal to

$$ulp/2 = B^{-p}/2.$$

4. The *maximum relative error* is equal to $ulp/(2.|x|) = 1/(2.|x|.B^p)$. If $x \ne 0$ then $|x| \ge B^{-p}$, so that the maximum relative error is less than or equal to $\frac{1}{2}$.

Example 3.6   The range of numbers x that can be represented in B's complement, with $B = 10$, $n = 9$ digits, and $ulp = 10^{-3}$ is

$$-10^6/2 \le x < 10^6/2.$$

The following numbers can be exactly represented:

$$-500000.000, \; -499999.999, \; -499999.998, \ldots, \; -0.001, 0.000,$$
$$0.001, \ldots, 499999.999.$$

The distance between them is equal to $ulp = 0.001$.

Definitions 3.10

1. In a *floating-point numeration system*, the representation consists of two numbers: a fixed-point number (the *significand*) $+s$ or $-s$, where s is a non-negative number, and an integer (the *exponent*) e. The corresponding number is $\pm s.b^e$, where b is the chosen base (not necessarily equal to B).

2. Let s_{min}, s_{max}, e_{min}, and e_{max} be the minimum and maximum values of s and e, respectively. The range of represented numbers is

$$-s_{max}.b^{e_{max}} \le x \le s_{max}.b^{e_{max}} \tag{3.22}$$

and the minimum absolute value of a represented number is

$$|x| \ge s_{min}.b^{e_{min}}. \tag{3.23}$$

3. Let ulp be the unit in the least representative position of the significand. Then the *distance* D between exactly represented numbers is $D = d.b^e$, where $d = ulp$ is the distance between two successive values of the significand. Thus the value of D depends on the exponent e. The *maximum error* is equal to

$$D_{max}/2 = ulp.b^{e_{max}}/2.$$

4. The *maximum relative error* is equal to $D/(2.|x|) = ulp.b^e/(2.s.b^e) = ulp/2.s$. As in the preceding case (Definition 3.9(4)) the maximum relative error is less than or equal to $\frac{1}{2}$.

Comment 3.6 In a floating-point system, with q digits for representing the absolute value s of the significand and t digits for representing the exponent, the range of positive numbers is

$$ulp.b^{e_{min}} \le x < ulp.B^q.b^{e_{max}},$$

the maximum error is equal to

$$max\ error_{floating} = ulp.b^{e_{max}}/2,$$

and the maximum relative error is equal to $\frac{1}{2}$.

In a fixed-point system with $q + t$ digits, the range of positive numbers is

$$ulp \le x < ulp.B^{q+t},$$

the maximum error is equal to

$$max\ error_{\text{fixed}} = ulp/2,$$

and the maximum relative error is equal to $\frac{1}{2}$.

In order to compare both systems, one can compute the quotient rr (*relative range*) between the maximum and the minimum value of x (x positive). In the floating-point system

$$rr_{\text{floating}} = B^q.b^{e_{\max}-e_{\min}}, \qquad (3.24)$$

and in the fixed point system

$$rr_{\text{fixed}} = B^{q+t}. \qquad (3.25)$$

Taking into account that $e_{\max} - e_{\min} \cong B^t$, it is obvious that

$$rr_{\text{floating}} >> rr_{\text{fixed}}.$$

Nevertheless, the maximum relative errors are equal. As regards the maximum errors, their values depend on the *ulp* (not necessarily the same value in both cases).

Example 3.7 In the *ANSI/IEEE* (*[ANS1985]*) *single-precision floating-point system*, the significand is a sign-magnitude integer

$$\pm s = \pm 1.s_{-1}s_{-2}\cdots s_{-23},$$

where $s_{-1}\,s_{-2}\cdots s_{-23}$ is called the *mantissa*, and the exponent is an excess $-\,127$ integer $e_7\,e_6\cdots e_0$. The 32-bit word

$$sign\ e_7 e_6 \cdots e_0 s_{-1} s_{-2} \cdots s_{-23}$$

represents the number

$$(-1)^{\text{sign}}.(1 + s_{-1}.2^{-1} + s_{-2}.2^{-2} + \cdots + s_{-23}.2^{-23}).2^e,$$

where $e = e_7.2^7 + e_6.2^6 + \cdots + e_0.2^0 - 127$.
    Thus

$$s_{\min} = 1,\ s_{\max} = 1.11\cdots 1 \cong 2,\ ulp = 2^{-23},\ e_{\min} = -127,\ e_{\max} = 128.$$

Nevertheless, e_{min} and e_{max} are not used for representing ordinary numbers; they are used for representing

$$0 \cong +1.0 \times 2^{-127}, \; -0 \cong -1.0 \times 2^{-127}, \; +\infty \cong 1.0 \times 2^{128}, \; -\infty \cong -1.0 \times 2^{128},$$

and other nonordinary numbers. The actual minimum and maximum values are

$$e_{min} = -126, \quad e_{max} = 127,$$

so that the range of represented numbers is $-2.2^{127} < x < 2.2^{127}$, that is

$$-2^{128} < x < 2^{128},$$

and the minimum positive represented number is 1.2^{-126}.

3.4 BIBLIOGRAPHY

[ANS1985] ANSI and IEEE, *IEEE Standard for Binary Floating-Point Arithmetic*, ANSI/ IEEE Standard, Std 754-1985, New York, 1985.

[BOO1951] A. D. Booth, A signed binary multiplication technique. *Q. J. Mechanics Appl. Math.*, **June**: 236–240 (1951).

[ERC2004] M. Ercegovac and T. Lang. *Digital Arithmetic*, Morgan Kaufmann Publishers, San Francisco, CA, 2004.

[PAR1999] B. Parhami, *Computer Arithmetic: Algorithms and Hardware Designs*, Oxford University Press, New York, 1999.

4

ARITHMETIC OPERATIONS: ADDITION AND SUBTRACTION

Addition is used as a primitive operation for computing most arithmetic functions, so that it deserves particular attention. The classical pencil and paper algorithm implies the sequential computation of a set of carries, each of them depending on the preceding one. As a consequence, the execution time of any program, or circuit, based on the classical algorithm is proportional to the number n of digits of the operands. In order to minimize the computation time, several general ideas have been proposed. One of them consists of modifying the classical algorithm in such a way that the computation time of each carry is minimal; the time complexity is still proportional to n, but the proportionality constant is smaller. Another approach rests on the use of a different numeration system; instead of adding two base-B n-digit numbers, two base-B^s (n/s)-digit numbers are considered. Several algorithms, different from the classical one and generally based on some kind of tree structure, have been proposed. If their implicit parallelism can be exploited, execution times proportional to $\log n$ are reached.

4.1 ADDITION OF NATURAL NUMBERS

4.1.1 Basic Algorithm

Consider the base-B representations of two n-digit numbers:

$$x = x_{n-1}.B^{n-1} + x_{n-2}.B^{n-2} + \cdots + x_0.B^0,$$

$$y = y_{n-1}.B^{n-1} + y_{n-2}.B^{n-2} + \cdots + y_0.B^0.$$

Synthesis of Arithmetic Circuits: FPGA, ASIC, and Embedded Systems
By Jean-Pierre Deschamps, Géry J. A. Bioul, and Gustavo D. Sutter
Copyright © 2006 John Wiley & Sons, Inc.

The following (pencil and paper) algorithm computes the $(n+1)$-digit representation of $z = x + y + c_{in}$ where c_{in} is an initial carry equal to 0 or 1.

Algorithm 4.1   Classic Addition

```
q(0):=c_in;
for i in 0..n-1 loop
    if x(i)+y(i)+q(i)>B-1 then q(i+1):=1; else q(i+1):=0;
    end if;
    z(i):=(x(i)+y(i)+q(i)) mod B;
end loop;
z(n):=q(n);
```

As $q(i+1)$ is a function of $q(i)$ the execution time of Algorithm 4.1 is proportional to n. In order to reduce the execution time of each iteration step, Algorithm 4.1 can be modified. First, define two binary functions of two B-valued variables, namely, the propagate (p) and generate (g) functions:

$$p(a, b) = 1 \text{ if } a + b = B - 1, \quad p(a, b) = 0 \text{ otherwise;}$$
$$g(a, b) = 1 \text{ if } a + b > B - 1, \quad g(a, b) = 0 \text{ otherwise.}$$

(4.1)

The next carry q_{i+1} can be calculated as follows:

```
if p(x(i), y(i))=1 then q(i+1):=q(i); else q(i+1):=g(x(i),
y(i)); end if;
```

The corresponding modified algorithm is the following one.

Algorithm 4.2   Carry-Chain Addition

```
--computation of the generation and propagation conditions:
for i in 0..n-1 loop g(i):=g(x(i),y(i)); p(i):=p(x(i),y(i));
end loop;
--carry computation:
q(0):=c_in;
for i in 0..n-1 loop
    if p(i)=1 then q(i+1):=q(i); else q(i+1):=g(i); end if;
end loop;
-sum computation
for i in 0..n-1 loop z(i):=(x(i)+y(i)+q(i)) mod B; end loop;
z(n):=q(n);
```

Comments 4.1

1. Observe that the first iteration includes $2.n$ B-ary operations (computation of $g(i)$ and $p(i)$) that could be executed in parallel. The second iteration is made up of n iteration steps that must be executed sequentially (as $q(i+1)$ is a function of $q(i)$) and consists of binary operations only. The last iteration includes

n B-ary operations (computation of $z(i)$) that could be executed in parallel. Algorithm 4.2 thus splits the operations into concurrent B-ary ones (first and third iterations) and sequential binary ones (second iteration). The sequential binary operations are the same whatever the base B. The expected computation time reduction is due to the substitution of the (relatively) complex instruction

```
if x(i)+y(i)+q(i)>B-1 then q(i+1):=1; else q(i+1):=0;
end if;
```

by the simpler one

```
if p(i)=1 then q(i+1):=q(i); else q(i+1):=g(i); end if;
```

2. The preceding instruction sentence is equivalent to the following Boolean equation:

$$q(i + 1) = p(i).q(i) \vee not(p(i)).g(i). \tag{4.2}$$

Furthermore, if the preceding relation is used, then the definition of the generate function can be modified:

$$g(a, b) = 1 \text{ if } a + b > B - 1, \quad g(a, b) = 0 \text{ if } a + b < B - 1,$$
$$g(a, b) = 0 \text{ or } 1 \text{ (don't care) otherwise.}$$

3. Another Boolean equation equivalent to (4.2) is

$$q(i + 1) = g(i) \vee p(i).q(i). \tag{4.3}$$

If the preceding relation is used, then the definition of the propagate function can be modified:

$$p(a, b) = 1 \text{ if } a + b = B - 1, \quad p(a, b) = 0 \text{ if } a + b < B - 1,$$
$$p(a, b) = 0 \text{ or } 1 \text{ (don't care) otherwise.}$$

4.1.2 Faster Algorithms

The values of $q(1), q(2), \ldots, q(n)$ could also be calculated in parallel:

Property 4.1

$$\forall i = 1, 2, \ldots, n:$$
$$q(i) = g(i - 1) \vee g(i - 2)..p(i - 1) \vee g(i - 3).p(i - 2).p(i - 1)$$
$$\vee g(i - 4).p(i - 3).p(i - 2).p(i - 1)\vee \tag{4.4}$$
$$\cdots \vee g(0).p(1).\cdots.p(i - 1) \vee q(0).p(0).p(1).\cdots.p(i - 1),$$

where symbol $\vee$ stands for the Boolean sum, $g(i) = g(x(i), y(i))$ and $p(i) = p(x(i), y(i))$.

Relation (4.4) is deduced from (4.3) by induction. The corresponding algorithm is the following one.

Algorithm 4.3

```
--computation of the generation and propagation conditions:
for i in 0..n-1 loop g(i):=g(x(i),y(i)); p(i):=p(x(i),y(i));
end loop;
--carry computation:
q(0):=c_in;
for i in 1..n loop
  q(i):=g(i-1) or g(i-2)*p(i-1) or...or g(0)*p(1)*...*
  p(i-1) or q(0)*p(0)*p(1)*...*p(i-1);
end loop;
--sum computation
for i in 0..n-1 loop z(i):=(x(i)+y(i)+q(i)) mod B; end loop;
z(n):=q(n);
```

The preceding algorithm is made up of three iterations whose operations could be executed in parallel as $q(i)$ just depends on the operands x, y, and c_in but not on the preceding carries. Nevertheless, the execution of

```
q(i):=g(i-1) or g(i-2)*p(i−1) or...or g(0)*p(1)*...*p(i-1)
  or q(0)*p(0)*p(1)*...*p(i-1);
```

implies the computation of a $(2.i + 1)$-variable switching function—a $(2.n + 1)$-variable function in the case of $q(n)$. Except for small values of n, specific algorithms must be defined for computing these functions. For that purpose two new concepts are introduced: the *dot operation and the generalized generate and propagate functions*:

Definitions 4.1

1. Given two 2-component binary vectors $a_i = (a_{i0}, a_{i1})$ and $a_k = (a_{k0}, a_{k1})$ the *dot operation* $\bullet$ defines an application from $B_2^2 \times B_2^2$ into B_2^2:

$$a_i \bullet a_k = (a_{i0} \vee a_{k0}.a_{i1}, a_{i1}.a_{k1}).$$

   It can easily be demonstrated that it is a noncommutative associative operation; $(0,1)$ is the neutral element and $(0,0)$ the left 0-element.

2. Given the generate and propagation functions $g(i)$ and $p(i)$, for $i \in \{0, 1, \ldots, n − 1\}$, the *generalized generate and propagate functions* $g(i:i-k)$ and $p(i:i-k)$, for $i \in \{0, 1, \ldots, n − 1\}$ and $k \in \{0, 1, \ldots, i\}$ are defined as follows:

$$(g(i:i − k), p(i:i − k)) = (g(i), p(i)) \bullet (g(i − 1), p(i − 1))$$
$$\bullet(g(i − 2), p(i − 2)) \bullet \cdots \bullet (g(i − k), p(i − k)). \tag{4.5}$$

The following property is deduced from (4.4) and from the preceding definitions.

Property 4.2

$$q(i+1) = g(i:i-k) \vee p(i:i-k).q(i-k). \qquad (4.6)$$

Then Algorithm 4.3 can be modified as follows.

Algorithm 4.4

```
--computation of the generation and propagation conditions:
for i in 0..n-1 loop g(i):=g(x(i),y(i)); p(i):=p(x(i),y(i));
 end loop;
--computation of the generalized generation and propagation
 conditions:
for i in 1..n loop
  (g(i-1:0), p(i-1:0)):=(g(i-1), p(i-1)) dot (g(i-2),
  p(i-2)) dot ... dot (g(0), p(0));
end loop;
--carry computation:
q(0):=c_in;
for i in 1..n loop q(i):=g(i-1:0) or p(i-1:0)*q(0); end loop;
--sum computation:
for i in 0..n-1 loop z(i):=(x(i)+y(i)+q(i)) mod B; end loop;
z(n):=q(n);
```

The second iteration of Algorithm 4.4, that is, the computation of all pairs $(g(i-1,0), p(i-1,0))$, can be performed in several ways. It is a particular case of a more general problem: Given a set of input data $a(0), a(1), \ldots, a(n-1)$ and an associative operator $\bullet$ (*dot*), compute

$$
\begin{aligned}
b(0) &= a(0), \\
b(1) &= a(1) \bullet a(0), \\
b(2) &= a(2) \bullet a(1) \bullet a(0), \\
&\cdots \\
b(n-1) &= a(n-1) \bullet \cdots \bullet a(1) \bullet a(0).
\end{aligned}
\qquad (4.7)
$$

The simplest (naïve) algorithm is

```
b(0):=a(0); for i in 1..n-1 loop b(i):=a(i) dot b(i-1); end loop;
```

whose execution time is proportional to n. Nevertheless, better algorithms have been proposed, among others ([BRE1982], [LAD1980], [KOG1973], [HAN1987], [SUG1990]). Two of them are described below; they are based on the definition of a procedure dot_procedure computing Equations (4.7); its input and output parameters are a natural number n (the number of input data), and two n-component vectors (the input data and the output result):

```
procedure dot_procedure (n:in natural;
a:in data_vector(0..n-1); b:out data_vector(0..n-1));
```

Assume that n is a power of 2 (0's should be added if necessary). A first algorithm consists of:

computing
$$c(0) = a(1) \bullet a(0),$$
$$c(1) = a(3) \bullet a(2),$$
$$\cdots$$
$$c((n/2) - 1) = a(n-1) \bullet a(n-2);$$

calling dot_procedure with parameters $n/2$, c, and d, so that

$$d(0) = b(1),$$
$$d(1) = b(3),$$
$$d(2) = b(5),$$
$$\cdots$$
$$d((n/2) - 1) = b(n-1);$$

computing the missing components of b;

$$b(2) = a(2) \bullet d(0), b(4) = a(4) \bullet d(1), \ldots, b(n-2) = a(n-2) \bullet d((n/2) - 2).$$

The *computation scheme* (or precedence graph, Chapter 10) is shown in Figure 4.1 (with $n = 16$). The corresponding recursive algorithm is the following.

Algorithm 4.5   Dot Procedure (1)

```
procedure dot_procedure (n:in natural; a:in data_vector(0..
n-1); b:out data_vector(0..n-1)) is
  c,d: data_vector(0..(n/2)-1);
begin
  if n=2 then b(0):=a(0); b(1):=a(1) dot a(0);
  else
  for i in 0..(n/2)-1 loop c(i):=a((2*i)+1) dot a(2*i);
  end loop;
  dot_procedure (n/2, c, d);
  b(0):=a(0);
  for i in 1..(n/2)-1 loop b(2*i):=a(2*i) dot d(i-1);
  b((2*i)+1):=d(i); end loop;
  end if;
end dot_procedure;
```

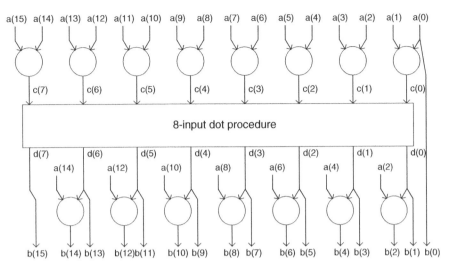

Figure 4.1 A 16-input dot procedure (first algorithm).

Both **for** loops are made up of dot operations that can be executed in parallel. The total execution time $T(n)$ is equal to $T_{dot} + T(n/2) + T_{dot}$, with $T(2) = T_{dot}$, so that

$$T_{\text{Algorithm } 4.5}(n) = (2.(\log_2 n) - 1).T_{dot}. \qquad (4.8)$$

The second algorithm consists of:

calling `dot_procedure` with parameters $n/2$, $a(0 .. (n/2) - 1)$, and $b(0 .. (n/2) - 1)$;
calling `dot_procedure` with parameters $n/2$, $a((n/2) .. n - 1)$, and c, so that

$$c(0) = a(n/2),$$
$$c(1) = a((n/2) + 1) \bullet a(n/2),$$
$$\cdots$$
$$c((n/2) - 1) = a(n - 1) \bullet a(n - 2) \bullet \cdots \bullet a(n/2);$$

computing the missing components of b,

$$b(n/2) = c(0) \bullet b((n/2) - 1),$$
$$b((n/2) + 1) = c(1) \bullet b((n/2) - 1),$$
$$\cdots$$
$$b(n - 1) = c((n/2) - 1) \bullet b((n/2) - 1).$$

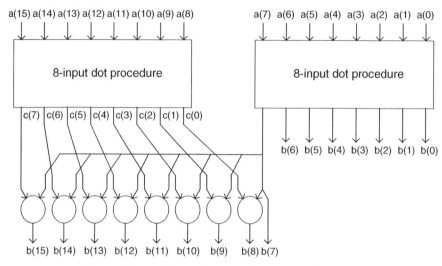

Figure 4.2 A 16-input dot procedure (second algorithm).

The computation scheme is shown in Figure 4.2 (with $n = 16$), and the corresponding recursive algorithm is the following.

Algorithm 4.6 Dot Procedure (2)

```
procedure dot_procedure (n:in natural; a:in data_vector(0..
n-1);b:out data_vector(0..n-1)) is
  c: data_vector(0..(n/2)-1);
begin
  if n=2 then b(0):=a(0); b(1):=a(1) dot a(0);
  else
    dot_procedure (n/2, a(0..(n/2)-1), b(0..(n/2)-1);
    dot_procedure (n/2, a((n/2)-1..n-1), c);
    for i in 0..n/2-1 loop b(i+(n/2)):=c(i) dot b((n/2)-1);
    end loop;
  end if;
end dot_procedure;
```

Both procedure calls can be executed in parallel, and the **for** loop is made up of dot operations that can also be executed in parallel. The total execution time $T(n)$ is equal to $T(n/2) + T_{\text{dot}}$, with $T(2) = T_{\text{dot}}$, so that

$$T_{\text{Algorithm 4.6}}(n) = (\log_2 n).T_{\text{dot}}. \qquad (4.9)$$

The following algorithm is deduced from Algorithm 4.4 and the definition of dot_procedure.

Algorithm 4.7   Parallel-Prefix Addition

```
a, b: data_vector(0..n-1, 0..1);
begin
  --computation of the generation and propagation conditions:
  for i in 0..n-1 loop a(i,0):=g(x(i),y(i)); a(i,1):=p(x(i),
  y(i))); end loop;
  --computation of the generalized generation and propagation
  --conditions:
  dot_procedure(n, a, b);
  --carry computation:
  q(0):=c_in;
  for i in 1..n loop q(i):=b(i,0) or b(i,1)*q(0); end loop;
  --sum computation
  for i in 0..n-1 loop z(i):=(x(i)+y(i)+q(i)) mod B; end loop;
  z(n):=q(n);
```

The preceding algorithm is made up of three iterations, whose operations can be executed in parallel, and a call to dot_procedure. The procedure execution time depends on the number of digits n; according to (4.8) or (4.9) it is proportional to $\log(n)$. The execution time of the iterations is independent of n. Thus for great values of n, the execution time of Algorithm 4.7 is practically proportional to $\log(n)$.

A logarithmic execution time can be obtained with a different algorithm using two new procedures. The first one,

procedure carry_lookahead_procedure (n:**in** natural; a:**in** data_
vector(0..n-1, 0..1); c_in: **in** bit; q:**out** bit_vector(1..n));

computes the n carries $q(1), q(2), \ldots, q(n)$, in function of the n generation and propagation conditions $g(t) = a(t,0)$ and $p(t) = a(t,1)$, and of c_in, that is,

$$q(t) = a(t-1,0) \vee a(t-2,0).a(t-1,1) \vee \cdots \vee a(0,0).a(1,1). \cdots .a(t-1,1) \vee$$
$$c\_in.a(0,1).a(1,1). \cdots .a(t-1,1), \forall t \in \{1, 2, \ldots, n\}.$$

The second one,

procedure carry_procedure (n:**in** natural; b:**in** data_vector(0..
n-1, 0..1); c_in: **in** bit; q:**out** bit_vector(1..n));

computes the n carries $q(1), q(2), \ldots, q(n)$, in function of the n generalized generation and propagation conditions $g(t{:}0) = b(t,0)$ and $p(t{:}0) = b(t,1)$, and of c_in, that is,

$$q(t) = b(t-1,0) \vee c\_in.b(t-1,1). \tag{4.10}$$

Assume that n can be factorized under the form $n = k.s$. The algorithm consists of:

calling k times the dot_procedure with parameters s, $a(j.s \,.. \,j.s + s - 1)$ and $c(j, 0..s - 1)$, where $j \in \{0, 1, \ldots, k - 1\}$, so that

$$c(j, 0) = a(j.s) = (g(j.s), p(j.s)),$$
$$c(j, 1) = a(j.s + 1) \bullet a(j.s) = (g(j.s + 1:j.s), p(j.s + 1:j.s)),$$
$$\cdots$$
$$c(j, s - 1) = a(j.s + s - 1) \bullet a(j.s + s - 2) \bullet \cdots \bullet a(j.s)$$
$$= (g(j.s + s - 1:j.s), p(j.s + s - 1:j.s));$$

calling carry_lookahead_procedure with parameters k, $c(0 .. k - 1, s - 1)$, c_in and d, so that

$$d(j) = g(j.s - 1:0) \vee c\_in.p(j.s - 1:0) = q(j.s);$$

calling k times the carry_procedure with parameters $s - 1$, $c(j, 0 .. s - 2)$, $d(j)$, and $e(j, 0 .. s - 2)$, where $j \in \{0, 1, \ldots, k - 1\}$, so that

$$e(j,i) = (g(j.s + i:j.s) \vee q(j.s).p(j.s + i:j.s) = q(j.s + i + 1).$$

The computation scheme is shown in Figure 4.3 (with $k = s = 4$), and the corresponding recursive algorithm is the following:

Algorithm 4.8

```
procedure carry_lookahead_procedure
(n:in natural; a:in data_vector(0..n-1, 0..1); c_in: in bit;
q:out bit_vector(1..n)) is
  c: data_vector(0..k-1, 0..s-1, 0..1); d: bit_vector(0..
  k-1);
begin
  for j in 0..k-1 loop dot_procedure(s, a(j*s..j*s+s-1), c(j,
  0..s-1)); end loop;
  carry_lookahead_procedure (k, c(0..k-1, s-1), c_in, d);
  for j in 0..k-1 loop
    carry_procedure(s-1, c(j, 0..s-2), d(j), e(j, 0..s-2));
  end loop;
  for j in 1..k-1 loop q(j*s):=d(j); end loop;
  for j in 0..k-1 loop
    for i in 0..s-2 loop q(j*s+i+1):=e(j, i); end loop;
  end loop;
end carry_lookahead_procedure;
```

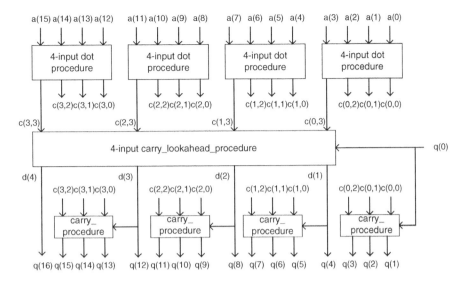

Figure 4.3 A 16-input `carry_lookahead_procedure`.

The procedure `carry_procedure` computes (4.10):

```
procedure carry_procedure
(n:in natural; b:in data_vector(0..n − 1, 0..1); c_in: in bit;
q:out bit_vector(1..n))
is begin
  for t in 1..n loop q(t)=b(t − 1,0) v c_in.b(t − 1,1); end loop;
end carry_procedure;
```

Let $T(n)$ be the execution time of `carry_lookahead_procedure`, $T_1(n)$ the execution time of `dot_procedure`, and T_2 the execution time of any one of the equations (4.10). The k calls to `dot_procedure` can be executed in parallel, and the same occurs with the k calls to `carry_procedure`. Furthermore, within `carry_procedure` the equations (4.10) can be calculated in parallel. Thus

$$T(k.s) = T_1(s) + T(k) + T_2.$$

Assume now that $n = s_1.s_2.\cdots.s_m$. The algorithm obtained by recursively calling the `carry_lookahead_procedure` has a computation time that can be calculated as follows:

$$T(s_1.s_2.\cdots.s_m) = T_1(s_1) + T(s_2.\cdots.s_m) + T_2,$$
$$T(s_2.\cdots.s_m) = T_1(s_2) + T(s_3.\cdots.s_m) + T_2,$$

$$\cdots$$

$$T(s_{m-1}.s_m) = T_1(s_{m-1}) + T(s_m) + T_2,$$
$$T(s_m) = T_1(s_m) + T_2,$$

so that

$$T(s_1.s_2.\cdots.s_m) = T_1(s_1) + T_1(s_2) + \cdots + T_1(s_m) + m.T_2. \tag{4.11}$$

In particular, if $n = s^m$ then

$$T(n) = m.(T_1(s) + T_2), \quad \text{where } m = \log_s n. \tag{4.12}$$

The complete addition algorithm is the following.

Algorithm 4.9   Carry-Lookahead Addition

```
a: data_vector(0..n-1, 0..1);
begin
  --computation of the generation and propagation conditions:
  for i in 0..n-1 loop a(i,0):=g(x(i),y(i)); a(i,1):=p(x(i),
  y(i))); end loop;
  --carry computation
  carry_lookahead_procedure(n, a, c_in, q);
  q(0):=c_in;
  --sum computation
  for i in 0..n-1 loop z(i):=(x(i)+y(i)+q(i)) mod B; end loop;
  z(n):=q(n);
```

4.1.3   Long-Operand Addition

In the case of *long-operand additions* it may be necessary to break down the n-digit operands into s-digit slices. A typical example is the implementation of n-bit arithmetic operations within an m-bit microprocessor, with $m < n$. Taking into account that an n-digit base-B number can also be considered as being an (n/s)-digit base-B^s number (Comment 3.1) a modified version of the basic algorithm 4.1 can be used. The iteration body of Algorithm 4.1 must be substituted by a procedure natural_addition, which computes the sum of two s-digit numbers:

```
procedure natural_addition (s: in natural; carry: in bit; x, y:
in digit_vector(0..s-1); next_carry: out bit; z: out digit_
vector(0..s-1);
```

Any one of the previously proposed algorithms (4.1, 4.2, 4.7, or 4.9) can be used for defining the `natural_addition` procedure. Then the following algorithm computes $x + y + c_{in}$.

Algorithm 4.10   Long-Operand Addition

```
q:=c_in;
for i in 0..n/s-1 loop
  natural_addition(s, q, x(i*s..(i*s)+s-1), y(i*s..(i*s)+s-1),
    q, z(i*s..(i*s)+s-1));
end loop;
z(n):=q;
```

Depending on the selection of the `natural_addition` procedure, the corresponding execution time is proportional to either $(n/s).s = n$ or $(n/s).\log s$.

Observe that modified versions of the other algorithms would not give shorter execution times: all of them include n sentences

```
z(i):=(x(i)+y(i)+q(i)) mod B;
```

equivalent, in base B^s, to n/s sentences

```
natural_addition(s, q(i), x(i*s..(i*s)+s-1), y(i*s..(i*s)+
s-1), not_used, z(i*s..(i*s)+s-1));
```

As the n/s preceding sentences must be executed sequentially (long-operand constraint), the execution time would still be proportional to either $(n/s).s = n$ or $(n/s).\log(s)$.

4.1.4   Multioperand Addition

Another important operation is the *multioperand addition*, that is, the computation of $z = x^{(0)} + x^{(1)} + \cdots + x^{(m-1)}$, where every $x^{(i)}$ is a natural number. Assume that the overall sum z does not exceed n digits and that all operands are expressed with n digits. The following algorithm computes z.

Algorithm 4.11   Basic Multioperand Addition

```
accumulator:=0;
for j in 0..m-1 loop
  natural_addition(n, 0, accumulator, x(j), not_used,
  accumulator);
end loop;
z:=accumulator;
```

Its execution time is proportional to $m.n$ or $m.\log n$ depending on the selected `natural_addition` procedure.

An interesting concept for executing multioperand additions is the *stored-carry form encoding* of the result of a 3-operand addition. Assume that a procedure

procedure three-to-two(w, x, y: **in** natural; u, v: **out** natural);

has been defined; it computes u and v such that

$$w + x + y = u + v.$$

Then the following algorithm computes the sum $z = x^{(0)} + x^{(1)} + \cdots + x^{(m-1)}$ of m natural numbers.

Algorithm 4.12

```
three-to-two(x(0), x(1), x(2), u(0), v(0));
for j in 3..m-1 loop
   three-to-two (x(j), u(j-3), v(j-3), u(j-2), v(j-2));
end loop;
natural_addition(n, 0, u(m-3), v(m-3), not_used, z);
```

The three-to-two procedure consists in expressing the sum z of three natural numbers (w, x, y) under the form of a pair (u, v) of two natural numbers in such a way that $z = u + v$. Assume now that w, x, and y are n-digit numbers, and q_in is a 1-digit number. The following algorithm computes two n-digit numbers u and v, and a 1-digit number q_out, such that

$$w + x + y + q\_in = q\_out.B^n + u + v: \qquad (4.13)$$

Algorithm 4.13   Stored-Carry Encoding

```
procedure stored-carry_encoding(w, x, y: in digit_vector(0..
n-1); q_in: in digit; u, v: out digit_vector(0..n-1); q_out:
out digit) is
begin
   q(0):=q_in;
   for i in 0..n-1 loop
      q(i+1):=(w(i)+x(i)+y(i))/B;
      u(i):=(w(i)+x(i)+y(i)) mod B;
   end loop;
   v:=q(0..n-1); q_out:=q(n);
end stored-carry_encoding;
```

Algorithm 4.13 is similar to the basic addition algorithm 4.1: two digits are computed at each step, and the first one, $q(i+1)$, can be considered as a B-ary carry (instead of a binary one when $B > 2$). Nevertheless, $q(i+1)$ does not depend on $q(i)$ so that the n iteration steps can be executed in parallel. In other words, at each step the carry $q(i+1)$ is stored instead of being transferred to the next iteration step. For that reason the pair (u, v) is said to be the *stored-carry form* of z.

The following multioperand algorithm, where $x(j, i)$ stands for $x^{(j)}(i)$, is deduced from Algorithms 4.12 and 4.13 (assuming that z is an n-digit number and that all operands are expressed with n digits).

Algorithm 4.14   Carry-Save Addition

```
stored-carry_encoding(x(0, 0..n-1), x(1, 0..n-1), x(2,
0..n-1), 0, u(0, 0..n-1), v(0, 0..n-1), not_used);
for j in 3..m-1 loop
  stored-carry_encoding (x(j,0..n-1), u(j-3, 0..n-1),
  v(j - 3, 0..n-1), 0, u(j-2, 0..n-1),
  v(j-2, 0..n-1), not_used);
end loop;
z(0):=u(m-3, 0);
natural_addition(n-1, 0, u(m-3, 1..n-1), v(m-3, 1..n-1),
 not_used, z(1..n-1));
```

The carry-save addition algorithm is made up of $m - 2$ calls to `stored-carry_encoding` and a call to an $(n - 1)$-digit addition procedure, so that the execution time is roughly proportional to $m + n$ or $m + \log(n)$, instead of $m.n$ or $m.\log(n)$.

Comments 4.2

1. Instead of the `three-to-two` procedure, more general *p-to-k* procedures could be defined, as well as multioperand addition algorithms in which $p - k$ new operands are added at each step. The generalized version of Algorithm 4.12 would include $m \cong (n - k)/(p - k)$ steps to reach k operands. Each step could be decomposed in a similar way as in the case of Algorithm 4.13. For instance, with $p = 7$ and $k = 3$, each step of the generalized algorithm 4.12 should compute the sum of seven numbers $w^{(0)}, w^{(1)}, \ldots, w^{(6)}$, and encode the result as a three-component vector; the generalized version of Algorithm 4.13 should compute

   ```
   q(i+2):=(w(0, i)+w(1, i)+···+w(6, i))/(B**2);
   r(i+1):=(w(0, i)+w(1, i)+···+w(6, i)-q(i+2)*(B**2))/B;
   u(i):=(w(0, i)+w(1, i)+···+w(6, i)) mod B;
   ```

   at each iteration step (observe that if $B \geq 2$, then $7.(B - 1) < B^2$). This idea, mainly applicable to the case of hardware implementations, will be developed in Chapter 11.

2. Another idea mainly applicable to hardware implementations is the substitution of the iterations (as in Algorithms 4.11, 4.12, and 4.14) by tree structures. It will also be developed in Chapter 11.

4.1.5   Long-Multioperand Addition

A *long-multioperand addition* can be executed by combining Algorithms 4.10 and 4.11.

Algorithm 4.15

```
accumulator:=0;
for j in 0..m-1 loop
  q:=0;
  for i in 0..n/s-1 loop
    natural_addition(s, q, accumulator(i*s..(i*s)+s-1),
      x(j,i*s..(i*s)+s-1), q, accumulator(i*s..(i*s)+s-1));
  end loop;
end loop;
z:=accumulator;
```

Its execution time is proportional to either $m.(n/s).s = m.n$ or $m.(n/s).\log(s)$.

The stored-carry encoding could be used too. The reduction of m n-digit operands to 2 n-digit operands can be performed by breaking down each n-digit operand into n/s s-digit ones and calling the stored-carry_encoding procedure $(m-2).(n/s)$ times. Then the so-obtained operands are added.

Algorithm 4.16   Carry-Save Long-Multioperand Addition

```
--m-to-2 reduction:
q:=0;
for i in 0..n/s-1 loop
  stored-carry_encoding (x(0, i*s..(i*s)+s-1),
  x(1, i*s..(i*s)+s-1), x(2, i*s..(i*s)+s-1),
  q, u(i*s..(i*s)+s-1), v(i*s..(i*s)+s-1), q);
end loop;
for j in 3..m-1 loop
  q:=0;
  for i in 0..n/s-1 loop
    stored-carry_encoding (x(j, i*s..(i*s)+s-1), u(i*s..(i*s)
    +s-1), v(i*s..(i*s)+s-1),q, u(i*s..(i*s)+s-1),
    v(i*s..(i*s)+s-1), q);
  end loop;
end loop;
--2-operand addition:
q:=0;
for i in 0..n/s-1 loop
  natural_addition(s, q, u(i*s..(i*s)+s-1), v(i*s..(i*s)+
  s-1), q, z(i*s..(i*s)+s-1));
end loop;
```

The m-to-2 reduction is performed in $(m-2).(n/s)$ steps, and the 2-operand addition execution time is proportional to either $(n/s).s = n$ or $(n/s).\log s$. The total execution time is roughly proportional to either $(n/s).(m + s)$ or $(n/s).(m + \log s)$ instead of $(n/s).m.s$ or $(n/s).m.\log s$.

4.2 SUBTRACTION OF NATURAL NUMBERS

The following (pencil and paper) algorithm computes the n-digit representation of $z = x - y - b_{in}$ where b_{in} is an initial *borrow* equal to 0 or 1; if z is negative—that means that z is not a natural number—the output borrow $q(n)$ is equal to 1.

Algorithm 4.17   Subtraction

```
q(0):=b_in;
for i in 0..n-1 loop
    if x(i)-y(i)-q(i)<0 then q(i+1):=1; else q(i+1):=0;
    end if; r(i):=(x(i)-y(i)-q(i)) mod B;
end loop;
negative:=q(n);
```

Another method consists in treating the subtraction of natural numbers as a particular case of the subtraction of integers (next section).

4.3 INTEGERS

In the case of integer numbers, the addition and subtraction algorithms depend on the particular representation. Three nonredundant representation methods are considered in what follows: B's complement, sign-magnitude, and excess-E (Chapter 3).

4.3.1 *B's Complement Addition*

Given two n-digit B's complement integers x and y, and an initial carry c_{in} equal to 0 or 1, then $z = x + y + c_{in}$ is an $(n + 1)$-digit B's complement integer. Assume that x and y are represented with $n+1$ digits. Then the natural numbers associated with x, y, and z are $R(x) = x \bmod B^{n+1}$, $R(y) = y \bmod B^{n+1}$, and $R(z) = z \bmod B^{n+1}$ (Definition 3.4), so that

$$R(z) = (x + y + c_{in}) \bmod B^{n+1} = (R(x) + R(y) + c_{in}) \bmod B^{n+1}.$$

Thus a straightforward addition algorithm consists in representing x and y with $n+1$ digits and adding the corresponding natural numbers, as well as the initial carry, modulo B^{n+1} (that means without taking into account the output carry). In order to represent x and y with one additional digit, Comment 3.2 is taken into account. As before, the procedure `natural_addition` computes the sum of two natural numbers.

Algorithm 4.18 B's Complement Addition

```
if x(n-1)<B/2 then x(n):=0; else x(n):=B-1; end if;
if y(n-1)<B/2 then y(n):=0; else y(n):=B-1; end if;
natural_addition(n+1, c_in, x, y, not_used, z);
```

Example 4.1   Assume that $B = 10$, $n = 4$, $c_{in} = 0$, $x = -2345$, and $y = -3674..$ Both x and y are negative so that they are represented by $R(x) = -2345 + 10^4 = 7655$ and $R(y) = -3674 + 10^4 = 6326$. First represent x and y with five digits: $R(x) = 97655$ and $R(y) = 96326$. Then add up $R(x)$ and $R(y)$ modulo 10^5: $(97655 + 96326) \bmod 10^5 = 93981$. As $9 \geq B/2 = 5$, the integer represented by 93981 is negative and equal to $93981 - 10^5 = -6019$, that is, the sum of x and y.

4.3.2 B's Complement Sign Change

Given an n-digit B's complement integer x, the inverse $z = -x$ of x is an $(n + 1)$-digit B's complement integer (actually the only case when $-x$ cannot be represented with n digits is when $x = -B^n/2$ and $-x = B^n/2$; i.e., $-x = -0.B^n + (B/2).B^{n-1} + 0.B^{n-2} + \cdots + 0.B^0$). The computation of the representation of $-x$ is based on the following property.

Property 4.3   Given two m-digit base-B natural numbers $a = a_{m-1}.B^{m-1} + a_{m-2}.B^{m-2} + \cdots + a_0. B^0$ and $b = (B - 1 - a_{m-1}).B^{m-1} + (B - 1 - a_{m-2}).B^{m-2} + \cdots + (B - 1 - a_0).B^0$, then

$$b = B^m - a - 1. \tag{4.14}$$

Assume that x is represented with $n + 1$ digits, and define x' as being the natural number deduced from $R(x)$ by substituting every digit x_i by $x_i' = B - 1 - x_i$. Then, according to Property 4.3,

$$x' = B^{n+1} - R(x) - 1$$

and

$$R(-x) = (-x) \bmod B^{n+1} = (B^{n+1} - R(x)) \bmod B^{n+1} = (x' + 1) \bmod B^{n+1}. \tag{4.15}$$

A straightforward inversion algorithm consists in representing x with $n + 1$ digits, complementing every digit to $B - 1$, then adding 1.

Algorithm 4.19 *B*'s Complement Sign Change

```
if x(n-1)<B/2 then x(n):=0; else x(n):=B-1; end if;
for i in 0..n loop x'(i):=B-1-y(i); end loop;
natural_addition(n+1, 1, x', 0, not_used, z);
```

Examples 4.2

1. Assume that $B = 10$, $n = 4$, $x = 2345$; x is nonnegative and is represented by $R(x) = x = 2345$. First represent x with five digits: $R(x) = 02345$. Then complement all digits to $B - 1 = 9$, and add 1: $(97654 + 1) \bmod 10^5 = 97655$. The integer represented by 97655 is $97655 - 10^5 = -2345$, that is, $-x$.

2. If $x = -5000$ then the four-digit representation of x is 5000 and its five-digit one is 95000. By complementing all digits and adding 1 the obtained result is $(04999 + 1) \bmod 10^5 = 05000$, which is the representation of the nonnegative number 5000.

3. If $x = 0$ then the four-digit representation of x is 0000 and its five-digit one is 00000. By complementing all digits and adding 1 the obtained result is $(99999 + 1) \bmod 10^5 = 00000$, which is the representation of the nonnegative number 0.

An alternative sign-change algorithm is based on the following observation: if x is represented under the form

$$R(x) = x_n.B^n + \cdots + x_{k+1}.B^{k+1} + x_k.B^k,$$

where $x_k > 0$, then the representation of $-x$ is

$$R(-x) = (B - 1 - x_n).B^n + \cdots + (B - 1 - x_{k+1}).B^{k+1}$$
$$+ (B - 1 - x_k).B^k + (B - 1).B^{k-1} + \cdots + (B - 1).B^0 + 1$$
$$= (B - 1 - x_n).B^n + \cdots + (B - 1 - x_{k+1}).B^{k+1} + (B - x_k).B^k.$$

In the following algorithm, the binary variable first_non_zero, initially equal to 0, is set to 1 as soon as the first nonzero digit of x is encountered.

Algorithm 4.20 *B*'s Complement Sign Change (Alternative Algorithm)

```
if x(n-1)<B/2 then x(n):=0; else x(n):=B-1; end if;
first_non_zero:=0;
for i in 0..n loop
  if first_non_zero=0 then
    if x(i)=0 then z(i):=0; else z(i):=B-x(i); first_non_zero
    :=1; end if;
  else z(i):=B-1-x(i);
  end if;
end loop;
```

4.3.3   *B*'s Complement Subtraction

Given two n-digit B's complement integers x and y, and an input borrow b_{in} equal to 0 or 1, then $z = x - y - b_{in}$ is an $(n+1)$-digit B's complement integer. Assume that x and y are represented with $n+1$ digits. Then the natural numbers associated with x, $-y$, and z are $R(x) = x \bmod B^{n+1}$, $R(-y) = (y' + 1) \bmod B^{n+1}$ (relation (4.15)) and $R(z) = z \bmod B^{n+1}$, so that

$$R(z) = (x - y - b_{in}) \bmod B^{n+1} = (R(x) + y' + (1 - b_{in})) \bmod B^{n+1}.$$

Thus a straightforward subtraction algorithm consists in representing x and y with $n+1$ digits, complementing the digits of y, and adding, modulo B^{n+1}, the corresponding natural numbers, as well as the inverted input borrow.

Algorithm 4.21   *B*'s Complement Subtraction

```
if x(n-1)<B/2 then x(n):=0; else x(n):=B-1; end if;
if y(n-1)<B/2 then y(n):=0; else y(n):=B-1; end if;
for i in 0..n loop y'(i):=B-1-y(i); end loop;
c_in:=1-b_in;
natural_addition(n+1, x, y', c_in, z, not_used);
```

Example 4.3   Assume that $B = 10$, $n = 4$, $b_{in} = 1$, $x = -2345$, and $y = 3674$; x is negative and y nonnegative, so that they are represented by $R(x) = -2345 + 10^4 = 7655$ and $R(y) = y = 3674$. First represent x and y with five digits: $R(x) = 97655$ and $R(y) = 03674$. Then compute $y' = 96325$, $c_{in} = 1 - b_{in} = 0$ and $(R(x) + y' + c_{in}) \bmod 10^5 = (97655 + 96325) \bmod 10^5 = 93980$. The integer represented by 93980 is equal to $93980 - 10^5 = -6020$, that is, $-2345 - 3674 - 1$.

4.3.4   *B*'s Complement Overflow Detection

In some cases it may be necessary to know whether the result of an operation actually is an $(n+1)$-digit number and not an n-digit one. A typical case is the arithmetic unit of a general-purpose computer: both the operands and the result are n-bit numbers, and an *overflow flag* is raised if the result does not fit within n bits. Assume that the previous algorithms (addition, inversion, and subtraction) are executed without extending the operands to $n+1$ bits:

1. Consider the case of addition. An overflow can occur when both operands have the same sign. First observe that if x and y belong to the interval $-B^n/2 \le x, y < B^n/2$, then $-B^n \le x + y + c_{in} \le 2.(B^n/2 - 1) + 1 = B^n - 1$; that is,

$$-B^n \le x + y + c_{in} < B^n. \tag{4.16}$$

So, if x and y are nonnegative, the sum $x + y + c_{in}$ could be greater than or equal to $B^n/2$. As $R(x) = x$ and $R(y) = y$, then $R(z) = (x + y + c_{in}) \bmod B^n$, and according to the previous hypothesis and to (4.16)

$$B^n/2 \leq x + y + c_{in} < B^n,$$

that is,

$$(B/2).B^{n-1} \leq x + y + c_{in} < B^n,$$

so that $z(n) = 0$ and $z(n-1) \geq B/2$.

The conclusion is that the sum of two nonnegative numbers, plus an initial carry, generates an apparently negative number if only n digits are available $(z(n-1) \geq B/2)$.

If x and y are negative the sum $x + y + c_{in}$ could be smaller than $-B^n/2$. As $R(x) = B^n + x$ and $R(y) = B^n + y$, then $R(z) = (2.B^n + x + y + c_{in}) \bmod B^n$, and according to the previous hypothesis and to (4.16)

$$2.B^n - B^n \leq 2.B^n + x + y + c_{in} < 2.B^n - B^n/2,$$

that is,

$$B^n \leq 2.B^n + x + y + c_{in} < B^n + (B/2).B^{n-1},$$

so that

$$z(n) = 1, \; z(n-1) < B/2.$$

The conclusion is that the sum of two negative numbers, plus an initial carry, generates an apparently nonnegative number if only n digits are available $(z(n-1) < B/2)$.

To summarize, the overflow detection is carried out just looking at the sign digits of the operands and the result. Under Boolean form:

$$\text{add\_ovf} = [(x(n-1) < B/2) \text{ and } (y(n-1) < B/2) \text{ and } (z(n-1) \geq B/2)]$$
$$\text{or } [(x(n-1) \geq B/2) \text{ and } (y(n-1) \geq B/2)$$
$$\text{and } (z(n-1) < B/2)]. \tag{4.17}$$

2. It has already been observed that, in the case of the sign-change operation, the only overflow situation is when $x = -B^n/2$, namely, $x(n-1) = B-1$ and $x(n-2) = \cdots = x(0) = 0$. The inversion algorithm, with n digits, generates $z = x$. Once again it's just a matter of looking at the sign digits of both the operand and the result:

$$\text{inv\_ovf} = (x(n-1) \geq B/2) \text{ and } (z(n-1) \geq B/2). \tag{4.18}$$

3. If a subtraction is performed, an overflow could occur if one operand is negative and the other one nonnegative. First observe that if x and y belong

to the interval $-B^n/2 \leq x, \ y < B^n/2$ then $-2.(B^n/2) - 1 < x - y - b_{in} < 2.(B^n/2)$, that is,

$$-B^n \leq x - y - b_{in} < B^n. \tag{4.19}$$

If $x \geq 0$ and $y < 0$ the difference $x - y - b_{in}$ could be greater than or equal to $B^n/2$. As $R(x) = x$, $R(y) = B^n + y$, $y' + 1 = B^n - R(y) = -y$, $R(z) = (R(x) + y' + 1 - b_{in})$ mod $B^n = (x - y - b_{in})$ mod B^n, then according to the previous hypothesis and to (4.19)

$$(B/2).B^{n-1} = B^n/2 \leq x - y - b_{in} < B^n,$$

so that $z(n) = 0$, $z(n - 1) \geq B/2$.

The conclusion is that the difference between a nonnegative number and a negative one, minus an initial borrow, generates an apparently negative number if only n digits are used ($z(n - 1) \geq B/2$).

If $x < 0$ and $y \geq 0$ the difference $x - y - b_{in}$ could be smaller than $-B^n/2$. As $R(x) = B^n + x$, $R(y) = y$, $y' + 1 = B^n - R(y) = B^n - y$, $R(z) = (R(x) + y' + 1 - b_{in})$ mod $B^n = (2.B^n + x - y - b_{in})$ mod B^n, then according to the previous hypothesis and to (4.19)

$$2.B^n - B^n = B^n \leq 2.B^n + x - y - b_{in} < 2.B^n - B^n/2 = B^n + (B/2).B^{n-1},$$

so that $z(n) = 1$, $z(n - 1) < B/2$.

The conclusion is that the difference between a negative number and a nonnegative one, minus an initial borrow, generates an apparently nonnegative number if only n digits are used ($z(n - 1) \geq B/2$).

As in the preceding cases the overflow detection is carried out just looking at the sign digits of the operands and the result. Under Boolean form:

$$\begin{aligned} \text{sub\_ovf} = &[(x(n - 1) < B/2) \text{ and } (y(n - 1) \geq B/2) \text{ and } (z(n - 1) \geq B/2)] \\ &\text{or } [(x(n - 1) \geq B/2) \text{ and } (y(n - 1) < B/2) \\ &\text{and } (z(n - 1) < B/2)]. \end{aligned} \tag{4.20}$$

Examples 4.4 $(B = 10, n = 4)$

1. Assume that $c_{in} = 0$, $x = 2345$, and $y = 4674$, and that the value of $x + y + c_{in}$ is computed. Then $R(x) = x = 2345$ and $R(y) = y = 4674$, so that $(R(x) + R(x) + c_{in})$ mod $10000 = 7019$, that is, the representation of the negative number -2981.

2. Assume now that $c_{in} = 0$, $x = -4726$, and $y = -2174$, and that the value of $x + y + c_{in}$ is computed. Then $R(x) = 10000 - 4726 = 5274$ and

$R(y) = 10000 - 2174 = 7826$, so that $(R(x) + R(x) + c_{in})$ mod $10000 = 3100$, that is, the representation of the nonnegative number 3100.

3. Compute the difference between $x = 2345$ and $y = -4726$, with $b_{in} = 0$. The corresponding representations are $R(x) = x = 2345$ and $R(y) = 10000 - 4726 = 5274$, so that $y' = 4725$ and $(2345 + 4725 + 1)$ mod $10000 = 7071$, that is, the representation of the negative number -2929.

Comments 4.3   About the reduced B's complement representation (Comment 3.2):

1. The sign extension just consists in duplicating the sign digit.
2. If $B = 2$, there is no difference between the reduced and the nonreduced 2's complement representation.
3. If x and y are n-digit reduced B's complement numbers, then $-B^{n-1} \le x$, $y < B^{n-1}$, so that

$$-2.B^{n-1} \le x + y + c_{in} < 2.B^{n-1} \quad \text{and} \quad -2.B^{n-1} \le x - y - b_{in} < 2.B^{n-1}.$$

If, furthermore, $B > 2$ and B is even, so that $B \ge 4$, then $2.B^{n-1} \le B^n/2$ and

$$-B^n/2 \le x + y + c_{in} < B^n/2 \quad \text{and} \quad -B^n/2 \le x - y - b_{in} < B^n/2.$$

Thus both $x + y + c_{in}$ and $x - y - b_{in}$ are n-digit B's complement numbers. There is an overflow if the result is not a reduced B's complement number, that is, if the sign digit does not belong to $\{0, B - 1\}$. Actually the sign digit is equal to $B-2$ in the case of a negative overflow (result $< -B^{n-1}$) and to 1 in the case of a positive one (result $\ge B^{n-1}$).

Examples 4.5 $(B = 10, n = 3, 10$'s complement reduced form)

1. Assume that $c_{in} = 0$, $x = 74$, and $y = 41$, and that the value of $x + y$ is computed. Then $R(x) = x = 074$ and $R(y) = y = 041$, so that $(R(x) + R(x) + c_{in})$ mod $1000 = 115$, a number whose sign digit does not belong to $\{0, 9\}$.
2. Assume now that $c_{in} = 0$, $x = -74$, and $y = -41$, and that the value of $x + y$ is computed. Then $R(x) = 1000 - 74 = 926$ and $R(y) = 1000 - 41 = 959$, so that $(R(x) + R(x) + c_{in})$ mod $1000 = 885$, a number whose sign digit does not belong to $\{0, 9\}$—actually the representation in nonreduced form of $885 - 1000 = -115$.
3. Compute the difference between $x = 74$ and $y = -41$, with $b_{in} = 0$. The corresponding representations are $R(x) = x = 074$ and $R(y) = 1000 - 41 = 959$, so that $y' = 040$ and $(074 + 040 + 1)$ mod $1000 = 115$, a number whose sign digit does not belong to $\{0, 9\}$—actually the representation in nonreduced form of 115.

4.3.5 Excess-*E* Addition and Subtraction

The addition and subtraction algorithms are based on the following properties.

Properties 4.4   Given two excess-*E* integers x and y, an initial binary carry c_{in} and an initial borrow b_{in}, then

$$R(x + y + c_{in}) = R(x) + R(y) + c_{in} - E,$$
$$R(x - y - b_{in}) = R(x) - R(y) - b_{in} + E,$$
$$R(-x) = -R(x) + 2.E.$$

Proof   According to Definition 3.3, $R(x) = x + E$, $R(y) = y + E$, and $R(x + y + c_{in}) = x + y + c_{in} + E$, so

$$R(x) + R(y) + c_{in} - E = (x + E) + (y + E) + c_{in} - E = x + y + c_{in} + E$$
$$= R(x + y + c_{in}),$$
$$R(x) - R(y) - b_{in} + E = (x + E) - (y + E) - b_{in} + E = x - y - b_{in} + E$$
$$= R(x - y - b_{in}),$$
$$-R(x) + 2.E = -(x + E) + 2.E = -x + E = R(-x).$$

If x and y are two n-digit excess-*E* integers, and if $z = x + y + c_{in}$ is also an n-digit excess-*E* integer, then a straightforward addition algorithm consists in representing x and y with $n + 1$ digits, adding them up with c_{in} and subtracting E. The result $R(z)$ is an $(n + 1)$-digit natural number whose first digit is 0.

Assume that a procedure `natural_subtraction` has been defined:

```
procedure natural_subtraction (s: in natural; borrow: in bit;
x, y: in digit_vector(0..s-1); next_borrow: out bit; z: out
digit_vector(0..s-1);
```

The following algorithms compute $z = x + y + c_{in}$.

Algorithm 4.22   Excess-*E* Addition

```
x(n):=0; y(n):=0;
natural_addition(n+1, x, y, c_in, w, not_used);
natural_subtraction(n+1, w, E, 0, z, not_used);
if z(n)>0 then overflow:=true; end if;
```

Similar algorithms can be defined for computing $z = x - y - b_{in}$ and $z = -x$:

Algorithm 4.23   Excess-*E* Subtraction

```
x(n):=0; y(n):=0;
natural_addition(n+1, x, E, 0, w, not_used);
natural_subtraction(n+1, w, y, b_in, z, not_used);
if z(n)>0 then overflow:=true; end if;
```

Algorithm 4.24 Excess-*E* Sign Change

```
x(n):=0;
E_by_2(0):=0;
for i in 1..n loop E_by_2(i):=E(i-1); end loop;
natural_subtraction(n+1, E_by_2, x, 0, z, not_used);
if z(n)>0 then overflow:=true; end if;
```

Examples 4.6 $(B = 10, n = 4, \text{excess } 5000)$

1. Assume that $c_{in} = 0$, $x = 2345$, and $y = 1674$, and that the value of $x + y + c_{in}$ is computed. Then $R(x) = 07345$ and $R(y) = 06674$, so that $R(x) + R(y) + c_{in} - 05000 = 09019$, that is, the representation of 4019.
2. Assume now that $c_{in} = 0$, $x = -2345$, and $y = 1674$, and that the value of $x + y + c_{in}$ is computed. Then $R(x) = 02655$ and $R(y) = 06674$, so that $R(x) + R(y) + c_{in} - 05000 = 4329$, that is, the representation of -671.
3. Compute the sum of $x = 2345$ and $y = 4726$, with $c_{in} = 0$. The corresponding representations are $R(x) = 07345$ and $R(y) = 09726$, so that $R(x) + R(y) + c_{in} - 05000 = 12071$ and the overflow flag is raised.
4. Compute the difference between $x = 2345$ and $y = 4726$, with $b_{in} = 0$. The corresponding representations are $R(x) = 07345$ and $R(y) = 09726$, so that $R(x) - R(y) - b_{in} + 05000 = 2619$, that is, the representation of -2381.
5. Compute the difference between $x = -2345$ and $y = 4726$, with $b_{in} = 0$. The corresponding representations are $R(x) = 02655$ and $R(y) = 09726$, so that $R(x) - R(y) - b_{in} + 05000 = 97929$ (modulo $B^{n+1} = 100000$) and the overflow flag is raised.
6. Compute the inverse of $x = -5000$. The corresponding representation is $R(x) = 00000$ so that $- R(x) + 2 \cdot 05000 = 10000$ and the overflow flag is raised.

4.3.6 Sign–Magnitude Addition and Subtraction

Given two n-digit sign-magnitude integers x and y, then $z = x + y + c_{in}$ is an $(n + 1)$-digit sign-magnitude integer. The following algorithm computes z.

Algorithm 4.25

```
if sign(x)=sign(y) then a:=abs(x)+abs(y); else a:=abs(x)-
abs(y);
end if;
if a<0 then sign(z):=sign(y); abs(z):=-a; else sign(z):=
sign(x); abs(z):=a; end if;
```

It is equivalent to the following algorithm (at the digit level).

Algorithm 4.26   Sign-Magnitude Addition

```
abs_x:=x(0..n-2)&0; abs_y:=y(0..n-2)&0;
if x(n-1)=y(n-1) then
  natural_addition(n, abs_x, abs_y, 0, a, not_used);
else
  for i in 0..n-1 loop abs_y'(i):=B-1-abs_y(i); end loop;
  natural_addition(n, abs_x, abs_y', 1, a, not_used);
end if;
if a(n-1)=B-1 then
  z(n):=y(n-1);
  for i in 0..n-1 loop a'(i):=B-1-a(i); end loop;
  natural_addition(n, 0, a', 1, z(0..n-1), not_used);
else
  z(n):=x(n-1); z(0..n-1):=a;
end if;
```

Example 4.7   ($B = 10$, $n = 5$). Assume that $x = +2345$ and $y = -7674$. First express the absolute values with five digits: $abs(x) = 02345$ and $abs(y) = 07674$. As the signs are different, compute $02345 + 92325 + 1 = 94671$. The first digit is equal to 9, indicating a negative value. The sign of the result is the same as the sign of y ($-$) and the absolute value of the result is $05328 + 1 = 05329$. So the final result is -5329.

Comment 4.4   With algorithm 4.26, if $sign(x) = 0$, $sign(y) = 1$, and $abs(x) = abs(y)$, the result is $+0$; if $sign(x) = 1$, $sign(y) = 0$, and $abs(x) = abs(y)$, the result is -0.

The subtraction $x - y$ is equivalent to the addition $x + minus\_y$ where $minus\_y = -y$, and the computation of $minus\_y$ is straightforward:

```
minus_y(n-1):=1-y(n-1); minus_y (0..n-2):=y(0..n-2);
```

4.4   BIBLIOGRAPHY

[BRE1982] R. Brent and H. T. Kung, A regular layout for parallel adders. *IEEE Trans. Comput.*, **C-31**(3): 260–264 (1982).

[HAN1987] T. Han and D. A. Carlson, Fast area-efficient VLSI adders. In: *Proceedings of the 8th Symposium on Computer Arithmetic*, 1987, pp. 49–56.

[KOG1973] P. M. Kogge and H. S. Stone, A parallel algorithm for the efficient solution of a general class of recurrence equations. *IEEE Trans. Comput.*, **C-22**(8): 786–793 (1973).

[LAD1980] R. E. Ladner and M. J. Fischer, Parallel prefix computation. *J. ACM* **27**(10): 831–838 (1980).

[SUG1990] B. Sugla and D. Carlson, Extreme area-time tradeoffs in VLS. *IEEE Trans. Comput.* **39**(2): 251–257 (1990).

5

ARITHMETIC OPERATIONS: MULTIPLICATION

Basically, multiplication is a very simple operation as it most often reduces to multi-operand addition. In early computers, multiplication was assumed too complex to receive a combinational implementation, typically considered too expensive at this time. For this historical reason, in most textbooks on computer arithmetic, multiplication algorithms are strongly biased by the sequential implementations. In this chapter, the authors attempt to remain consistent with their general philosophy, presenting the algorithms in a way that never settles on a specific implementation technique. Although the Ada-like language, utilized in the algorithm descriptions, could suggest some kind of sequential implementations, the actual interpretations cannot involve any choice between space or time iteration of the presented step-by-step processes. This approach is particularly well suited to provide the designer, with a range of options, based on the diversity of technologies at hand, speed–cost compromises, and other constraints to be dealt with. Actually, it is important to realize that the algorithmic complexity is not necessarily tied to the actual required performance of some practical application.

Base-B is generally assumed, while base-2 is extensively treated whenever the specificity of the binary system results in prominent features or allows significant algorithmic simplifications. Most multiplication algorithms share a common feature: they produce, in one way or another, all the digitwise partial products of the operands. The complexity of the corresponding cell or procedure is thus a key point to be considered by the designer when selecting the base. As quoted in Chapter 3,

Synthesis of Arithmetic Circuits: FPGA, ASIC, and Embedded Systems
By Jean-Pierre Deschamps, Géry J. A. Bioul, and Gustavo D. Sutter
Copyright © 2006 John Wiley & Sons, Inc.

the most used bases are 2 (binary), 4 (quaternary or radix-4), 8 (octal or radix-8), 16 (hexadecimal or radix-16), and 10 (decimal). The examples treated in this chapter will be limited to those bases, although most theorems hold for any base B. As far as the technology deals with two-level signals and devices, binary coding is assumed in most practical implementations. Nevertheless, from the algorithmic point of view, the base coding aspect is not relevant.

Logarithmic techniques for multiplication are not generally used because logarithm computation algorithms do not exhibit a better complexity behavior than multiplication itself. Actually, if look-up tables (LUTs) are available, the process is interesting because it reduces to a simple addition. Nevertheless, the cost of look-up tables is formidable except for small operand sizes.

Let us point out, finally, that the fast evolution of technology may change the optimization criteria and the performance factors of some types of physical implementations. So it is quite difficult to forecast future interest in the respective algorithm options. This chapter presents the most used multiplication algorithms while Chapter 12 is devoted to multiplier design with some typical FPGA and IC implementations.

5.1   NATURAL NUMBERS MULTIPLICATION

5.1.1   Introduction

The most basic multiplication algorithms for n-digit $\times$ m-digit B-ary natural numbers (shift and add algorithms) proceed in two phases:

1. Digitwise partial products ($n \times m$),
2. *Multioperand* addition.

The classic computation scheme to introduce multiplication is given in Figure 5.1a, where partial products appear lined up according to their respective weight. This scheme is historically related to the pencil and paper implementation of the operation. This simple scheme is easily built up by noting that the partial products $x_i y_j$ are lined up in the column whose index $k = i + j$ corresponds to the weight B^k. Observe that whenever $B > 2$, the partial products may need two base-B digits. For $B > 2$, a possible multiplication scheme is displayed at Figure 5.1b, where $X_i Y_j$ and $x_i y_j$ stand, respectively, for the integer product

$$X_i Y_j = (x_i . y_j)/B \tag{5.1}$$

and the mod B product

$$x_i y_j = (x_i . y_j) \bmod B \tag{5.2}$$

Observe that the column index k remains $i + j$ for products (5.2) but is computed as $i + j + 1$ for products (5.1).

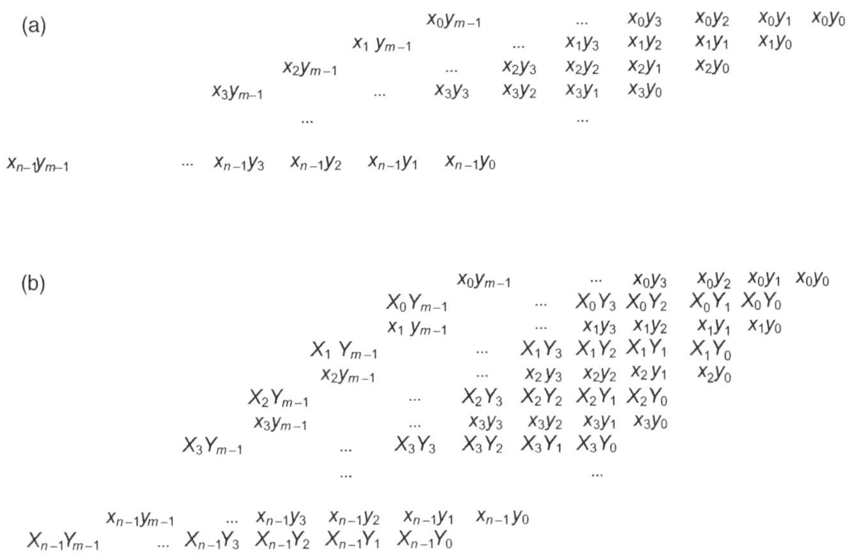

(a)
$$
\begin{array}{ccccccccc}
 & & & & x_0y_{m-1} & \cdots & x_0y_3 & x_0y_2 & x_0y_1 & x_0y_0 \\
 & & & x_1y_{m-1} & & \cdots & x_1y_3 & x_1y_2 & x_1y_1 & x_1y_0 \\
 & & x_2y_{m-1} & & & \cdots & x_2y_3 & x_2y_2 & x_2y_1 & x_2y_0 \\
 & x_3y_{m-1} & & & & \cdots & x_3y_3 & x_3y_2 & x_3y_1 & x_3y_0 \\
 & & \cdots & & & & \cdots & & & \\
x_{n-1}y_{m-1} & & & \cdots & x_{n-1}y_3 & x_{n-1}y_2 & x_{n-1}y_1 & x_{n-1}y_0 & &
\end{array}
$$

(b)
$$
\begin{array}{ccccccc}
 & & & x_0y_{m-1} & \cdots & x_0y_3 & x_0y_2 & x_0y_1 & x_0y_0 \\
 & & X_0Y_{m-1} & & \cdots & X_0Y_3 & X_0Y_2 & X_0Y_1 & X_0Y_0 \\
 & & x_1y_{m-1} & & \cdots & x_1y_3 & x_1y_2 & x_1y_1 & x_1y_0 \\
 & X_1Y_{m-1} & & \cdots & X_1Y_3 & X_1Y_2 & X_1Y_1 & X_1Y_0 \\
 & x_2y_{m-1} & & \cdots & x_2y_3 & x_2y_2 & x_2y_1 & x_2y_0 \\
X_2Y_{m-1} & & \cdots & X_2Y_3 & X_2Y_2 & X_2Y_1 & X_2Y_0 \\
 x_3y_{m-1} & & \cdots & x_3y_3 & x_3y_2 & x_3y_1 & x_3y_0 \\
X_3Y_{m-1} & & \cdots & X_3Y_3 & X_3Y_2 & X_3Y_1 & X_3Y_0 \\
 \cdots & & & \cdots & & & \\
x_{n-1}y_{m-1} & \cdots & x_{n-1}y_3 & x_{n-1}y_2 & x_{n-1}y_1 & x_{n-1}y_0 \\
X_{n-1}Y_{m-1} & \cdots & X_{n-1}Y_3 & X_{n-1}Y_2 & X_{n-1}Y_1 & X_{n-1}Y_0
\end{array}
$$

Figure 5.1   (a) Multiplication scheme and (b) multiplication scheme for $B > 2$.

The cost/speed constraints are key factors to set trade-offs between combinational parallel schemes and sequential implementations. The most popular algorithms, with a number of implementation schemes, are proposed in what follows, where n-digit by m-digit operands are considered for generality.

5.1.2 Shift and Add Algorithms

5.1.2.1 Shift and Add 1   Multiplication is known as a commutative operation in which both operands play the same mathematical role. At the algorithmic point of view, the situation is somewhat different. Actually, in most algorithm descriptions, one of the operands, called the *multiplicator*, is viewed as some kind of parameter set, while the other one, the *multiplicand*, is viewed as a data set.

Let the multiplicator X and the multiplicand Y be given by

$$X = x_{n-1}.B^{n-1} + x_{n-2}.B^{n-2} + \cdots + x_0.B^0,$$

$$Y = y_{m-1}.B^{m-1} + y_{m-2}.B^{n-2} + \cdots + y_0.B^0, \quad x_i, y_i \in \{0, 1, \ldots, B-1\},$$

Let

$$Z = X.Y \tag{5.3}$$

with

$$Z = z_{n+m-1}.B^{n+m-1} + z_{n+m-2}.B^{n+m-2} + \cdots + z_0.B^0.$$

Since

$$0 \leq X \leq B^n - 1 \quad \text{and} \quad 0 \leq Y \leq B^m - 1,$$

then

$$0 \leq Z \leq (B^n - 1).(B^m - 1).$$

Equation (5.3) can be written

$$Z = x_{n-1}.Y.B^{n-1} + x_{n-2}.Y.B^{n-2} + \cdots + x_2.Y.B^2 + x_1.Y.B + x_0.Y, \tag{5.4}$$

then expanded as

$$Z = ((\ldots((0.B + x_{n-1}.Y).B + x_{n-2}.Y).B + \cdots + x_2.Y).B + x_1.Y).B + x_0.Y, \tag{5.5}$$

called the *Hörner expansion.* This suggests the following algorithm.[1]

Algorithm 5.1   Hörner Shift and Add 1

```
P(n):=0;
for i in 0..n-1 loop
  P(n-1-i):=P(n-i)*B+X(n-1-i)*Y;
end loop;
Z:=P(0);
```

5.1.2.2 Shift and Add 2 It will be shown in Chapter 12 that a right to left factorization (5.6) reduces by half the adder length.

$$Z/B^n = B^{-1}.(x_{n-1}.Y + B^{-1}.(x_{n-2}.Y + \cdots + B^{-1}.(x_1.Y + B^{-1}.(x_0.Y + 0))\cdots)). \tag{5.6}$$

Algorithm 5.2   Hörner Shift and Add 2

```
P(0):=0;
for i in 0..n-1 loop
  P(i+1):=(P(i)+X(i)*Y)/B;
end loop;
Z=P(n)*(B**n);
```

The difference between the roles of X and Y clearly appears in the recurrence formula of the Hörner Algorithms 5.1 and 5.2. At each step, the multiplicand Y is

---

[1]In the used Ada-like language, $*$ stands for multiplication, $**$ stands for exponentiation, and $/$ stands without ambiguity for division integer or not.

multiplied by $X(k)$ then either (Algorithm 5.1) added to the (left-)shifted result of the preceding step or (Algorithm 5.2) added to the result of the preceding step and then (right-)shifted. The multiplicator component $X(k)$ sets how many times the multiplicand Y has to be added. For $B = 2$, the process is quite simple because, at each step, $X(k) \in \{0,1\}$ just sets if Y has to be added or not, while, whenever $B \geq 3$, nontrivial partial products have to be generated.

Example 5.1   Let $X = 367169$ and $Y = 24512$ be two decimal numbers: $n = 6$, $m = 5$

The Hörner expressions (5.5) and (5.6) backing Algorithms 5.1 and 5.2 respectively, are:

$$Z = ((\cdots(3.24512).10 + 6.24512).10 + 7.24512).10 + 1.24512).10$$
$$+ 6.24512).10 + 9.24512$$

and

$$Z/10^6 = ((\cdots(9.24512).10^{-1} + 6.24512).10^{-1} + 1.24512).10^{-1}$$
$$+ 7.24512).10^{-1} + 6.24512).10^{-1} + 3.24512).10^{-1}$$

Let us compute step by step.

Hörner Algorithm 5.1

Step 1: $P(5) =$ $(3*24512) =$ 73536
Step 2: $P(4) = (P(5) *10 + 6*24512)) =$ 882432
Step 3: $P(3) = (P(4) *10 + 7*24512)) =$ 8995904
Step 4: $P(2) = (P(3) *10 + 1*24512)) =$ 89983552
Step 5: $P(1) = (P(2) *10 + 6*24512)) =$ 899982592
Step 6: $P(0) = (P(1) *10 + 9*24512)) = 9000046528 = Z$

Hörner Algorithm 5.2

Step 1: $P(1) =$ $(9*24512) *10^{-1} =$ 22060.8
Step 2: $P(2) = (P(1) + (6*24512)) *10^{-1} =$ 16913.28
Step 3: $P(3) = (P(2) + (1*24512)) *10^{-1} =$ 04142.528
Step 4: $P(4) = (P(3) + (7*24512)) *10^{-1} =$ 17572.6528
Step 5: $P(5) = (P(4) + (6*24512)) *10^{-1} =$ 16464.46528
Step 6: $P(6) = (P(5) + (3*24512)) *10^{-1} = 09000.046528 = Z/10^6$
$Z = 9000046528$

This example illustrates the fact that in Algorithm 5.2, a digit is extracted at each step, so the complexity of the sum is always limited to n digits. In Algorithm 5.1,

$n + m$ digits are involved, and no digit is extracted before the end of the process. The same would occur in a standard addition process, whenever adding from left to right instead of right to left.

5.1.2.3   *Extended Shift and Add Algorithm: XY + C + D*   Observe that

$$Z = X.Y \le (B^n - 1).(B^m - 1) = B^{n+m} - B^n - B^m + 1,$$

in such a way that the full digit capacity of Z is not used, namely, $n + m$ digits that would allow one to represent numbers up to $(B^{n+m} - 1)$. So, for the design of a device using this capacity, it is convenient to define the function

$$Z = X.Y + C + D,$$

where C and D are n-digit and m-digit numbers, respectively, in such a way that

$$Z \le (B^n - 1).(B^m - 1) + (B^n - 1) + (B^m - 1) = B^{n+m} - 1.$$

The modified Hörner Algorithm 5.2 is expressed as follows.

Algorithm 5.3   Extended Shift and Add

```
P(0):=D;
for i in 0..n−1 loop
  P(i+1):=(P(i)+X(i)*Y+C(i))/B;
end loop;
Z:=P(n)*(B**n);
```

Property 5.1   In the preceding algorithm, $P(i)$ is an $m + i$ number whose integer and fractional parts contain m and i digits, respectively

5.1.2.4   *Cellular Shift and Add*
Cellular Ripple-Carry Algorithm   The recurrence operation of Algorithm 5.3 can be written in the form

$$P(i) + x_i.(y_0 + y_1.B + \cdots + y_{m-1}.B^{m-1}) + c_i$$

where, according to Property 5.1,

$$P(i) = p_{m-1+i}.B^{m-1} + p_{m-2+i}.B^{m-2} + \cdots + p_i.B^0 + p_{i-1}.B^{-1}$$
$$+ p_{i-2}.B^{-2} + \cdots + p_0.B^{-i}$$

so that

$$P(i + 1) = ((p_{m-1+i} + x_i.y_{m-1}).B^{m-1} + \cdots + (p_i + x_i.y_0).B^0 + c_i$$
$$+ p_{i-1}.B^{-1} + \cdots + p_0.B^{-i})/B. \tag{5.7}$$

In what follows, $p(i, j)$ stands for the jth digit of $P(i)$. Expression (5.7) can be computed by an algorithm for adding natural numbers (basic addition Algorithm

4.1). Assuming that $p(i, m+i-1), \ldots, p(i, 0)$ are the $m+i$ digits of $P(i)$, then the $m+i+1$ digits $p(i+1, m+i), \ldots, p(i+1, 0)$ of $P(i+1)$ are computed as follows (symbol / stands for integer division):

```
for j in 0..m-1 loop p(0, j):=D(j); end loop;
carry(i, 0):=C(i);
for j in 0..i-1 loop p(i+1, j):=p(i, j); end loop;
for j in 0..m-1 loop
  p(i+1, i+j):=(p(i, i+j)+X(i)*Y(j)+carry(i, j)) mod B;
  carry(i, j+1):=(p(i, i+j)+X(i)*Y(j)+carry(i, j))/B;
end loop;
p(i+1, m+i):=carry(i, m);
```

Observe that $p(i, i+j) \leq B - 1$ and $x_i.y_j \leq (B-1)^2$, so that if $carry(i,j) \leq B - 1$ then $carry(i, j) + p(i, i+j) + x_i. \ y_j$ is a two-digit number $(\leq B^2 - 1)$, namely, $[carry(i, j+1), sum(i, j)]$ or $carry(i, j+1).B + sum(i, j)$. Thus an essential difference with the basic addition algorithm 4.1 (in base $B > 2$) comes from the range of the carry $[0, B - 1]$ instead of $[0,1]$.

The precedence graph (Chapter 10) of the multiplication algorithm for $n = 3$, $m = 4$, is shown in Figure 5.2.

Algorithm 5.4 Cellular Ripple-Carry

(Symbol / stands for integer division)

```
for j in 0..m-1 loop p(0, j):=D(j); end loop,
for i in 0..n-1 loop
```

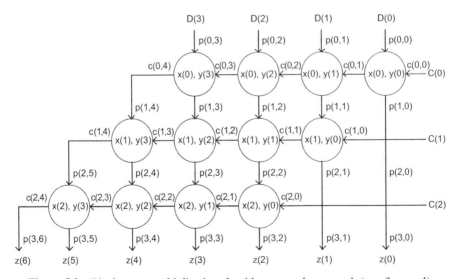

Figure 5.2 Ripple-carry multiplication algorithm–precedence graph ($n = 3$; $m = 4$).

```
  carry(i, 0):=C(i);
  for j in 0..i-1 loop p(i+1, j):=p(i, j); end loop;
  for j in 0..m-1 loop
    p(i+1, i+j):=(p(i, i+j)+X(i)*Y(j)+carry(i, j)) mod B;
    carry(i, j+1):=(p(i, i+j)+X(i)*Y(j)+carry(i, j))/B;
  end loop;
  p(i+1, m+i):=carry(i, m);
end loop;
for j in 0..m+n-1 loop Z(j):=p(n, j); end loop;
```

Comment 5.1 The main loop, named (i, j)-cell loop, computes the functions $carry(i, j + 1)$ and $sum(i, j)$. The other loops are used for indexing purposes, such as assigning $p(0, j)$ to $D(j)$ or $carry(i, 0)$ to $C(i)$. The name *cell* takes its origin from the full-combinational cellular array implementation (Chapter 12, Section 12.2.3) of this algorithm. Actually, the (i, j)-cell can be implemented by any mix of hardware and firmware, where the choice can be made by the designer according to the resources at hand. As it will be shown in Chapter 12, the indexing is not trivial because the time performances can be directly and significantly affected. In a full-hardware implementation (Chapter 12), indexing $p(i, j)$ corresponds to connection assignments between cells, as it already appears in Figure 5.2, where clear relationships come out between input and output indexes of (i, j)-cells.

Cellular Carry–Save Algorithm The following carry–save algorithm differs from the ripple-carry algorithm 5.4 by the indexing loops and a final adding stage loop. The basic concepts of carry-save adders (CSAs, Chapter 4) are applied in the way carries are saved from one loop to the other one, allowing more parallelism in the cell computation. The precedence graph of the carry-save multiplication algorithm for $n = 3$, $m = 4$, is shown in Figure 5.3.

Algorithm 5.5   Carry-Save

(Symbol / stands for integer division)

```
for j in 0..m-1 loop p(0, j):=D(j); carry(0, j):=C(j); end
loop;
for i in 0..n-2 loop p(i+1, m+i):=0; end loop;
for i in 0..n-1 loop
    for j in 0..i-1 loop p(i+1, j):=p(i, j); end loop;
    for j in 0..m-1 loop
      p(i+1, i+j):=(p(i, i+j)+X(i)*Y(j)+carry(i, j)) mod B;
      carry(i+1, j):=(p(i, i+j)+X(i)*Y(j)+carry(i, j))/B;
    end loop;
end loop;
for j in 0..n-1 loop Z(j):=p(n, j); end loop;
k(0):=0; p(n, n+m-1):=0;
for j in 0..m-1 loop
  Z(j+n):=(p(n, n+j)+c(n, j)+k(j)) mod B;
  k(j+1):=(p(n, n+j)+c(n, j)+k(j))/B;
end loop;
```

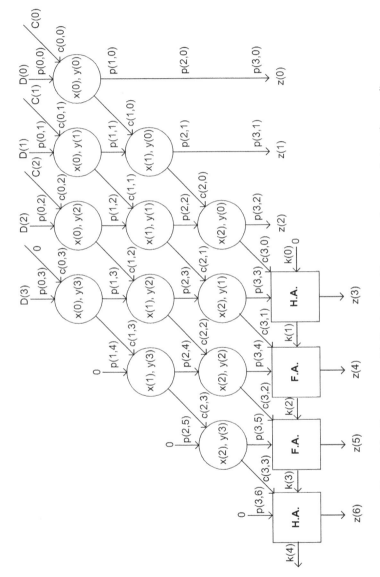

Figure 5.3 Carry-save multiplication algorithm–precedence graph ($n = 3$; $m = 4$).

Comment 5.2   The (i, j)-cell loop computation scheme of the carry-save algorithm is much the same as that of the ripple-carry one. But a final adding loop has to be executed. Nevertheless, thanks to the reindexing, significant time is saved by taking a better profit of parallelism, either in the software execution or in hardware implementation. In particular, a data-flow machine with m processing units could implement the program within n steps whose elementary time delay would be that of one (i, j)-cell. Despite their aspect, the precedence graphs of Figures 5.2 and 5.3 are not circuits, because no assumption is made about the way cells are implemented. Nevertheless, assuming a full combinational circuit implementation for such cells would convert those figures to explicit circuit schemes (Chapter 12).

Whenever $B = 2$, the practical implementation of the (i, j)-cell is quite simple, because the partial products $x_i.y_j$ are expressed by a single bit while two B-ary digits are generally needed for higher base values.

Let us point out that the precedence graph of Figure 5.3 presents several 0 inputs. Therefore the related border cells may be accordingly simplified.

5.1.3   Long-Operand Algorithm

In the case of *long-operand multiplications*, it may be necessary to break down the n-digit (or m-digit) operands into s-digit slices. Assume that a procedure multiplier has been defined; it computes the product of two s-digit numbers (symbol / stands for integer division):

```
procedure multiplier (s: in natural; carry, w, x, y: in
digit_vector (0..s - 1); next_carry, z: out digit_vector
(0..s-1)) is
begin
  z:=(carry+w+x*y)mod(B**s);
  next_carry:=(carry+w+x*y)/(B**s);
end multiplier;
```

The procedure multiplier generates two s-digit numbers ($\leq B^s - 1$). The next algorithm computes the product $Z = X.Y$, where X is an n-digit number and Y is an m-digit one.

Algorithm 5.6   Long-Operand

```
for k in 0..s-1 loop zero(k):=0; end loop;
for j in 0..m/s-1 loop p(0, j):=zero; end loop;
for i in 0..n/s-1 loop
  carry(i,0):=zero;
  for j in 0..m/s-1 loop
    multiplier(s, carry(i, j), p(i, i+j), X(i*s..i*s+s-1),
    Y(i*s..i*s+s-1), carry(i, j+1), sum(i, j));
  end loop;
```

```
 p(i+1, m/s+i):=carry(i, m);
   for j in 0..i-1 loop p(i+1, j):=p(i, j); end loop;
   for j in 0..m-1 loop p(i+1, j+i):=sum(i, j); end loop;
end loop;
for j in 0..m/s+n/s-1 loop Z(j*s..j*s+s - 1):=p(n,j);
end loop;
```

5.2 INTEGERS

Conceptually, the simplest method for multiplying signed integers consists in computing, from the representations of the operands X and Y, the corresponding absolute value of the product together with the appropriate sign. The absolute value of the product would be obtained by one of the methods described before, while the product sign would be readily given through the exclusive OR function applied to a suitable binary representation of the respective signs. The process would then be completed by the computation of the signed representation of the result. This method is appropriate whenever the operands are provided in sign-magnitude form. In this case, signed multiplication appears trivial. In the case of B's complement or signed-digit representations, other methods have to be recommended.

5.2.1 *B*'s Complement Multiplication

Let X and Y be two integers represented in the reduced B's complement numeration system (Chapter 3, Comment 3.2):

$$R(X) = x_{n-1}.B^{n-1} + x_{n-2}.B^{n-2} + \cdots + x_0.B^0,$$

$$R(Y) = y_{m-1}.B^{m-1} + y_{m-2}.B^{m-2} + \cdots + y_0.B^0,$$

$$x_{n-1}, y_{m-1} \in \{0, B-1\} \quad \text{and} \quad x_i, y_j \in \{0, 1, \ldots, B-1\}, \tag{5.8}$$

$$\forall i \neq n-1, \forall j \neq m-1;$$

x_{n-1} and y_{m-1} are called sign-digits (sign-bits for $B = 2$). The respective weights of X and Y can be positive or negative according to the following definitions.

X and Y can be expressed in the form

$$X = x'_{n-1}.B^{n-1} + x_{n-2}.B^{n-2} + \cdots + x_0.B^0,$$

$$Y = y'_{m-1}.B^{m-1} + y_{m-2}.B^{m-2} + \cdots + y_0.B^0, \tag{5.9}$$

where $x'_{n-1} = 0$ if $x_{n-1} = 0$ and $x'_{n-1} = -1$ if $x_{n-1} = B - 1$; $y'_{n-1} = 0$ if $y_{n-1} = 0$ and $y'_{n-1} = -1$ if $y_{n-1} = B - 1$.

It can easily be shown that this sign-digit convention is well suited for the sign extension operation. Actually, under the above sign-digit interpretations, x_{n-1} (resp. y_{m-1}) can be indefinitely reproduced on the left side of the digit string without changing the numerical value of the expressed number. This takes for granted that the leftmost digit of the new string is viewed as the new sign-digit while the other digits remain positively weighted.

So X and Y, respectively belong to the ranges

$$-B^{n-1} \leq X \leq B^{n-1} - 1 \quad \text{and} \quad -B^{m-1} \leq Y \leq B^{m-1} - 1,$$

and the product $Z = X.Y$ belongs to the range

$$-B^{n+m-2} < Z \leq B^{n+m-2}$$

and can be represented in the form

$$R(Z) = z'_{n+m-1} . B^{n+m-1} + z_{n+m-2} . B^{n+m-2} + \cdots + z_0 . B^0.$$

5.2.1.1 *Mod B^{n+m} B's Complement Multiplication*   A straightforward multiplication algorithm can be deduced from the fact that the representation of a number x in the B's complement system can be mapped to a number $R(x) = x \bmod B^n$, n being the number of digits of the representation. X and Y are first represented with $n + m$ digits. Then

$$X \equiv R(X) \bmod B^{n+m} \quad \text{and} \quad Y \equiv R(Y) \bmod B^{n+m},$$

so that

$$Z = X.Y \equiv R(X).R(Y) \bmod B^{n+m}$$

and

$$R(Z) = R(X).R(Y) \bmod B^{n+m}$$

Assume that a `truncated_multiplication` procedure has been defined, which computes $c = a.b \bmod B^{n+m}$, a and b being two $(n + m)$-digit base-B natural numbers:

```
procedure truncated_multiplication
(a, b: in digit_vector(0..n+m-1); c: out digit_vector(0..
n+m-1);
```

Any natural-number multiplication algorithm may be used, as the mod B^{n+m} reduction is just a matter of truncating the results.

Algorithm 5.7   Mod B^{n+m} B's Complement Multiplication

```
for i in n..m+n-1 loop x(i):=x(n-1); end loop;
for i in m..m+n-1 loop y(i):=y(m-1); end loop;
truncated_multiplication(x, y, z);
```

The first two loops consist of sign extension up to $n + m$ digits.

Example 5.2

$B = 10 \quad n = m = 3, \quad X = -53, \quad Y = 65,$
$R(X) = 1000 - 53 = 947, \quad R(Y) = 065,$
6-digit representations: $R(X) = 999947, R(Y) = 000065,$
$R(X). R(Y) = 999947 \times 65 = 64996555$
$R(Z) = 64996555 \bmod 10^6 = 996555$
$Z = -10^5 + 96555 = -3445$

5.2.1.2 Signed Shift and Add Another method is based on a modification of the Hörner algorithm 5.2. It consists in subtracting Y from the last partial result $P(n-1)$ whenever the digit $X(n-1) = B-1$ (X negative).

Algorithm 5.8 Signed Shift and Add

```
P(0):=0;
for i in 0..n-2 loop
  P(i+1):=(P(i)+X(i)*Y)/B;
end loop;
if X(n-1)=0 then P(n):=P(n-1)/B; else P(n):=(P(n-1)-Y)/B;
end if;
Z:=P(n)*(B**n);
```

Example 5.3 Let

$X = -53 \quad \text{and} \quad Y = 65; \quad B = 10, \quad n = m = 3,$
$R(X) = 1000 - 53 = 947, R(Y) = 065, R(-65) = 9935$

> *Step 1:* $P(1) = \qquad (7^*65)\,{}^*10^{-1} = 0045.5$
> *Step 2:* $P(2) = (P(1) + (4^*65))\,{}^*10^{-1} = 0030.55$
> *Step 3:* $P(3) = (P(2) + (9935))\,{}^*10^{-1} = 996.555$
> $996.555 \times 10^3 = 996555$
> $Z = -10^5 + 96555 = -3445$

5.2.1.3 Postcorrection B's Complement Multiplication A third method, introduced by Baugh and Wooley ([BAU73]) for array multipliers in base 2, is deduced from relations (5.9). The negative (sign-digit) and positive parts of the multiplicator multiply the negative and positive parts of the multiplicand. Actually, the result of multiplication of the positive parts is then corrected by three supplementary terms depending on the respective sign-digits. The computation scheme is described in what follows.

The product $Z = X.Y$ can be expressed in the form

$$Z = (x'_{n-1}.B^{n-1} + X_0).(y'_{m-1}.B^{m-1} + Y_0)$$

where x'_{n-1} and y_{m-1} are the sign-digits while X_0 and Y_0, are, respectively, $(n-1)$- and $(m-1)$-digit natural numbers:

$$X_0 = x_{n-2}.B^{n-2} + \cdots + x_0.B^0 \quad \text{and} \quad Y_0 = y_{m-2}.B^{m-2} + \cdots + y_0.B^0.$$

Thus

$$Z = x'_{n-1}.y'_{m-1}.B^{n+m-2} + x'_{n-1}.Y_0.B^{n-1} + y'_{m-1}.X_0.B^{m-1} + X_0.Y_0. \quad (5.10)$$

The product $X_0.Y_0$ is a straight multiplication of natural numbers. The sign-digit product is different from zero only when both operands are negative, and consists of a (positive) shifted 1, $(n+m-2)$ positions to the left. The other two terms are, respectively, $(n-1)$-shifted Y_0 and $(m-1)$-shifted X_0, to be subtracted only when $x'_{n-1} \neq 0$ or $y'_{m-1} \neq 0$, respectively. Suitable sign extensions have to be performed on each of the four terms.

Assume that the following procedures have been previously defined: a multiplication procedure that computes $c = a.b$, a being an $(n-1)$-digit base-B natural number and b an $(m-1)$-digit one:

procedure multiplication (a: **in** digit_vector(0..n-2); b: **in** digit_vector(0..m-2); c: **out** digit_vector(0..n+m-3);

an addition procedure that computes $c = a + b \mod B^{n+m}$, a, b, and c being $(n+m)$-digit B's complement numbers:

procedure addition (a, b: **in** digit_vector(0..n+m-1); c: **out** digit_vector(0..n+m-1);

a subtraction procedure that computes $c = a - b \mod B^{n+m}$, a, b, and c being $(n+m)$-digit B's complement numbers:

procedure subtraction (a, b: **in** digit_vector(0..n+m-1); c: **out** digit_vector(0..n+m-1).

In the following algorithm, shifted_one stands for the $(n+m)$-digit representation of B^{n+m-2}:

Algorithm 5.9   Postcorrection B's Complement Multiplication

```
multiplication (X(0..n-2), Y(0..m-2), a(0..n+m-3));
a(m+n-1):=0; a(m+n-2):=0;
if x(n-1)=B-1 then
   for i in 0..n-2 loop b(i):=0; end loop;
   for i in 0..m-2 loop b(i+n-1):=y(i); end loop;
```

```
  b(n+m-1):=0; b(n+m-2):=0;
  subtraction (a, b, c);
else c:=a; end if;
if y(m-1)=B-1 then
  for i in 0..m-2 loop d(i):=0; end loop;
  for i in 0..n-2 loop d(i+m-1):=x(i); end loop;
  d(n+m-1):=0; d(n+m-2):=0;
  subtraction (c, d, e);
else e:=c; end if;
if x(n-1)=B-1 and y(n-1)=B-1 then addition (e, shifted_one,
z); else z:=e; end if;
```

The above program is structured in four parts: a multiplication, two conditional shift-and-subtract procedures, and a conditional shifted-one addition. According to the resources at hand, it may be suitable to implement the shift-and-subtract procedure by a `sign-change` procedure followed by a shift, and then an `addition` procedure. In this case as well, the sign-digit of each operand is conditioning the procedure execution. The `sign-change` procedure on an n-digit number X is described as follows:

```
for i in 0..n-1 loop x(i):=B-1-x(i);
end loop;
X:=X+1;
```

Example 5.4   Let

$$X = -53 \quad \text{and} \quad Y = -65; \quad B = 10, \quad n = m = 3,$$
$$R(X) = 1000 - 53 = 947, \quad R(Y) = 1000 - 65 = 935$$
$$\textit{multiplication procedure } X_0 = 47, Y_0 = 35, 47 \times 35 = 1645$$
$$a = 001645 \text{(6-digit representation)}$$
$$b = 003500$$
$$c = (001645 + 996499 + 1) \bmod 10^6 = 998145$$
$$d = 004700$$
$$e = (998145 + 995299 + 1) \bmod 10^6 = 993445$$
$$z = (993445 + 010000) \bmod 10^6 = 003445$$

Comment 5.3   The only case where Z cannot be expressed as an $(m+n-1)$-digit number is when $X = -B^{n-1}$ and $Y = -B^{m-1}$, so that $Z = B^{n+m-2}$ and its representation is $010\ldots0$. The previous algorithms could be modified accordingly: first detect whether $R(X) = R(-B^{n-1}) = (B-1)00\ldots0$ and $R(Y) = R(-B^{m-1}) = (B-1)\,00\ldots0$; if so, $R(Z) = 010\ \ldots\ 0$. In the contrary case the computation can be performed with $n+m-1$ digits. Then the obtained result $Z(n+m-2..0)$ is extended to $n+m$ digits by defining $Z(n+m-1) = Z(n+m-2)$.

5.2.2 Postcorrection 2's Complement Multiplication

In base 2, the postcorrection 2's complement multiplication[2] algorithm is simplified thanks to the carry-free feature of the partial products $x_i.y_j$ computation (BAU1973). Since the base is 2, equation (5.10) can be written

$$Z = X_0.Y_0 + x_{n-1}.y_{m-1}.B^{n+m-2} - x_{n-1}.Y_0.B^{n-1} - y_{m-1}.X_0.B^{m-1}. \tag{5.11}$$

As well as in the general case, the first two terms may be computed through a straight natural number multiplication procedure and a correction by a $(m+n-2)$-shifted positive bit $(x_{n-1}.y_{m-1})$. This shifted-one correction occurs whenever x_{n-1} and y_{m-1} are both 1. The two terms $-x_{n-1}.Y_0.B^{n-1}$ and $-y_{n-1}.X_0.B^{m-1}$ can be computed through (sign-change procedure) a bitwise complementation of the shifted products $x_{n-1}.(0Y_0)$ and $y_{m-1}.(0X_0)$, then adding 1 at levels $n-1$ and $m-1$, respectively. Observe that the sign-bit of $(n-1)$-shifted $x_{n-1}.(0Y_0)$ and $(m-1)$ shifted $y_{m-1}.(0X_0)$, share the same position: $n+m-2$. So a single negative-weight 1 at level $n+m-1$ will replace the two negative-weight 1's at level $n+m-2$. Whenever $n = m$, adding 1 twice at level $n-1$ can be done through adding 1 once at level n. Let's point out that, whenever x_{n-1} and/or y_{n-1} are zero, changing sign of $x_{n-1}.Y_0$ and/or $y_{n-1}.X_0$ doesn't affect the result. Figures 5.4 and 5.5 illustrate the computation scheme. Although this scheme suggests a

				1		1			
						not(x_0y_3)	x_0y_2	x_0y_1	x_0y_0
					not(x_1y_3)	x_1y_2	x_1y_1	x_1y_0	
				not(x_2y_3)	x_2y_2	x_2y_1	x_2y_0		
			not(x_3y_3)	x_3y_2	x_3y_1	x_3y_0			
sign bit		not(x_4y_3)	x_4y_2	x_4y_1	x_4y_0				
−1	x_5y_3	not(x_5y_2)	not(x_5y_1)	not(x_5y_0)					

Figure 5.4 A 2's complement multiplication scheme ($n = 6$, $m = 4$)

			1				
				not(x_0y_3)	x_0y_2	x_0y_1	x_0y_0
			not(x_1y_3)	x_1y_2	x_1y_1	x_1y_0	
sign bit		not(x_2y_3)	x_2y_2	x_2y_1	x_2y_0		
−1	x_3y_3	not(x_3y_2)	not(x_3y_1)	not(x_3y_0)			

Figure 5.5 A 2's complement multiplication scheme ($n = m = 4$).

[2] $\wedge$, dot, or no-symbol stand for Boolean AND, while + stands for real sum. $\bar{x}$ or *not* stand for the Boolean complementation (*not* function).

combinational circuit (Chapter 12), any sequential or programmed implementation can profit from the described features. The sign-bit 1, set at level $n + m - 1$, will vanish whenever a positive carry will occur at this level. An equivalent option, without assigning the initial sign-bit 1, would be to complement the final bit z_{n+m-1}.

5.2.3 Booth Multiplication for Binary Numbers

The *Booth algorithm* and the *modified Booth algorithm* ([BOO1951], [DAV1977]), have been and still are very popular ([KOR1993], [OBE1964], [PAR1999]) in a number of implementations of signed multiplication for 2's complement binary numbers. In most applications, the size of the slices (parameter r) does not exceed 2, in order to limit the complexity of the coded-digit products. Actually, for $r = 2$, one has to deal with shift, add, and subtract operations only. For greater values of r, the complexity of partial products is a significant part of the speed/cost compromises. Observe that, thanks to the sign extension availability, n/r may be considered integer without loss of generality. For B's complement representations, the method doesn't seem attractive, not only because of the partial products complexity but because of the coding itself. Nevertheless, from the theoretical point of view, the extension is straightforward, as will be shown in Section 5.2.4.

5.2.3.1 Booth-r Algorithms Assume that the *Booth-r representation* of X is used (Chapter 3, Definition 3.6):

$$X = x'_{k-1}.B^{k-1} + x'_{k-2}.B^{k-2} + \cdots + x'_0.B^0,$$

with

$$n = r.k, \quad B = 2^r$$

and (Booth-r coding)

$$x'_i = -x_{i.r+r-1}.2^{r-1} + x_{i.r+r-2}.2^{r-2} + x_{i.r+r-3}.2^{r-3}$$
$$+ \cdots + x_{i.r+2}.2^2 + x_{i.r+1}.2 + x_{i.r} + x_{i.r-1} \tag{5.12}$$

Then, Hörner Algorithm 5.2 (shift and add 2) can readily be used for computing $Z = X.Y$.

Algorithm 5.10 Booth-r Multiplication

```
P(0):=0;
for i in 0..k-1 loop
   P(i+1):=(P(i)+X'(i)*Y)/(2**r);
end loop;
Z:=P(n)*(2**n);
```

The product $x'_i.Y$ is equal to

$$
\begin{aligned}
x'_i.Y = &-x_{i.r+r-1}.Y.2^{r-1} + x_{i.r+r-2}.Y.2^{r-2} + \cdots + x_{i.r+2}.Y.2^2 \\
&+ x_{i.r+1}.Y.2 + x_{i.r}.Y + x_{i.r-1}.Y
\end{aligned}
\tag{5.13}
$$

so that the computation can be performed as follows.

Algorithm 5.11   Booth-r Multiplication

```
P(0):=0;
for i in 0..k-1 loop
  if X((i*r)-1)=1 then sum(-1):=Y; else sum(-1):=0; end if;
  for j in 0..r-2 loop
    if X((i*r)+j)=1 then sum(j):=sum(j-1)+Y*(2**j); else
    sum(j):=sum(j-1); end if;
  end loop;
  if X((i*r)+r-1)=1 then sum(r-1):=sum(r-2)-Y*(2**(r-1)); else
  sum(r-1):=sum(r-2); end if;
  P(i+1):=(P(i)+sum(r-1))/(2**r);
end loop;
Z:=P(n)*(2**n);
```

5.2.3.2   *Per Gelosia Signed-Digit Algorithm*   Another method, the *Per Gelosia* ([DAV1977]), consists in encoding both operands according to the Booth-r coding formula (5.12) and modifying the cellular shift and add Algorithm 5.4. It can be proved that, given integers a, b, c, and d belonging to the interval $[-2^{r-1}, 2^{r-1}]$, $p = a.b + c + d$ can be decomposed in a unique way under the form

$$
p_1.2^r + p_0
\tag{5.14}
$$

where both p_1 and p_0 belong to the interval $[-2^{r-1}, 2^{r-1} - 1]$. Digits p_1 and p_0 can thus be represented by r-tuples in 2's complement notation. Booth-r coded digits $x'(i)$ or $y'(j)$, according to formula (5.12), can be proved to belong to the interval $[-2^{r-1}, 2^{r-1}]$. So the products of Booth-r coded digits $x'(i)$ and $y'(j)$ can be specified in the form (5.14). Assume that two functions G and H have been defined which compute $p_1 = G(a, b, c, d)$ and $p_0 = H(a, b, c, d)$. Assume moreover that a `Booth_encode` procedure has been defined to compute the Booth-r digits of X and Y, respectively. An extra `Booth_decode` procedure is needed to express the final result, given in r-bit signed digits, in the classical 2's complement binary form.

Algorithm 5.12   Per Gelosia Booth-r Signed Digit Multiplication

```
XX:=Booth_encode(X); YY:=Booth_encode(Y);
for j in 0..m/r-1 loop p(0, j):=0; end loop;
for i in 0..n/r-1 loop
  carry(i, 0):=0;
  for j in 0..m/r-1 loop
    sum(i, j):=H(XX(i), YY(j), carry(i, j), p(i, i+j));
    carry(i, j+1):=G(XX(i), YY(j), carry(i, j), p(i, i+j));
  end loop;
```

```
p(i+1, m/r+i):=carry(i, m/r);
  for j in 0..i-1 loop p(i+1, j):=p(i, j); end loop;
  for j in 0..m/r-1 loop p(i+1, j+i):=sum(i, j); end loop;
end loop;
for j in 0..n/r+m/r-1 loop ZZ(j):=p(n/r, j); end loop;
Z:=Booth_decode(ZZ);
```

Figure 5.6 displays the (i, j)-cell, to illustrate the indexing suggested in the above algorithm. The carry-save Algorithm 5.5 could have been used to provide a better overall algorithmic complexity.

Example 5.5 Let

$$n = m = 12; \quad r = 3;$$
$$X = 101\ 011\ 110\ 001(\text{decimal} - 1295)$$
$$Y = 011\ 100\ 010\ 011\ (\text{decimal } 1811)$$

Booth-3 coding (expressed in decimal for convenience) is performed as follows

$XX\ (0) = 1 + 0 = 1$	$YY\ (0) = 3 + 0 = 3$
$XX\ (1) = -2 + 0 = -2$	$YY\ (1) = 2 + 0 = 2$
$XX\ (2) = 3 + 1 = 4$	$YY\ (2) = -4 + 0 = -4$
$XX\ (3) = -3 + 0 = -3$	$YY\ (3) = 3 + 1 = 4$
$XX = -3\ 4\ -2\ 1$	$YY = 4\ \ -4\ 2\ 3$

The operations are illustrated in Figure 5.7 where the result is expressed as

$$Z = -1,\ -1, 1,\ -4,\ -4,\ -4,\ -4, 3$$

or

$$-1.8^7 - 1.8^6 + 1.8^5 - 4.8^4 - 4.8^3 - 4.8^2 - 4.8 + 3;\ \text{in decimal } Z = -2345245.$$

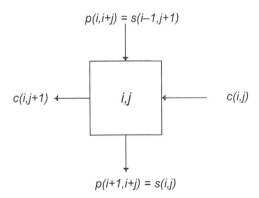

Figure 5.6 Booth-r (i, j)-cell.

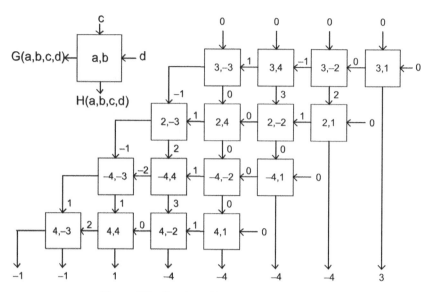

Figure 5.7   Booth multiplication, Example 5.5.

Let us point out that, although Booth-coded operands are in the range $[-2^{r-1},$ $+2^{r-1}]$, signed-digit results $Z(i)$ are in the range $[-2^{r-1}, +2^{r-1}-1]$. So in the example $(r=3)$, 4 will never appear as a digit result. This means that the Booth_encode procedure successively generates $(r+1)$-bit signed digits while the Booth_decode will generate, step-by-step, r-bit-digit outputs from r-bit inputs and carry-in.

The Booth_encode and Booth_decode functions are defined as follows:

Algorithm 5.13   Booth_encode

```
X(-1):=0;
for i in 0..n/r-1 loop
  a:=X((i*r)-1);
  for j in 0..r-2 loop a:=(a+X((i*r)+j))/2; end loop;
  a:=(a-X((i*r)+r-1))/2;
  XX(i):=a*(2**r);
end loop;
```

Algorithm 5.14.a   Booth_decode 1

```
a:=0;
for i in 0..n/r-1 loop
a:=(a+XX(i))/(2**r);
end loop;
X:=a*(2**n);
```

Example 5.6 Let

$$n = 15; \quad r = 3; \quad YY = 1 -4 -4\,2\,3 = 001, 100, 100, 010, 011$$

The result is built up in 2's complement representation,

Step 1: $(000 + 011)/2^3 = 000011/2^3 = 000.011$
Step 2: $(000.011 + 010)/2^3 = 010.011/2^3 = 000.010011$
Step 3: $(000.010011 - 100)/2^3 = 100.010011/2^3$-the first bit (sign-bit) is negative
  so the shift involves sign extension: $100.010011/2^3 = 111.100010011$
Step 4: $(111.100010011 - 100)/2^3 = 1011.100010011/2^3 = 111.011100010011$
Step 5: $(111.011100010011 +001)/2^3 = 0.000011100010011$
Step 6: $YY = 0.000011100010011 * 2^{15} = 011100010011$ (decimal 1811).

Another way to decode the signed-digit expression of $Z(i)$ rests on a procedure that roughly corresponds to the reverse application of the coding process described in Algorithm 5.13 (Booth_encode). Assuming

$$
\begin{aligned}
z_i = &-z_{i.r+r-1}.2^{r-1} + z_{i.r+r-2}.2^{r-2} + z_{i.r+r-3}.2^{r-3} + \cdots + z_{i.r+2}.2^2 \\
&+ z_{i.r+1}.2 + z_{i.r} \\
z_i^+ = &z_{i.r+r-1}.2^{r-1} + z_{i.r+r-2}.2^{r-2} + z_{i.r+r-3}.2^{r-3} + \cdots + z_{i.r+2}.2^2 \\
&+ z_{i.r+1}.2 + z_{i.r}, \quad i = 0,1,\ldots, k-1
\end{aligned}
\tag{5.15}
$$

and assuming moreover

$$z_{0.-1} = c_0\_in = 0; \; c_k\_in = z_{k.r} \text{ (sign-bit)},$$

The 2's complement expression Z'' of the result will be given by the following algorithm.

Algorithm 5.14.b **Booth_decode 2**

($\bar{z}$ stands for the Boolean complement of z)

```
for i in 0,1,..k-2 loop
if c_i_in=0
then
z''_i=z_i^+; c_{i+1}_in=z_{i.r+r-1}
else
z''_i=(z_i^+-c_i_in) mod 2^r; c_{i+1}_in=z_{i.r+r-1} ∨ (z̄_{i.r+r-2}.z̄_{i.r+r-3}.
...z̄_{i.r}); end if;
end loop;
if c_{k-1}_in=0
then
z''_{k-1} =z_{k-1}^+; z_{k.r}=z_{k.r-1}
else
z''_{k-1}=(z_{k-1}^+-c_{k-1}_in) mod 2^r; z_{k.r}=z_{kr-1} ∨ (z̄_{k.r-2}.z̄_{k.r-3}.
...z̄_{(k-1).r}); end if;
```

The above algorithm actually sets all signed digits to a positive value carrying a negative correction bit to the following digit (on the left) whenever necessary. The last correction bit $z_{k,r}$ is the sign bit.

Example 5.7   Let

$$n = 15; r = 3; k = 5; Z = 1 \ -4 \ -4 \ 2 \ 3 = 001, 100, 100, 010, 011$$

The result is built up in 2's complement representation,

Step 1: $c_0\_in = 0 \Longrightarrow z_0'' = 011; c_1\_in = z_2 = 0.$
Step 2: $c_1\_in = 0 \Longrightarrow z_1'' = 010; c_2\_in = z_5 = 0.$
Step 3: $c_2\_in = 0 \Longrightarrow z_2'' = 100; c_3\_in = z_8 = 1.$
Step 4: $c_3\_in = 1 \Longrightarrow z_3'' = (100 - 001) \bmod 8 = 011; c_4\_in = z_{11} \ \vee \ (\bar{z}_{10}.\bar{z}_9)$
      $= 1.$
Step 5: $c_4\_in = 1 \Longrightarrow z_4'' = (001 - 001) \bmod 8 = 000; z_{15} = z_{14} \ \vee \ (\bar{z}_{13}.\bar{z}_{12})$
      $= 0;$

$Z'' = z_4'' \ z_3'' \ z_2'' \ z_1'' \ z_0'' = 0 \ 000011100010011$

5.2.4   Booth Multiplication for Base-B Numbers
(Booth-r Algorithm in Base B)

The above algorithms may be extended to base $B > 2$, within the following conditions.

B is assumed even, and the sign-digit $x_{n-1} \in \{0,1,\ldots, B-1\}$ is valued $+ x_{n-1}$ whenever $x_{n-1} < B/2$ and $x_{n-1} - B$, otherwise. The sign function $\mathrm{sign}(x)$ is defined as

$$\mathrm{sign}(x) = \lfloor 2.x/B \rfloor. \tag{5.16}$$

This means that

$$\mathrm{sign}(x) = 1 \text{ if } x \geq B/2 \quad (X \text{ negative})$$

and

$$\mathrm{sign}(x) = 0 \text{ if } x < B/2 \quad (X \text{ positive}).$$

A base-B signed number $X = x_{n-1}, x_{n-2}, \ldots, x_1, x_0$ is thus given by the expression

$$X = (x_{n-1} - B.\mathrm{sign}(x_{n-1})).B^{n-1} + x_{n-2}.B^{n-2} + \cdots + x_2.B^2 + x_1.B + x_0 \tag{5.17}$$

Then, given an *n-digit* base-B integer shredded into n/r r-digit slices, the generalized Booth coding can be performed according to

$$x_i' = (x_{i.r+r-1} - B.\mathrm{sign}(x_{i.r+r-1})).B^{r-1} + x_{i.r+r-2}.B^{r-2} + \cdots + x_{i.r+2}.B^2$$
$$+ x_{i.r+1}.B + x_{i.r} + \mathrm{sign}(x_{i.r-1}) \tag{5.18}$$

This means that each slice is viewed as a B's complement r-digit number. So whenever $x_{i.r+r-1} \geq B/2$, a correction $+ \mathrm{sign}(x_{i.r+r-1})$ has to be made at the next left digit $x_{i.r+r}$.

Example 5.8 Let

$$n = 24; r = 4; B = 10$$
$$X = 0000\,9124\,6458\,2123\,5252\,5632\,2145$$

Booth-4 coding

$$X = 0001\,9125\,6458\,2124\,5253\,5632\,2145 = x_6', x_5', x_4', x_3', x_2', x_1', x_0'$$

1 has to be added at the end of every slice whose next digit (on the right) is greater than 4.

The decimal values of Booth digits are given by

$$x_6' = 1$$
$$x_5' = -1000 + 125 = -875$$
$$x_4' = -4000 + 458 = -3542$$
$$x_3' = 2124$$
$$x_2' = -5000 + 253 = -4747$$
$$x_1' = -5000 + 632 = -4368$$
$$x_0' = 2145$$

Assuming $C = 10^4$,

$$X = 1.C^6 - 875.C^5 - 3542.C^4 + 2124.C^3 - 4747.C^2 - 4368.C + 2145$$
$$= 9124\,6458\,2123\,5252\,5632\,2145$$

Algorithm 5.15   Booth-r in Base B

The Booth-r multiplication, Algorithm 5.11, can be modified as follows.

```
P(0):=0;
for i in 0..k-1 loop
  if X((i*r)-1) >= B/2 then sum(-1):=Y; else sum(-1):=0; end if;
  for j in 0..r-2 loop
    sum(j):=sum(j-1)+X((i*r)+j)*Y*(B**j);
  end loop;
  if X((i*r)+r-1) >= B/2 then sum(r-1):=sum(r-2)+(X((i*r)+r-1)
  -B)*Y*(B**(r-1));
    else sum(r-1):=sum(r-2)+X((i*r)+r-1)*Y*(B**(r-1));
    end if; P(i+1):=(P(i)+sum(r-1))/(B**r);
end loop;
Z:=P(n)*(B**n);
```

The complexity of the partial products computation is an important drawback for the above algorithm. Moreover, increasing the base or increasing the side of the slices in Booth coding are approaches conceptually similar. Both techniques reduce the number of computational steps, but the rise of the step complexity is the price to be paid. The Booth algorithm in base-B could be more suitable for some Per Gelosia inspired applications with suitable hardware resources such as look-up table (LUT) to implement the partial products within acceptable times. Obviously, increasing B and/or r would quickly make the LUT size unmanageable.

5.3 SQUARING

5.3.1 Base-B Squaring

Although any multiplication system could readily be used for squaring, specialized systems can provide both time and cost savings. Actually, normal multiplication requires n^2 partial products while the maximum depth of the adding tree reaches n (maximum column depth in Figure 5.1a). Squaring only needs $n(n - 1)/2$ partial products, plus n digit-squares. Moreover, the adding process is simplified by the fact that each nonsquare partial product digit appears twice. As far as partial products are multiplied by two beforehand, the maximum depth of the adding tree is reduced to $\lceil n/2 \rceil$. The complexity of squaring in base $B > 2$ comes from the fact that digit double products could need up to three digits while squaring only needs two. So the saving in column depth is consumed by second-order carries. For the particular case $B = 2$, the digit squaring is trivial as double products are performed through straight shifts. Moreover, Boolean simplification provides further reductions in the depth of the adding tree.

5.3.1.1 Cellular Carry–Save Squaring Algorithm Algorithm 5.16 is a modification of the cellular carry-save Algorithm 5.5. It computes $X^2 + C + D$; the main difference rests upon the (i, j)-cell loops, adding squares whenever $i = j$, and adding double products otherwise, to the successive carries and partial sums. What is saved in computing the double products is somewhat consumed by the carries, to be handled by additional adding procedures. Actually, to cope with the basic scheme suggested by Algorithm 5.5, the only modification (besides that of the partial products) rests on the substitution of some (i, j)-cell procedures by half-adder ones.

Algorithm 5.16 Carry-Save Squaring

(Symbol / stands for integer division)

```
for j in 0..n-1 loop p(0, j):=D(j); carry1(0, j):=C(j);
  carry2(0, j):=0;end loop;
  p(1, n):=0;
  sum(0, 0):=(carry1(0, 0)+p(0, 0)+X(0)**2) mod B;
  carry1(1, 0):=((carry1(0, 0)+p(0, 0)+X(0)**2)/B) mod B;
  carry2(1, 1):=(carry1(0, 0)+p(0, 0)+X(0)**2)/B**2;
for j in 1..n-1 loop
```

```
  sum(0, j):=(carry1(0, j)+carry2(0, j)+p(0, j)+2*X(0)*X(j))
   mod B;
  carry1(1,j):=((carry1(0,j)+carry2(0,j)+p(0,j)+2*X(0)*
   X(j))/B) mod B;
  carry2(1,j+1):=(carry1(0,j)+carry2(0,j)+p(0,j)+2*X(0)*X(j))/B**2;
end loop;
for j in 0..n-1 loop p(1,j):=sum(0,j);end loop;
p(2, n+1):=carry2(1, n);
for i in 1..n-2 loop
  for j in 0..i-1 loop
    sum(i, j):=(carry1(i,j)+p(i, i+j)) mod B;
    carry1(i+1,j):=(carry1(i,j)+p(i, i+j))/B;
  end loop;
    sum(i,i):=(carry1(i,i)+carry2(i,i)+p(i,i+i)+X(i)**2) mod B;
    carry1(i+1,i):=((carry1(i,i)+carry2(i,i)+p(i,i+i)+X(i)**2)
     /B) mod B;
    carry2(i+1,i+1):=(carry1(i,i)+carry2(i,i)+p(i, i+i)+X(i)**2)
     /B**2;
  for j in i+1..n-1 loop
    sum(i,j):=(carry1(i,j)+carry2(i,j)+p(i,i+j)+2*X(i)*X(j))
     mod B;
    carry1(i+1, j):=((carry1(i, j)+carry2(i, j)+p(i,i+j)+2*
     X(i)*X(j))/B) mod B;
    carry2(i+1, j+1):=(carry1(i, j)+carry2(i, j)+p(i, i+j)+2*
     X(i)*X(j))/B**2;
  end loop;
  for j in 0..i-1 loop p(i+1, j):=p(i, j); end loop;
  for j in 0..n-1 loop p(i+1, j+i):=sum(i, j); end loop;
  p(i+2, n+i+1):=carry2(i+1, n);
end loop;

for j in 0..n-2 loop
  sum(n-1, j):=(carry1(n-1, j)+p(n-1, n-1+j))mod B;
  carry1(n, j):=(carry1(n-1, j)+p(n-1, n-1+j))/B;
end loop;
sum(n-1, n-1):=(carry1(n-1, n-1)+carry2(n-1, n-1)+p(n-1,
2*n-2)+X(n-1)**2) mod B;
carry1(n, n-1):=((carry1(n-1, n-1)+carry2(n-1, n-1)+p(n-1,
2*n-2)+X(n-1)**2)/B) mod B;
carry2(n, n):=(carry1(n-1, n-1)+carry2(n-1, n-1)+p(n-1,
2*n-2)+X(n-1)**2)/B**2;
for j in 0..n-2 loop p(n, j):=p(n-1, j); end loop;
for j in 0..n-1 loop p(n, n-1+j):=sum(n-1, j); end loop;
p(n, 2*n):= carry2(n, n);

for j in 0..n-1 loop Z(j):=p(n, j); end loop;
k(0):=0;
for j in 0..n-1 loop
  Z(j+n):=(p(n, n+j)+carry1(n, j)+k(j)) mod B;
  k(j+1):=(p(n, n+j)+carry1(n, j)+k(j))/B;
end loop;
```

It is straightforward to figure out the relative computational complexity of a specialized squaring procedure, with respect to classic multiplication. Let us point out that the $n(n+1)/2$ base-B partial products (either double products or squares) will generate less than $3n(n+1)/2$ base-B digits to be added according to some selected algorithm. Using the common multiplication algorithm, this quantity is $2n^2$. Asymptotically, a 25% saving can be expected, through the best use of the adding tree reduction (multioperand addition; Chapter 11). Although some extra benefit could be taken from the fact that the upper carry digit belongs to $\{0, 1\}$, the potential time/cost saving doesn't look attractive for general-purpose computers. Some combinational implementations are presented in Chapter 12. As a matter of fact, squaring is not statistically frequent in most applications where suitable multiplication resources are generally available. So the interest of designing special devices for base-B squaring remains limited. On the contrary, significant advantages can be taken from specific features of base-2 representation, as shown in the following section.

5.3.2 Base-2 Squaring

The following three Boolean relations are key properties allowing important simplifications to the squaring computation scheme.

1. $x_0 \wedge x_0 = x_0, \quad x_0 \in \{0, 1\}$ (5.19)

   the square of a binary digit is itself (carry-free).

2. $x_i x_j + x_i x_j = 2x_i x_j = (x_i x_j, 0)_{\text{base } 2}, \quad x_i, x_j \in \{0, 1\}$ (5.20)

   the operation of adding a partial product to itself may be replaced by a left-shift.

3. $x_i x_j + x_j = (x_i x_j, \bar{x}_i x_j)_{\text{base } 2}.$ (5.21)

Thanks to these properties, the computation scheme of Figure 5.1 can be reduced as shown in Figure 5.8.

This scheme is easily built up by noting that columns of index k even are made up with products $x_i x_j$ such that $i < j$ and $i + j = k - 1$ and $\bar{x}_{k/2-1} x_{k/2}$, while columns of index k odd follow the same rule but with the additional term $x_{(k-1)/2-1} x_{(k-1)/2}$ instead of $x_{k/2-1} x_{k/2}$. In total, $n(n+1)/2$ partial products have to be considered

$x_{n-2}x_{n-1}$	$\bar{x}_{n-2}x_{n-1}$	$\cdots$	x_0x_7	x_0x_6	x_0x_5	x_0x_4	x_0x_3	x_0x_2	$\bar{x}_0x_1$	$-$	x_0
			x_1x_6	x_1x_5	x_1x_4	x_1x_3	$\bar{x}_1x_2$	x_0x_1			
			x_2x_5	x_2x_4	$\bar{x}_2x_3$	x_1x_2					
			$\bar{x}_3x_4$	x_2x_3							

Figure 5.8 Squaring computation scheme.

instead of n^2 for the classical multiplication scheme. With respect to a full multiplication scheme, the saving is now 50% (asymptotically). The most popular techniques to deal with this problem are related to multioperand addition procedures (Chapter 11) using various methods for reducing the number of operands ([DAD1965], [OBE1964], [WAL1964]).

Whenever methods for squaring can be fast enough to compete with multiplication, an interesting alternative for multiplying is inferred from the following formula:

$$4X.Y = [(X + Y)^2 - (X - Y)^2]$$ (5.22)

So binary multiplication can be made from two squares, three signed additions, and one shift. For small numbers, a look-up table inspired technique may be an interesting approach for fast multiplication using squares. We conclude this section by observing that general integer exponentiation can be performed using squaring and multiplication as primitive operations. A suitable factorization of the exponent would minimize the quantity of multiplying/squaring steps.

5.4 BIBLIOGRAPHY

[BAU1973] C. R. Baugh and B. A. Wooley, A two's complement parallel array multiplication algorithm. *IEEE Trans. Comput.*, **C-22**: 1045–1047 (1973).

[BOO1951] A. D. Booth, A signed binary multiplication technique. *Q. J. Mech. Appl. Math.* **4**: 236–240 (1951).

[DAD1965] L. Dadda, Some schemes for parallel multipliers. *Alta Frequenza* **34**: 349–356 (1965).

[DAV1977] M. Davio and G. Bioul, Fast parallel multiplication. *Philips Res. Rpts.* **32**: 44–70 (1977).

[KOR1993] I. Koren, *Computer Arithmetic Algorithms*, Prentice Hall, Englewood Cliffs, NJ, 1993.

[MAS1990] M. Nagamatsu, et al., A 15-ns 32 × 32-b CMOS multiplier with an improved parallel structure. *IEEE J. Solid-State Circuits* **25**(2): 494–499 (1990).

[OBE1964] S. F. Oberman and M. Flynn, *Advanced Computer Arithmetic Design*, Wiley-Interscience, Hoboken, NJ, 2001.

[OKL1996] V. G. Oklobdzija, D. Villeger, and S. S. Liu, A method for speed optimized partial product reduction and generation of fast parallel multipliers using an algorithmic approach. *IEEE Trans. Comput.* **45**(3): 294–305 (1996).

[PAR1999] Behrooz Parhami, *Computer Arithmetic, Algorithms and Hardware Designs*, Oxford University Press, New York, 1999.

[WAN1995] Z. Wang, G. A. Jullien, and W. C. Miller, A new design technique for column compression multipliers. *IEEE Trans. Comput.* **44**(8): 962–970 (1995).

[WAL1964] C. S. Wallace, A suggestion for fast multipliers. *IEEE Trans. Electron. Comput.* **EC-13**(Feb): 14–17 (1964).

6

ARITHMETIC OPERATIONS: DIVISION

Integer or finite length fractional numbers can be multiplied exactly, whenever sufficient length is allowed for the result. Division doesn't share this feature. As a matter of fact, division generally does not provide a finite length result. The accuracy must be defined beforehand by setting the unit in the least significant position (*ulp*) of the result. The number of algorithmic cycles will therefore depend on the desired accuracy, not on the operand length.

Digit recurrence algorithms represent the most common class of division techniques: a single quotient-digit is produced at each computation step. The classic pencil and paper method belongs to this class. The time complexity is thus a linear function of the desired number of quotient-digits. SRT division ([SWE1957], [ROB1958], [TOC1958]) has been widely used in computer applications. The *digit recurrence* algorithmic step mainly consists in an estimation of the greatest multiple of the divisor to be subtracted from the remainder.

Functional iteration is another class of algorithms. These algorithms use function-solving techniques to converge, from an initial estimation, toward the quotient with the required precision. The main feature of this method rests on the faster than linear convergence, typically quadratic. The main drawbacks are the step complexity and the need for additional computations to provide the final remainder, thus increasing the rounding complexity. The most used *functional iteration* techniques are based on *Newton–Raphson* convergence equations and *Taylor–MacLaurin* expansions (*Goldschmidt's algorithm*). This technique has been used in several commercial applications.

Synthesis of Arithmetic Circuits: FPGA, ASIC, and Embedded Systems
By Jean-Pierre Deschamps, Géry J. A. Bioul, and Gustavo D. Sutter
Copyright © 2006 John Wiley & Sons, Inc.

Very high radix and *variable latency* classes of algorithms are described in the literature ([OBE1995]). Their practical use is more limited and more justified for specific applications ([OBE1995], [BRI1993]).

A number of division methods, more or less related to the above four main classes, are described in the literature ([OBE1997], [PAR1999], [FLY2001], [ERC2004]).

6.1 NATURAL NUMBERS

Let X and Y be two natural numbers with $Y > 0$. Define Q and R, respectively, as the *quotient* and the *remainder* of the division of X by Y, with an accuracy of p fractional base-B digits:

$$B^p.X = Q.Y + R,$$

where Q and R are natural numbers, and $R < Y$. In other words,

$$X = (Q.B^{-p}).Y + (R.B^{-p}), \quad \text{with } R.B^{-p} < Y.B^{-p}, \tag{6.1}$$

so that the unit in the least significant position (*ulp*) of $Q.B^{-p}$ and $R.B^{-p}$ is equal to B^{-p}. In the particular case where $p = 0$, that is,

$$X = Q.Y + R, \quad \text{with } R < Y, \tag{6.2}$$

Q and R are the *quotient* and the *remainder* of the *integer division* of X by Y.

The basic algorithm applies to operands X and Y such that

$$X < Y. \tag{6.3}$$

In the general case, to ensure that $X < Y$, a previous alignment step is necessary. Assume that X is an m-digit base-B number, that is, $X < B^m$; then

substitute Y by $Y' = B^m.Y$, so that $Y' \geq B^m.1 > X$;

compute the quotient Q and the remainder R' of the division of X by Y', with an accuracy of $p + m$ fractional base-B digits, that is,

$$B^{p+m}.X = Q.Y' + R', \quad \text{with } R' < Y'$$

so that

$$B^p.X = Q.Y + R, \quad \text{with } R = R'/B^m < Y.$$

The next theorem constitutes the justification of the basic division algorithm.

Theorem 6.1 Fundamental Equation of Division Given two natural numbers a and b such that $a < b$, there exists two natural numbers q and r satisfying $Ba = q.b + r$, with $q \in \{0, 1, \ldots, B - 1\}$ and $r < b$.

The iterative application of the preceding theorem, that is,

$$B.r(0) = q(1).Y + r(1), \quad r(1) < Y,$$
$$B.r(1) = q(2).Y + r(2), \quad r(2) < Y,$$
$$\cdots$$
$$B.r(p-1) = q(p).Y + r(p), \quad r(p) < Y,$$

(6.4)

with $r(0) = X$, generates the following relation,

$$X.B^p = (q(1).B^{p-1} + q(2).B^{p-2} + \cdots + q(p).B^0).Y + r(p),$$

(6.5)

so that

$$Q = q(1).B^{p-1} + q(2).B^{p-2} + \cdots + q(p).B^0 \text{ and } R = r(p).$$

(6.6)

Assume that a procedure `division_step` has been defined

```
procedure division_step (a, b: in natural; q, r: out
natural);
```

that computes q and r such that $B.a = q.b + r$, with $q \in \{0, 1, \ldots, B-1\}$ and $r < b$. Then the following basic division algorithm is a straightforward application of (6.4) and (6.5).

Algorithm 6.1   Basic Division

```
r(0):=X;
for i in 1..p loop
   division_step (r(i − 1), Y, q(i), r(i));
end loop,
```

It generates the base-B representation $q(1)q(2) \cdots q(p)$ of Q and the remainder $R = r(p)$. If $B = 2$ the `division_step` procedure is very simple.

Algorithm 6.2   Base-2 Division Step

```
z:=2*a-b;
if z<0 then q:=0; r:=2*a; else q:=1; r:=z; end if;
```

If B is greater than 2 the division step is more complex, namely:

Algorithm 6.3   Base-B Division Step

```
if B*a<b then q:=0; r:=B*a;
elsif B*a<2*b then q:=1; r:=B*a-b;
elsif B*a<3*b then q:=2; r:=B*a-(2*b);
..
elsif B*a<(B-1)*b then q:=B-2; r:=B*a-((B-2)*b);
else q:=B-1; r:=B*a-((B-1)*b);
```

Observe that at every step of Algorithm 6.1, the new remainder $r(i)$ is equal to $B.r(i-1) - q(i).Y$. If $q(i) = 0$ then $r(i) = B.r(i-1)$ and the preceding remainder $r(i-1)$ is said to be *restored* (actually, restored and shifted). For that reason, the basic division algorithm is also known as the *restoring division algorithm*.

Comment 6.1 If a previous alignment is necessary, instead of substituting Y by $Y' = B^m.Y$, an alternative option is to substitute X by $X' = X/B$ and Y by $Y' = B^{m-1}.Y$, and to compute the quotient Q and the remainder R' of the division of X' by Y', with an accuracy of $p+m$ fractional base-B digits, that is,

$$B^{p+m}.X' = B^{p+m-1}.X = Q.Y' + R', \quad \text{with} \quad R' < Y',$$

so that

$$B^p.X = Q.Y + R, \quad \text{with} \quad R = R'/B^{m-1} < Y.$$

Observe that the substitution of X by X' is equivalent to the substitution of the first algorithm step, namely,

$$B.X = q(1).Y + r(1), \quad r(1) < Y$$

by

$$X = q(1).Y + r(1), \quad r(1) < Y.$$

Examples 6.1

1. Compute $12/15$ with an accuracy of 8 fractional bits:

$$r(0) = 12,$$

$$2.12 - 15 \geq 0 \rightarrow q(1) = 1, \quad r(1) = 24 - 15 = 9,$$
$$2.9 - 15 \geq 0 \rightarrow q(2) = 1, \quad r(2) = 18 - 15 = 3,$$
$$2.3 - 15 < 0 \rightarrow q(3) = 0, \quad r(3) = 6,$$
$$2.6 - 15 < 0 \rightarrow q(4) = 0, \quad r(4) = 12,$$
$$2.12 - 15 \geq 0 \rightarrow q(5) = 1, \quad r(5) = 24 - 15 = 9,$$
$$2.9 - 15 \geq 0 \rightarrow q(6) = 1, \quad r(6) = 18 - 15 = 3,$$
$$2.3 - 15 < 0 \rightarrow q(7) = 0, \quad r(7) = 6,$$
$$2.6 - 15 < 0 \rightarrow q(8) = 0, \quad r(8) = 12.$$

So $Q = 11001100 = 204$, $R = 12$, and $2^8.12 = 204.15 + 12$.

2. (Integer division) Given an 8-bit natural number X ($X < 256$) and a positive integer Y, the integer division of X by Y is computed as follows. The divisor Y is substituted by $Y' = Y.256$, the accuracy is equal to $p + m = 0 + 8 = 8$ bits, and the final remainder R' will be substituted by $R = R'/256$. As an example, assume

that $X = 124$ and $Y = 15$:

$$Y' = 15.256 = 3840,$$
$$r(0) = 124,$$

$$2.124 - 3840 < 0 \rightarrow q(1) = 0, \quad r(1) = 248,$$
$$2.248 - 3840 < 0 \rightarrow q(2) = 0, \quad r(2) = 496,$$
$$2.496 - 3840 < 0 \rightarrow q(3) = 0, \quad r(3) = 992,$$
$$2.992 - 3840 < 0 \rightarrow q(4) = 0, \quad r(4) = 1984,$$
$$2.1984 - 3840 \geq 0 \rightarrow q(5) = 1, \quad r(5) = 3968 - 3840 = 128,$$
$$2.128 - 3840 < 0 \rightarrow q(6) = 0, \quad r(6) = 256,$$
$$2.256 - 3840 < 0 \rightarrow q(7) = 0, \quad r(7) = 512,$$
$$2.512 - 3840 < 0 \rightarrow q(8) = 0, \quad r(8) = 1024.$$

Thus $Q = 00001000 = 8$, $R = 1024/256 = 4$, and $124 = 8.15 + 4$.

The same operation can be performed taking into account Comment 6.1:

$$Y'15.128 = 1920,$$
$$r(0) = 124/2,$$

$$124 - 1920 < 0 \rightarrow q(1) = 0, \quad r(1) = 124,$$
$$2.124 - 1920 < 0 \rightarrow q(2) = 0, \quad r(2) = 248,$$
$$2.248 - 1920 < 0 \rightarrow q(3) = 0, \quad r(3) = 496,$$
$$2.496 - 1920 < 0 \rightarrow q(4) = 0, \quad r(4) = 992,$$
$$2.992 - 1920 \geq 0 \rightarrow q(5) = 1, \quad r(5) = 1984 - 1920 = 64,$$
$$2.64 - 1920 < 0 \rightarrow q(6) = 0, \quad r(6) = 128,$$
$$2.128 - 1920 < 0 \rightarrow q(7) = 0, \quad r(7) = 256,$$
$$2.256 - 1920 < 0 \rightarrow q(8) = 0, \quad r(8) = 512.$$

Thus $Q = 00001000 = 8$, $R = 512/128 = 4$, and $124 = 8.15 + 4$.

3. Given a 6-bit natural number X and a 3-bit positive integer Y, compute X/Y with an accuracy of $p = 4$. To ensure that $X < Y$, the divisor is substituted by $Y' = Y.2^6$, the division is performed with an accuracy of $p = 4 + 6 = 10$, and the final remainder will be divided by 2^6. Assume that $X = 101011$ (43) and $Y = 111$ (7), so that $Y' = 111000000$:

Initial step, $i = 0$
$r(0) = X = 000101011;$
$i = 1; 2.r(0) - Y'$
0001010110 − **0111**000000 $< 0 \rightarrow q(1) = 0; \quad r(1) = 001010110$ (restoring)
$i = 2; 2.r(1) - Y'$

$0010101100 - 0111000000 < 0 \rightarrow q(2) = 0; \quad r(2) = 010101100$ (restoring)

$i = 3; 2.r(2) - Y'$

$0101011000 - 0111000000 < 0 \rightarrow q(3) = 0; \quad r(3) = 101011000$ (restoring)

$i = 4; 2.r(3) - Y'$

$1010110000 - 0111000000 = 011110000 \geq 0 \rightarrow q(4) = 1;$

$r(4) = 011110000$

$i = 5; 2.r(4) - Y'$

$0111100000 - 0111000000 = 000100000 \geq 0 \rightarrow q(5) = 1;$

$r(5) = 000100000$

$i = 6; 2.r(5) - Y'$

$0001000000 - 0111000000 < 0 \rightarrow q(6) = 0; \quad r(6) = 001000000$ (restoring)

$i = 7; 2.r(6) - Y'$

$0010000000 - 0111000000 < 0 \rightarrow q(7) = 0; \quad r(7) = 010000000$ (restoring)

$i = 8; 2.r(7) - Y'$

$0100000000 - 0111000000 < 0 \rightarrow q(8) = 0; \quad r(8) = 100000000$ (restoring)

$i = 9; 2.r(8) - Y'$

$1000000000 - 0111000000 = 001000000 \geq 0 \rightarrow q(9) = 1;$

$r(9) = 001000000$

$i = 10; 2.r(9) - Y'$

$0010000000 - 0111000000 < 0 \rightarrow q(10) = 0; \quad r(10) = 010000000$

Thus

$$Q = 0001100010 \ (98), R = 010000000 \ (128),$$

so that

$$43.2^4 = 98.7 + 128/2^6, \quad \text{that is, } 43 = (98/16).7 + (2/16).$$

Observe that Y' is a multiple of 2^6, so that the subtraction $2.r(i) - Y'$ is performed with the four (highlighted) most significant bits of $2.r(i)$ and Y'; the other bits of $2.r(i)$ are just propagated to the next step.

Another observation is that all numbers are even. A better solution is to substitute Y by $Y' = Y.2^5$ and X by $X/2$ (Comment 6.1). Then $2.r(0) = X = 000101011$. The computation is the same as before without the final bit (always equal to 0). The final result is

$$Q = 0001100010 \ (98), \quad R = 01000000 \ (64),$$

so that

$$43.2^4 = 98.7 + 64/2^5, \quad \text{that is, } 43 = (98/16).7 + (2/16).$$

Let us point out that most scientific hand calculators feature integer binary operations. Therefore, to produce a nonzero quotient, the initial multiplication of the dividend by 2^p is compulsory.

Algorithm 6.3 is quite unpractical. A better solution consists in looking first for a tentative value q_t of q. This process assumes that a normalization procedure sets both a and b as n-digit natural numbers such that $a < b$ and $B^n > b \geq B^{n-1}$; that is, the leftmost digit of b is non-zero. The input parameters of the `division_step` procedure are a and b. One defines the truncated values a_t and b_t of a and b as follows (/ stands for integer division):

$$a_t = a/B^{n-3} \quad \text{and} \quad b_t = b/B^{n-2}. \tag{6.7}$$

Observe that, as $a < b$, $a_t < b/B^{n-3} < B^n/B^{n-3} = B^3$; $b_t < B^n/B^{n-2} = B^2$ and $b_t \geq B^{n-1}/B^{n-2} = B$.

The tentative value of q, that is, q_t, is computed from

$$a_t = q_t.b_t + r_t, \text{ with } r_t < b_t.$$

A 5-input (B-ary) look-up table (LUT) may be used as a fast computing device for that purpose.

Lemma 6.1 Given an n-digit dividend X, an n-digit divisor Y, and a remainder $r(i)$ such that (/ stands for integer division)

$$r(i) < Y, \quad r(0) = X, \quad Y_{n-1} \neq 0,$$
$$r_t(i) = B.r(i)/B^{n-2} = r(i)/B^{n-3}, \tag{6.8}$$
$$Y_t = Y/B^{n-2},$$

and defining q_t as

$$q_t = r_t(i)/Y_t, \tag{6.9}$$

then the correct value of $q = B.r(i)/Y$ is either q_t or $q_t - 1$, that is,

$$q_t - 1 \leq B.r(i)/Y \leq q_t. \tag{6.10}$$

From relations (6.8) and (6.9), it is obvious that $B^{n-1} \leq Y \leq B^n - 1$, while $q_t \leq B - 1$.

If the k most significant digits of a given divisor Y are equal to 0, then a previous (normalizing) step will require substituting Y by $Y.B^k$ so that $Y.B^k \geq B^{n-1}$, while $r(i)$ may be assumed an n-digit number with any number of zeros upfront. If an n-digit dividend X is given greater then the n-digit divisor Y, then a previous (normalizing) step will require substituting X by X/B.

Proof The rightmost part of inequality (6.10) (/ stands for real division) is derived from

$$B.r(i)/Y \leq (r_t(i).B^{n-2} + \alpha)/Y_t.B^{n-2}, \tag{6.11}$$

where α stands for $B.r(i) - r_t(i).B^{n-2}$; so $\alpha < B^{n-2}$ and

$$B.r(i)/Y < (r_t(i) + 1)/Y_t \quad \text{and} \quad \lfloor B.r(i)/Y \rfloor \le q_t. \tag{6.12}$$

The leftmost part of inequality (6.10) is proven as follows.
    Equation (6.8) can be written

$$r_t(i).B^{n-2} \le B.r(i), \tag{6.13}$$

while (6.9) can be written

$$r_t(i) - q_t.Y_t \ge 0. \tag{6.14}$$

Using (6.13) in (6.14),

$$B.r(i) - q_t.(Y - \beta) \ge 0, \tag{6.15}$$

where β stands for $Y - Y_t.B^{n-2}$; so $\beta < B^{n-2}$, then, as $q_t \le B - 1$

$$q_t.\beta < B^{n-1} \le Y, \tag{6.16}$$

allowing (6.15) to be written

$$B.r(i) - (q_t - 1).Y > 0, \tag{6.17}$$

or

$$B.r(i)/Y > q_t - 1 \text{ then } \lfloor B.r(i)/Y \rfloor \ge q_t - 1, \tag{6.18}$$

which completes the proof.

Algorithm 6.4   Restoring Base-B Division Step

```
at:=a/B**(n-3); bt:=b/B**(n-2);
qt:=at/bt;
remainder:=B*a-qt*b;
if remainder<0 then q:=qt-1; r:=remainder+b;
else q:=qt; r:=remainder; end if;
```

    In this case, the restoring operation consists of adding the divisor to the remainder whenever the latter is negative. Then the remainder is iteratively shifted as in the basic division algorithm.

Example 6.2 Compute $752024/876544$ (base 10):

$$q(0) = 0, \, r(0) = 752024;$$
$$i = 1$$
$$a_t = 752, \, b_t = 87, \, q_t = 752/87 = 8,$$
$$\text{remainder} = 10 \times 752024 - 8 \times 876544 = 507888 > 0,$$
$$q(1) = 8, \, r(1) = 507888;$$
$$i = 2$$
$$a_t = 507, \, b_t = 87, \, q_t = 507/87 = 5,$$
$$\text{remainder} = 10 \times 507888 - 5 \times 876544 = 696160 > 0,$$
$$q(2) = 5, \, r(2) = 696160;$$
$$i = 3$$
$$a_t = 696, \, b_t = 87, \, q_t = 696/87 = 8,$$
$$\text{remainder} = 10 \times 696160 - 8 \times 876544 = -50752 < 0,$$
$$q(3) = 7, \, r(3) = -50752 + 876544 = 825792;$$
$$i = 4$$
$$a_t = 825, \, b_t = 87, \, q_t = 825/87 = 9,$$
$$\text{remainder} = 10 \times 825792 - 9 \times 876544 = 369024 > 0,$$
$$q(4) = 9, \, r(4) = 369024;$$
$$i = 5$$
$$a_t = 369, \, b_t = 87, \, q_t = 369/87 = 4,$$
$$\text{remainder} = 10 \times 369024 - 4 \times 876544 = 184064 > 0,$$
$$q(5) = 4, \, r(5) = 184064$$

So

$$752024.10^5 = 85794 \times 876544 + 184064 \text{ or}$$
$$752024 = 0.85794 \times 876544 + 1.84064.$$

6.2 INTEGERS

6.2.1 General Algorithm

Let X be an integer and Y a natural number with $Y > 0$. Define Q and R respectively as the *quotient* and the *remainder* of the division of X by Y, with an accuracy of p fractional base-B digits:

$$B^p.X = Q.Y + R,$$

where Q and R are integers, $-Y \leq R < Y$ and $sign(R) = sign(X)$. In other words,

$$X = (Q.B^{-p}).Y + (R.B^{-p}), \quad \text{with}$$
$$-Y.B^{-p} \leq R.B^{-p} < Y.B^{-p} \quad \text{and} \quad sign(R.B^{-p}) = sign(X), \tag{6.19}$$

so that the unit in the least significant position (*ulp*) of $Q.B^{-p}$ and $R.B^{-p}$ is equal to B^{-p}. In the particular case where $p = 0$, that is,

$$X = Q.Y + R, \quad \text{with} - Y \leq R < Y \quad \text{and} \quad sign(R) = sign(X), \tag{6.20}$$

Q and R are the *quotient* and the *remainder* of the *integer division* of X by Y.
    The basic algorithm applies to operands X and Y such that

$$-Y \leq X < Y. \tag{6.21}$$

In the general case, to ensure that $-Y \leq X < Y$, a previous alignment step is necessary. Assume that X is an m-digit reduced B's complement number, that is, $-B^{m-1} \leq X < B^{m-1}$; then

substitute Y by $Y' = B^{m-1}.Y$, so that $Y' \geq B^{m-1}.1 > X$ and $-Y' \leq -B^{m-1}.1 \leq X$;
compute the quotient Q and the remainder R' of the division of X by Y', with an accuracy of $p + m - 1$ fractional base-B digits, that is,

$$B^{p+m-1}.X = Q.Y' + R', \quad \text{with} - Y' \leq R' < Y' \quad \text{and} \quad sign(R') = sign(X),$$

that is,

$$B^p.X = Q.Y + R, \quad \text{with} -Y \leq R = R'/B^{m-1} < Y \quad \text{and} \quad sign(R) = sign(X).$$

Comment 6.2 If X is negative and $X.B^p$ is an exact multiple of Y, for example, $X.B^p = K.Y$, then the remainder must be negative, so that

$$Q = (K + 1) \quad \text{and} \quad R = -Y.$$

Nevertheless, the desired result was (probably)

$$Q = K \quad \text{and} \quad R = 0.$$

An alternative definition of the quotient and remainder could be

$$B^p.X = Q.Y + R, \quad \text{with} -Y < R < Y \quad \text{and} \quad sign(R) = sign(X) \text{ if } R \neq 0,$$

that is, Definition 2.4.

The digit-recurrence algorithms are based on the following sequence of equations:

$$X = r(0),$$
$$B.r(0) = q(1).Y + r(1), \quad -Y \le r(1) < Y,$$
$$B.r(1) = q(2).Y + r(2), \quad -Y \le r(2) < Y, \quad\quad (6.22)$$
$$\cdots$$
$$B.r(p-1) = q(p).Y + r(p), \quad -Y \le r(p) < Y.$$

Multiply the first equation by B^p, the second by B^{p-1}, the third by B^{p-2}, and so on. Then sum up the $p+1$ equations. The following relation is obtained:

$$B^p.X = (q(1).B^{p-1} + q(2).B^{p-2} + \cdots + q(p).B^0).Y + r(p),$$
$$\text{with} \quad -Y \le r(p) < Y. \quad\quad (6.23)$$

So

$$Q = q(1).B^{p-1} + q(2).B^{p-2} + \cdots + q(p).B^0 \quad \text{and} \quad R = r(p). \quad\quad (6.24)$$

As a matter of fact, the 2-unknowns $(q(i+1), r(i+1))$ system

$$B.r(i) = q(i+1).Y + r(i+1), \quad \text{with} \quad -Y \le r(i+1) < Y,$$

has two solutions as the interval $-Y \le r(i+1) < Y$ includes two values of $r(i+1)$ such that $r(i+1) \equiv B.r(i) \bmod Y$. Thus different algorithms can be defined according to the way the values of $r(i+1)$ and $q(i+1)$ are chosen. The range of $q(i+1)$ is given by the following inequalities:

$$q(i+1) = (B.r(i) - r(i+1))/Y < (B.Y + Y)/Y = (B+1),$$
$$q(i+1) = (B.r(i) - r(i+1))/Y > (-B.Y - Y)/Y = -(B+1);$$

thus

$$-B \le q(i+1) \le B.. \quad\quad (6.25)$$

The diagram of Figure 6.1 (the *Robertson diagram*) gives, for every value of $B.r(i)$, the two possible values of $r(i+1)$ along with the corresponding value of $q(i+1)$.

According to (6.25) the values of $q(i)$ computed with (6.22) are not necessarily base-B digits. If the minimum and maximum values $-B$ and B are not used, then $q(i)$ is a signed base-B digit. An easy way to transform the obtained signed-digit

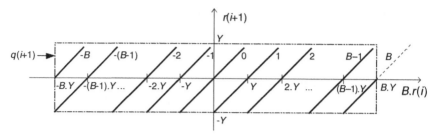

Figure 6.1   Robertson diagram: computation of $r(i + 1)$ and $q(i + 1)$.

representation into a nonsigned one (e.g., a B's complement representation) consists in defining, at each step, the ith component of two new variables q_pos and q_neg:

if $q(i) > 0$ then q_pos $= q(i)$, q_neg $= 0$,
if $q(i) = 0$ then q_pos $= 0$, q_neg $= 0$,
if $q(i) < 0$ then q_pos $= 0$, q_neg $= -q(i)$,

so that both q_pos and q_neg are p-digit base-B numbers. It remains to compute

$$q = \text{q\_pos} - \text{q\_neg}.$$

As a matter of fact, the conversion from the signed-digit representation to a B's complement one can also be performed on the fly ([ERC1987], [ERC1992], [OBE1997]).

A final step is necessary in order that the condition $sign(R) = sign(X)$ be satisfied:

if $R < 0$ and $X \geq 0$ then substitute R by $R + Y$ and Q by $Q - 1$;
if $R \geq 0$ and $X < 0$ then substitute R by $R - Y$ and Q by $Q + 1$.

Comment 6.3   If a previous alignment is necessary, instead of substituting Y by $Y' = B^{m-1} . Y$, an alternative option is to substitute X by $X' = X/B$ and Y by $Y' = B^{m-2} . Y$, and to compute the quotient Q and the remainder R' of the division of X' by Y', with an accuracy of $p + m - 1$ fractional base-B digits, that is,

$$B^{p+m-1} . X' = B^{p+m-2} . X = Q.Y' + R', \quad \text{with} -Y' \leq R' < Y' \quad \text{and}$$
$$sign(R') = sign(X')$$

so that

$$B^p . X = Q.Y + R, \quad \text{with} \quad -Y \leq R = R'/B^{m-2} < Y \quad \text{and} \quad sign(R) = sign(X).$$

Observe that the substitution of X by X' is equivalent to the substitution of the first algorithm step, namely,

$$B.X = q(1).Y + r(1), \quad -Y \leq r(1) < Y$$

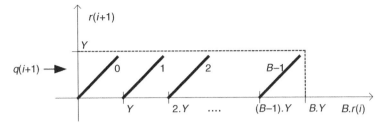

Figure 6.2 Restoring algorithm.

by

$$X = q(1).Y + r(1), \quad -Y \le r(1) < Y.$$

6.2.2 Restoring Division Algorithm

A simple way of choosing between the two possible values of $r(i+1)$ is to add the condition

$$sign(r(i+1)) = sign(r(i)),$$

so that all remainders have the same sign as X. Assume that X is nonnegative. Then the diagram of Figure 6.1 is reduced to the right upper quarter (Figure 6.2). The corresponding algorithm is the classical restoring algorithm of Section 6.1.

6.2.3 Base-2 Nonrestoring Division Algorithm

In the binary case, the diagram of Figure 6.1 (without the minimum and maximum values -2 and 2) is reduced to the diagram of Figure 6.3a. The way the values of $r(i+1)$ and $q(i+1)$ are chosen is shown in Figure 6.3b. The main difference with respect to the recovering algorithm is that the decision about the values of $r(i+1)$ and $q(i+1)$ only depends on the sign of $r(i)$, and it is no longer necessary to compare $2.r(i)$ with Y:

if $2.r(i)$ is nonnegative, then choose $q(i+1) = 1$ and $r(i+1) = 2.r(i) - Y$;
if $2.r(i)$ is negative, then choose $q(i+1) = -1$ and $r(i+1) = 2.r(i) + Y$.

The corresponding algorithm is the following:

Algorithm 6.5 Nonrestoring Division, First Version

```
r(0):=X;
for i in 0..p-1 loop
  if r(i)<0 then q_pos(i+1):=0; q_neg(i+1):=1;
  r(i+1):=2*r(i)+Y;
  else q_pos(i+1):=1; q_neg(i+1):=0; r(i+1):=2*r(i)-Y;
  end if;
end loop;
```

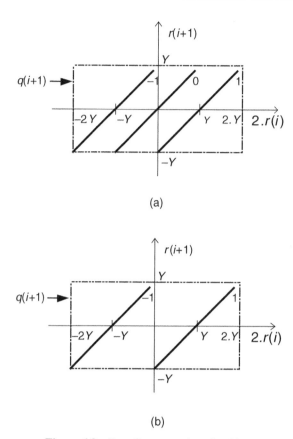

Figure 6.3   Base-2 nonrestoring algorithm.

```
Q:=q_pos-q_neg;
R:=r(p);
if X>=0 and R<0 then R:=R+Y; Q:=Q-1;
elsif X<0 and R>=0 then R:=R-Y; Q:=Q+1;
end if;
```

Nevertheless, a better algorithm, including an implicit on-the-fly conversion, can be used. At each step, instead of computing $q(1)$, $q(2)$, and so on, the following values are computed:

$$q'(0) = (1 + q(1))/2, \, q'(1) = (1 + q(2))/2, \ldots, q'(p - 1) = (1 + q(p))/2. \quad (6.26)$$

Observe that these new coefficients are bits:

    if $q(i) = 1$, then $q'(i - 1) = 1$;
    if $q(i) = -1$, then $q'(i - 1) = 0$.

According to (6.24) and (6.26),

$$
\begin{aligned}
Q &= q(1).2^{p-1} + q(2).2^{p-2} + \cdots + q(p).2^0 \\
&= (2.q'(0) - 1).2^{p-1} + (2.q'(1) - 1).2^{p-2} + \cdots + (2.q'(p-1) - 1).2^0 \\
&= 2(q'(0).2 + q'(1).2^{p-2} + \cdots + q'(p-1).2^0) - (2^p - 1) \\
&= -(1 - q'(0)).2^p + q'(1).2^{p-1} + \cdots + q'(p-1).2^1 + 1.2^0.
\end{aligned}
\tag{6.27}
$$

Thus the vector

$$
(1 - q'(0))\, q'(1) \cdots q'(p-1)\; 1
$$

is the 2's complement representation of Q.

The corresponding algorithm is the following:

Algorithm 6.6   Nonrestoring Division, Second Version

```
r(0):=X;
for i in 0..p-1 loop
  if r(i)<0 then Q(i):=0; r(i+1):=2*r(i)+Y;
  else Q(i):=1; r(i+1):=2*r(i)-Y;
  end if;
end loop;
Q(0):=1-Q(0); Q(p):=1; R:=r(p);
if X>=0 and R<0 then R:=R+Y; Q:=Q-1;
elsif X<0 and R>=0 then R:=R-Y; Q:=Q+1;
end if;
```

Observe that, before the final sign correction, the obtained quotient Q is always odd.

Examples 6.3

1. Compute $-12/15$ with an accuracy of 8 fractional bits:

$$
\begin{aligned}
&r(0) = -12, \\
q(0) = 0, \quad &r(1) = -24 + 15 = -9, \\
q(1) = 0, \quad &r(2) = -18 + 15 = -3, \\
q(2) = 0, \quad &r(3) = -6 + 15 = 9, \\
q(3) = 1, \quad &r(4) = 18 - 15 = 3, \\
q(4) = 1, \quad &r(5) = 6 - 15 = -9, \\
q(5) = 0, \quad &r(6) = -18 + 15 = -3, \\
q(6) = 0, \quad &r(7) = -6 + 15 = 9, \\
q(7) = 1, \quad &r(8) = 18 - 15 = 3.
\end{aligned}
$$

So $Q = 100110011 = -205$, $R = 3$, and $2^8.(-12) = (-205).15 + 3$. The sign correction generates the final result: $Q = (-205) + 1 = -204$, $R = 3 - 15 = -12$, so that $2^8.(-12) = (-204).15 + (-12)$.

2. (Integer division) Given a 9-bit integer X ($-256 \leq X < 256$) and a positive integer Y, the integer division of X by Y is computed as follows. The divisor Y is substituted by $Y' = Y.256$, the accuracy is equal to $p + m - 1 = 0 + 9 - 1 = 8$ bits, and the final remainder R' will be substituted by $R = R'/256$. As an example, assume that $X = -247$ and $Y = 15$:

$$Y' = 15.256 = 3840,$$
$$r(0) = -247,$$
$$q(0) = 0, \quad r(1) = -494 + 3840 = 3346,$$
$$q(1) = 1, \quad r(2) = 6692 - 3840 = 2852,$$
$$q(2) = 1, \quad r(3) = 5704 - 3840 = 1864,$$
$$q(3) = 1, \quad r(4) = 3728 - 3840 = -112,$$
$$q(4) = 0, \quad r(5) = -224 + 3840 = 3616,$$
$$q(5) = 1, \quad r(6) = 7232 - 3840 = 3392,$$
$$q(6) = 1, \quad r(7) = 6784 - 3840 = 2944,$$
$$q(7) = 1, \quad r(8) = 5888 - 3840 = 2048.$$

So, $Q = 111101111 = -17$, $R = 2048/256 = 8$, and $-247 = (-17).15 + 8$. The sign correction generates the final result: $Q = (-17) + 1 = -16$, $R = 8 - 15 = -7$, so that $-247 = (-16).15 + (-7)$.

The same operation can be performed taking into account Comment 6.3:

$$Y' = 15.128 = 1920,$$
$$r(0) = -247/2,$$
$$q(0) = 0, \quad r(1) = -247 + 1920 = 1673,$$
$$q(1) = 1, \quad r(2) = 3346 - 1920 = 1426,$$
$$q(2) = 1, \quad r(3) = 2852 - 1920 = 932,$$
$$q(3) = 1, \quad r(4) = 1864 - 1920 = -56,$$
$$q(4) = 0, \quad r(5) = -112 + 1920 = 1808,$$
$$q(5) = 1, \quad r(6) = 3616 - 1920 = 1696,$$
$$q(6) = 1, \quad r(7) = 3392 - 1920 = 1472,$$
$$q(7) = 1, \quad r(8) = 2944 - 1920 = -1024.$$

So, $Q = 111101111 = -17$, $R = 1024/128 = 8$, and $-247 = (-17).15 + 8$. The sign correction generates the final result: $Q = (-17) + 1 = -16$, $R = 8 - 15 = -7$, so that $-247 = (-16).15 + (-7)$.

3. Given a 7-bit 2's complement integer X and a 6-bit natural number Y belonging to the interval $-2^3.Y \leq X < 2^3.Y$, compute X/Y with an accuracy of $p = 5$.

To ensure that $-Y \leq X < Y$, the divisor is substituted by $Y' = 2^2.Y$, the dividend by $X' = X/2$, the division is performed with an accuracy of $p = 5 + 3 = 8$, and the final remainder will be divided by 2^2. Assume that $X = 1010101$ (-43) and $Y = 000111$ (7):

$r(0) = X'; \quad 2.r(0) = X = 1010101 < 0;$

$i = 0$

$q(0) = 0; \quad 2.r(0) + Y' = r(1) = \mathbf{10101}01 + \mathbf{00111}00 = 1110001 < 0$

$i = 1$

$q(1) = 0; \quad 2.r(1) + Y' = r(2) = \mathbf{11000}10 + \mathbf{00111}00 = 1111110 < 0$

$i = 2$

$q(2) = 0; \quad 2.r(2) + Y' = r(3) = \mathbf{11111}00 + \mathbf{00111}00 = 0011000 \geq 0$

$i = 3$

$q(3) = 1; \quad 2.r(3) - Y' = r(4) = \mathbf{01100}00 - \mathbf{00111}00 = 0010100 \geq 0$

$i = 4$

$q(4) = 1; \quad 2.r(4) - Y' = r(5) = \mathbf{01010}00 - \mathbf{00111}00 = 0001100 \geq 0$

$i = 5$

$q(5) = 1; \quad 2.r(5) - Y' = r(6) = \mathbf{00110}00 - \mathbf{00111}00 = 1111100 < 0$

$i = 6$

$q(6) = 0; \quad 2.r(6) + Y' = r(7) = \mathbf{11110}00 + \mathbf{00111}00 = 0010100 \geq 0$

$i = 7$

$q(7) = 1; \quad 2.r(7) - Y' = r(8) = \mathbf{01010}00 - \mathbf{00111}00 = 0001100 \geq 0$

$i = 8$

$q(8) = 1$

Thus $Q = 100111011$ and $R = 0001100$. Since $X < 0$ and $r(8) > 0$, a correction has to be made:

$Q = 100111011 + 1 = 100111100 \; (= -196); \quad R = 0001100 - 0011100$

$= 1110000 \; (= -16)$

so that $(-43).32 = (-196).7 + (-16/4)$, that is, $-43 = (-196/32).7 + (-4/32)$.

Observe that Y' is a multiple of 2^2, so that the operation $2.r(i) \pm Y'$ is performed with the five (boldface) most significant bits of $2.r(i)$ and Y'; the other bits of $2.r(i)$ are just propagated to the next step.

4. Given a 7-bit 2's complement integer X and a 6-bit natural number Y belonging to the interval $-2^4.Y \leq X < 2^4.Y$, compute X/Y with an accuracy of $p = 4$. To ensure that $-Y \leq X < Y$, the divisor is substituted by $Y' = 2^3.Y$, the dividend by $X' = X/2$, the division is performed with an accuracy of $p = 4 + 4 = 8$,

and the final remainder will be divided by 2^3. Assume that $X = 00101011$ (43) and $Y = 000101$ (5):

$r(0) = X'$; $2.r(0) = X = 0101011 \geq 0$;

$i = 0$

$q(0) = 1$; $2.r(0) - Y' = r(1) = \mathbf{0101}011 - \mathbf{0101}000 = 0000011 \geq 0$

$i = 1$

$q(1) = 1$; $2.r(1) - Y' = r(2) = \mathbf{0000}110 - \mathbf{0101}000 = 1011110 < 0$

$i = 2$

$q(2) = 0$; $2.r(2) + Y' = r(3) = \mathbf{1011}1100 + \mathbf{0101}000 = 1100100 < 0$

$i = 3$

$q(3) = 0$; $2.r(3) + Y' = r(4) = \mathbf{1001}000 + \mathbf{0101}000 = 1110000 < 0$

$i = 4$

$q(4) = 0$; $2.r(4) + Y' = r(5) = \mathbf{1100}000 + \mathbf{0101}000 = 0001000 \geq 0$

$i = 5$

$q(5) = 1$; $2.r(5) + Y' = r(6) = \mathbf{0010}000 - \mathbf{0101}000 = 1101000 < 0$

$i = 6$

$q(6) = 0$; $2.r(6) + Y' = r(7) = \mathbf{1010}000 + \mathbf{0101}000 = 1111000 < 0$

$i = 7$

$q(7) = 0$; $2.r(7) + Y' = r(8) = \mathbf{1110}000 + \mathbf{0101}000 = 0011000 \geq 0$

$i = 8$

$q(8) = 1$

Thus $Q = 010001001$ ($=$ 137) and $R = 0011000$ ($=$ 24). Since $X \geq 0$ and $r(8) \geq 0$, no corrections have to be made, so that $43.16 = 137.5 + 24/2^3$, that is, $43 = (137/16).5 + (3/16)$.

6.2.4 SRT Radix-2 Division

Consider again the diagram of Figure 6.3a. There are two overlapping areas such that

if $-Y \leq 2.r(i) < 0$ then $q(i + 1)$ can be chosen equal to either -1 or 0;
if $0 \leq 2.r(i) < Y$ then $q(i + 1)$ can be chosen equal to either 0 or 1.

The initial goal of the SRT-2 procedure was to reduce the number of additions or subtractions, choosing $q(i + 1) = 0$ and $r(i + 1) = 2.r(i)$ as often as possible. Another advantage of the existence of overlapping areas is that, within particular conditions of allowed range on the remainder, the choice of $q(i + 1)$ can be done as a function of the truncated values of Y and $r(i)$. The drawback is the nonunique form of the final quotient because of the use of a redundant quotient-digit set, namely, $\{-1, 0, 1\}$. This is solved through a final conversion process.

If the strategy consists in selecting $q(i+1) = 0$ whenever possible, the selection is achieved according to the following rule (Figure 6.3a):

if $2.r(i) < -Y$, then $q(i+1) = -1$;
if $-Y \leq 2.r(i) < Y$, then $q(i+1) = 0$;
if $2.r(i) \geq Y$, then $q(i+1) = 1$.

In practice, detecting the situation that allows $q(i+1) = 0$ (i.e., $-Y \leq 2.r(i) < Y$) is not straightforward, unless through a trial subtraction, an operation to be avoided whenever possible to make the savings effective. Sweeney, Robertson, and Tocher ([SWE1957], [ROB1958], [TOC1958]) suggest an alternative solution to this question. Instead of allowing the range $-Y \leq r(i) < Y$ for the partial remainders, a restricted range is enforced.

Let Y be an n-bit natural number whose most significant bit is equal to 1, that is,

$$2^{n-1} \leq Y < 2^n,$$

and X an integer belonging to the range

$$-2^{n-1} \leq X < 2^{n-1},$$

(an n-bit 2's complement number) so that $-Y \leq X < Y$. Then the system (6.22), with $B = 2$, is substituted by the following:

$$X = r(0),$$
$$2.r(0) = q(1).Y + r(1), \quad -2^{n-1} \leq r(1) < 2^{n-1},$$
$$2.r(1) = q(2).Y + r(2), \quad -2^{n-1} \leq r(2) < 2^{n-1}, \tag{6.28}$$
$$\cdots$$
$$2.r(p-1) = q(p).Y + r(p), \quad -2^{n-1} \leq r(p) < 2^{n-1}.$$

The corresponding graphical representation is shown in Figure 6.4a.
    The selection of $q(i+1)$ and $r(i+1)$ is done as follows (Figure 6.4b):

if $2.r(i) < -2^{n-1}$, then $q(i+1) = -1$, $r(i+1) = 2.r(i) + Y$;
if $-2^{n-1} \leq 2.r(i) < 2^{n-1}$, then $q(i+1) = 0$, $r(i+1) = 2.r(i)$;
if $2.r(i) \geq 2^{n-1}$, then $q(i+1) = 1$, $r(i+1) = 2.r(i) - Y$.

Observe that $w = 2.r(i)$ belongs to the interval $-2^n \leq w < 2^n$. If it is represented as an $(n+1)$-bit 2's complement number, the comparison with 2^{n-1} is very simple and can be done with the two most significant bits: See Table 6.1.

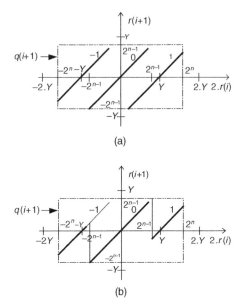

Figure 6.4   SRT-2 division algorithm.

The corresponding algorithm is the following:

Algorithm 6.7   SRT-2 Division

```
r(0):=X;
for i in 0..p-1 loop
  if  2.r(i)<-(2**(n-1))  then  q_pos(i+1):=0;  q_neg(i+1):=1;
  r(i+1):=2*r(i)+Y;
    elsif 2.r(i)>=2**(n-1) then q_pos(i+1):=1; q_neg(i+1):=0;
    r(i+1):=2*r(i)-Y;
    else q_pos(i+1):=0; q_neg(i+1):=0; r(i+1):=2*r(i);
    end if;
end loop;
Q:=q_pos-q_neg;
R:=r(p);
if X>=0 and R<0 then R:=R+Y; Q:=Q-1;
elsif X<0 and R>=0 then R:=R-Y; Q:=Q+1;
end if;
```

TABLE 6.1   SRT-2 Algorithm: Selection of $q(i + 1)$

w_n	w_{n-1}	$q(i + 1)$
0	0	0
0	1	1
1	0	-1
1	1	0

Examples 6.4

1. Given an 8-bit 2's complement integer X ($-128 \leq X < 128$) and an 8-bit positive integer Y whose most significant bit is equal to 1 ($128 \leq Y < 256$), compute the quotient and the remainder of the division of X by Y. The range of possible values of w is $-256 \leq w < 256$. This range is partitioned into three intervals: $w < -128$, $-128 \leq w < 128$ and $128 \leq w$, to which correspond the values -1, 0, and 1 for $q(i+1)$. Assume that $X = -84$, $Y = 247$ and $p = 8$:

$r(0) = -84$,

$2.r(0) < -128$	$q(1) = -1$,	$r(1) = 2.r(0) + Y$	$= -168 + 247 = 79$,
$128 < 2.r(1)$	$q(2) = 1$,	$r(2) = 2.r(1) - Y$	$= 158 - 247 = -89$,
$2.r(2) < -128$	$q(3) = -1$,	$r(3) = 2.r(2) + Y$	$= -178 + 247 = 69$,
$-128 \leq 2.r(3) < 128$	$q(4) = 0$,	$r(4) = 2.r(3)$	$= 138$,
$128 < 2.r(4)$	$q(5) = 1$,	$r(5) = 2.r(4) - Y$	$= 276 - 247 = 29$,
$-128 \leq 2.r(5) < 128$	$q(6) = 0$,	$r(6) = 2.r(5)$	$= 58$
$-128 \leq 2.r(6) < 128$	$q(7) = 0$,	$r(7) = 2.r(6)$	$= 116$,
$-128 \leq 2.r(7) < 128$	$q(8) = 0$,	$r(8) = 2.r(7)$	$= 232$,

So q_pos $= 01001000$ ($= 72$), q_neg $= 10100000$ ($= 160$), $Q = 72 - 160 = -88$, $R = 232$. The sign correction generates the final result: $Q = (-88) + 1 = -87$, $R = 232 - 247 = -15$, so that $(-84).256 = (-87).247 + (-15)$.

2. Let $X = 0101001101$ ($= 333$) be a 2's complement 10-bit integer and $Y = 1101000000$ ($= 832$) a 10-bit natural number whose most significant bit is equal to 1. Compute X/Y with an accuracy of $p = 8$. At each step w will be represented in the form of a 2's complement number with $10 + 1 = 11$ bits. For adding or subtracting Y, a 2's complement 11-bit representations will be used:

$$Y = 01101000000,$$
$$-Y = 10011000000.$$

The step-by-step procedure is described as follows.

Step #	Remainder Computation	q_pos	q_neg
$i = 0$	$w = 2.X = 01010011010$		
	$w_{10} \, w_9 = 01 \rightarrow q(1) = 1$	1	0
	$w \quad 01010011010$		
	$-Y \quad \underline{10011000000}$		
	$r(1) \quad 1101011010$		
$i = 1$	$w = 2.r(1) = 11010110100$		
	$w_{10} \, w_9 = 11 \rightarrow q(2) = 0$	0	0

Step #	Remainder Computation	q_pos	q_neg
	w 11010110100		
	0 00000000000		
	$r(2)$ 1010110100		
$i = 2$	$w = 2.r(2) = 10101101000$		
	$w_{10}\, w_9 = 10 \rightarrow q(3) = -1$	0	1
	w 10101101000		
	Y 01101000000		
	$r(3)$ 0010101000		
$i = 3$	$w = 2.r(3) = 00101010000$		
	$w_{10}\, w_9 = 00 \rightarrow q(4) = 0$	0	0
	w 00101010000		
	0 00000000000		
	$r(4)$ 0101010000		
$i = 4$	$w = 2.r(4) = 01010100000$		
	$w_{10}\, w_9 = 01 \rightarrow q(5) = 1$	1	0
	w 01010100000		
	$-Y$ 10011000000		
	$r(5)$ 1101100000		
$i = 5$	$w = 2.r(5) = 11011000000$		
	$w_{10}\, w_9 = 11 \rightarrow q(6) = 0$	0	0
	w 11011000000		
	0 00000000000		
	$r(6)$ 1011000000		
$i = 6$	$w = 2.r(6) = 10110000000$	0	1
	$w_{10}\, w_9 = 10 \rightarrow q(7) = -1$		
	w 10110000000		
	Y 01101000000		
	$r(7)$ 0011000000		
$i = 7$	$w = 2.r(7) = 00110000000$		
	$w_{10}\, w_9 = 00 \rightarrow q(8) = 0$	0	0
	w 00110000000		
	0 00000000000		
	$r(8)$ 0110000000	$(= 384)$	

As the final remainder is positive, Y doesn't need to be added, and Q is given by:

$$Q = \text{Q\_pos} - \text{Q\_neg} = 10001000 - 00100010 = 01100110 \ (= 102).$$

The overall operation can be resumed as

$$333.2^8 = 102.832 + 384.$$

6.2.5 SRT Radix-2 Division with Stored-Carry Encoding

The most time-consuming operation, at each step of algorithm 6.7, is clearly the computation of the new remainder $r(i+1)$: n-bit addition. The key idea for saving this time is to perform a carry-save sum, that is, a reduction of three operands to two (stored-carry encoding, Chapter 4). So every remainder $r(i)$ will be expressed in the form of a sum of two 2's complement numbers. The SRT-2 carry-save algorithm departs from Algorithm 6.7 by the range allowed for the successive remainders and by the way the values of $q(i+1)$ and $r(i+1)$ are chosen as functions of $r(i)$.

Let Y be an n-bit natural number whose most significant bit is equal to 1, that is,

$$2^{n-1} \leq Y < 2^n, \tag{6.29}$$

and let X be an integer belonging to the range $-Y \leq X < Y$, so that

$$-2^n < X < 2^n$$

(an $(n+1)$-bit 2's complement number). Then the system (6.22), with $B = 2$, generates the quotient Q and the remainder R of the division of X by Y with an accuracy of p fractional bits. The selection of $q(i+1)$ and $r(i+1)$ must be done as shown in Figure 6.3a.

Every remainder $r(i)$ belongs to the range $-Y \leq r(i) < Y$, so that $r(i)$ satisfies the following inequalities

$$-2^n < -Y \leq r(i) < Y < 2^n,$$

and $w = 2.r(i)$ belongs to the interval

$$-2^{n+1} < w < 2^{n+1}. \tag{6.30}$$

In 2's complement, with $n+3$ bits,

$$110000\cdots0 < w < 010000\cdots0.$$

All along the algorithm execution, w will be represented in stored-carry form, that is, in the form

$$w = s + c$$

where s and c are $(n+3)$-bit numbers (Figure 6.5a).

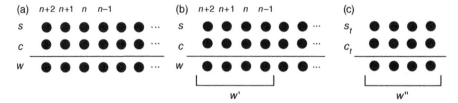

Figure 6.5   Stored-carry representation of $w = 2.r(i)$.

Define w' as being the truncated value of $w = 2.r(j)$, namely (Figure 6.5b)

$$w' = \lfloor w/2^{n-1} \rfloor.$$

According to (6.30),

$$-4 < w/2^{n-1} < 4.$$

Thus

$$-4 \leq w' \leq 3. \tag{6.31}$$

The maximum difference between w and $w'.2^{n-1}$ is smaller than 2^{n-1}, that is,

$$w - 2^{n-1} < w'.2^{n-1} \leq w. \tag{6.32}$$

Define the truncated values s_t and c_t of s and c as

$$s_t = \lfloor s/2^{n-1} \rfloor \quad \text{and} \quad c_t = \lfloor c/2^{n-1} \rfloor,$$

and w'' as being the result of adding s_t and c_t (Figure 6.5c). The difference between w' and w'' is the possible carry from the rightmost positions, so that

$$w' - 1 \leq w'' \leq w'. \tag{6.33}$$

Thus, from (6.31) and (6.33),

$$-5 \leq w'' \leq 3, \tag{6.34}$$

that is, in 2's complement,

$$1011 \leq w'' \leq 0011,$$

and from (6.32) and (6.33),

$$w - 2^n < w''.2^{n-1} \leq w, \tag{6.35}$$

that is,

$$w''.2^{n-1} \leq w < w''.2^{n-1} + 2^n. \tag{6.36}$$

The selection of $q(i+1)$ is done as follows (see Figure 6.3a):

if $-5 \leq w'' < -1$, that is, $-5 \leq w'' \leq -2$, then (6.36) $w < 0$ and $q(i+1) = -1$;

if $-1 \leq w'' < 0$, that is, $w'' = -1$, then (6.36) and (6.29) $-Y \leq -2^{n-1} \leq w < 2^{n-1} \leq Y$, and $q(i+1) = 0$;

if $0 \leq w'' \leq 3$, then (6.36) $0 \leq w$, and $q(i+1) = 1$.

The corresponding selection rules are show in Table 6.2.

Assume that `carry_save` is a function that expresses the sum of three integers in the form of two integers (stored-carry encoding, Chapter 4). Then define an `srt_step` procedure:

```
procedure srt_step (s, c, Y: in integer; q_pos, q_neg: out
bit; next_s, next_c: out integer) is
begin
w":=s(n+2..n-1)+c(n+2..n-1);
case w" is
  when 0000|0001|0010|0011|=>
    q_pos:=1; q_neg:=0; (next_s, next_c):=carry_save(s, c,
    -Y);
  when 1011|1100|1101|1110=>
    q_pos:=0; q_neg:=1; (next_s, next_c):=carry_save(s, c, Y);
  when others=>
    q_pos:=0; q_neg:=0; (next_s, next_c):=carry_save(s, c, 0);
  end case;
end srt_step;
```

The SRT-2 algorithm can now be stated as follows (s' and c' stand for $s/2$ and $c/2$):

Algorithm 6.8   SRT-2 Division with Stored-Carry Encoding

```
s'(0):=x; c'(0):=0;
for i in 0..p-1 loop
  srt_step(2*s'(i), 2*c'(i), y, q_pos(p-1-i), q_neg(p-1-i),
  s'(i+1), c'(i+1));
end loop;
r:=s'(p)+c'(p);
q:=q_pos-q_neg;
if x>=0 and r<0 then r:=r+y; q:=q-1;
elsif x<0 and r>=0 then r:=r-y; q:=q+1;
end if;
```

TABLE 6.2   SRT-2 Algorithm with Stored-Carry Encoding: Selection of $q(i + 1)$

w''	$q(i+1)$
0000	1
0001	1
0010	1
0011	1
0100	—
0101	—
0110	—
0111	—
1000	—
1001	—
1010	—
1011	-1
1100	-1
1101	-1
1110	-1
1111	0

Comment 6.4 When X belongs to the interval $-Y \leq X < 2^{n-1}$, it has been observed ([SUT2004]) that w'' never reaches 3, that is, $-5 \leq w'' \leq 2$. Then Table 6.2 can be modified as shown Table 6.3.

The most significant bit of w'', that is, w_3'', is no longer necessary and Table 6.3 is reduced Table 6.4.

TABLE 6.3   Modified $q(i + 1)$ Selection Table

w''	$q(i+1)$
0000	1
0001	1
0010	1
0011	—
0100	—
0101	—
0110	—
0111	—
1000	—
1001	—
1010	—
1011	-1
1100	-1
1101	-1
1110	-1
1111	0

TABLE 6.4 Reduced $q(i + 1)$ Selection Table

$w_2'' \, w_1'' \, w_0''$	$q(i + 1)$
000	1
001	1
010	1
011	-1
100	-1
101	-1
110	-1
111	0

Examples 6.5

1. Let $X = 001110111011$ ($=955$) be a 12-bit 2's complement integer and $Y = 11010101101$ ($=1709$) an 11-bit positive integer. Compute X/Y with an accuracy of $p = 8$. At each step s and c will be represented in the form of 2's complement integers with $11 + 3 = 14$ bits. For adding or subtracting Y, 2's complement 13-bit representations will be used:

$$Y = 0011010101101,$$
$$-Y = 1100101010011.$$

The step-by-step procedure is described as follows (*nop* stands for *no operation*):

Step #	Carry-Save Remainder Computation	q_pos	q_neg
$i = 0$	$s = 2.X = 00011101110110;\ s_t = 0001$		
	$c = 00000000000000;\ c_t = 0000$		
	$w'' = 0001 \rightarrow q(0) = 1 \rightarrow$ subtract Y	1	0
	$\quad\quad s \quad\quad 00011101110110$		
	$\quad\quad c \quad\quad 00000000000000$		
	$\quad\ -Y \quad\ \underline{1100101010011}$		
	$\ next\_s' \quad 1111000100101$		
	$\ next\_c' \quad 0001010100100$		
$i = 1$	$s = 11110001001010;\ s_t = 1111$		
	$c = 00010101001000;\ c_t = 0001$		
	$w'' = 0000 \rightarrow q(1) = 1 \rightarrow$ subtract Y	1	0
	$\quad\quad s \quad\quad 11110001001010$		
	$\quad\quad c \quad\quad 00010101001000$		
	$\quad\ -Y \quad\ \underline{1100101010011}$		
	$\ next\_s' \quad 0000001010001$		
	$\ next\_c' \quad 1101010010100$		

Step #	Carry-Save Remainder Computation	q_pos	q_neg
$i = 2$	$s = 00000010100010$; $s_t = 0000$		
	$c = 11010100101000$; $c_t = 1101$		
	$w'' = 1101 \rightarrow q(2) = -1 \rightarrow$ add Y	0	1

$$
\begin{array}{ll}
s & 00000010100010 \\
c & 11010100101000 \\
Y & \underline{0011010101101} \\
next\_s' & 1001100100111 \\
next\_c' & 0100101010000
\end{array}
$$

$i = 3$	$s = 10011001001110$; $s_t = 1001$		
	$c = 01001010100000$; $c_t = 0100$		
	$w'' = 1101 \rightarrow q(3) = -1 \rightarrow$ add Y	0	1

$$
\begin{array}{ll}
s & 10011001001110 \\
c & 01001010100000 \\
Y & \underline{0011010101101} \\
next\_s' & 1001001000011 \\
next\_c' & 0110101011000
\end{array}
$$

$i = 4$	$s = 10010010000110$; $s_t = 1001$		
	$c = 01101010110000$; $c_t = 0110$		
	$w'' = 1111 \rightarrow q(4) = 0 \rightarrow nop$	0	0

$$
\begin{array}{ll}
s & 10010010000110 \\
c & \underline{01101010110000} \\
next\_s' & 1111000110110 \\
next\_c' & 0000100000000
\end{array}
$$

$i = 5$	$s = 11110001101100$; $s_t = 1111$		
	$c = 00001000000000$; $c_t = 0000$		
	$w'' = 1111 \rightarrow q(5) = 0 \rightarrow nop$	0	0

$$
\begin{array}{ll}
s & 11110001101100 \\
c & \underline{00001000000000} \\
next\_s' & 1111001101100 \\
next\_c' & 0000000000000
\end{array}
$$

$i = 6$	$s = 11110011011000$; $s_t = 1111$		
	$c = 00000000000000$; $c_t = 0000$		
	$w'' = 1111 \rightarrow q(6) = 0 \rightarrow nop$	0	0

$$
\begin{array}{ll}
s & 11110011011000 \\
c & \underline{00000000000000} \\
next\_s' & 1110011011000 \\
next\_c' & 0000000000000
\end{array}
$$

$i = 7$ $s = 11100110110000;\ s_t = 1110$

$c = 00000000000000;\ c_t = 0000$

$w'' = 1110 \to q(7) = \text{-}1 \to \text{add } Y$ 0   1

s	11100110110000	
c	00000000000000	
Y	0011010101101	
$next\_s'$	1111100011101	
$next\_c'$	0000101000000	
$r(8)$	0000001011101	$(= 93)$

As the final remainder is positive, the divisor Y doesn't need to be added to the final remainder $r(8)$, and Q is given by

$$Q = Q\_pos - Q\_neg = 11000000 - 00110001;$$

In 2's complement:

$$Q = 011000000 + 111001111 = 010001111\ (= 143).$$

The overall operation can be resumed as

$$955.256 = 143.1709 + 93.$$

2. Let Y be an 8-bit positive integer whose most significant bit is equal to 1 ($128 \le Y < 256$) and let X be a 9-bit 2's complement integer ($-256 \le X < 255$). In order to compute the X/Y with an accuracy of 4 fractional bits, X is first divided by 2 (Comment 6.3) and the division is performed with an accuracy of 5 fractional bits. Assume that $X = 100000111$ (-249) and $Y = 10010011$ (147). At each step s and c will be represented in the form of 2's complement integers with $8 + 3 = 11$ bits. For adding or subtracting Y, 2's complement 10-bit representations will be used:

$$Y = 0010010011,$$

$$-Y = 1101101101.$$

The step-by-step procedure is described as follows (*nop* stands for *no operation*):

Step #	Carry-Save Remainder Computation	q_pos	q_neg
$i = 0$	$s = X = 11100000111;\ s_t = 1110$		
	$c = 00000000000;\ c_t = 0000$		
	$w'' = 1110 \to q(0) = -1 \to \text{add } Y$	0	1

s	11100000111	
c	00000000000	
Y	0010010011	
$next\_s'$	1110010100	
$next\_c'$	0000000110	

Step #	Carry-Save Remainder Computation	q_pos	q_neg
$i = 1$	$s = 11100101000;\ s_t = 1110$		
	$c\ =\ 00000001100;\ c_t = 0000$		
	$w'' = 1110 \to q(1) = -1 \to$ add Y	0	1
	$\quad\quad s\quad\quad\quad 11100101000$		
	$\quad\quad c\quad\quad\quad 00000001100$		
	$\quad\quad Y\quad\quad\quad \underline{0010010011}$		
	$\quad\quad next\_s'\quad 1110110111$		
	$\quad\quad next\_c'\quad 0000010000$		
$i = 2$	$s = 11101101110;\ s_t = 1110$		
	$c\ =\ 00000100000;\ c_t = 0000$		
	$w'' = 1110 \to q(2) = -1 \to$ add Y	0	1
	$\quad\quad s\quad\quad\quad 11101101110$		
	$\quad\quad c\quad\quad\quad 00000100000$		
	$\quad\quad Y\quad\quad\quad \underline{0010010011}$		
	$\quad\quad next\_s'\quad 1111011101$		
	$\quad\quad next\_c'\quad 0001000100$		
$i = 3$	$s = 11110111010;\ s_t = 1111$		
	$c\ =\ 00010001000;\ c_t = 0001$		
	$w'' = 0000 \to q(3) = 1 \to$ subtract Y	1	0
	$\quad\quad s\quad\quad\quad 11110111010$		
	$\quad\quad c\quad\quad\quad 00010001000$		
	$\quad\quad -Y\quad\quad\quad \underline{1101101101}$		
	$\quad\quad next\_s'\quad 0001011111$		
	$\quad\quad next\_c'\quad 1101010000$		
$i = 4$	$s = 00010111110;\ s_t = 0001$		
	$c\ =\ 11010100000;\ c_t = 1101$		
	$w'' = 1110 \to q(4) = -1 \to$ add Y	0	1
	$\quad\quad s\quad\quad\quad 00010111110$		
	$\quad\quad c\quad\quad\quad 11010100000$		
	$\quad\quad Y\quad\quad\quad \underline{0010010011}$		
	$\quad\quad next\_s'\quad 1010001101$		
	$\quad\quad next\_c'\quad \underline{0101100100}$		
	$\quad\quad r(5)\quad\quad\quad 1111110001$	$(= -15)$	

As the final remainder is negative, the divisor Y doesn't need to be subtracted from the final remainder $r(5)$, and Q is given by

$$Q = Q\_pos - Q\_neg = 00010 - 11101;$$

In 2's complement

$$Q = 000010 + 100011 = 100101(= -27).$$

The overall operation can be resumed as

$$(-249/2).32 = (-27).147 + (-15)$$

that is,

$$-249 = (-27/16).147 + (-15/16).$$

6.2.6 P–D Diagram

Apart from the Robertson diagram, another popular graphical tool used to illustrate the quotient selection problem is the P–D (*partial remainder–divisor*) plot diagram ([FRE1961]). It is a representation of the domain where possible values of the quotient-digit may be assigned. First define normalized values of Y, $r(i)$, $r(i+1)$, w, w', and w'':

$$d = Y/2^n, \quad r = r(i)/2^n, \quad r^+ = r(i+1)/2^n, \quad y = w/2^n, \tag{6.37}$$

so that

$$r = y/2 \tag{6.38}$$

and the normalized values of w' and w'' are

$$r' = w'/2, \quad r'' = w''/2. \tag{6.39}$$

Thus

$$\tfrac{1}{2} \le d < 1, \quad -d \le r < d, \quad -2 \le r' \le \tfrac{3}{2}, \quad \text{and} \quad -\tfrac{5}{2} \le r'' \le \tfrac{3}{2}. \tag{6.40}$$

With these normalized values, the selection of $q = q(i+1)$ corresponding to Algorithm 6.8 (SRT division with stored-carry encoding) is done as follows (Figure 6.6):

if $-\tfrac{5}{2} \le r'' < -\tfrac{1}{2}$, then $q = -1$;
if $-\tfrac{1}{2} \le r'' < 0$, then $q = 0$;
if $0 \le r'' \le \tfrac{3}{2}$, then $q = 1$.

In the P–D diagram, the coordinates $(2.r, d)$ are linked to none, one, or several acceptable values of q. Figure 6.7 displays the P–D diagram associated to the

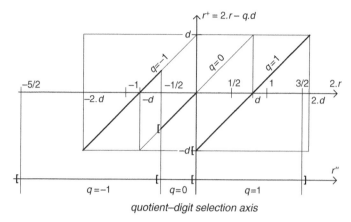

Figure 6.6 Robertson diagram.

partial remainder diagram of Figure 6.6. The domain presented in Figure 6.7 covers the $(2.r, d)$ coordinates compatible with the conditions $\frac{1}{2} \le d < 1$ and $-2d \le 2.r < 2.d$. This domain is divided into six zones separated by the lines $2.r = -2.d$, $2.r = -d$, $2.r = 0$, $2.r = d$, and $2.r = 2.d$. Above the line $2.r = 2.d$ and under the line $2.r = -2.d$ lie two *never reached* zones; they correspond to coordinates out of range. Between the lines $2.r = -2.d$ and $2.r = -d$, the value -1 has to be assigned to q; this zone corresponds to the coordinates $2.r$ in the interval $[-2.d, -d[$ of the diagram in Figure 6.6. Between the lines $2.r = -d$ and $2.r = 0$, either value -1 or 0 may be assigned to q; this zone corresponds to the coordinates $2.r$ in the interval $[-d, 0[$ of the diagram in Figure 6.6. In the same way, one can show that in the positive field, the zone between the lines $2.r = 0$ and $2.r = d$ corresponds to q in $\{0,1\}$, while the zone between the lines $2.r = d$ and $2.r = 2.d$ correspond to $q = 1$.

Now the q selection strategy, defined in Figure 6.6, can be mapped on the P–D plot diagram: the lines $r'' = -\frac{1}{2}$ and $r'' = 0$, highlighted in Figure 6.7, set the limits for choosing $q = -1, 0$, or 1. As it appears in the following section, the P–D plot diagram is particularly well suited to deal with high-radix (bases 2^k with $k \ge 2$) because of more complex quotient-digit selection rules.

Actually, the P–D plot diagram as shown in Figure 6.7 is the level zero of a tri-dimensional (3-D) diagram whose third (vertical) axis would be r^+, assuming that the horizontal axes are $2.r$ and d. Equations $r^+ = 2.r - q.d$ for $q = -1, 0$, and 1 now represent planes crossing level zero at lines $2.r = -d$, $2.r = 0$, and $2.r = d$, respectively. The diagram of Figure 6.6 is the intersection, at some allowed coordinate of d, of this tridimensional figure with a plane parallel to axes r^+, $2.r$, that is, parallel to plane $d = 0$. Figure 6.8 shows the 3-D Robertson/P–D diagram for the plane corresponding to $q = -1$. Lines $2.r = -d$, $2.r = 0$, and $2.r = d$ (from P–D plot diagram at $r^+ = 0$) are highlighted in Figure 6.8, together with line $r^+ = 2.r - q.d$ (from the Robertson diagram at $d = \frac{3}{4}$).

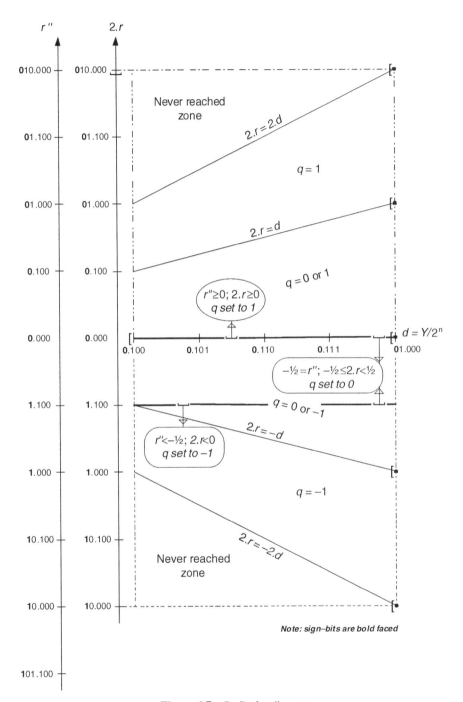

Figure 6.7   *P–D* plot diagram.

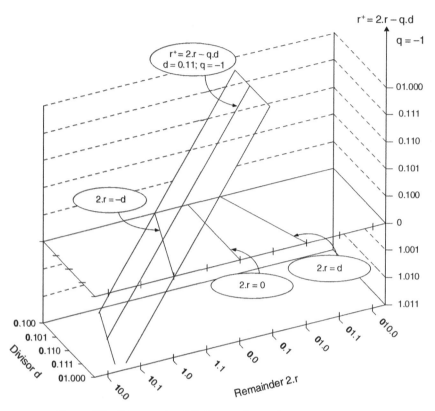

Figure 6.8   Tridimensional Robertson/P–D diagram for $q = -1$.

6.2.7   SRT-4 Division

The SRT method can be extended to any base 2^k with a variety of quotient-digit sets. Nevertheless, the step complexity increases with k as more comparisons are involved in the quotient-digit selection process and more divisor multiples have to be computed. The designer will consider trade-offs between cycle time and the number of cycles. A lot of alternatives are proposed in the literature ([ERC1990], [ERC2004], [FAN1989], [MON1994], [QUA1992], [SRI1995], [TAY1985]). An SRT-4 ($B = 2^2$) algorithm, with redundant quotient-digit sets, is presented in the following.

One assumes that X and Y are n-bit 2's complement integers with

$$2^{n-2} \leq Y < 2^{n-1},$$
$$-Y \leq X < Y. \qquad (6.41)$$

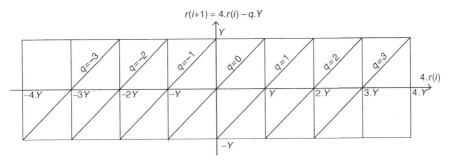

Figure 6.9   SRT-4 partial remainder diagram.

A first version is given with the quotient-digit set $[-3, 3]$. This means that the multiples $\pm 2.Y$ and $\pm 3.Y$ have to be generated according to the possible selection of $q = q(i + 1) = \pm 2$ or ± 3. The Robertson diagram is given in Figure 6.9.

The new remainder $r(i + 1)$ is now computed as $4.r(i) - q(i + 1).Y$, where $q(i + 1)$ is selected in such a way that $r(i + 1)$ can be kept in the range $[-Y, Y[$; the maximum allowed range for $4.r(i)$ is now $[-2^{n+1}, 2^{n+1}[$. As it will appear later, it is no longer possible to locate a point $(4.r(i), Y)$ in some zone $[k.Y, (k + 1).Y]$ of Figure 6.9, only from the numerical value $4.r(i)$. One also needs extra information on Y. The $P-D$ plot diagram presented in Figure 6.10 emphasizes this point and illustrates a possible quotient-digit selection strategy (r stands for $r(i)$ and q for $q(i + 1)$). Although the q-*select* zones are symmetric with respect to the Y-axis, the way to select q is not. This comes from the fact that, in 2's complement representation, truncated nonsign bits are always positive whatever the sign is. As in Figure 6.7, lines $4.r = q.Y$ are defining q-*select* zones. From Figure 6.9 one can check that between the vertical lines $4.r = k.Y$ and $4.r = (k + 1).Y$, q can be either k or $k + 1$; this does not hold for the outermost zones where q has to be -3 or 3, according to the side; these zones are shown in Figure 6.10.

In Figure 6.10, the values on axis $4.r$ and Y have been truncated by $n - 2$ and $n - 4$ bits, respectively. For clarity, sign-bits appear as bold characters. The remaining bits are actually the only ones needed for algorithmic purposes. The q-selection strategy is illustrated in Figure 6.10 where heavy lines stand for border limit lines between consecutive options for q. The symbols "number & arrows" and "[", lying on a border-line, respectively, highlight the choice for q, the zone covered by this choice, and the inclusion of the limit line in that choice. At coordinates $4.r = \mathbf{0}10$, $\mathbf{0}1$, $\mathbf{1}1$, and $\mathbf{1}0$, horizontal border-lines still hold between regions of different options for q. This is no longer valid for values between $[\mathbf{1}00, \mathbf{1}01,[$ and $\mathbf{0}11$, $\mathbf{0}100[$, where so-called *staircase* lines emphasize that the selection in that case depends on Y too. Take a closer look at those border-lines to make sure that the selections proposed in Figure 6.10 are correct. First, remember that the bits deleted are always positive, so the head bits of $4.r$ or Y (truncated coordinates $4.r$, Y) stand for the minimum of the nonintegers $4.r/2^{n-2}$ and $Y/2^{n-4}$; notice that the rightmost point of any border-line is never reached as Y never reaches $\mathbf{0}1000$. For

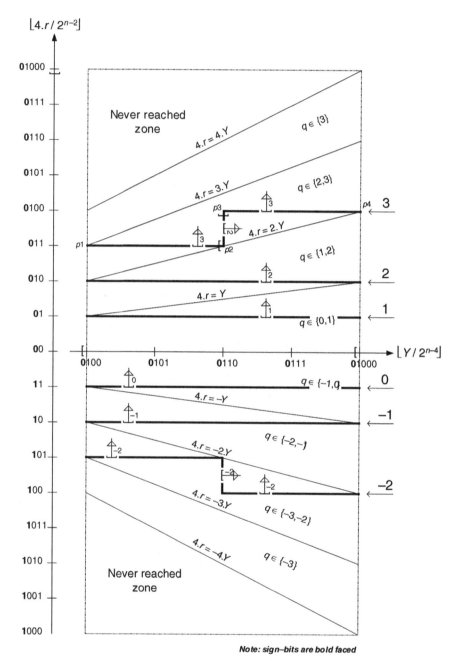

Figure 6.10   *P–D* plot diagram for SRT-4 division.

TABLE 6.5   SRT-4 Quotient-Digit Selection Table

$4r(i)$	Y	$q(i+1)$
≥ 4	—	3
$[3, 4[$	$[4, 6[$	3
$[3, 4[$	$[6, 8[$	2
$[2, 3[$	—	2
$[1, 2[$	—	1
$[-1, 1[$	—	0
$[-2, -1[$	—	-1
$[-3, -2[$	—	-2
$[-4, -3[$	$[6, 8[$	-2
$[-4, -3[$	$[4, 6[$	-3
< -4	—	-3

these two reasons, the border-lines labeled -2, -1, 0, 1, and 2, in Figure 6.10, may include all the edge line points in the choice for q, extremities and corners included. The situation of the line labeled 3 is somewhat different: the choice $q = 3$ for the point labeled $p1$ is valid because $p1$ stands between two zones where this choice is allowed. The same principle holds for point $p4$ that moreover is never reached because of the above-mentioned limit condition on Y. Points $p2$ and $p3$ appear more critical as they are corner points in between border-lines with different choices. $p3 = 3$ is valid for standing inside a zone where this choice is allowed. $p2$ has to be excluded from the choice $q = 3$ (line $p1 \rightarrow p2$) because, with full coordinates, this point could fall either in the zone $q \in \{1,2\}$ or in the zone $q \in \{2,3\}$; thus $p2$ is given the value 2 of line $p2 \rightarrow p3$.

The complete selection process can be resumed in Table 6.5.

Comment 6.5   For the sake of simplicity the truncated part of the exact n-bit remainder $r(i)$ has been considered as coordinate references on Figure 6.10; four bits of the shifted remainder are required by the SRT-4 division algorithm 6.9. The carry-save techniques are applicable in this case, as far as a sufficient quantity of bits are saved in the s_t, c_t representation of the remainder; moreover, the flexibility in the choice of the quotient-digit set, as well as that of the q-selection strategy, allows one to consider a number of alternatives in the algorithms. As in the SRT-2 case, the carry-save computation still has to be implemented in a way that prevents each of the carry-save components s' and c' from exceeding the $(n + 2)$-bit length.

Algorithm 6.9   SRT-4 Division

(/ Stands for integer division)

```
yt:=y/2**n-4
r(0):=X
for i in 0..p-1 loop
```

```
      4.rt(i):=4.r(i)/2**n-2
      if 4.rt(i)<-4 then q(i+1):=-3; q_pos(i+1):=0; q_neg
         (i+1):=3; r(i+1):=4.r(i)+3*Y;
      elsif 4.rt(i)<-3 then
            if yt<6 then q(i+1):=-3; q_pos(i+1):=0; q_neg(i+1):=3;
            r(i+1):=4.r(i)+3*Y; else q(i+1):=-2; q_pos(i+1):=0;
            q_neg(i+1):=2; r(i+1):=4.r(i)+2*Y; end if;
      elsif 4.rt(i)<-2 then q(i+1):=-2; q_pos(i+1):=0;
         q_neg(i+1):=2; r(i+1):=4.r(i)+2*Y;
      elsif 4.rt(i)<-1 then q(i+1):=-1; q_pos(i+1):=0;
         q_neg(i+1):=1; r(i+1):=4.r(i)+Y;
      elsif 4.rt(i)<1 then q(i+1):=0; q_pos(i+1):=0;
         q_neg(i+1):=0; r(i+1):=4.r(i);
      elsif 4.rt(i)<2 then q(i+1):=1; q_pos(i+1):=1;
         q_neg(i+1):=0; r(i+1):=4.r(i)-Y;
      elsif 4.rt(i)<3 then q(i+1):=2; q_pos(i+1):=2;
         q_neg(i+1):=0; r(i+1):=4.r(i)-2*Y;
      elsif 4.rt(i)<4 then
      if yt>=6 then q(i+1):=2; q_pos(i+1):=2; q_neg(i+1):=0;
         r(i+1):=4.r(i)-2*Y; else q(i+1):=3; q_pos(i+1):=0;
         q_neg(i+1):=3; r(i+1):=4.r(i)+3*Y; end if;
    else q(i+1):=3; q_pos(i+1):=0; q_neg(i+1):=3;
       r(i+1):=4.r(i)+3*Y; end if;
end loop;
q:=q_pos - q_neg;
If r(p)>=0 then R:=r(p); else R:=r(p)+Y; q:=q-1;
```

Example 6.6   Let $X = 010110111100101100110$ and $Y = 011010110$ be 2's complement numbers Compute X/Y with an accuracy of $p = 6$.

According to the conditions X in $[-Y, Y[$, and $2^{n-2} \leq Y < 2^{n-1}$, a first preliminary scaling of Y is necessary: $Y = 011010110000000000000$. So, $p = 6 + 12 = 18$, $n = 21$, and two more bits will be necessary to express the multiples of Y.

First compute the multiples of Y as follows

$$Y = 000110101100000000000000$$
$$2.Y = 001101011000000000000000$$
$$3.Y = 010100001000000000000000$$
$$-Y = 111001010100000000000000$$
$$-2.Y = 110010101000000000000000$$
$$-3.Y = 101011111100000000000000$$

For q-selection purposes one has to compute the integer division

$$Y/2^{17} = 0110 \, (\text{decimal } 6)$$

The step-by-step procedure is described as follows (sign-bits are bold face; *nop* stands for *no operation*; *symbol*/stands for integer division).

Step i	Remainder Computation	$4.rt\,(i)$ (decimal)	q-select (decimal)	q_pos	q_neg
0	$r(0) = \mathbf{0}0010110111100101100110$ $4.r(0) = \mathbf{0}101101111001011001100$ $-3.Y = \mathbf{1}01011111110000000000000$ $r(1) = \mathbf{0}0001011100010110011000$	$4.r(0)/2^{19} = 5$ $5 \geq 4$	3	3	0
1	$4.r(1) = \mathbf{0}0101110001011001100100$ $-2.Y = \mathbf{1}10010101000000000000000$ $r(2) = \mathbf{1}111100010101100100000$	$4.r(1)/2^{19} = 2$ $2 \in [2, 3[$	2	2	0
2	$4.r(2) = \mathbf{1}11000101011001100000000$ $+1.Y = \mathbf{0}00110101100000000000000$ $r(3) = \mathbf{1}1111101011100110000000$	$4.r(2)/2^{19} = -2$ $-2 \in [-2, -1[$	-1	0	1
3	$4.r(3) = \mathbf{1}111010111001100000000$ $nop = \mathbf{0}000000000000000000000000$ $r(4) = \mathbf{1}111010111001100000000$	$4.r(3)/2^{19} = -1$ $-1 \in [-1, 1[$	0	0	0
4	$4.r(4) = \mathbf{1}1010111001100000000000$ $+2.Y = \mathbf{0}0110101100000000000000$ $r(5) = \mathbf{0}0001100101100000000000$	$4.r(4)/2^{19} = -3$ $-3 \in [-3, -2[$	-2	0	2
5	$4.r(5) = \mathbf{0}01100101100000000000000$ $-2.Y = \mathbf{1}10010101000000000000000$ $r(6) = \mathbf{1}111110101000000000000$	$4.r(5)/2^{19} = 3$ $3 \in [3, 4[$ $Y \in [6, 8[$	2	2	0
6	$4.r(6) = \mathbf{1}111010100000000000000000$ $nop = \mathbf{0}000000000000000000000000$ $r(7) = \mathbf{1}111010100000000000000$	$4.r(6)/2^{19} = -1$ $-1 \in [-1, 1[$	0	0	0
7	$4.r(7) = \mathbf{1}1010100000000000000000$ $+2.Y = \mathbf{0}0110101100000000000000$ $r(8) = \mathbf{0}0001001100000000000000$	$4.r(7)/2^{19} = -3$ $-3 \in [-3, -2[$	-2	0	2
8	$4.r(8) = \mathbf{0}0100110000000000000000$ $-2.Y = \mathbf{1}10010101000000000000000$ $r(9) = \mathbf{1}1110000100000000000000$	$4.r(8)/2^{19} = 2$ $2 \in [2, 3[$	2	2	0

As $r(9) < 0$, Y has to be added to the final remainder,

$$R = \mathbf{1}1110000100000000000000$$
$$+\ \underline{\mathbf{0}0011010110000000000000}$$
$$= \mathbf{0}0001011010000000000000,$$

and the quotient has to be reduced by one unit. This can be done by reducing q_pos(8) to 1.

So the final quotient is given by

$$Q = \text{Q\_pos} - \text{Q\_neg},$$

that is,

$$Q = (3, 2, 0, 0, 0, 2, 0, 0, 1) - (0, 0, 1, 0, 2, 0, 0, 2, 0,) = (3, 1, 2, 3, 2, 1, 3, 2, 1).$$

In binary, $Q = \mathbf{0}110110111001111001$, and the overall operation (accuracy 18 and pre-scaling 12) can be resumed as

$$\mathbf{0}101101111001011001110.2^{18} = \mathbf{0}110110111001111001.011010110.2^{12}$$

$$+ \, 00001011010.2^{12}.$$

Comment 6.6   As far as the final computation time for $Q = \text{Q\_pos} - \text{Q\_neg}$ may be considered negligible with respect to the whole process, it may not be necessary to speed up the final conversion process. The influence of the conversion on the overall time will mainly depend on the required accuracy and the size of operands. On the other hand, the final correction on the last remainder, whenever negative, may be processed in parallel with the conversion process. To minimize the impact of the conversion time, several on-the-fly conversion algorithms have been proposed in the literature ([ERC1987], [ERC1992], [OBE1997]). Basically these algorithms perform the conversion as the digits are produced. Prescaling the divisor or both operands is another idea proposed in the literature ([ERC1983], [SVO1963]); as far as an efficient method can be used to make the divisor close enough to 1, the quotient-digit selection depends on the dividend only. As time has to be consumed for the scaling operation, the advantages are not clear in general cases. The SRT-4 algorithm has been used in the early Pentium processors and was the origin of the famous *Pentium bug*; the error, due to flaws in the look-up tables, was not detected on the test bench because the probability of addressing the tables at those incorrect entries was actually very weak ([EDE1997]).

6.2.8   Base-*B* Nonrestoring Division Algorithm

In bases other than 2 or 2^k (high-radix), nonrestoring division algorithms have received little attention in the literature. Nevertheless, for the sake of generality, a more general approach on division algorithms with extended quotient-digit sets will be developed in the following.

Definitions 6.1

1. A system of digits (weights) $\{d_i\}$ is said to be complete when it is able to represent any number as a weighted sum of powers of the base B; a *nonredundant* quotient-digit system in base B is defined as a complete system of exactly B digits, that is, the minimum quantity of needed digit values to express the

quotient (or any number) as a base-B number. The set $\{0, 1, \ldots, B - 1\}$ is most commonly used.

2. In a *redundant* quotient-digit set, the number of allowed quotient-digits is greater than B. The most used sets are of the form $\{-\alpha, -\alpha + 1, \ldots, 0, \ldots, \alpha - 1, \alpha\}$ with $\alpha \geq \lceil B/2 \rceil$, that is, a symmetric set of consecutive integers.

3. The *redundancy factor* ρ is defined as $\rho = \alpha/(B - 1)$.

4. A quotient-digit set with $\alpha = \lceil B/2 \rceil$ is said to be *minimally redundant*.

5. A quotient-digit set with $\alpha = B - 1$ ($\rho = 1$) is said to be *maximally redundant*.

6. A quotient-digit set with $\alpha > B - 1$ ($\rho > 1$) is said to be *over-redundant*.

The first algorithm to be presented for base-B nonrestoring division uses a redundant quotient-digit set $\{0, 1, 2, \ldots, B\}$. A tentative quotient estimation is made from the truncated operands-look-up tables (LUTs) can be used to speed up this phase – then a possible correction has to be carried out in order to convert the obtained quotient into a final nonredundant representation. The following lemmas and theorems justify the algorithm.

Lemmas 6.2   Let

$$B.R = r_{n+1}, r_n, r_{n-1}, \ldots, r_1, r_0,$$
$$Y = y_{n-1}, y_{n-2}, \ldots, y_1, y_0 \tag{6.42}$$

be two base-B positive integers ($B \geq 3, n \geq 3$). $B.R$, the shifted remainder, and Y, the divisor, comply with the following conditions:

$$B.R/Y < (B + 1),$$
$$B^{n-1} \leq Y \leq B^n - 1. \tag{6.43}$$

Define moreover

$$R_t = \lfloor B.R/B^{n-3} \rfloor; \quad Y_t = \lfloor Y/B^{n-3} \rfloor; \quad q = \lfloor B.R/Y \rfloor;$$
$$q_t = \lfloor R_t/Y_t \rfloor; \quad q_t^* = \lfloor R_t/(Y_t + 1) \rfloor. \tag{6.44}$$

Lemma 6.2.1

$$0 \leq q_t \leq B + 1,$$
$$0 \leq q_t^* \leq B, \tag{6.45}$$

Proof   The inequalities $0 \leq q_t$ and $0 \leq q_t^*$, are trivial.

Condition (6.43) can be written

$$B.R < (B+1).Y;$$

as

$$R_t.B^{n-3} \leq B.R \quad \text{and} \quad Y < (Y_t + 1).B^{n-3},$$

then

$$R_t.B^{n-3} < (B+1).(Y_t + 1).B^{n-3} \rightarrow R_t < (B+1).(Y_t + 1). \tag{6.46}$$

Definitions (6.44) imply that $B^2 \leq Y_t \leq B^3 - 1 \rightarrow B + 1 < Y_t$ or $(B+1)/Y_t < 1$; thus

$$R_t/Y_t < (B+1).(Y_t + 1)/Y_t = (B+1) + (B+1)/Y_t,$$

and

$$q_t = \lfloor R_t/Y_t \rfloor \leq B + 1,$$

which completes the proof of the first inequality (6.45).

The second inequality (6.45) is deduced from (6.46), written as

$$R_t.B^{n-3}/(Y_t + 1).B^{n-3} < (B+1) \rightarrow R_t/(Y_t + 1) < (B+1).$$

Then

$$q_t^* = \lfloor R_t/(Y_t + 1) \rfloor \leq B,$$

which completes the proof of the second inequality (6.45).

Lemma 6.2.2

$$q_t - 1 \leq q_t^* \leq q_t. \tag{6.47}$$

Proof The fundamental equation of division for R_t/Y_t may be written

$$R_t = q_t.Y_t + \rho \quad \text{with} \quad \rho < Y_t, \tag{6.48}$$

Equation (6.48) may be expressed in any of the following two forms:

$$R_t = q_t.(Y_t + 1) + \rho - q_t, \tag{6.49}$$

or

$$R_t = (q_t - 1).(Y_t + 1) + \rho - q_t + (Y_t + 1), \tag{6.50}$$

If $\rho - q_t \geq 0$, (6.49) may be written

$$R_t = q_t^*.(Y_t + 1) + \rho^*, \quad \text{with} \quad q_t^* = q_t \quad \text{and} \quad \rho^* = \rho - q_t < (Y_t + 1). \tag{6.51}$$

Otherwise $\rho - q_t < 0$, and (6.50) is written

$$R_t = q_t^*.(Y_t + 1) + \rho^*, \quad \text{with} \quad q_t^* = q_t - 1 \quad \text{and}$$
$$\rho^* = \rho - q_t + (Y_t + 1) < (Y_t + 1). \tag{6.52}$$

Observe that, as $q_t \leq B + 1$, and $B^2 \leq Y_t$, then $q_t < Y_t$, and ρ^* cannot be negative. The proof is now complete.

Lemma 6.2.3

$$q_t^* \leq q \leq q_t. \tag{6.53}$$

Proof   The following inequality is straightforward

$$R_t.B^{n-3}/(Y_t + 1).B^{n-3} < B.R/Y \leq (R_t + \varepsilon).B^{n-3}/Y_t.B^{n-3}, \quad \varepsilon = 1 - B^{3-n},$$

or

$$R_t/(Y_t + 1) < B.R/Y \leq (R_t + \varepsilon)/Y_t,$$

then

$$\lfloor R_t/(Y_t + 1) \rfloor \leq \lfloor B.R/Y \rfloor \leq \lfloor (R_t + \varepsilon)/Y_t \rfloor \rightarrow q_t^* \leq q \leq q_t.$$

Lemmas 6.2.2 and 6.2.3 may be merged and expressed in the following theorem.

Theorem 6.2   Assuming (6.44),

$$R_t = \lfloor B.R/B^{n-3} \rfloor; \quad Y_t = \lfloor Y/B^{n-3} \rfloor; \quad q = \lfloor B.R/Y \rfloor;$$
$$q_t = \lfloor R_t/Y_t \rfloor; \quad q_t^* = \lfloor R_t/(Y_t + 1) \rfloor,$$

then

$$q_t - 1 \le q_t^* \le q \le q_t. \tag{6.54}$$

Theorem 6.3 Let $R(i)$ and Y be the ith remainder and the divisor, respectively, as defined in (6.42). Definitions given in (6.44) hold. The initial dividend is denoted $B \cdot R(0)$. Define moreover

$$R(i + 1) = B.R(i) - q.Y; \quad R^*(i + 1) = B.R(i) - q_t^*.Y. \tag{6.55}$$

If

$$0 \le B.R(i) < (B + 1).Y,$$

then

$$0 \le B.R^*(i + 1) < (B + 1).Y. \tag{6.56}$$

Proof If $q_t^* = q$, then

$$R^*(i + 1) = R(i + 1) = B.R(i) - q.Y < Y \rightarrow B.R^*(i + 1) < (B + 1).Y.$$

Otherwise, using Theorem 6.2 (6.54)

$$q = q_t \quad \text{and} \quad q_t^* = q_t - 1 = q - 1.$$

This situation corresponds to the conditions of (6.52), where

$$\rho = R_t(i) - q_t.Y_t < q_t < B - 1 \rightarrow \rho + 1 = R_t(i) + 1 - q_t.Y_t < B \le Y_t/B.$$

On the other hand,

$$R(i + 1) < (R_t(i) + 1).B^{n-3} - q_t.Y_t.B^{n-3} = (\rho + 1).B^{n-3} < Y_t.B^{n-3}/B$$

$$\le Y/B. \tag{6.57}$$

As $q_t^* = q - 1$,

$$R(i + 1) = B.R(i) - q.Y \quad \text{and} \quad R^*(i + 1) = B.R(i) - (q - 1).Y,$$

then

$$R^*(i + 1) = R(i + 1) + Y < (B + 1).Y/B,$$

which completes the proof.

The base-B nonrestoring division algorithm, presented below, applies to the n-digit base-B numbers X (dividend) and Y (divisor). A normalizing operation has to set Y to comply with the second condition (6.43), namely, $B^{n-1} \leq Y \leq B^n - 1$. The first condition (6.43), assuming $X = B.R(0)$, is written

$$B.R(0)/Y < (B + 1) \tag{6.58}$$

and is always true. Nevertheless, if floating-point representation standards are used, the dividend X is set (shifted) to the greatest value such that $X < Y$. The tentative quotient-digits q_t^* are computed by dividing the truncated remainder by $(Y_t + 1)$, that is, the truncated divisor augmented by 1. The redundant set of quotient-digits is in the range $[0, B]$. Whenever B is selected a correction $(+1)$ has to be carried out to the preceding quotient-digit, generating a possible carry propagation. This correction may be done *on-the-fly*, or at the end of the process, as a conversion step. The following algorithm generates two quotient-digit vectors Q and $Q1$ whose final sum $(Q + B.Q1)$ is the actual quotient.

Algorithm 6.10   Nonrestoring Base-B Division Step

```
rt:=B*r/B**(n-3); yt:=y/B**(n-3);
qt:=rt/yt+1; qt1:=qt/B; qt0:=qt modB;
remainder:=B*r-qt*y;
```

Example 6.7 X and Y are positive 5-digit decimal numbers Compute X/Y as $45598/45522$ with an accuracy of $p = 7$. To ensure that $X < Y$, a preliminary normalization procedure sets the operands to 045598 and 455220, respectively ($n = 6$, $p = 7 + 1 = 8$).

Step #	Remainder Computation	qt1	qt0
$i = 0$	$Y_t + 1 = 456, B.R(0) = 045598, (B.R(0))_t = 045,$		
	$q_t(0) = 045/456 = 0,$	0	0
	Remainder $= 45598 - 0 = 45598$		
$i = 1$	$B.R(1) = 455980, (B.R(1))_t = 455,$		
	$q_t(1) = 455/456 = 0,$	0	0
	Remainder $= 455980 - 0 = 455980$		

Step #	Remainder Computation	qt1	qt0
$i = 2$	$B.R(2) = 4559800$, $(B.R(2))_t = 4559$, $q_t(2) = 4559/456 = 9$, Remainder $= 4559800 - 4096980 = 462820$	0	9
$i = 3$	$B.R(3) = 4628200$, $(B.R(3))_t = 4628$, $q_t(3) = 4628/456 = 10$, Remainder $= 4628200 - 4552200 = 76000$	1	0
$i = 4$	$B.R(4) = 760000$, $(B.R(4))_t = 760$, $q_t(4) = 760/456 = 1$, Remainder $= 760000 - 455220 = 304780$	0	1
$i = 5$	$B.R(5) = 3047800$, $(B.R(5))_t = 3047$, $q_t(5) = 3047/456 = 6$, Remainder $= 3047800 - 2731320 = 316480$	0	6
$i = 6$	$B.R(6) = 3164800$, $(B.R(6))_t = 3164$, $q_t(6) = 3164/456 = 6$, Remainder $= 3164800 - 2731320 = 433480$	0	6
$i = 7$	$B.R(7) = 4334800$, $(B.R(7))_t = 4334$, $q_t(7) = 4334/456 = 9$, Remainder $= 4334800 - 4096980 = 237820$	0	9
$i = 8$	$B.R(8) = 2378200$, $(B.R(8))_t = 2378$, $q_t(8) = 2378/456 = 5$, Remainder $= 2378200 - 2276100 = 102100$	0	5

The final correction is expressed as

$$009016695 + 001000000 = 010016695,$$

and the overall operation can be resumed as

$$45598.10^8 = 455220.10016695 + 102100.$$

The decimal system has been used without ambiguity to represent the accuracy in the multiplicative factor 10^8.

The minimum redundancy of the quotient-digit set allows an easy correction procedure: adding 1 at level $i - 1$ whenever a quotient-digit $q(i)$ reaches B. The remainder computation is aided by the fact that the remainder is always positive. The tentative quotient-digits may be extracted from look-up tables, or computed by a specific circuit. Instead of computing the remainder as a full k-digit subtraction, a carry-save technique may be applied to store the remainder as the signed sum of two k-digit numbers.

6.3   CONVERGENCE (FUNCTIONAL ITERATION) ALGORITHMS

6.3.1   Introduction

Functional iteration algorithms represent division as a function. Numerical calculus techniques are used to solve, for example, Newton–Raphson equations or Taylor–MacLaurin expansions. These methods provide better than linear convergence, but the step complexity is somewhat more important than the one involved in digit recurrence algorithms. Since functional iteration algorithms use multiplication as a basic operation, the step complexity will mainly depend on the performance of the multiplication resources at hand ([FER1967], [FLY1970]). In practice, whenever the division process is integrated in a general-purpose arithmetic processor, one of the advantages comes from the availability of multiplication without additional hardware cost. It has been reported ([OBE1994], [FLY1997]) that, in typical floating-point applications, sharing multipliers does not significantly affect the overall performances of the arithmetic unit. In what follows, the divisor d is assumed to be a positive and normalized number such that, for example, $1/B \leq d < 1$. In most practical binary applications, IEEE normalization standards are used: $1 \leq d < 2$.

6.3.2   Newton–Raphson Iteration Technique

Coming back to the general division equation written

$$D = d.Q + r, \tag{6.59}$$

the theoretical exact quotient ($r = 0$) may be written

$$Q = D/d = D.(1/d).$$

Actually, the Newton–Raphson method first computes the reciprocal x of the divisor d, with the required precision, and then the result is multiplied by the dividend D. Using $x = 1/d$ as a root target, the priming function

$$f(x) = 1/x - d, \tag{6.60}$$

may be considered for root extraction $\{x | f(x) = 0\}$. To solve $f(x) = 0$, the following equation is iteratively used to evaluate x:

$$x_{i+1} = x_i - f(x_i)/f'(x_i), \tag{6.61}$$

where $f'(x_i)$ stands for $(df/dx)_i$, that is, the first derivative of $f(x)$ at point x_i.

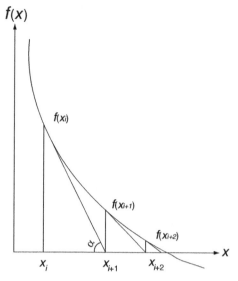

Figure 6.11   Newton–Raphson convergence graph.

The first steps of (6.61) are depicted in Figure 6.11. Equality (6.61) is readily inferred from

$$f'(x_i) = \tan(\alpha) = f(x_i)/(x_i - x_{i+1}),$$

that points to x_{i+1} as the intersection of the x-axis with the tangent to $f(x)$ at $x = x_i$. Function (6.60) is continuous and derivable in the area (close enough to the root) where the successive approximations are processed. In base B, assuming d in $[1/B, 1[$, any value of x in $]1, B]$ could be chosen as a first approximation. To speed up the convergence process, a look-up table (LUT) is generally used for a first approximation. Let x_0 ($\neq 0$) be this first approximation; then

$$f(x_0) = 1/x_0 - d,$$

$$[df/dx]_0 = -1/x_0^2.$$

Using (6.61):

$$x_1 = x_0.(2 - d.x_0),$$

and recursively

$$x_2 = x_1.(2 - d.x_1),$$

$$\cdots \tag{6.62}$$

$$x_{i+1} = x_i.(2 - d.x_i).$$

In base 2, if d is initially set to

$$\tfrac{1}{2} \leq d < 1, \tag{6.63}$$

then

$$1 < 1/d \leq 2.$$

Selecting x_0 within the range $1 < x_0 \leq 2$ will ensure a quadratic convergence. Actually, assuming $x_i = 1/d + \varepsilon_i$, error ε_{i+1} can be evaluated, according to (6.62), as

$$1/d + \varepsilon_{i+1} = (1/d + \varepsilon_i).(2 - d.(1/d + \varepsilon_i)), \tag{6.64}$$

ending at

$$|\varepsilon_{i+1}| = d.\varepsilon_i^2, \tag{6.65}$$

which means that, at each step, the number of relevant digits is multiplied by two. If x_0 is drawn from a t-digit precision LUT, the minimum precision p will be reached after $k = \lceil \log_2 p/t \rceil$ steps.

Algorithm 6.11 Reciprocal Computation

```
x(0):=LUT (d);
for i in 0..k-1 loop
x(i+1):=x(i)*(2-d*x(i));
end loop;
Q:=x(k);
```

Example 6.8 Let $d = 0.1618280$ (base 10) and compute $1/d$ with precision 32.

Assume a look-up table with 4-digit precision.

$$
\begin{aligned}
x_0 &= 6.179 (LUT),\\
d.x_0 &= 0.1618280 \times 6.179 = 0.99993521,\\
x_1 &= 6.179 \times (2 - 0.99993521) = 6.1794003,\\
d.x_1 &= 0.1618280 \times 6.1794003 = 0.9999999917484,\\
x_2 &= 6.1794003 \times (2 - 0.9999999917484) = 6.179400350989939,\\
d.x_2 &= 0.1618280 \times 6.179400350989939 = 0.999999999999999848492,\\
x_3 &= 6.179400350989939 \times (2 - 0.999999999999999848492),\\
&= 6.179400350989939362285883777837 \pm 10^{-31}.
\end{aligned}
$$

In Example 6.8, one can observe that the error is always by default (as $2 - dx_i > 1$). This is in agreement with the monotony of the convergence (Figure 6.11). Nevertheless, the practical convergence could appear different according to the way the rounding on the last digit is made: by excess or by default.

Two dependent multiplications and one subtraction are needed at each computation step. In base B the subtraction can be substituted by (i) setting the integer part to 1 and (ii) a digitwise $(B-1)$'s complement operation then (iii) adding 1 at the last digit level; the add-1 operation can be skipped to speed up the process at the cost of a slower convergence rate. In base 2 the subtraction is actually a 2's complement operation, which in turn can be replaced by a bitwise complementation at the cost of a slower convergence rate. Example 6.9 illustrates the algorithm in base 2.

Example 6.9   Let $d = 0.1010111$ in base 2; compute $1/d$ with precision 32.

Assume a look-up table with 4-digit (rounded) precision.

$$x_0 \quad = 1.100\,(LUT);$$
$$d.x_0 = 0.1010111 \times 1.100 = 1.0000010100;$$
$$2 - d.x_0 = 0.1111101100;$$
$$x_1 \quad = 1.100 \times 0.1111101100 = 1.0111100 - 8\,\text{bits rounded};$$
$$d.x_1 = 0.1010111 \times 1.0111100 = 0.11111111100100;$$
$$2 - d.x_1 = 1.00000000011100;$$
$$x_2 \quad = 1.0111100 \times 1.00000000011100 = 1.011110001010010$$
$$\quad\quad - 16\,\text{bits rounded};$$
$$d.x_2 = 0.1010111 \times 1.011110001010010 = 0.1111111111111111011110;$$
$$2 - d.x_2 = 1.0000000000000000100010;$$
$$x_3 \quad = 1.011110001010010 \times 1.0000000000000000100010$$
$$\quad\quad = 1.0111100010100100110010000001011 \pm 2^{-31}$$

The same example is now treated replacing the 2's complement operation by bitwise complementation.

$$x_0 \quad = 1.100\,(LUT),$$
$$d.x_0 = 0.1010111 \times 1.100 = 1.0000010100;$$
$$\text{not } d.x_0 = 0.1111101011;$$
$$x_1 \quad = 1.100 \times 0.1111101011 = 1.0111100 - 8\,\text{bits rounded};$$
$$d.x_1 = 0.1010111 \times 1.0111100 = 0.11111111100100;$$
$$\text{not } d.x_1 = 1.00000000011011;$$
$$x_2 \quad = 1.0111100 \times 1.00000000011011$$
$$\quad\quad = 1.011110001001111 - 16\,\text{bits rounded};$$
$$d.x_2 = 0.1010111 \times 1.011110001001111 = 0.1111111111111011011001;$$
$$\text{not } d.x_2 = 1.0000000000000100100110;$$
$$x_3 \quad = 1.011110001001111 \times 1.0000000000000100100110$$
$$\quad\quad = 1.01111000101001001100 \pm 2^{-20} - \text{other next bits irrelevant}$$

6.3.3 MacLaurin Expansion—Goldschmidt's Algorithm

Taylor expansions are well-known mathematical tools for the numerical calculus of functions. The MacLaurin series, as the Taylor expansion of the function $1/(1+x)$, is the base of a division algorithm called *Goldschmidt's algorithm* ([GOL1964]).

Consider the classical Taylor expansion at point p

$$f(x) = \Sigma_{0 \le i \le m}[(x-p)^i/i!].(d^i f(x)/dx^i)_{x=p}, \tag{6.66}$$

applied to the function

$$f(x) = 1/(1+x). \tag{6.67}$$

At the point $x = 0$, the following MacLaurin expansion is established:

$$1/(1+x) = 1 - x + x^2 - x^3 + x^4 - x^5 + \cdots, \tag{6.68}$$

or equivalently,

$$1/(1+x) = (1-x).(1+x^2).(1+x^4).(1+x^8).(1+x^{16}) \cdots, \tag{6.69}$$

This is the foundation of Goldschmidt's algorithm. As the Newton–Raphson method does, the series expansion technique consists of computing the reciprocal $1/d$ of the divisor d. The series $1/(1+x)$ at $x = 0$ is easier to compute than $1/x$ at $x = 1$. The problem at hand is now to let D/d converge toward $Q/1$, which is readily achieved through successive multiplications by the factors of MacLaurin expression (6.69).

For this purpose, the following is assumed:

$$x = d - 1,$$
$$1/B \le d < 1 \text{ (base-}B \text{ normalized form).} \tag{6.70}$$

The procedure is then carried out as follows:

$$Q = D/d = D.[1/(1+x)]$$
$$= D.(1-x).(1+x^2).(1+x^4).(1+x^8).(1+x^{16}) \cdots, \tag{6.71}$$

$d(0)$ and $D(0)$ are set to d and D respectively; a look-up table evaluation procedure could refine this first approximation to get a better convergence rate. Then

$$d(1) = d(0).(1 - x) = d(0).(2 - d(0)),$$
$$D(1) = D(0).(2 - d(0)),$$
$$d(2) = d(1).(1 + x^2) = d(1).(2 - d(1)),$$
$$D(2) = D(1).(2 - d(1)), \tag{6.72}$$
$$\cdots$$
$$d(i) = d(i - 1).(2 - d(i - 1)),$$
$$D(i) = D(i - 1).(2 - d(i - 1)).$$

Algorithm 6.12   Goldschmidt's Algorithm

```
d(0):=LUT (divisor); DD(0):=LUT (dividend);
for i in 1..k-1 loop
d(i):=d(i-1)*(2-d(i-1));
DD(i):=DD(i-1)*(2-d(i-1));
end loop;
Q:=DD(k-1);
```

Comments 6.7

1. Each step is made up of one subtraction (2's complement in base 2) and two multiplications.

2. At step i, the computed value of $d(i)$ is given by $d(i) = 1 - x^{\exp(i)}$ where $\exp(i) = 2^i$; since (6.70) $1/B \le d < 1$, then $(1 - B)/B \le x < 0$, as initial pre-scaling of d and D is carried out, a fast quadratic convergence is ensured: each correction factor $(1 + x^{\exp(i)})$ will duplicate the precision.

3. The complexity of the Newton–Raphson method is similar to the one involved in Goldschmidt's algorithm. Nevertheless, Goldschmidt's method directly provides the quotient while Newton–Raphson computes the reciprocal first, then a further multiplication is needed. Moreover, in Newton–Raphson, the two multiplications are dependent so they have to be processed sequentially while parallelism is allowed in Goldschmidt's method.

4. Newton Raphson's convergence method is self-correcting in the sense that any error (i.e., taking bitwise complement instead of 2's complement) is corrected in the following steps. On the contrary any error in Goldschmidt's algorithm will never be corrected. The final result will be the quotient of $D(j)/d(j)$, assuming step j is the step where the last errors have been committed.

5. Goldschmidt's algorithm has been used in the IBM System/360 model 91 ([AND1967]) and, more recently, in the IBM System/390 ([SCHW1999])

and AMD-K7 microprocessor ([OBE1999]). Another algorithm based on series expansions has been implemented in IBM's Power3 ([SCHM1999]).

6. Goldschmidt's algorithm is suited for combinatorial implementations; factors of (6.72) can be calculated in parallel.

Example 6.10   Let $D = 0.152525$, $d = 0.161828$ (base 10); compute D/d with precision 32.

A look-up table with four-digit precision would approximate $1/d$ at 6.179, so multiplying both d and D by this value allows presetting $d(0)$ and $D(0)$ while reducing the number of steps:

$$D(0) = 0.152525 \times 6.179 = 0.942451975,$$

$$d(0) = 0.161828 \times 6.179 = 0.999935212,$$

$$d(1) = 0.999935212 \times (2 - 0.999935212) = 0.9999999995802515056,$$

$$D(1) = 0.942451975 \times (2 - 0.999935212) = 0.9425130345785563$$

$$d(2) = 0.9999999995802515056 \times (2 - 0.9999999995802515056)$$
$$= 0.99999999999999998238112014489332,$$

$$D(2) = 0.9425130345785563 \times (2 - 0.9999999995802515056)$$
$$= 0.94251303853474057216724145450635$$

$$d(3) = 0.99999999999999998238112014489332$$
$$\times (2 - 0.99999999999999998238112014489332)$$
$$= 0.9999999999999999999999999999999,$$

$$D(3) = 0.94251303853474057216724145450635$$
$$\times (2 - 0.99999999999999998238112014489332)$$
$$= 0.94251303853474058877326544232143$$

Comment 6.8   An important question about convergence algorithms is the evaluation of the exact (i.e., minimum) amount of bits necessary for any prescaling or intermediate calculations, to ensure a correct result within the desired precision. Actually, the accurate calculus of rounding errors (LUT data as well as outputs from intermediate operations) is not a straightforward matter. This mathematical problem has been treated extensively in the literature ([DAS1995], [COR1999], [EVE2003]). Using extrabits is a safe and easy way to ensure correctness; nevertheless, a careful error computation can lead to significant savings ([EVE2003]).

6.4   BIBLIOGRAPHY

[AND1967] S. F. Anderson, J. G. Earle, R. E. Goldschmidt, and D. M. Powers, The IBM System/360 Model 91: Floating-point execution unit. *IBM J. Res. Dev.*, **11** (Jan.): 34–53 (1967).

[ATK1968] D. E. Atkins, Higher-radix division using estimates of the divisor and partial remainders. *IEEE Trans. Comput.*, **C-17**(10): 925–934 (1968).

[BRI1993] W. S. Briggs and D.W. Matula, A 17 × 69 bit multiply and add unit with redundant binary feedback and single cycle latency. In: *Proceedings of the 11th IEEE Symposium on Computer Arithmetic*, July 1993, pp. 163–170.

[COR1999] M. A. Cornea-Hasegan, R. A. Golliver, and P. Markstein, Correctness proofs outline for Newton–Raphson based floating-point divide and square root algorithms: In: *Proceedings of the 14th IEEE Symposium on Computer Arithmetic*, 1999, pp. 96–105.

[DAS1995] D. DasSarma and D. W. Matula, Faithfull bipartite ROM reciprocal tables. In *Proceedings of the 12th IEEE Symposium Computer Arithmetic*, 1995 pp. 17–28.

[EDE1997] A. Edelman, The mathematics of the Pentium division bug. *SIAM Rev.* **39**(1): 54–67 (1997).

[ERC1983] M. D. Ercegovac, A higher radix division with simple selection of quotient digits. In *Proceedings of the 6th IEEE Symposium Computer Arithmetic*, July 1983, pp. 94–98.

[ERC1987] M. D. Ercegovac and T. Lang, On-the-fly conversion of redundant into conventional representations. *IEEE Trans. Comput.*, **36**(7): 895–897 (1987).

[ERC1990] M. D. Ercegovac and T. Lang, Simple radix-4 division with operands scaling. *IEEE Trans. Comput.*, **39**(9): 1204–1207 (1990).

[ERC1992] M. D. Ercegovac and T. Lang, On-the-fly rounding. *IEEE Trans. Comput.* **41**, (12):1497–1503 (1992).

[ERC2004] M. D. Ercegovac and T. Lang, *Digital Arithmetic*, Morgan Kaufmann, San Francisco, CA, 2004.

[EVE2003] G. Even, P.-M. Seidel, and W. E. Ferguson, A parametric error analysis of Goldschmidt's division algorithm. In: *Proceedings of the 16th IEEE Symposium on Computer Arithmetic*, June 2003.

[FAN1989] J. Fandrianto, Algorithm for high-speed shared radix-8 division and radix-8 square root. In *Proceedings of the Ninth IEEE Symposium on Computer Arithmetic*, July 1989, pp. 68–75.

[FER 1967] D. Ferrari, A division method using a parallel multiplier. *IEEE Trans. Comput.*, **16**: 224–226, (April 1967).

[FLY1970] M. J. Flynn, On division by functional iteration. *IEEE Trans. Comput.* **19**: (August 1970).

[FLY1997] M. J. Flynn, *Modern Research in Computer Arithmetic*, Stanford Architecture and Arithmetic Group, Stanford Technical Report, 1997.

[FLY2001] M. J. Flynn and S. F. Oberman, *Advanced Computer Arithmetic Design*, Wiley-Interscience, Hoboken, NJ, 2001.

[FRE1961] C. V. Freiman, Statistical analysis of certain binary division algorithms. *IRE Proc.* **49**: 91–103 (1961).

[GOL1964] R. Goldschmidt, *Applications of Division by Convergence.* Master's thesis, MIT, June 1964.

[MON1994] P. Montuschi and L. Ciminiera, Radix-8 division with over-redundant digit set. *J. VLSI Signal Processing* **7** (3): 259–270 (1994).

[NAD1956] M. Nadler, A high speed electronic arithmetic unit for automatic computing machines. *Acta Tech. (Prague)*, **6**: 464–478 (1956).

[OBE1994] S. F. Oberman and M. J. Flynn, Design issues in floating point division. Technical Report No. CSL-TR-94-647, Computer Systems Laboratory, Stanford University, December 1994.

[OBE1995] S. F. Oberman and M. J. Flynn, An analysis of division algorithms and implementations. Stanford Technical Report CSL-TR-95-675, July 1995.

[OBE1997] S. F. Oberman and M. J. Flynn, Division algorithms and implementations. *IEEE Trans. Comput.* **46**: 833–854 (Aug. 1997).

[OBE1999] S. F. Oberman, Floating point division and square root algorithms and implementation in the AMD-K7 microprocessor. In: *Proceedings of the 14th IEEE Symposium on Computer Arithmetic*, April 1999, pp. 106–115.

[PAR1999] B. Parhami, *Computer Arithmetic, Algorithms and Hardware Designs*, Oxford University Press, New York, 1999.

[QUA1992] N. Quach and M. Flynn, A radix-64 floating-point divider. Technical Report CSL-TR-92–529, Computer Systems Laboratory, Stanford University, June 1992.

[ROB1958] J. E. Robertson, A new class of digital division methods. *IRE Trans. Electron. Comput.*, **EC-7**: 218–222 (Sept. 1958).

[SCHM1999] M. S. Schmookler, R. C. Agarwal, and F. G. Gustavson, Series approximation methods, for divide and square root in the Power3 Microprocessor. In: *Proceedings of the 14th IEEE Symposium Computer Arithmetic*, April 1999, pp. 116–123.

[SCHW1999] E. M. Schwarz, R.M. Smith, and C. A. Krygowski, The S/390 G5 floating point unit supporting hex and binary architectures. In: *Proceedings of the 14th IEEE Symposium on Computer Arithmetic*, April 1999, pp. 258–265.

[SRI1995] H. Srinivas and K. Parhi, A fast radix-4 division algorithm and its architecture. *IEEE Trans. Comput.* **44**(6): 826–831 (1995).

[SUT2004] G. Sutter, G. Bioul, and J.-P. Deschamps, Comparative study of SRT-dividers in FPGA. In: *Lecture Notes in Computer Science*, Vol. 3203, Springer-Verlag, 2004, pp. 209–220.

[SVO1963] A. Svoboda, An algorithm for division. *Stroje na Zpracovani Informaci*, **9**: 25–34 (1963).

[SWE1957] J. Cocke and D. W. Sweeney, High speed arithmetic in a parallel device. IBM Technical Report, February 1957.

[TAY1985] G. S. Taylor, Radix 16 SRT dividers with overlapped quotient selection stages. In: *Proceedings of the Seventh IEEE Symposium on Computer Arithmetic*, June 1985, pp. 64–71.

[TOC1958] T. D. Tocher, Techniques of multiplication and division for automatic binary computer. *Q. J. Mech. Appl. Math.*, **2**: 364–384 (1958).

7

OTHER ARITHMETIC OPERATIONS

This chapter is devoted to arithmetic functions and operations other than the four basic ones. Number representation systems conversion procedures are first analyzed; they play a prominent role in arithmetic processes since a variety of algorithms are designed for a wide-ranging number of systems and/or bases (radices). Further on, this chapter reviews classical methods for approximating logarithmic, exponential, and trigonometric functions. Polynomial approximation, Taylor–MacLaurin series, and convergence methods are described with a special attention to CORDIC algorithms and their applications to trigonometric functions. Square rooting algorithms founded on digit recurrence and convergence methods are finally surveyed.

A common feature of a number of modern algorithm implementations is the increased use of look-up tables (LUTs), a practice fully compatible with the evolution of the ROM technology toward larger size and lower cost. The main consequence of LUT-based techniques is then a low-cost speed-up of the overall procedures.

7.1   BASE CONVERSION

Given the representation of a number in a specific system, conversion operations consist of finding the representation of this number in another system. Arithmetic

Synthesis of Arithmetic Circuits: FPGA, ASIC, and Embedded Systems
By Jean-Pierre Deschamps, Géry J. A. Bioul, and Gustavo D. Sutter
Copyright © 2006 John Wiley & Sons, Inc.

algorithms deal with a diversity of systems such as base-B for naturals or signed systems for integers; the most common signed systems are sign-magnitude, B's complement, excess-k, or signed-digit systems such as Booth coding. Finally, special attention is paid to floating-point representations for computer applications. Redundant systems are also important in arithmetic operations, among others in multiplication (Booth algorithm, Chapter 5) or division (SRT, Chapter 6). Booth coding and redundant base-B coding are generally related to specific algorithms; the conversion techniques are therefore developed in the sections devoted to the respective algorithms. Floating-point conversion is reviewed in Chapter 16. The most classic problem to deal with is the base conversion for base-B unsigned representations: given a number by its representation in base B_1, find its corresponding representation in base B_2.

Let the base-B_1 (source system) representation of a natural number x be given by

$$x = x_{n-1}.B_1^{n-1} + x_{n-2}.B_1^{n-2} + \cdots + x_0.B_1^0, \tag{7.1}$$

weighted sum of powers of B. The problem at hand is to compute the base-B_2 (target system) representation of x

$$x = y_{m-1}.B_2^{m-1} + y_{m-2}.B_2^{m-2} + \cdots + y_0.B_2^0. \tag{7.2}$$

A first simple solution consists of computing (7.1) in the target system. Expression (7.1) may be written under the Hörner form (Chapter 5) as

$$x = ((\cdots(0.B_1 + x_{n-1}).B_1 + x_{n-2}).B_1 + \cdots).B_1 + x_0, \tag{7.3}$$

then iteratively computed in base B_2. The following algorithm generates the value of x in base-B_2 from its base-B_1 representation $(x_{n-1}, x_{n-2}, \ldots, x_0)$:

```
acc:=0; for i in 0..n-1 loop acc:=acc*B1+x(n-1-i);
end loop; x:=acc;
```

In order to generate the base-B_2 representation of x, the preceding algorithm is executed in base B_2. The base-B_2 representation of $acc.B_1 + x_{n-1-i}$ is generated by the following set of integer divisions:

$$acc_0.B_1 + x_{n-1-i} = B_2.q_1 + r_0,$$
$$acc_1.B_1 + q_1 = B_2.q_2 + r_1,$$
$$acc_2.B_1 + q_2 = B_2.q_3 + r_2,$$

$$\cdots$$

$$acc_{m-1}.B_1 + q_{m-1} = B_2.q_m + r_{m-1}.$$

where acc_i is the digit i of acc expressed in base B_2.

From the preceding system the following equality is deduced:

$$acc.B_1 + x_{n-1-i} = B_2^m.q_m + r_{m-1}.B_2^{m-1} + r_{m-2}.B_2^{m-2} + \cdots + r_0.B_2^0.$$

Choose the value of m in such a way that $B_1^n \le B_2^m$. In the last equality $q_m = 0$. The following algorithm executes the assignation $acc := acc.B_1 + x_{n-1-i}$ in base B_2:

```
q:=x(n-1-i);
for j in 0..m-1 loop q:=(acc(j)*B1+q)/B2;
  acc(j):=(acc(j)*B1+q) mod B2; end loop;
```

The complete conversion algorithm is then given as follows:

Algorithm 7.1   Base Conversion Algorithm

```
-acc:=0;
for j in 0..m-1 loop acc(j):=0; end loop;
for i in 0..n-1 loop
    --acc:=acc*B1+x(n-1-i);
    q:=x(n-1-i);
    for j in 0..m-1 loop q:=(acc(j)*B1+q)/B2;
       acc(j):=(acc(j)*B1+q) mod B2; end loop;
end loop;
--x:=acc;
for j in 0..m-1 loop x(j):=acc(j); end loop;
```

The basic computation primitive calculates

$$q^+ = (acc(j).B_1 + q)/B_2, \ acc^+(j) = (acc(j).B_1 + q) \bmod B_2 \qquad (7.4)$$

where

$$q, \ q^+ \in \{0, 1, \ldots, B_1 - 1\}, \quad acc(j), acc^+(j) \in \{0, 1, \ldots, B_2 - 1\}.$$

Examples 7.1

1. Compute the hexadecimal representation of the decimal number $(9128)_{10}$. Observing that $16^3 < 10^4 < 16^4$, one selects $m = 4$.
2. Compute the decimal representation of the hexadecimal number $(23A8)_{16}$. Observing that $10^4 < 16^4 < 10^5$, one should select $m = 5$; actually, $m = 4$ is sufficient for this particular case so $m = 4$ is selected to reduce the number of useful steps.

Problem 2 is the back-conversion of the result of problem 1; both examples are presented in parallel columns of the following table.

$(9128)_{10} \rightarrow$ base 16	$(23A8)_{16} \rightarrow$ base 10
$acc = (0\ 0\ 0\ 0)$	$acc = (0\ 0\ 0\ 0)$
$0.10 + 9 = 16.0 + 9$	$0.16 + 2 = 10.0 + 2$
$0.10 + 0 = 16.0 + 0$	$0.16 + 0 = 10.0 + 0$
$acc = (0\ 0\ 0\ 9)$	$acc = (0\ 0\ 0\ 2)$
$9.10 + 1 = 16.5 + 11$	$2.16 + 3 = 10.3 + 5$
$0.10 + 5 = 16.0 + 5$	$0.16 + 3 = 10.0 + 3$
$0.10 + 0 = 16.0 + 0$	$0.16 + 0 = 10.0 + 0$
$acc = (0\ 0\ 5\ 11)$	$acc = (0\ 0\ 3\ 5)$
$11.10 + 2 = 16.7 + 0$	$5.16 + 10 = 10.9 + 0$
$5.10 + 7 = 16.3 + 9$	$3.16 + 9 = 10.5 + 7$
$0.10 + 3 = 16.0 + 3$	$0.16 + 5 = 10.0 + 5$
$0.10 + 0 = 16.0 + 0$	$0.16 + 0 = 10.0 + 0$
$acc = (0\ 3\ 9\ 0)$	$acc = (0\ 5\ 7\ 0)$
$0.10 + 8 = 16.0 + 8$	$0.16 + 8 = 10.0 + 8$
$9.10 + 0 = 16.5 + 10$	$7.16 + 0 = 10.11 + 2$
$3.10 + 5 = 16.2 + 3$	$5.16 + 11 = 10.9 + 1$
$0.10 + 2 = 16.0 + 2$	$0.16 + 9 = 10.0 + 9$
$acc = (2\ 3\ 10\ 8)$	$acc = (9\ 1\ 2\ 8)$
$(9128)_{10} = (23A8)_{16}$	$(23A8)_{16} = (9128)_{10}$

Decimal-to-binary and binary-to-decimal conversions may readily be handled by Algorithm 7.1. According to the prominent role of those particular cases in digital system implementations, special attention has been given to them in the literature. Basically three methods are more commonly described for decimal-to-binary conversion. The first one consists of subtracting from the successive remainders R the greatest power of 2 (2^p) inferior or equal to R. Successive exponents p identify the bit-1 positions of the desired binary expression. Step i of a nonrestoring version of this algorithm would add or subtract $2^{(n-1-i)}$ according to the sign of the preceding remainder: adding $2^{(n-1-i)}$ to a negative remainder and conversely. The second method consists of computing the remainders (parity) of the integer division by 2 of the successive decimal quotients, starting from the decimal number to be converted. The third method is the adaptation of Algorithm 7.1. At the implementation level, the complexity would depend on the form decimal digits are provided. Decimal numbers are generally assumed represented in BCD form (binary coded decimal, Chapter 3). Then algorithms are implemented as a sequence of binary operations. A close look at Example 7.2(3) shows that the first step of the third method (Algorithm 7.2) actually sets the leftmost decimal digit to BCD; so this step can be skipped if the decimal number to convert is already BCD coded. In this algorithm, the successive steps iteratively perform the multiplication of the intermediate result (stored in acc) by the base $(10)_{10} = (1010)_2$, and then add the next (binary coded) decimal digit. Let us point out moreover that multiplying a number N by $(1010)_2$ can readily be implemented as $2.N + 8.N$ or $2.(N + 4.N)$, that is, a simple 1-*shifted*($N + 2$-*shiftedN*) operation. Assuming m and n to be the number of digits of the decimal and binary representations, respectively, the time complexity (in number of steps) of the first two methods is n while the third one is m. The step complexity has to be evaluated in the context of the implementation resources. The BCD-to-binary conversion algorithm can be stated as follows.

Algorithm 7.2   BCD-to-Binary Conversion Algorithm

```
-acc:=(bcd_m-1, bcd_m-2,..bcd_0);
for j in 0..m-1 loop acc(j):=bcd(j); end loop;
bin(m-1):=acc(m-1);
for j in 1..m-1 loop bin(m-1-j):=(bin(m-j)+bin(m-j)*100)*10
  +acc(m-j-1); end loop;
x:=bin(0);
```

Examples 7.2   Convert in binary $(922)_{10}$

1. Nonrestoring 2^p subtracting.

$$922 - 2^9 = \quad 410 \rightarrow x_9 = 1$$
$$410 - 2^8 = \quad 154 \rightarrow x_8 = 1$$
$$154 - 2^7 = \quad 26 \rightarrow x_7 = 1$$
$$26 - 2^6 = -38 \rightarrow x_6 = 0$$
$$-38 + 2^5 = \quad -6 \rightarrow x_5 = 0$$
$$-6 + 2^4 = \quad 10 \rightarrow x_4 = 1$$
$$10 - 2^3 = \quad 2 \rightarrow x_3 = 1$$
$$2 - 2^2 = \quad -2 \rightarrow x_2 = 0$$
$$-2 + 2^1 = \quad 0 \rightarrow x_1 = 1$$
$$0 - 2^0 = \quad -1 \rightarrow x_0 = 0$$
$$(922)_{10} = (1110011010)_2$$

2. Division-by-2.

$$
\begin{array}{rcl}
922/2 & = & 461 + \mathbf{0} \\
461/2 & = & 230 + \mathbf{1} \\
230/2 & = & 115 + \mathbf{0} \\
115/2 & = & 57 + \mathbf{1} \\
57/2 & = & 28 + \mathbf{1} \\
28/2 & = & 14 + \mathbf{0} \\
14/2 & = & 7 + \mathbf{0} \\
7/2 & = & 3 + \mathbf{1} \\
3/2 & = & 1 + \mathbf{1} \\
1/2 & = & 0 + \mathbf{1}
\end{array}
$$

$$(922)_{10} = (1110011010)_2$$

3. Algorithm 7.2.

$acc = 1001, 0010, 0010; \ m = 3;$
$bin(2) = 1001$

Step 1

$bin(1) = (1001 + 100100).10 + 0010 = 1011100$

Step 2

$bin(0) = (1011100 + 101110000).10 + 0010 = (1110011010)_2$

$(922)_{10} = (1110011010)_2$

The most intuitive method to achieve a binary-to-decimal conversion is to compute the decimal representations of the relevant powers of 2, and then add them together. This corresponds to the decimal computation of the expression $\Sigma_i x_i.2^i$. An algorithm consisting of computing the remainders of successive divisions by ten doesn't appear easy to implement. Algorithm 7.1 looks the most attractive as it still reduces to a straight application of the Hörner expansion (7.3). Assuming that a BCD representation is desired for the converted result, the binary-to-BCD algorithmic step performs the multiplication of the intermediate result (stored in *acc*) by the base $2 = (10)_2$, and then adds the next bit. For this purpose, define a procedure to multiply a BCD number by 2 (BCDx2 procedure). Note that a shifted (multiplied by 2) 4-bit binary number $(x_3, x_2, x_1, x_0) \le (1001)$ can be coded in BCD form $(y(1)_3, y(1)_2, y(1)_1, y(1)_0; y(0)_3, y(0)_2, y(0)_1, y(0)_0)$, applying the following simple rules:

$$y(1)_3 = y(1)_2 = y(1)_1 = y(0)_0 = 0, \tag{7.5}$$

$$y(1)_0 = x_3 \lor (x_2.(x_1 \lor x_0)), \tag{7.6}$$

$$(y(0)_3, y(0)_2, y(0)_1, y(0)_0) = [(x_2, x_1, x_0, 0) + (0, y(1)_0, y(1)_0, 0)] \bmod 16. \tag{7.7}$$

Observe that the x2 operation on a single BCD digit leaves $y(i)_0$ at 0; on a full BCD number $y(i)_0$ will assume the value of the carry (7.6) coming from the right neighbor digit; this carry will not propagate further. This feature allows a parallel digit processing in the BCDx2 procedure.

Steps i and $i + 1$ of the procedure BCDx_step are defined according to (7.6) to (7.7):

Step i

$$y(i + 1)_0: = x(i)_3 \lor (x(i)_2.(x(i)_1 \lor x(i)_0)), \text{ carry to step } (i + 1)$$

$$y(i): = (x(i)_2, x(i)_1, x(i)_0, y(i)_0) + (0, y(i + 1)_0, y(i + 1)_0, 0)] \bmod 16.$$

Step i + 1

$$y(i + 2)_0: = x(i + 1)_3 \lor (x(i + 1)_2.(x(i + 1)_1 \lor x(i + 1)_0)), \text{ carry to step } (i + 2),$$

$$y(i + 1): = (x(i + 1)_2, x(i + 1)_1, x(i + 1)_0, y(i + 1)_0)$$
$$+ (0, y(i + 2)_0, y(i + 2)_0, 0)] \bmod 16.$$

As quoted above, all the steps may be executed in parallel. The above equations define the procedure BCDx2_step computing the BCD expression of its BCD input multiplied by two:

procedure BCDx2_step (bcd(i): **in** bcd; 2bcd (i): **out** bcd)

Algorithm 7.3 describes the binary-to-BCD version of Algorithm 7.1.

Algorithm 7.3   Binary-to-BCD Conversion Algorithm

```
--acc:=(bin_n-1, bin_n-2,..., bin_0);
for j in 0..n-1 loop acc(j):=bin(j); end loop;
bcd(n-1):=acc(n-1);
for j in 1..n-1 loop
procedure BCDx2_step (bcd(n-j));
bcd(n-1-j):=2bcd(n-j)+acc(n-j-1); end loop;
x:=bcd(0);
```

Example 7.3   Convert in BCD $(1011101011101001)_2$; $n = 16$

$acc = 1011101011101001$

Initial step
      $bcd(15) = 0001$
Step 1
      $2bcd(15) = 0010$
      $bcd(14) = 0010 + 0 = 0010$
Step 2
      $2bcd(14) = 0100$
      $bcd(13) = 0100 + 1 = 0101$
Step 3
      $2bcd(13) = 1\ 0000$
      $bcd(12) = 1\ 0000 + 1 = 1\ 0001$

Step 4

   $2bcd(12) = 10\ 0010$
   $bcd(11)\ = 10\ 0010 + 1 = 10\ 0011$

Step 5

   $2bcd(11) = 100\ 0110$
   $bcd(10)\ = 100\ 0110 + 0 = 100\ 0110$

Step 6

   $2bcd(10) = 1001\ 0010$
   $bcd(9)\ = 1001\ 0010 + 1 = 1001\ 0011$

Step 7

   $2bcd(9) = 1\ 1000\ 0110$
   $bcd(8)\ = 1\ 1000\ 0110 + 0 = 1\ 1000\ 0110$

Step 8

   $2bcd(8) = 11\ 0111\ 0010$
   $bcd(7)\ = 11\ 0111\ 0010 + 1 = 11\ 0111\ 0011$

Step 9

   $2bcd(7) = 111\ 0100\ 0110$
   $bcd(6)\ = 111\ 0100\ 0110 + 1 = 111\ 0100\ 0111$

Step 10

   $2bcd(6) = 1\ 0100\ 1001\ 0100$
   $bcd(5)\ = 1\ 0100\ 1001\ 0100 + 1 = 1\ 0100\ 1001\ 0101$

Step 11

   $2bcd(5) = 10\ 1001\ 1001\ 0000$
   $bcd(4)\ = 10\ 1001\ 1001\ 0000 + 0 = 10\ 1001\ 1001\ 0000$

Step 12

   $2bcd(4) = 101\ 1001\ 1000\ 0000$
   $bcd(3)\ = 101\ 1001\ 1000\ 0000 + 1 = 101\ 1001\ 1000\ 0001$

Step 13

   $2bcd(3) = 1\ 0001\ 1001\ 0110\ 0010$
   $bcd(2)\ = 1\ 0001\ 1001\ 0110\ 0010 + 0 = 1\ 0001\ 1001\ 0110\ 0010$

Step 14

   $2bcd(2) = 10\ 0011\ 1001\ 0010\ 0100$
   $bcd(1)\ = 10\ 0011\ 1001\ 0010\ 0100 + 0 = 10\ 0011\ 1001\ 0010\ 0100$

Step 15

   $2bcd(1) = 100\ 0111\ 1000\ 0100\ 1000$
   $bcd(0)\ = 100\ 0111\ 1000\ 0100\ 1000 + 1 = 100\ 0111\ 1000\ 0100\ 1001$

$(1011101011101001)_2 = (100\ 0111\ 1000\ 0100\ 1001)_{\text{BCD}} = (47849)_{10}$

Comments 7.1 Algorithm 7.3 needs n steps to convert an n-bit binary number into BCD. This figure can go down by processing 2-bit, 3-bit, or 4-bit slices per step, that is, handling the binary vector as a base-4, octal, or hexadecimal number, respectively. Nevertheless, the step complexity will be significantly increased because the adding part is no longer carry-propagation free, as it is in the binary case: adding one unit to an even number ($2bcd$) never generates a carry.

7.2 RESIDUE NUMBER SYSTEM CONVERSION

7.2.1 Introduction

The residue number system (RNS) is not widely used in practice; nevertheless, in some specific classes of algorithms, the RNS can provide an important speed-up by replacing operations on large numbers by parallel processing on small size operands ([GAR1959], [SZA1967]). A residue number system is defined as a system of s natural numbers $\{m_i\}$ called *moduli*, the greatest of which is generally denoted m. The *RNS representation* of a given integer N, with respect to $\{m_i\}$, is the vector r, the components of which, called *residues modulo-m_i*, are the s successive remainders of the integer divisions N/m_i. Residues are denoted

$$r_i = |N|m_i. \tag{7.8}$$

The least common multiple (*lcm*) of a system $\{m_i\}$ is the range, denoted M, of the related RNS system, that is, the quantity of different integers that can be represented in the system. Whenever a system $\{m_i\}$ consists of pairwise prime moduli, the associated RNS is said to be *nonredundant*. Selection of moduli, as small as possible, maximizes the speed of RNS arithmetic operations, as operand sizes never exceed the size of greatest modulus m. The most important drawbacks of the RNS are the complexity of overflow detection and the conversion process. Whenever those operations are not critical in the overall complexity of the application at hand, the RNS is fully justified.

7.2.2 Base-*B* to RNS Conversion

The most intuitive method for base-B to RNS coding consists in performing the successive mod m_i reductions as shown in the following example (see Chapter 8 for reduction mod m).

Example 7.4 Let $\{m_i\} = \{31, 17, 7, 5, 3\}$ and $N = (789)_{10}$; compute $\{r_i\}$

$$s = 5;$$
$$r_1 = 789 \bmod 3 = 0$$
$$r_2 = 789 \bmod 5 = 4$$
$$r_3 = 789 \bmod 7 = 5$$
$$r_4 = 789 \bmod 17 = 7$$
$$r_5 = 789 \bmod 31 = 14$$
$$(789)_{10} = (14, 7, 5, 4, 0)_{\text{RNS}}$$

A faster method consists in precomputing modulo m_i the B multiples of the successive powers of B; for n base-B digits, $n \times B \times s$ modulo m_i reductions are needed. The RNS expressions of the relevant powers of B are first extracted from the look-up table, and then added modulo m_i componentwise to achieve the RNS conversion. Table 7.1 shows the precomputed residues related to the first 5 powers of 10 with the set of 5 moduli $\{23, 17, 7, 5, 6\}$.

TABLE 7.1   RNS {23, 17, 7, 5, 6}: Precomputed Residues for Multiples of the First 5 Powers of 10

$a_i.10^0$ ↓	23	17	7	5	6	$a_i.10^2$ ↓	23	17	7	5	6	$a_i.10^4$ ↓	23	17	7	5	6
0	0	0	0	0	0	0	0	0	0	0	0	0	0	0	0	0	0
1	1	1	1	1	1	**100**	**8**	**15**	**2**	**0**	**4**	10000	18	4	4	0	4
2	2	2	2	2	2	200	16	13	4	0	2	20000	13	8	1	0	2
3	3	3	3	3	3	300	1	11	6	0	0	30000	8	12	5	0	0
4	4	4	4	4	4	400	9	9	1	0	4	40000	3	16	2	0	4
5	5	5	5	0	5	500	17	7	3	0	2	50000	21	3	6	0	2
6	6	6	6	1	0	600	2	5	5	0	0	60000	16	7	3	0	0
7	7	7	0	2	1	700	10	3	0	0	4	70000	11	11	0	0	4
8	**8**	**8**	**1**	**3**	**2**	800	18	1	2	0	2	**80000**	**6**	**15**	**4**	**0**	**2**
9	9	9	2	4	3	900	3	16	4	0	0	90000	1	2	1	0	0

$a_i.10^1$ ↓	23	17	7	5	6	$a_i.10^3$ ↓	23	17	7	5	6
0	0	0	0	0	0	0	0	0	0	0	0
10	10	10	3	0	4	**1000**	**11**	**14**	**6**	**0**	**4**
20	20	3	6	0	2	2000	22	11	5	0	2
30	7	13	2	0	0	3000	10	8	4	0	0
40	**17**	**6**	**5**	**0**	**4**	4000	21	5	3	0	4
50	4	16	1	0	2	5000	9	2	2	0	2
60	14	9	4	0	0	6000	20	16	1	0	0
70	1	2	0	0	4	7000	8	13	0	0	4
80	11	12	3	0	2	8000	19	10	6	0	2
90	21	5	6	0	0	9000	7	7	5	0	0

Example 7.5   Let $\{m_i\} = \{23, 17, 7, 5, 6\}$ and $N = (81148)_{10}$; compute $\{r_i\}$.

$s = 5$ and the range is computed as $23.17.7.5.6 = 82110$. From Table 7.1, the following RNS expressions (boldface lighted in the table) are extracted:

$$80000 = (06, 15, 04, 00, 02)_{RNS}$$
$$1000 = (11, 14, 06, 00, 04)_{RNS}$$
$$100 = (08, 15, 02, 00, 04)_{RNS}$$
$$40 = (17, 06, 05, 00, 04)_{RNS}$$
$$8 = (08, 08, 01, 03, 02)_{RNS}$$

TABLE 7.2   RNS {23, 17, 7, 5, 6}: Precomputed Residues for the First 22 Powers of 2

i	Mod $\rightarrow$	23	17	7	5	6	i	Mod $\rightarrow$	23	17	7	5	6
	$2^i \downarrow$							$2^i \downarrow$					
0	1	1	1	1	1	1	11	2048	1	8	4	3	2
1	2	2	2	2	2	2	12	4096	2	16	1	1	4
2	4	4	4	4	4	4	13	8192	4	15	2	2	2
3	8	8	8	1	3	2	14	16384	8	13	4	4	4
4	16	16	16	2	1	4	15	32768	16	9	1	3	2
5	32	9	15	4	2	2	16	65536	9	1	2	1	4
6	64	18	13	1	4	4	17	131072	18	2	4	2	2
7	128	13	9	2	3	2	18	262144	13	4	1	4	4
8	256	3	1	4	1	4	19	524288	3	8	2	3	2
9	512	6	2	1	2	2	20	1048576	6	16	4	1	4
10	1024	12	4	2	4	4	21	2097152	12	15	1	2	2

The component wise sums are computed as

$$r_1 = (02 + 04 + 04 + 04 + 02) \bmod 6 = 4$$
$$r_2 = (00 + 00 + 00 + 00 + 03) \bmod 5 = 3$$
$$r_3 = (04 + 06 + 02 + 05 + 01) \bmod 7 = 4$$
$$r_4 = (15 + 14 + 15 + 06 + 08) \bmod 17 = 7$$
$$r_5 = (06 + 11 + 08 + 17 + 08) \bmod 23 = 4$$
$$(81148)_{10} = (4, 7, 4, 3, 4)_{\text{RNS}}$$

For $B = 2$, the table is simplified because just one residue expression is needed per power of 2. Table 7.2 shows the residues of powers of 2 in the same RNS used for Table 7.1. The complexity of Table 7.2 is lower than that of Table 7.1. This is paid for by an increased average quantity of power-of-2 residues to be added during the second phase of the conversion process. Observe that both Tables 7.1 and 7.2 are easily reduced thanks to the cyclic feature of the remainder sequences. Those tables are thus readily compacted. It is worth pointing out that a careful choice of moduli can lower the computational complexity for a number of applications ([GAR 1959], [SZA 1967], [SOD1986]).

Example 7.6   Let $\{m_i\} = \{23, 17, 7, 5, 6\}$ and $N = (10011110011111100)_2$; compute $\{r_i\}$.

$s = 5$ and the range is computed as $23.17.7.5.6 = 82110$. From Table 7.2, the following RNS expressions are extracted

$$2^{16} = (09, 01, 02, 01, 04)$$
$$2^{13} = (04, 15, 02, 02, 02)$$
$$2^{12} = (02, 16, 01, 01, 04)$$
$$2^{11} = (01, 08, 04, 03, 02)$$
$$2^{10} = (12, 04, 02, 04, 04)$$
$$2^{7} = (13, 09, 02, 03, 02)$$
$$2^{6} = (18, 13, 01, 04, 04)$$
$$2^{5} = (09, 15, 04, 02, 02)$$
$$2^{4} = (16, 16, 02, 01, 04)$$
$$2^{3} = (08, 08, 01, 03, 02)$$
$$2^{2} = (04, 04, 04, 04, 04)$$

The componentwise sums are computed as

$r_1 = (04 + 02 + 04 + 02 + 04 + 02 + 04 + 02 + 04 + 02 + 04) \bmod 6 = 4$
$r_2 = (01 + 02 + 01 + 03 + 04 + 03 + 04 + 02 + 01 + 03 + 04) \bmod 5 = 3$
$r_3 = (02 + 02 + 01 + 04 + 02 + 02 + 01 + 04 + 02 + 01 + 04) \bmod 7 = 4$
$r_4 = (01 + 15 + 16 + 08 + 04 + 09 + 13 + 15 + 16 + 08 + 04) \bmod 17 = 7$
$r_5 = (09 + 04 + 02 + 01 + 12 + 13 + 18 + 09 + 16 + 08 + 04) \bmod 23 = 4$

$$(10011110011111100)_2 = (4, 7, 4, 3, 4)_{RNS}$$

Algorithm 7.4 converts an n-digit $\{x_i\}$ base-B number N into an s-moduli $\{m_j\}$ RNS representation R. One assumes given a n-input $n.s$-output LUT to store the $B \times n \times s$ possible residues $r_j(i)$ of $x_i.B^i$; $i \in \{0, 1, \ldots, n-1\}, j \in \{1, 2, \ldots, s\}$, $x_i \in \{0, 1, \ldots, B-1\}$. Let LUT$(x(i), i, j)$ be the RNS component $r_j(i)$ of $x_i.B^i$. The RNS expression of N is then computed as the RNS sum $r(1) + r(2) + \cdots + r(s)$.

Algorithm 7.4   Base-B to RNS Conversion

```
for i in 0..n-1 loop
  for j in 1..s loop
  acc (i,j):=LUT(x(i), i, j); end loop;
end loop;
for j in 1..s loop
  R:=0;
  for i in 0..n-1 loop
```

```
  R:=(R+acc(i,j)) mod m(j); end loop;
  R(j):=R;
end loop;
```

7.2.3   RNS to Base-*B* Conversion

Theorem 7.1   Chinese Remainder Theorem (CRT)

The Chinese remainder theorem (7.9), presented in Chapter 2 (Properties 2.4), is the key theorem for some RNS decoding algorithms. It can be stated as follows. Let $\{m_i\}$ be a nonredundant system of s moduli, while $\{r_i\}$ is the RNS representation of an integer N in this system. Then ($|a|_{\text{mod } m}$ stands for a mod m)

$$|N|_M = \left| \Sigma_{1 \leq i \leq s} \ m_i^* |r_i/m_i^*|_{m_i} \right|_{\text{mod } M}, \tag{7.9}$$

where

$$M = \Pi_{1 \leq i \leq s} \ m_i; \quad m_i^* = M/m_i.$$

The following method for RNS to base-*B* conversion is a straightforward application of (7.9). As in the preceding case, some precomputations are assumed, namely, $\{m_i^*\}$ and $\{(1/m_i^*) \bmod m_i\}$. The algorithm consists of computing, in base *B*, s products $[r_i.(1/m_i^*)] \bmod m_i$, and one mod *M* sum of s 2-operand products. Algorithms for mod *m* operations are detailed in Chapter 8. Algorithms 7.5 and 7.6 convert a RNS expression into a base-*B* one. Define the following procedures for computing *M*, $\{m_i^*\}$, and $\{(1/m_i^*) \bmod m_i\}$: M_procedure and star_procedure.

Algorithms 7.5

Algorithm 7.5.1

```
procedure M_procedure (s:in natural; m:in data_vector(1..s);
  M:out natural)
is
  m: data_vector(1..s);
begin
  M:=1;
  for i in 1..s loop M:=M*m(i); end loop;
```

Algorithm 7.5.2

```
procedure star_procedure (s:in natural; M:in data; m:in
  data_vector(1..s); mstar:out data_vector(1..s);
  invmstar:out data_vector(1..s)
is
  m: data_vector(1..s); M: data
begin
  for i in 1..s loop mstar(i):=M/m(i);
  invmstar(i):=(m(i)/M)mod m(i); end loop;
end star_procedure;
```

Algorithm 7.6   CRT RNS to Base-B Conversion

Let (r_i) be the RNS representation of a number N in a system defined by the moduli $\{m_i\}$. After calling M_procedure and star_procedure, the base-B representation of N is computed according to (7.9) as follows:

```
for i in 1..s loop
   acc:=0; acc:=acc+(mstar(i)*((r(i)*invmstar(i)) mod m(i)))
   mod M; end loop;
   N:=acc;
```

Another method to convert the RNS to base-B consists in converting the RNS into an intermediate representation system called a *mixed-radix unsigned digit system* ([HUA1983]) or a *mixed numeration system*. According to Chapter 3, a mixed numeration system is a weighted positional integer representation system that can be defined by

1. a set of n positions or ranks numbered from $n - 1$ to 0 when written from left to right;
2. a set of natural numbers $\{b_{n-1}, \ldots, b_1, b_0\}$, called radices, in one-to-one correspondence with the ranks;
3. a set of natural numbers $\{B_n, B_{n-1}, \ldots, B_1, B_0\}$, called weights, computed as

$$B_0 = 1$$
$$B_1 = b_0.B_0$$
$$\cdots$$
$$B_n = b_{n-1}.B_{n-1} = \Pi_{0 \leq i \leq n-1}\, b_i.$$

It can be shown that any integer X, such that $(B_n - 1) \leq X \leq 0$, has a unique representation

$$x_{n-1}, \ldots, x_1, x_0$$

such that

$$X = B_{n-1}.x_{n-1} + \cdots + B_1.x_1 + B_0.x_0, \quad 0 \leq x_i \leq b_i - 1 \tag{7.10}$$

It is straightforward to note that the range of a mixed-radix system is the same as the one of a RNS system with radix set $\{b_i\}$ as moduli. The RNS to mixed-radix conversion algorithm consists of an iterative extraction of the remainders modulo b_i (substep 1), subtracting this remainder from the RNS expression (substep 2) and dividing the new RNS expression by b_i (substep 3). The sequence of remainders is the desired mixed-radix representation. This procedure is resumed in Algorithm 7.7, where $b(i)$ stands for the modulus i, $r(j)$ stands for the RNS component j of number N, and $mr(k)$ stands for the mixed-radix component k.

Algorithm 7.7   Mixed-Radix `Digit_extraction_step` **Procedure**

```
--Step 1
  for j in 1..s loop
    acc(j):=r(j); end loop;
  --substep1
  mr(0):=acc(1);
  --substep2
  for j in 1..s loop
    r1(j):=(acc(1))mod b(j); acc(j):=(acc(j)-r1(j))mod b(j);
    end loop;
  --substep3
  for j in 2..s loop
    acc(j):=(acc(j)/b(1))mod b(j);
...
--step i
  --substep1
  mr(i-1):=acc(i);
  --substep2
  for j in i..s loop
    ri(j):=(acc(i))mod b(j); acc(j):=(acc(j)-ri(j))mod b(j);
    end loop;
  --substep3
  for j in i+1..s loop
    acc(j):=(acc(j)/b(i))mod b(j);
```

In order to speed-up the division procedure, the inverses of b_i modulo $b_{j \neq i}$ are precomputed; doing so, divisions are replaced by multiplications. The process is explained through the following simple numeric example.

Example 7.7   Let $\{7, 3, 5\}$ be the set of moduli defining a source RNS and $(6, 2, 3)$ be a number N expressed in this system. The corresponding target mixed-radix system is defined by the set of radices $\{7, 3, 5\}$ leading to the set of weights $\{105, 15, 5, 1\}$. The range is thus $[0, 104]$, equal to that of the source RNS. Observe that the moduli (resp. radices) do not need to be ordered by size, but the same order has to be respected for both systems. Strictly needed precomputed inverses are

$$|1/3|_7 = 5; \ |1/5|_3 = 2; \ |1/5|_7 = 3.$$

Substep 1 of step 1 extracts **3** as the remainder modulo 5 of N: the first rightmost mixed-radix digit of N is thus 3; substep 2 consists of subtracting 3, expressed in the RNS as $(3, 0, 3)$, from N

$$(6, 2, 3) - (3, 0, 3) = (3, 2, 0),$$

and substep 3 consists of dividing by 5,

$$(3, 2, 0)/5 = (3.|1/5|_7, 2.|1/5|_3, 0) = (|3.3|_7, |2.2|_3, 0) = (2, 1, 0).$$

The problem is now reduced to the moduli $\{7, 3\}$ and the number $(2, 1)$. The next step 2 is described as follows.

Step 2:

    Substep 1: **1** is the second mixed radix digit as the remainder modulo 3;

    Substep 2: $(2, 1) - (1, 1) = (1, 0)$;

    Substep 3: $(1, 0)/3 = (1.|1/3|_7, 0) = (1.5, 0) = (5, 0)$.

Step 3: Extract **5** as the last mixed-radix digit.

The desired mixed-radix representation is $(5, 1, 3)_{\text{M-R}}$. Using the above computed weights, the base-B expression can readily be computed; for example, in base 10

$$N = 5.15 + 1.5 + 3 = (83)_{10}.$$

Observe that the inverse of the leftmost modulus is not needed while the other inverses are computed modulo the left side moduli only.

Comment 7.2   Garner's algorithm ([GAR1959]), for RNS to base-B conversion (Chapter 2), actually computes the mixed-radix digits within a preliminary step of a procedure computing the base-B digits through a mixed-radix to base-B conversion.

7.3 LOGARITHMIC, EXPONENTIAL, AND TRIGONOMETRIC FUNCTIONS

Most often the computation of functions such as logarithms and exponential or trigonometric functions are made through software-implemented algorithms applied to floating-point representations. Hardware or microprogrammed systems are mainly justified for special-purpose computing devices such as ASIC or embedded systems. As it is generally not possible to get an exact result, approximation methods have to be used together with error estimation techniques. Newton–Raphson, Taylor–MacLaurin series, or polynomial approximations are the most common approaches to compute these functions. For trigonometric functions, CORDIC (linear convergence) algorithms are well suited. Arguments included in the range [1, 2[(floating-point IEEE standard) are suitable for most approximation methods that need to limit the range of the argument. Whenever a specific range is imposed on the operands, a prescaling operation may be necessary: so an initial step may be included in the algorithmic procedure. Crucial questions for approximation methods are error estimation and effective rounding techniques; these problems start from table design (first approximation LUT) up to the final result. Numerous methods, algorithms, and implementations are proposed in the literature ([SPE1965], [CHE1972], [ERC1977], [COD1980], [KOS1991], [MAR1990], [TAN1991], [ERC1995], [PAL2000], [CAO2001]). As for the basic operations, the choice will depend on the speed/cost compromises and other constraints imposed on the designer. Approximation methods usually assume the availability of the four basic operations as arithmetic primitives at hand, together with look-up tables for a first "reasonably good" approximation to start from.

7.3.1 Taylor–MacLaurin Series

The classical *Taylor expansion* (at point a) formula can be written

$$\Sigma_{0 \le i \le n}((x - a)^i / i!).d^i f(x)/dx^i)_{x=a} + R_n. \tag{7.11}$$

When this expansion converges over a certain range of x, that is, when

$$\lim_{n \to \infty} R_n = 0, \tag{7.12}$$

the *expansion* is referred to as the *Taylor series*; if $a = 0$, it is called the *MacLaurin series* (Chapter 2).

Consider the convergence behavior of the exponential functions expressed by the following series at $x = 0$:

$$e^x = \Sigma_{0 \le i \le n}(x^i / i!) + R_n, \tag{7.13}$$

$$e^{-x} = \Sigma_{0 \le i \le n}((-x)^i / i!) + R_n, \tag{7.14}$$

$$a^x = e^{x. \ln (a)} = \Sigma_{0 \le i \le n}((x. \ln (a))^i / i!) + R_n. \tag{7.15}$$

The above expressions can be factorized (Hörner scheme) to suggest a simple iterative computational step; for example, (7.13) for $n = 8$ can be written

$$((\cdots (x/8 + 1).x/7 + 1).x/6 + 1).x/5 + 1).x/4 + 1).x/3 + 1).$$
$$x/2 + 1).x + 1 + R_8. \tag{7.16}$$

Formula (7.16) can be computed in 8 divide-and-add steps with an error inferior to $(1 + 0.1).x^9 / 9! > R_8$, that is, for $0 \le x \le 1/2$, an error ε (maximum for $x = 1/2$) given by

$$\varepsilon_{max} < (1.1).(1/2)^9 / 9! < 6.10^{-9}. \tag{7.17}$$

Actually, (7.16), computed at $x = 1/2$, yields (rounded down) 1.6487212650; ten-decimal precision computation of $e^{1/2}$ yields (rounded up) 1.6487212708, that is, an error inferior to the value computed at (7.17). Obviously, the convergence is faster as x becomes smaller, so a prescaling operation could be necessary if the number of algorithmic steps is fixed. Expression (7.14) can be factorized (Hörner scheme) as

$$((\cdots (x/8 - 1).x/7 + 1).x/6 - 1).x/5 + 1).x/4 - 1).x/3 + 1).$$
$$x/2 - 1).x + 1 + R_8. \tag{7.18}$$

Formula (7.18) is computed in the same way as (7.16), but because of the alternating signs, the absolute value of the error ε_{max} is somewhat smaller, actually $x^9 / 9!$.

Expression (7.15) is quite similar to (7.14) excepted for the prescaling: the range of $x.\ln(a)$ has to be considered instead of just that of x.

Logarithmic function series are also easy to compute, in particular, $\ln(x)$ can be written

$$\ln(x) = \Sigma_{1 \le i \le n}(-)^{i+1}.(x-1)^i/i + R_n, \quad 0 < x \le 2 \tag{7.19}$$

or, for $n = 10$ (Hörner scheme),

$$((\cdots((1-x)/10 + 1/9).(x-1) - 1/8).(x-1)$$
$$+ 1/7).(x-1)\cdots - 1/2).(x-1) + 1).(x-1) + R_8. \tag{7.20}$$

A close analysis of (7.19) shows that in the range $[1, 2]$ the convergence rate can be very slow; typically, for $x = 2$, the error ε_{max} is expressed as

$$|\varepsilon_{max}| < 1/2.(n+1). \tag{7.21}$$

If the range is restricted to $[0, 1/2]$, the maximum error after n steps will be

$$|\varepsilon_{max}| < 1/2^n.(n+1), \tag{7.22}$$

that is, for $x = 1/2$, less than 10^{-4} after 10 steps.

Prescaling the argument of logarithmic functions, together with the corresponding result adjustment, does not involve a significant increase in the overall complexity. Actually, if the argument has been initially divided by k, a correction term $+\ln(k)$ has to be added to adjust the final result. For exponential functions the general prescaling problem is generally not straightforward. The particular case of the function 2^N, with N expressed in floating point, can be processed as follows. Let $m.2^k$ be the floating-point representation of N; m is the mantissa and k the exponent. The scaling process consists of reducing N as

$$Ns = m.2^k - \lfloor m.2^k \rfloor = 0.m_{-(k+1)}m_{-(k+2)}\cdots \tag{7.23}$$

So, assuming $E = \lfloor m.2^k \rfloor$, one can write

$$2^N = 2^{Ns+E} = 2^{Ns}.2^E$$

where 2^{Ns} is the mantissa of 2^N and E the exponent. So the scaling process reduces the computation of 2^N to that of 2^{Ns}.

Trigonometric functions are also readily expressed as series, in particular,

$$\sin(x) = \Sigma_{0 \le i \le n}(-)^i.(x)^{2.i+1}/(2.i+1)! + R_n, \quad -\infty < x \le \infty \tag{7.24}$$

or, for $n = 6$ (Hörner scheme)

$$((\cdots(x^2/156 - 1).x^2/110 + 1).x^2/72 - 1).x^2/42 + 1).x^2/20 - 1)x^2/6$$
$$+ 1).x + R_8. \tag{7.25}$$

To get a fast convergence rate, scaling is also required in this case. Since trigonometric functions are periodic, the scaling can be a straight reduction of the argument to, for example, the range $[-\pi/2, +\pi/2]$, using trigonometric equivalence relations; this can thus be processed without calling for any postcorrection to the final result. If the range $[-\pi/2, +\pi/2]$ holds, the convergence rate is fair: ε_{max} (for $|x| = \pi/2$) can be expressed as

$$|\varepsilon_{max}| < (\pi/2)^{2.n+3}/(2.n + 3)!. \tag{7.26}$$

That is, for $n = 6$, $|\varepsilon_{max}| < 10^{-9}$.

7.3.2 Polynomial Approximation

Basically, polynomial approximation methods consist of building a polynomial $P(x)$ of degree n, that matches the target function at n points. The principle is general for continuous functions. Taylor–MacLaurin series are particular cases. The choice of the matching points is not fixed, neither is the degree of the polynomial nor the quantity of them (e.g., several polynomials may approximate the same function in separate ranges). A number of methods and algorithms have been proposed in the literature. Moreover, prescaling is often recommended to speed-up the convergence. A classical numerical calculus technique is used to compute the coefficients of the approximation polynomial, once the degree and the desired matching points $\{x_k\}$ are selected; it basically consists of resolving $n+1$ equations, linear in $\{c_i\}$, of the form

$$P(x_k) = f(x_k), \tag{7.27}$$

where $P(x_k) = \Sigma c_i . x_k^i$ and $f(x_k)$ are, respectively, the degree-n polynomial and the function to approximate, both computed at x_k. Several alternative methods exist to obtain these coefficients without actually solving (7.27). The polynomial coefficients are computed once, then stored; the way the polynomial is then computed for a given value of the argument defines the algorithm computing the approximated function at this argument. A Hörner-type factorization is a first approach to define an iterative computation step.

Comment 7.3 Hörner Scheme Revisited In Sections 7.3.1 and 7.3.2, the Hörner scheme was mentioned as a possible iterative way to implement algorithms. Whenever pipelined or parallel arithmetic operators are available, this approach could be significantly improved; for instance, if one can take advantage of multiple multiply-and-add and squaring units, a useful generalized Hörner scheme may be defined as follows.
Let

$$P(x) = ((\cdots(x.c_{k-1} + c_{k-2}).x + c_{k-3}).x + \cdots + c_1).x + c_0 \tag{7.28}$$

be the Hörner expansion of the polynomial $P(x)$.
Let

$$C_i^2(x) = x.c_{2i+1} + c_{2i}. \tag{7.29}$$

An extended Hörner expression can be written

$$P(x) = ((\cdots(C_{k-1}^2(x).x^2 + C_{k-2}^2(x)).x^2 + C_{k-3}^2(x)).x^2 + \cdots + C_1^2).x^2 + C_0^2(x).$$
(7.30)

Assuming available operators computing (7.29) and squaring units, the number of steps is roughly divided by two. Expressions (7.29) and (7.30) can be generalized further; assuming

$$C_i^m(x) = x^{m-1}.c_{mi+m-1} + \cdots + x.c_{mi+1} + c_{mi}$$
$$= (\cdots(c_{mi+m-1}.x + c_{mi+m-2}).x + \cdots + c_{mi+1}).x + c_{mi},$$
(7.31)

one can write

$$P(x) = ((\cdots(C_{k-1}^m(x).x^m + C_{k-2}^m(x)).x^m + C_{k-3}^m(x)).x^m$$
$$+ \cdots + C_1^m).x^m + C_0^m(x),$$
(7.32)

called the *generalized Hörner expansion (GHE)*, which takes advantage of higher level primitive polynomial operators. Expansion (7.32) corresponds to a polynomial of degree $k.m - 1$. Therefore, thanks to the availability of parallel and/or pipelined operators, multiple level Hörner computation schemes can be built to speed-up the overall process. For instance, a degree-63 polynomial could be computed in 9 multiply-and-add steps if a 3-level scheme is used. First, 16 degree-3 polynomials can be computed (3 steps); four degree-15 polynomials are then worked out using the degree-3 polynomials as primitives (3 steps), and another 3 steps are finally needed to compute the degree-63 polynomial using the degree-16 ones as primitives. A practical implementation of this circuit is shown in Chapter 14.

7.3.3 Logarithm and Exponential Functions Approximation by Convergence Methods

Convergence methods consist of two parallel processes on two related sequences; typically, one sequence converges to 1 (*multiplicative normalization*) or 0 (*additive normalization*) while the other one converges to the function to approximate. Division using Goldschmidt's algorithm is an example of multiplicative normalization: while the divisor sequence converges to 1, the dividend converges to the desired quotient.

7.3.3.1 Logarithm Function Approximation by Multiplicative Normalization
Define

$$c(i) = 1 + a_i.2^{-i}, \quad a_i \in \{-1, 0, 1\}$$
(7.33)

as the *multiplicative normalizing function*, where a_i is selected in such a way that the sequence

$$x(i+1) = x(i).c(i) \quad \text{(auxiliary sequence)}, \quad x(i) \in B(2^n)$$
(7.34)

converges toward 1. Then the sequence

$$y(i + 1) = y(i) - \ln c(i) \tag{7.35}$$

can be set to converge toward the result $\ln (x)$. If $y(0)$ and $x(0)$ are, respectively, set to 0 and to the argument x, and assuming $x(p) \cong 1$, one can write

$$x(p) = x.\Pi_i c(i) \cong 1 \rightarrow 1/x \cong \Pi_i c(i);$$
$$y(p) = y - \Sigma_i \ln c(i) = -\ln \Pi_i c(i) = \ln (x). \tag{7.36}$$

To make the convergence of (7.34) possible, the argument x needs to be in a range such that

$$x.\min\left(\lim_{p\to\infty} \Pi_{1\le i\le p} c(i)\right) \le 1 \quad \text{and} \quad x.\max\left(\lim_{p\to\infty} \Pi_{1\le i\le p} c(i)\right) \ge 1$$

that is,

$$x \le 1 \Big/ \lim_{p\to\infty} \Pi_{1\le i\le p}(1 - 2^{-i}) \quad \text{and} \quad x \ge 1 \Big/ \lim_{p\to\infty} \Pi_{1\le i\le p}(1 + 2^{-i}),$$

that is, $0.42 \le x \le 3.45$. \tag{7.37}

This means that the argument x could need to be prescaled to fall in the range (7.37). An argument x in the range $[1, 2[$ (such as, e.g., a floating-point mantissa) fits perfectly; otherwise use a straightforward prescaling operation that replaces x by x' such that $x = x'.2^s$ (x' in $[1, 2[$). The algorithm computes $\ln (x')$, then a final additive correction of $s.\ln (2)$ is completed. Observe that the lower bound of (7.37) can be lowered to 0.21, as $(1 + 2^0)$ can be accepted as a first normalizing factor for computing $x(1)$.

In practical implementations of this algorithm, look-up tables are used to read out the successive values of $\ln (1 \pm 2^{-i})$, needed to compute $y(i + 1)$ of (7.35). For x in $[1/2, 2[$, a_i can be selected according to the following rules:

$$a_0 = 0, \tag{7.38}$$
$$\text{if } x(i) > 1, a_i = -x_{-i}(i), \quad i \ge 1, \tag{7.39}$$
$$\text{if } x(i) < 1, a_i = +x_{-i}(i).not(x_{-i-1}(i)), \quad i \ge 1. \tag{7.40}$$

The above rules are justified by the following two lemmas, also showing that the convergence rate reaches precision p after p steps (linear convergence).

Lemma 7.1 Let

$$x(k) = 1 + 2^{-k} + \varepsilon, \quad 0 \le \varepsilon \le 2^{-k} - 2^{-n}, \quad k \le n, \tag{7.41}$$

be the n-bit auxiliary sequence vector at step k; then

$$1 - 2^{-2k} \leq x(k).(1 - 2^{-k}) < 1 + 2^{-k}. \tag{7.42}$$

Proof The left inequality is trivial, it corresponds to $\varepsilon = 0$. The right inequality is deduced from the computation of $x(k).(1 - 2^{-k})$ for ε maximum, $2^{-k} - 2^{-n}$.

The practical interpretation of (7.42) is the impact of rule (7.39) on $x(k+1)$ whenever $x(k)$ is greater than one with a fractional part made up of a $(k - 1)$-zero string and a one at position k. Then $x(k+1)$ will be either greater than one, exhibiting a similar pattern with at least one zero more, or inferior to one $(x_0(k+1) = 0)$ with at least $2k$ ones as the header of the fractional part. In both cases, the target value $x(p) = 1$ is approximated by $x(k+1)$ with at least one bit more.

Lemma 7.2 Let

$$x(k) = 1 - 2^{-k} + \varepsilon, \quad 0 \leq \varepsilon \leq 2^{-k} - 2^{-n}, \quad k \leq n, \tag{7.43}$$

be the n-bit auxiliary sequence vector at step k, then

$$1 - 2^{-2k} \leq x(k).(1 + 2^{-k}) < 1 + 2^{-k}. \tag{7.44}$$

Proof The right inequality is trivial, it corresponds to $\varepsilon = 0$. The left inequality is deduced from the computation of $x(k).(1 + 2^{-k})$ for ε maximum, that is, $2^{-k} - 2^{-n}$.

The practical interpretation of (7.44) is the impact of rule (7.40) on $x(k+1)$ whenever $x(k)$ is less than one with a fractional part made up of a k-one string and a zero at position $k + 1$. Then $x(k+1)$ will be either less than one, exhibiting a similar pattern with at least $2k$ ones as the header of the fractional part, or greater than one $(x_0(k+1) = 1)$ with at least $k + 1$ zeros as the header of the fractional part. In both cases, the target value $x(p) = 1$ is approximated by $x(k+1)$ with at least one bit more.

Comment 7.4

1. The selection (7.38) is justified by the fact that a decision about multiplying by $a_i.2^{-i} + 1$ (7.33) cannot be made before knowing the next bit. Actually, considering bit x_0 only (either 1 or 0) one cannot know whether the sequence $x(i)$ is already 1 (end of convergence process) or not.

2. When $x(i) > 1$, the strategy described by (7.39) consists of detecting the first nonzero bit of $x(i)$ then multiplying by $(-2^{-i} + 1)$. When $x(i) > 1$, Lemma 7.1 shows that, at step i, bits $x_{-k > -i}(i)$ are all zeros.

3. When $x(i) < 1$, the strategy described by (7.40) consists of detecting the last nonzero bit of $x(i)$ then multiplying by $(2^{-i} + 1)$. When $x(i) \leq 1$, Lemma 7.2 shows that, at step i, bits $x_{-k > -i}(i)$ are all ones.

Algorithm 7.8 Logarithm Computation by Multiplicative Normalization

The argument x is in $[1/2, 2[$: $x = x(0).x(1)\ x(2) \cdots x(n)$. Let $xx(i, j)$ be the component j of $xx(i) = xx(i,\ 0).xx(i,\ 1)\ xx(i,\ 2) \cdots xx(i,\ n)$. Let $lut(i) = \ln(1 + a(i).2^{-i})$ read from the table.

```
a(0):=0; c(0):=1; xx(1):=x; yy(1):=0;
for i in 1..p-1 loop
  if xx(i)=1 then exit; end if;
  if xx(i)>1 then a(i):=-xx(i,i) else
  a(i):=xx(i,i)*not(xx(i,i+1)); end if;
  c(i):=1+a(i)*2**(-i); xx(i+1):=xx(i)*c(i);
  yy(i+1):=yy(i)-lut(i);
end loop;
```

Example 7.8   In this example the auxiliary sequence $x(i)$ is computed in the binary system, while, for readability, the sequence $y(i)$ is computed in decimal; the precision is then readily verified. The functional values $\ln(1 \pm 2^{-i})$ are assumed given by look-up tables. x is in $[1, 2[$.

Let

$$x = x(0) = x_0.x_{-1}x_{-2}x_{-3}x_{-4}x_{-5} = 1.10111 = (1.71875)_{10}$$
$$y(0) = 0$$

Compute $\ln(x)$ with precision $p = 8$.

i	a_i	$c(i)$ $a_i.2^{-i} + 1$	$x(i+1)$ $x(i).c(i)$	$\ln c(i)$	$y(i+1)$ $y(i) - \ln c(i)$
—	—	—	$x(0) = 1.10111$	—	$y(0) = 0$
0	$a_0 = 0$	$0.2^{-0} + 1$ $c(0) = 1$	$(1.10111).(1)$ $x(1) = 1.1011100$	0	0
1	$a_1 = -1$	$-2^{-1} + 1$ $c(1) = 0.1$	$(1.1011100).(0.1)$ $x(2) = 0.11011100$	-0.69314718	0.69314718
2	$a_2 = 1$	$2^{-2} + 1$ $c(2) = 1.01$	$(0.11011100).(1.01)$ $x(3) = 1.00010011$	0.223143551	0.470003628
3	$a_3 = 0$	$0.2^{-3} + 1$ $c(3) = 1$	$(1.00010011).1$ $x(4) = 1.00010011$	0	0.470003628
4	$a_4 = -1$	$-2^{-4} + 1$ $c(4) = 0.1111$	$(1.00010011).(0.1111)$ $x(5) = 1.00000010$	-0.064538521	0.534542149
5	$a_5 = 0$	$0.2^{-5} + 1$ $c(5) = 1$	$(1.00000010).1$ $x(6) = 1.00000010$	0	0.534542149
6	$a_6 = 0$	$0.2^{-6} + 1$ $c(6) = 1$	$(1.00000010).1$ $x(7) = 1.00000010$	0	0.534542149
7	$a_7 = -1$	$-2^{-7} + 1$ $c(7) = 0.1111111$	$(1.00000010).(0.1111111)$ $x(8) = 1$ (rounded up)	-0.007843177	0.542385326

The actual decimal value of ln (1.71875) is 0.541597282 ± 10^{-9}, the difference from the computed result is less than $8.10^{-4} < 2^{-10}$.

As it appears in the preceding example, whenever $a_i = 0$, the only effect of step i on the computation process consists of incrementing the step number; both sequences $x(i)$ and $y(i)$ remain unchanged. So, by detecting strings of 0 or 1 in $x(i)$, one could readily jump to the next nontrivial computation step. The following example illustrates this feature.

Example 7.9   As in the preceding example the auxiliary sequence $x(i)$ is computed in the binary system, while sequence $y(i)$ is computed in decimal. The functional values ln (1 ± 2^{-i}) are given by look-up tables. x is now in $[1/2, 2[$. Strings **00** .. or **11** .. are highlighted by bold type.

Let

$$x = x(0) = x_0.x_{-1}x_{-2}x_{-3}x_{-4}x_{-5} = 0.10011 = (0.59375)_{10}$$
$$y(0) = 0$$

Compute ln (x) with precision $p = 10$ (see Table 7.3).

The actual decimal value of ln (0.59375) is -0.521296923 ± 10^{-9}, the difference from the computed result is less than $4.10^{-6} < 2^{-10}$.

7.3.3.2   Exponential Function Approximation by Additive Normalization

Define

$$\ln c(i) = \ln (1 + a_i.2^{-i}), \quad a_i \in \{-1, 0, 1\} \tag{7.45}$$

as the *additive normalizing function*, where a_i is selected in such a way that the sequences

$$x(i+1) = x(i) - \ln c(i) \quad \text{(auxiliary sequence)} \tag{7.46}$$

converges toward 0. Then the sequence (*main sequence*)

$$y(i+1) = y(i).c(i) \tag{7.47}$$

can be set to converge toward the result e^x. If $y(0)$ and $x(0)$ are, respectively, set to 1 and to the argument x, and assuming $x(p) \cong 0$, one can write

$$x(p) = x - \Sigma_i \ln c(i) \cong 0 \rightarrow x \cong \Sigma_i \ln c(i) = \ln \Pi_i c(i);$$

$$y(p) = y.\Pi_i c(i) = \Pi_i c(i) = e^{\ln (\Pi c(i))} = e^x.$$

To make the convergence of (7.46) possible, the argument x needs to be in a range such that

$$x - \min\left(\lim_{p \to \infty} \Sigma_{1 \le i \le p} \ln c(i) \right) \ge 0 \quad \text{and} \quad x - \max\left(\lim_{p \to \infty} \Sigma_{1 \le i \le p} \ln c(i) \right) \le 0,$$

that is,

$$x \ge \left(\lim_{p \to \infty} \Sigma_{1 \le i \le p} \ln (1 - 2^{-i}) \right) \quad \text{and} \quad x \le \left(\lim_{p \to \infty} \Sigma_{1 \le i \le p} \ln (1 + 2^{-i}) \right) \le 0, \tag{7.48}$$

TABLE 7.3

i	a_i	$\dfrac{c(i)}{a_i.2^{-i}+1}$	$\dfrac{x(i+1)}{x(i).c(i)}$	$\ln c\,(i)$	$\dfrac{y(i+1)}{y(i)-\ln c\,(i)}$
—	—	—	$x(0)=0.1001100000$	—	$y(0)=0$
0	$a_0=0$	$0.2^{-0}+1$ $c(0)=1$	$(0.10011).(1)$ $x(1)=0.1001100000$	0	0
1	$a_1=1$	$2^{-1}+1$ $c(1)=1.1$	$(0.1001100000).(1.1)$ $x(2)=0.1110010000$	0.405465108	−0.405465108
2	$a_2=0$	—	$x(3)=x(2)$	—	—
3	$a_3=1$	$1.2^{-3}+1$ $c(3)=1.001$	$(0.111001).(1.001)$ $x(4)=1.0000000010$	0.117783035	−0.523248143
$4\to8$	$\begin{array}{c}a_4\to a_8\\=0\end{array}$	—	$x(9)=x(4)$	—	—
9	$a_9=-1$	$-2^{-9}+1$ $c(9)=0.111111111$	$(1.0000000010).(0.111111111)$ $x(10)=1$ (rounded up)	−0.001955035	−0.521293108

189

or

$$-1.242 \le x \le 0.869. \tag{7.49}$$

Observe that the upper bound of (7.48) can be raised to 1.562, as $\ln(1 + 2^0)$ can be accepted as the first normalizing factor for computing $x(1)$. This extends the range (7.49) to $]-1.242, 1.562]$. Here again, the argument x could need to be prescaled to fall in the (extended) range (7.49). The range $[1, 2[$ (floating-point mantissa) doesn't fit any more; the range $[-1, 1[$ does. A possible prescaling operation can replace x by x' such that $x = x'.2^s$ (x' in $[-1, 1[$). The algorithm first computes $e^{x'}$, then e^x is provided after an s-time squaring (square rooting if $s < 0$) correction on $e^{x'}$. In particular, a single shift reduces the range $[1, 2[$ to $[1/2, 1[$: $x = x'.2 \to x'$ in $[1/2, 1[$; squaring $e^{x'}$ provides e^x.

In the practical implementations of this algorithm, look-up tables are used to read out the successive values of $\ln(1 \pm 2^{-i})$, needed to compute $x(i + 1)$ of (7.46). For x in $[-1, 1[$, a_i can be selected according to the following rule:

whenever nonzero, a_i holds the same sign as $x(i)$, then

$$|a_i| = 1 \text{ if } |\ln(1 + a_i.2^{-i})| \le 2.|x(i)|, \quad |a_i| = 0 \text{ otherwise.} \tag{7.50}$$

This simple rules ensures that, at each step, $|x(i)|$ will either decrease or stay unchanged; $a_i = 0$ prevents $x(i)$ from increasing. Precision p is reached after p steps.

An interesting feature of the method comes from the following relation:

$$\lim_{i \to \infty} \ln(1 + 2^{-i}) = 2^{-i}. \tag{7.51}$$

2^{-i} is the first term of the Taylor–MacLaurin series (7.19). The approximation $\ln(1 + 2^{-i}) \cong 2^{-i}$ produces an error bounded by (2^{-2i}), that is, less than 2^{-20} after 10 steps. This very fast convergence behavior (see Table 7.3) can be used to speed-up the algorithm. If the desired precision p is great enough, the algorithmic procedure can be stopped after $p/2$ iteration steps; at this point one can write

$$x(p/2) = 0.00 \cdots 0 x_{p/2+1} x_{p/2+2} \cdots x_n = \Sigma_{p/2+1 \le i \le n} x_i.2^{-i},$$

$$x(p) \cong x(p/2) - \Sigma_{p/2+1 \le i \le p} x_i.2^{-i} \cong x(p/2) - \Sigma_{p/2+1 \le i \le p} \ln(1 + x_i.2^{-i}) \cong 0, \tag{7.52}$$

corresponding to the selection $a_i = x_i, \forall i \ge p/2$.

Actually (7.52), the convergence of $x(p)$ to zero, is settled by subtracting $\Sigma_{p/2+1 \le i \le p} x_i.2^{-i}$ from $x(p/2)$; for $p/2$ great enough $x_i.2^{-i}$ can be approximated by $\ln(1 + x_i.2^{-i})$, $(i \ge p/2)$.

On the other hand, for p great enough (2^{-i} small enough),

$$\Pi_{p/2+1 \le i \le p}(1 + x_i.2^{-i}) \cong (1 + \Sigma_{p/2+1 \le i \le p} x_i.2^{-i}) \cong (1 + x(p/2)). \tag{7.53}$$

As a consequence, the final $p/2$ iteration steps can be replaced by

$$y(p) = y(p/2).(1 + x(p/2)). \tag{7.54}$$

The auxiliary sequence has been built up to $x(p/2)$, then settled to $x(p) = 0$ by (7.52); the main sequence $y(p)$ can be computed from $y(p/2)$ in just one computation step (7.54). Algorithm 7.9 resumes the general procedure. Chapter 14 describes an implementation including the above-mentioned one-step simplification for the last $p/2$ steps.

Algorithm 7.9 Exponential Computation by Additive Normalization

The argument x is in $[-1, 1[$. Let $lutn(i) = (1 - 2^{-i})$, $lutp(i) = (1 + 2^{-i})$, $lutlnn(i) = \ln (1\text{-}2^{-i})$ and $lutlnp(i) = \ln (1 + 2^{-i})$ be read from look-up tables.

```
xx(0):=x; yy(0):=1;
for i in 0..p-1 loop
  if xx(i)=0 then exit, end if;
  if xx(i)<0 then
    if abs(lutlnn(i))<=abs(2*xx(i)) then
    xx(i+1):=xx(i)-lutlnn(i); yy(i+1):=yy(i)*lutn(i);
    else xx(i+1):=xx(i); yy(i+1):=yy(i); end if;
  else
    if abs(lutlnp(i))<=abs(2*xx(i)) then
    xx(i+1):=xx(i)-lutlnp(i); yy(i+1):=yy(i)*lutp(i);
    else xx(i+1):=xx(i); yy(i+1):=yy(i); end if;
  end if;
end loop;
```

Example 7.10 For the sake of readability, the following example is treated in decimal. Table 7.4 exhibits the values of $\ln (1 + 2^{-i})$ and $\ln (1 - 2^{-i})$ for $i = 0, 1, \ldots, 15$. This table is used to compute the successive values of x (i).

The problem at hand is to compute $e^{0.65625}$ with precision 2^{-32}. The first 16 steps are displayed in Table 7.5. The desired precision is reached through a final single step computing (7.54) for $p = 16$. The result is 1.92755045011, while the exact result rounded to the 11th decimal is $e^{0.65625} = 1.92755045016$. This makes a difference around $5.10^{-11} < 2^{-32}$.

Comment 7.5 The general exponentiation function x^y may be computed as

$$x^y = e^{y \ln (x)}, \tag{7.55}$$

which may be calculated by logarithm computation and multiplication. The particular cases of integer powers (squaring, cubing, etc.) may be calculated through customized multiplication schemes. Convergence methods, such as the one described above, are most often used to implement exponentiation algorithms. An important particular case is the square root (power $\frac{1}{2}$) for which a classical digit recurrence algorithm, very similar to division, can easily be adapted. Square rooting is reviewed in Section 7.4.

TABLE 7.4 Look-Up Table $\ln(1 \pm 2^{-i})$

i	$1 + 2^{-i}$	$1 - 2^{-i}$	$lutp(i) = ln(1 + 2^{-i})$	$lutn(i) = ln(1 - 2^{-i})$
0	2	0	0.6931471805599945	—
1	1.5	0.5	0.405465108108164	−0.6931471805599945
2	1.25	0.75	0.2231435513142100	−0.287682072451781
3	1.125	0.875	0.117783035656383	−0.133531392624523
4	1.0625	0.9375	0.060624621816435	−0.064538521137571
5	1.03125	0.96875	0.030771658666754	−0.031748698314580
6	1.015625	0.984375	0.015504186535965	−0.015748356968139
7	1.0078125	0.9921875	0.007782140442055	−0.007843177461026
8	1.00390625	0.99609375	0.003898640415657	−0.003913899321136
9	1.001953125	0.998046875	0.001951220131262	−0.001955034835803
10	1.0009765625	0.9990234375	0.000976085973055	−0.000977039647827
11	1.00048828125	0.99951171875	0.000488162079501	−0.000488400498109
12	1.000244140625	0.999755859375	0.000244110827527	−0.000244170432174
13	1.0001220703125	0.9998779296875	0.000122062862526	−0.000122077763687
14	1.00006103515625	0.99993896484375	0.000061033293681	−0.000061037018971
15	1.000030517578125	0.999969482421875	0.000030517112473	−0.000030518043796

TABLE 7.5 $e^{0.65625}$

i	$\ln(1+2^{-i})\ (a^i=1)$	$\ln(1-2^{-i})\ (a^i=-1)$	a_i	$x(i+1)=x(i)-\ln(1+a_i.2^{-i})$	$y(i+1)$
				$x(0)=0.6562500000$	$y(0)=1$
0	**0.693147180559945**	—	1	-0.036897180559945	$y(1)=y(0).(1+2^{-0})$
1	0.405465108108164	-0.693147180559945	0	-0.036897180559945	$y(2)=y(1)$
2	0.223143551314210	-0.287682072451781	0	-0.036897180559945	$y(3)=y(2)$
3	0.117783035656383	-0.133531392624523	0	-0.036897180559945	$y(4)=y(3)$
4	0.060624621816435	**-0.064538521137571**	-1	0.027641340577626	$y(5)=y(4).(1-2^{-4})$
5	**0.030771658666754**	-0.031748698314580	1	-0.003130318089128	$y(6)=y(5).(1+2^{-5})$
6	0.015504186535965	-0.015504835968139	0	-0.003130318089128	$y(7)=y(6)$
7	0.007782140442055	-0.007843177461026	0	-0.003130318089128	$y(8)=y(7)$
8	0.003898640415657	**-0.003913899321136**	-1	0.000783858123208	$y(9)=y(8).(1-2^{-8})$
9	0.001951220131262	-0.001955034835803	0	0.000783858123208	$y(10)=y(9)$
10	**0.000976085973055**	-0.000977039647827	1	-0.000192504741047	$y(11)=y(10).(1+2^{-10})$
11	0.000488162079501	-0.000488400498109	0	-0.000192504741047	$y(12)=y(11)$
12	0.000244110827527	**-0.000244170432174**	-1	0.000051665691127	$y(13)=y(12).(1-2^{-12})$
13	0.000122062862526	-0.000122077763687	0	0.000051665691127	$y(14)=y(13)$
14	**0.000061033293681**	-0.000061037018971	1	-0.000009367602554	$y(15)=y(14).(1+2^{-14})$
15	0.000030517112473	-0.000030518043796	0	-0.000009367602554	$y(16)=y(15)=1.9275685068$

$y(32)=y(16).(1+x(16))=1.9275685068 \times 0.9999906324 = 1.92755045011$

Exact result rounded to the 11th decimal: $e^{0.65625}=1.92755045016$

193

7.3.4   Trigonometric Functions—CORDIC Algorithms

CORDIC algorithms ([VOL1959], [ERC1987], [DEL1989], [VOL2000]) belong to
the class of linear convergence methods. They are particularly well suited to evaluation of trigonometric functions, although they apply to a number of other functions
such as the logarithm, exponential, or square root. Historically, the Coordinate
Rotation Digital Computer had been designed as a special-purpose digital computer
for airborne computation in real-time. The CORDIC techniques were then extended
for solving trigonometric relationships involved in plane coordinate rotation and
conversion from Cartesian to polar coordinate systems. The CORDIC algorithms
converge linearly (one bit per cycle) and are rather simple to implement in practice.

Conceptually, the method consists of computing the new coordinates of a given
vector after rotation by a given angle α; after rotation, the unit vector $(0, 0) - (1, 0)$
would get its new end point coordinates as $(\cos \alpha, \sin \alpha)$.

Let $(0, 0) - (x_1, y_1)$ be the vector $(0, 0) - (x_0, y_0)$ rotated by an angle α
(Figure 7.1), the basic rotation formulas are given by

$$x_1 = x_0 \cos \alpha - y_0 \sin \alpha \tag{7.56}$$

$$y_1 = x_0 \sin \alpha + y_0 \cos \alpha \tag{7.57}$$

The CORDIC algorithm defines, as the auxiliary sequence, the residual rotation
remaining to be completed after step i, namely,

$$a_{i+1} = a_i - \alpha_i. \tag{7.58}$$

If a_0 is the desired rotation angle, the auxiliary sequence will be set to converge
to zero while the main sequence converges toward the function to evaluate.
The method replaces rotations as defined by formulas (7.56) and (7.57) by
pseudorotations as illustrated in Figure 7.2. Pseudorotations expand coordinates
by a factor of $1/\cos \alpha_i = (1 + \tan^2 \alpha_i)^{1/2}$; (7.56) and (7.57) for step i are replaced by

$$x_{i+1} = x_i - y_i \tan \alpha_i = (x_i \cos \alpha_i - y_i \sin \alpha_i).(1 + \tan^2 \alpha_i)^{1/2} \tag{7.59}$$

and

$$y_{i+1} = x_i \tan \alpha_i + y_i = (x_i \sin \alpha_i + y_i \cos \alpha_i).(1 + \tan^2 \alpha_i)^{1/2}. \tag{7.60}$$

Assuming x_0 and y_0 to be the initial coordinates x and y, after p rotations by angle
α_i, the recursive use of (7.59) and (7.60) leads to

$$x_p = (x \cos (\Sigma \alpha_i) - y \sin (\Sigma \alpha_i)).\Pi(1 + \tan^2 \alpha_i)^{1/2} \tag{7.61}$$

and

$$y_p = (x \sin (\Sigma \alpha_i) + y \cos (\Sigma \alpha_i)).\Pi(1 + \tan^2 \alpha_i)^{1/2}. \tag{7.62}$$

On the other hand, assuming a to be the overall rotation angle, that is, the argument
of the trigonometric function to be evaluated, then the auxiliary sequence is written

$$a_p = a - \Sigma \alpha_i. \tag{7.63}$$

The set $\{\alpha_i\}$ is selected in order to make (7.61) and (7.62) easy to calculate and, in particular, to be able to precompute the factor $\Pi(1 + \tan^2 \alpha_i)^{1/2}$. For this sake, $\{\alpha_i\}$ is pre-set except for the sign; then, convergence will be achieved by a suitable strategy on sign selection. In the following, an algorithm is developed for computing sine and cosine functions.

Let the set $\{\alpha_i\}$ be such that

$$\tan \alpha_i = s_i 2^{-i}, \quad s \in \{-1, 1\}. \tag{7.64}$$

Equations (7.59) and (7.60) can then be written

$$x_{i+1} = x_i - y_i.s_i 2^{-i} \tag{7.65}$$

and

$$y_{i+1} = x_i.s_i 2^{-i} + y_i, \tag{7.66}$$

while (7.58) becomes

$$a_{i+1} = a_i - \tan^{-1}(s_i 2^{-i}). \tag{7.67}$$

After p iterations, a_p is assumed close enough to zero, and (7.63) $a = \Sigma \alpha_i$; y_p and x_p, computed iteratively by (7.65) and (7.66), are introduced in (7.61) and (7.62), respectively, leading to

$$x_p = k.(x \cos a - y \sin a) \tag{7.68}$$

and

$$y_p = k.(x \sin a + y \cos a), \tag{7.69}$$

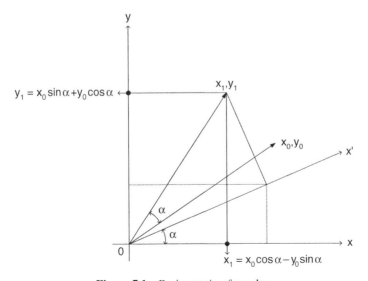

Figure 7.1 Basic rotation formulas.

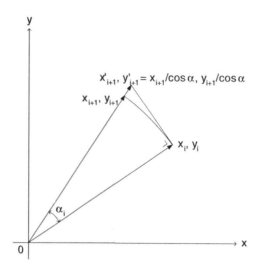

Figure 7.2 Pseudorotations.

where the factor $k = \Pi(1 + \tan^2\alpha_i)^{1/2}$ can be precomputed since $\tan^2\alpha_i$ is not dependent on the sign of α_i. Then, setting the initial values $x_0 = x = 1/k$ and $y_0 = y = 0$, the cosine and sine functions are given by x_p and y_p, respectively.

Precomputed values of $\tan^{-1}(s_i\, 2^{-i})$ are stored in a look-up table. The size of the look-up table sets the precision p. Table 7.6 shows the first 25 values of $\tan^{-1}(s_i\, 2^{-i})$ together with the fast converging values of $k = \Pi(1 + \tan^2 \alpha_i)^{1/2}$: for $i \geq 20$, $k = 1.646760258121$ and $x_0 = x = 1/k = 0.607252935009$. The strategy for selecting the sign s_i is trivial. Since each angle α_i (Table 7.6) is smaller than half the preceding angle α_{i-1}, the convergence may be achieved by selecting $s_i = \text{sign}$ (a_{i-1}). The domain of convergence is greater than $[-90°, 90°]$, so any trigonometric value can be computed (with a prospective use of trigonometric relations for angle reduction). More precisely, for 25-step precision, the range is $[-99.88°, 99.88°]$, $\pm$ the sum of the α_i listed in Table 7.6. Algorithm 7.10 describes the CORDIC procedure for computing $\sin a$ and $\cos a$. Without loss of generality, the decimal system has been used for a better readability.

Algorithm 7.10 Sine and Cosine CORDIC Procedure

Let *lut(i)* be $\alpha_i = \tan^{-1}(2^{-i})$ read from the table.

```
x(0):=0.607252935009; y(0):=0; a(0):=a;
if a(0)=0 then x(0):=1; y(0):=0; exit; end if;
for i in 0..p-1 loop
  if a(i)>0 then s(i):=1 else s(i):=-1; end if;
  x(i+1):=x(i)-(y(i)*s(i)*2**(-i));
  y(i+1):=y(i)+(x(i)*s(i)*2**(-i));
  a(i+1):=a(i)-(s(i)*lut(i));
  end loop;
```

TABLE 7.6 Look-Up Table for $\tan^{-1}(2^{-i})$

i	2^{-i}	$\alpha_i = \tan^{-1}(2^{-i})$ (rad)	$\alpha_i = \tan^{-1}(2^{-i})$ (deg)	$(1 + \tan^2 \alpha_i)^{1/2}$	$k = \Pi(1 + \tan^2 \alpha_i)^{1/2}$
0	1.0000000000	0.7853981634	45.0000000000	1.41421356	1.4142135623731
1	0.5000000000	0.4636476090	26.5650511771	1.11803399	1.5811388300842
2	0.2500000000	0.2449786631	14.0362434679	1.03077641	1.6298006013007
3	0.1250000000	0.1243549945	7.1250163489	1.00778222	1.6424840657522
4	0.0625000000	0.0624188100	3.5763343750	1.00195122	1.6456889157573
5	0.0312500000	0.0312398334	1.7899106082	1.00048816	1.6464922787125
6	0.0156250000	0.0156237286	0.8951737102	1.00012206	1.6466932542736
7	0.0078125000	0.0078123411	0.4476141709	1.00003052	1.6467435065969
8	0.0039062500	0.0039062301	0.2238105004	1.00000763	1.6467560702049
9	0.0019531250	0.0019531225	0.1119056771	1.00000191	1.6467592111398
10	0.0009765625	0.0009765622	0.0559528919	1.00000048	1.6467599963756
11	0.0004882813	0.0004882812	0.0279764526	1.00000012	1.6467601926847
12	0.0002441406	0.0002441406	0.0139882271	1.00000003	1.6467602417620
13	0.0001220703	0.0001220703	0.0069941137	1.00000001	1.6467602540313
14	0.0000610352	0.0000610352	0.0034970569	1.00000000	1.6467602570986
15	0.0000305176	0.0000305176	0.0017485284	1.00000000	1.6467602578655
16	0.0000152588	0.0000152588	0.0008742642	1.00000000	1.6467602580572
17	0.0000076294	0.0000076294	0.0004371321	1.00000000	1.6467602581051
18	0.0000038147	0.0000038147	0.0002185661	1.00000000	1.6467602581171
19	0.0000019073	0.0000019073	0.0001092830	1.00000000	1.6467602581201
20	0.0000009537	0.0000009537	0.0000546415	1.00000000	1.6467602581208
21	0.0000004768	0.0000004768	0.0000273208	1.00000000	1.6467602581210
22	0.0000002384	0.0000002384	0.0000136604	1.00000000	1.6467602581211
23	0.0000001192	0.0000001192	0.0000068302	1.00000000	1.6467602581211
24	0.0000000596	0.0000000596	0.0000034151	1.00000000	1.6467602581211
25	0.0000000298	0.0000000298	0.0000017075	1.00000000	1.6467602581211

In binary, the operations involved in one step of Algorithm 7.10 are: two i-bit shifts, one table look-up, and three signed additions. Computation of $x(i+1)$, $y(i+1)$ and $a(i+1)$ are independent; hardware implementations can thus profit from the potential parallelism. In this case, the elementary step delay is that of a table look-up (or a shift—whatever operation is slower) plus a signed addition. Algorithm 7.10 is illustrated in Example 7.11.

Example 7.11 Let $a = 75°$; compute the binary expression of $\cos a$ and $\sin a$ Table 7.6 displays the first steps of the algorithm. Since the binary system is at hand, the initial value of x, namely $x(0) = 0.607252935009_{10}$, has to be expressed in binary, with the appropriate precision. In the example, 13 bits have been used $(x(0) = 0.1001101101110)$ to set the procedure for a 12-bit precision. The results after 5 steps (Table 7.7) are $\cos 75° = x(5) = 0.0100001011011$ and $\sin 75° = y(5) = 0.1111011011111$. The first seven bits are correct but this precision is irrelevant because too few steps have been completed. The average convergence

TABLE 7.7   cos 75° sin 75°: First Five Computational Steps

				$x(0) =$ 0.1001101101110	$y(0) = 0$	$a = 75°$
i	2^{-i}	$lut(i) = \tan^{-1} 2^{-i}$ (deg)	$s(i)$	$x(i+1) =$ $x(i) - y(i).s(i).2^{-i}$	$y(i+1) =$ $y(i) + x(i).s(i).2^{-i}$	$a(i+1) =$ $a(i) - s(i).lut(i)$
0	1.0000000000	45.000000000	1	0.1001101101110	0.1001101101110	30.000000000
1	0.1000000000	26.565051177	1	0.0100110110111	0.1110100100110	3.434948823
2	0.0100000000	14.036243468	1	0.0001001101101	0.1111110010011	−10.601294645
3	0.0010000000	7.125016349	−1	0.0011001100110	0.1111101000110	−3.476278296
4	0.0001000000	3.576334375	−1	0.0100001011011	0.1111011011111	0.100056079

rate (1 bit per step) is reached after a sufficient number of steps, namely, whenever $\tan^{-1} 2^{-i}$ is close enough to 2^{-i}; Table 7.6 shows that for $i = 10$, $\tan^{-1} 2^{-i}$ may be set to 2^i with an error inferior to 4.10^{-7}.

The next decimal example runs 28 computational steps (Table 7.8). In this example, the results are $\cos 30° = x(28) = 0.8660254031$ and $\sin 30° = y(28) = 0.5000000012$. In both cases, the error is $O(10^{-9})$, that is, $O(2^{-28})$.

7.4   SQUARE ROOTING

Square rooting deserves special attention because of its frequent use in a number of applications ([ERC1994], [OBE1999]). Although square rooting can be viewed as a particular case of the exponential operation, the similarity with division is a more important consideration for the choice of algorithms. Several techniques in base-B and in binary systems are reviewed in this section.

7.4.1   Digit Recurrence Algorithm—Base-B Integers

Let

$$X = x_{2n-1}, x_{2n-2}, \ldots, x_1, x_0, \quad x_i \in \{0, 1, \ldots, B - 1\} \qquad (7.70)$$

be the $2n$-digit base-B radicand.

The square root Q and the remainder R are denoted

$$Q = q_{n-1}, q_{n-2}, \ldots, q_1, q_0 \quad q_i \in \{0, 1, \ldots, B - 1\} \qquad (7.71)$$

and

$$R = r_n, r_{n-1}, \ldots, r_1, r_0 \quad r_i \in \{0, 1, \ldots, B - 1\} \qquad (7.72)$$

respectively.

The remainder

$$R = X - Q^2 \qquad (7.73)$$

TABLE 7.8 cos 30°, sind 30°

i	2^{-i}	$lut(i) = \tan^{-1}2^{-i}$ (deg)	$s(i)$	$x(0) = 0.607252935$ $x(i+1) =$ $x(i) - y(i).s(i).2^{-i}$	$y(0) = 0$ $y(i+1) =$ $y(i) + x(i).s(i).2^{-i}$	$a = 30°$ $a(i+1) =$ $a(i) - s(i).lut(i)$
0	1.0000000000	45.0000000000	1	0.6072529350	0.6072529350	-15
1	0.5000000000	26.5650511771	-1	0.9108794025	0.3036264675	11.56505118
2	0.2500000000	14.0362434679	1	0.8349727856	0.5313463181	-2.471192291
3	0.1250000000	7.12501634890	1	0.9013910754	0.4269747199	4.653824058
4	0.0625000000	3.57633437500	1	0.8747051554	0.4833116621	1.077489683
5	0.0312500000	1.78991060825	1	0.8596016660	0.5106461982	$-7.124209252 \cdot 10^{-1}$
6	0.0156250000	$8.9517371021 \cdot 10^{-1}$	-1	0.8675805128	0.4972149222	$1.827527850 \cdot 10^{-1}$
7	0.0078125000	$4.4761417086 \cdot 10^{-1}$	1	0.8636960212	0.5039928950	$-2.648613858 \cdot 10^{-1}$
8	0.0039062500	$2.2381050037 \cdot 10^{-1}$	-1	0.8656647435	0.5006190824	$-4.105088547 \cdot 10^{-2}$
9	0.0019531250	$1.1190567707 \cdot 10^{-1}$	-1	0.8666425151	0.4989283309	$7.085479160 \cdot 10^{-2}$
10	0.0009765625	$5.5952891894 \cdot 10^{-2}$	-1	0.8661552804	0.4997746615	$1.490189970 \cdot 10^{-2}$
11	0.00048828125	$2.7976452617 \cdot 10^{-2}$	1	0.8659112498	0.5001975889	$-1.307455292 \cdot 10^{-2}$
12	0.000244140625	$1.3988227142 \cdot 10^{-2}$	-1	0.8660333684	0.4999861848	$9.136742266 \cdot 10^{-4}$
13	$1.220703125 \cdot 10^{-4}$	$6.9941136754 \cdot 10^{-3}$	-1	0.8659723349	0.5000919018	$-6.080439449 \cdot 10^{-3}$
14	$6.103515625 \cdot 10^{-5}$	$3.4970568507 \cdot 10^{-3}$	1	0.8660028581	0.5000390470	$-2.583382598 \cdot 10^{-3}$
15	$3.0517578125 \cdot 10^{-5}$	$1.7485284270 \cdot 10^{-3}$	-1	0.8660181181	0.5000126187	$-8.348541711 \cdot 10^{-4}$
16	$1.5258789063 \cdot 10^{-5}$	$8.7426421369 \cdot 10^{-4}$	-1	0.8660257477	0.4999994043	$3.941004264 \cdot 10^{-5}$
17	$7.6293945313 \cdot 10^{-6}$	$4.3713210687 \cdot 10^{-4}$	1	0.8660219330	0.5000060116	$-3.977220642 \cdot 10^{-4}$
18	$3.8146972656 \cdot 10^{-6}$	$2.1856605344 \cdot 10^{-4}$	1	0.8660238403	0.5000027079	$-1.791560108 \cdot 10^{-4}$
19	$1.9073486328 \cdot 10^{-6}$	$1.0928302672 \cdot 10^{-4}$	-1	0.8660247940	0.5000010561	$-6.987298407 \cdot 10^{-5}$
20	$9.5367431641 \cdot 10^{-7}$	$5.4641513360 \cdot 10^{-5}$	1	0.8660252709	0.5000002302	$-1.523147071 \cdot 10^{-5}$
21	$4.7683715820 \cdot 10^{-7}$	$2.7320756680 \cdot 10^{-5}$	1	0.8660255093	0.4999998173	$1.208928597 \cdot 10^{-5}$
22	$2.3841857910 \cdot 10^{-7}$	$1.3660378340 \cdot 10^{-5}$	-1	0.8660253901	0.5000000237	$-1.571092374 \cdot 10^{-6}$
23	$1.1920928955 \cdot 10^{-7}$	$6.8301891700 \cdot 10^{-6}$	1	0.8660254497	0.4999999205	$5.259096796 \cdot 10^{-6}$
24	$5.9604644775 \cdot 10^{-8}$	$3.4150945850 \cdot 10^{-6}$	1	0.8660254497	0.4999999721	$1.844002211 \cdot 10^{-6}$
25	$2.9802322388 \cdot 10^{-8}$	$1.7075472925 \cdot 10^{-6}$	1	0.8660254050	0.4999999979	$1.364549189 \cdot 10^{-7}$
26	$1.4901161194 \cdot 10^{-8}$	$8.5377364625 \cdot 10^{-7}$	1	0.8660253975	0.5000000108	$-7.173187274 \cdot 10^{-7}$
27	$7.4505805969 \cdot 10^{-9}$	$4.2688682313 \cdot 10^{-7}$	-1	0.8660254013	0.5000000044	$-2.904319042 \cdot 10^{-7}$
28	$3.7252902985 \cdot 10^{-9}$	$2.1344341156 \cdot 10^{-7}$	-1	0.8660254031	0.5000000012	$-7.698849268 \cdot 10^{-8}$

complies with the condition

$$R \le 2Q, \tag{7.74}$$

ensuring that $Q = \lfloor X^{1/2} \rfloor$.

The classical pencil and paper method, described in what follows, assumes that all roots of 2-digit numbers are available ($(B^2 - 1) \times (B - 1)$ look-up table). Allowing the first digit of X to be zero, the radicand can always be sliced into n 2-digit groups. Algorithm 7.11 computes the square root of the $2n$-digit integer X. The symbol * stands for integer multiplication, while ** and $/$ stand for integer exponentiation and division respectively. One defines the function $P(i, k)$ as

$$P(i, k) = (2^*B^*Q(n - i) + k)^*k^*B^{**}(2^*(n - i)) \tag{7.75}$$

Algorithm 7.11   Integer Square Rooting in Base B

```
R(0):=X;  Q(n-1):=0;
q(n-1):=SQR (R(0)/B**(2*n-2));
Q(n-2):=q(n-1);
R(1):=R(0)-(q(n-1)**2)*(B**(2*n-2));

for i in 2..n loop
  if P(i, B-1)<=R(i-1) then R(i):=R(i-1)-P(i, B-1);
  q(n-i):=B-1;
  Q(n-i-1):=B*Q(n-i)+q(n-i);
  elsif P(i, B-2)<=R(i) then R(i):=R(i-1)-P(i, B-2);
  q(n-i):=B-2;
  Q(n-i-1):=B*Q(n-i)+q(n-i);
  elsif
  ...
  elsif P(i, 1) <=R(i) then R(i):=R(i-1)-P(i, 1); q(n-i):=1;
  Q(n-i-1):=B*Q(n-i)+q(n-i);
  else R(i):=R(i-1); q(n-i):=0; Q(n-i-1):=B*Q(n-i)+q(n-i);
  end if;
end loop;
```

Example 7.12   Compute the square root of $X = (591865472)_{10}$

Slicing ($n = 5$) gives $X = R(0) = 05$ '91 '86 '54 '72

Step 1

$$Q(4) = 0$$

$$Q(3) = q(4) = \lfloor 5^{1/2} \rfloor = 2$$
$$R(1) = \quad 0591865472$$
$$\underline{-0400000000}$$
$$= \quad 191865472$$

Step 2

$$P(2,9) = 49.9 \cdot 10^6 > R(1)$$

$$P(2,8) = 48.8 \cdot 10^6 > R(1)$$

$$\ldots$$

$$P(2,4) = 44.4 \cdot 10^6 = 176000000 < R(1)$$

$$R(2) = \quad 191865472$$

$$-176000000$$

$$= \quad 015865472$$

$$q(3) = 4; \ Q(2) = 10 \cdot 2 + 4 = 24$$

Step 3

$$\ldots$$

$$P(3,3) = 483.3 \cdot 10^4 = 14490000 < R(2)$$

$$R(3) = \quad 15865472$$

$$-14490000$$

$$= \quad 01375472$$

$$q(2) = 3; \quad Q(1) = 10 \cdot 24 + 3 = 243$$

Step 4

$$\ldots$$

$$P(4,2) = 4862.2 \cdot 10^2 = 972400 < R(3)$$

$$R(4) = \quad 1375472$$

$$-972400$$

$$= \quad 0403072$$

$$q(1) = 2; \quad Q(0) = 10 \cdot 243 + 2 = 2432$$

Step 5

$$\ldots$$

$$P(4,8) = 48648.8 \cdot 10^0 = 389184 < R(4)$$

$$R(5) = \quad 403072$$

$$-389184$$

$$= \quad 013888 : \text{final remainder}$$

$$q(0) = 8; \quad Q(-1) = Q = 24328 : \text{integer square root of } 591865472.$$

$$X = Q^2 + R(5) : 591865472 = 24328^2 + 13888.$$

Comments 7.6 The process can be carried on further, up to the desired quantity of digits after the decimal point.

The function $P(i, k)$ (7.75) of Algorithm 7.11 is used to compute the greatest value of k verifying $P(i, k) \leq R(i - 1)$.

Actually, k can be defined algebraically by the formula

$$k = \lfloor -B.Q(n - i) + (B^2.Q^2(n - i) + R(i - 1)/B^{2(n-i)})^{1/2} \rfloor, \qquad (7.76)$$

the integer part of the solution to $k' \in R$ such that

$$R(i - 1) = P(i, k'), \qquad (7.77)$$

where

$$P(i, k) \leq R(i - 1) < P(i, k + 1). \qquad (7.78)$$

Obviously formula (7.76), using the square root, is useless for algorithmic implementation. Other methods must therefore be derived to get k verifying (7.78).

Techniques for square rooting in base $B > 2$ are quite similar to those for base-B division. The recurrence formula for the remainder is given by

$$R(i) = R(i - 1) - P(i, q(n - i)), \qquad (7.79)$$

with

$$P(i, q(n - i)) = (2B.Q(n - i) + q(n - i)).q(n - i).B^{2(n-i)}, \qquad (7.80)$$

for $2n$-digit integers.

7.4.2   Restoring Binary Shift-and-Subtract Square Rooting Algorithm

In base 2, the computation of the function $P(i, k)$ is obviously more straightforward, since just 0 or 1 have to be considered for k. Moreover, since $P(i, 0) = 0$, just $P(i, 1)$, hereafter denoted $P(i)$, is computed,

$$P(i) = (4.Q(n - i) + 1).2^{2(n-i)}. \qquad (7.81)$$

Defining

```
P(i)=(4*Q(n-i)+1)*2**(2*(n-i)),                                 (7.82)
```

Algorithm 7.11 can be simplified as follows.

Algorithm 7.12   Integer Binary Square Rooting; Restoring

```
R(0):=X; Q(n-1):=0; q(n):=0; P(1):=2**(2*(n-1));
for i in 1..n, loop
  if R(i-1)-P(i)>=0
  then R(i):=R(i-1)-P(i); q(n-i):=1;
  Q(n-i-1):=2*Q(n-i)+q(n-i);
  else R(i):=R(i-1); q(n-i):=0; Q(n-i-1):=2*Q(n-i)+
  q(n-i);
```

end if;
end loop;

Example 7.13   Compute the square root of

$$X = 101101100011 \qquad (n = 6)$$

Step 1

$$
\begin{aligned}
P(1) \qquad &= \qquad 10000000000 \\
R(0) - P(1) &= \quad 101101100011 \\
&\qquad \underline{-10000000000} \\
&= \quad 011101100011 = \; R(1) \geq 0 \\
q(5) = 1; \; & Q(4) = 1
\end{aligned}
$$

Step 2

$$
\begin{aligned}
P(2) \qquad &= \qquad 10100000000 \\
R(1) - P(2) &= \quad 011101100011 \\
&\qquad \underline{-10100000000} \\
&= \quad 001001100011 = \; R(2) \geq 0 \\
q(4) = 1; \; & Q(3) = 11
\end{aligned}
$$

Step 3

$$
\begin{aligned}
P(3) \qquad &= \qquad 1101000000 \\
R(2) - P(3) &= \quad 1001100011 \\
&\qquad \underline{-1101000000} \\
&\qquad \qquad \quad < 0 \\
R(3) = R(2) &= 1001100011 \\
q(3) = 0; \; & Q(2) = 110
\end{aligned}
$$

Step 4

$$
\begin{aligned}
P(4) \qquad &= \qquad 110010000 \\
R(3) - P(4) &= \quad 1001100011 \\
&\qquad \underline{-110010000} \\
&= \quad 0011010011 = \; R(4) \geq 0 \\
q(2) = 1; \; & Q(1) = 1101
\end{aligned}
$$

Step 5

$$P(5) \quad = \quad 11010100$$
$$R(4) - P(5) = \quad 11010011$$
$$\underline{-11010100}$$
$$< 0$$

$$R(5) = R(4)$$
$$q(1) = 0; \quad Q(0) = 11010$$

Step 6

$$P(6) \quad = \quad 1101001$$
$$R(5) - P(6) = \quad 11010011$$
$$= \quad -1101001$$
$$= \quad 01101010 = R(6) \geq 0$$

$$q(0) = 1; Q(-1) = Q = 110101$$

$$X = Q^2 + R(6):101101100011 = (110101)^2 + 1101010$$

Comments 7.7

1. Generality is not lost assuming $2n$ digits for the radicand X; a first digit zero can always be assumed.
2. Step 1 of Algorithm 7.12 actually computes $q(n\text{-}1)$ as the integer square root of the leftmost two bits of the radicand X. If X is not headed by "00", $q(n-1)$ will always be one.

7.4.3   Nonrestoring Binary Add-and-Subtract Square Rooting Algorithm

As well as in the classical binary division algorithm, restoring the remainder, whenever negative, is not necessary since the following operation can cope with a negative remainder and perform in a single operation the restoring process together with the next step operation.

Considering a current step i, notice that when

$$R(i-1) - P(i) \geq 0, \tag{7.83}$$

the nonrestoring algorithm proceeds like the restoring one.

Whenever

$$R(i-1) - P(i) < 0, \tag{7.84}$$

the restoring process would add back $P(i)$ to the partial result (7.84), set $q(n-i)$ to zero, then subtract $P(i+1)$ from restored $R(i-1)$. These operations can be substituted by adding, to the partial result (7.84), the following expression

$$P^*(i+1) = P(i) - P(i+1) \tag{7.85}$$

at the next step $i+1$.

Since, from step i (7.84), $q(n-i) = q_{n-i} = 0$, one can write:

$$P(i) \qquad = q_{n-1} \quad \cdots \quad q_{n-i+2} \quad q_{n-i+1} \quad 0 \qquad 1 \qquad \overbrace{000\cdots0}^{2(n-i)} \qquad (7.86)$$

and

$$P(i+1) \quad = 0 \qquad q_{n-1} \cdots \qquad q_{n-i+2} \quad q_{n-i+1} \qquad q_{n-1} \underset{\underset{\text{set to } 0}{\downarrow}}{0\,1}\, \overbrace{0\cdots0}^{2(n-i-1)} \qquad (7.87)$$

then

$$P^*(i+1) = P(i) - P(i+1) = q_{n-1} \quad \cdots \quad q_{n-i+2} \quad q_{n-i+1} \quad 0\,1\,1\,\overbrace{0\cdots0}^{2(n-i-1)} \quad (7.88)$$

Assuming

$$Q_{n-i-1} = q_{n-1} \quad \cdots \quad q_{n-i},$$

(7.88) can be written

$$P^*(i+1) = P(i) - P(i+1) = [(4.Q_{n-i-1}) + 11].2^{2(n-i-1)} \quad i \geq 2. \qquad (7.89)$$

The nonrestoring binary square rooting method is described in Algorithm 7.13, where $P(i)$ and Pstar(i) are defined as follows

$$P(i) := (4*Q)n-i)+1)*2**(2*(n-i)), \qquad (7.90)$$
$$\text{Pstar}(i) := (4*Q(n-i)+11)*2**)2*(n-i)). \qquad (7.91)$$

Algorithm 7.13   Integer Binary Square Rooting; Nonrestoring

```
R(0):=X; Q(n-1):=0; q(n):=0; P(1):=2**(2*(n-1));
R(1):=R(0)-P(1);
if R(1)>=0
then q(n-1):=1; Q(n-2):=1;
else q(n-1):=0; Q(n-2):=0;
end if;
for i in 2..n loop
  if R(i-1)>=0
  then R(i):=R(i-1)-P(i);
  else R(i):=R(i-1)+Pstar(i);
  end if;
  if R(i)>=0
  then q(n-i):=1; Q(n-i-1):=2*Q(n-i)+q(n-i);
  else q(n-i):=0; Q(n-i-1):=2*Q(n-i)+q(n-i);
  end if;
end loop;
```

As well as for nonrestoring division algorithms, a final correction may be necessary. Whenever the final remainder is negative, it needs to be adjusted to the previous positive one.

Example 7.14   Note: 2's complement notation is used; the sign bit is in boldface type

Compute the square root of

$$X = \mathbf{0}\ \ 1\ \ 1\ \ 0\ \ 0\ \ 1\ \ 1\ \ 0\ \ 1\ \ 0\ \ 0\ \ 0\ \ 0\ \ 1\quad (n = 7)$$

Step 1

$$P(1) = \mathbf{0}\ \ 1\ \ 0\ \ 0\ \ 0\ \ 0\ \ 0\ \ 0\ \ 0\ \ 0\ \ 0\ \ 0\ \ 0\ \ 0;$$
$$-P(1) = \mathbf{1}\ \ 1\ \ 0\ \ 0\ \ 0\ \ 0\ \ 0\ \ 0\ \ 0\ \ 0\ \ 0\ \ 0\ \ 0\ \ 0$$
$$R(1) = R(0) - P(1) = \mathbf{0}\ \ 1\ \ 1\ \ 0\ \ 0\ \ 1\ \ 1\ \ 0\ \ 1\ \ 0\ \ 0\ \ 0\ \ 0\ \ 1$$

$$\underline{\qquad \mathbf{1}\ \ 1\ \ 0\ \ 0\ \ 0\ \ 0\ \ 0\ \ 0\ \ 0\ \ 0\ \ 0\ \ 0\ \ 0\ \ 0}$$
$$\mathbf{0}\ \ 1\ \ 0\ \ 0\ \ 1\ \ 1\ \ 0\ \ 1\ \ 0\ \ 0\ \ 0\ \ 0\ \ 1 = R(1) \geq 0$$

$$R(1) \geq 0 \Rightarrow q(6) = 1\,;\ Q(5) = 1$$

Step 2

$$P(2) = \mathbf{0}\ \ 1\ \ 0\ \ 1\ \ 0\ \ 0\ \ 0\ \ 0\ \ 0\ \ 0\ \ 0\ \ 0\ \ 0\ \ 0;$$
$$- P(2) = \mathbf{1}\ \ 0\ \ 1\ \ 1\ \ 0\ \ 0\ \ 0\ \ 0\ \ 0\ \ 0\ \ 0\ \ 0\ \ 0\ \ 0$$
$$R(2) = R(1) - P(2) = \mathbf{0}\ \ 1\ \ 0\ \ 0\ \ 1\ \ 1\ \ 0\ \ 1\ \ 0\ \ 0\ \ 0\ \ 0\ \ 1$$

$$\underline{\qquad \mathbf{1}\ \ 0\ \ 1\ \ 1\ \ 0\ \ 0\ \ 0\ \ 0\ \ 0\ \ 0\ \ 0\ \ 0\ \ 0\ \ 0}$$
$$\mathbf{1}\ \ 0\ \ 1\ \ 0\ \ 1\ \ 1\ \ 0\ \ 1\ \ 0\ \ 0\ \ 0\ \ 0\ \ 1 = R(2) < 0$$

$$R(2) < 0 \Rightarrow q(5) = 0\,;\ Q(4) = 10$$

Step 3

$$P^*(3) = \mathbf{0}\ \ 1\ \ 0\ \ 1\ \ 1\ \ 0\ \ 0\ \ 0\ \ 0\ \ 0\ \ 0\ \ 0\ \ 0$$
$$R(3) = R(2) + P^*(3) = \mathbf{1}\ \ 0\ \ 1\ \ 0\ \ 1\ \ 1\ \ 0\ \ 1\ \ 0\ \ 0\ \ 0\ \ 0\ \ 1$$

$$\underline{\qquad \mathbf{0}\ \ 1\ \ 0\ \ 1\ \ 1\ \ 0\ \ 0\ \ 0\ \ 0\ \ 0\ \ 0\ \ 0\ \ 0}$$
$$\mathbf{0}\ \ 1\ \ 0\ \ 1\ \ 0\ \ 0\ \ 0\ \ 0\ \ 1 = R(3) \geq 0$$

$$R(3) \geq 0 \Rightarrow q(4) = 1\,;\ Q(3) = 101$$

Step 4

$$P(4) = \mathbf{0} \ 1 \ 0 \ 1 \ 0 \ 1 \ 0 \ 0 \ 0 \ 0 \ 0 \ 0 \ ;$$
$$-P(4) = \mathbf{1} \ 0 \ 1 \ 0 \ 1 \ 1 \ 0 \ 0 \ 0 \ 0 \ 0 \ 0$$
$$R(4) = R(3) - P(4) = \mathbf{0} \ 0 \ 0 \ 0 \ 1 \ 0 \ 1 \ 0 \ 0 \ 0 \ 0 \ 1$$

$$\begin{array}{c} \mathbf{1} \ 0 \ 1 \ 0 \ 1 \ 1 \ 0 \ 0 \ 0 \ 0 \ 0 \ 0 \\ \hline \mathbf{1} \ 0 \ 1 \ 1 \ 0 \ 1 \ 1 \ 0 \ 0 \ 0 \ 0 \ 1 = R(4) < 0 \end{array}$$

$$R(4) < 0 \Rightarrow q(3) = 0; \ Q(2) = 1010$$

Step 5

$$P^*(5) = \mathbf{0} \ 1 \ 0 \ 1 \ 0 \ 1 \ 1 \ 0 \ 0 \ 0 \ 0$$
$$R(5) = R(4) + P^*(5) = \mathbf{1} \ 0 \ 1 \ 1 \ 0 \ 1 \ 1 \ 0 \ 0 \ 0 \ 0 \ 1$$

$$\begin{array}{c} \mathbf{0} \ 0 \ 1 \ 0 \ 1 \ 0 \ 1 \ 1 \ 0 \ 0 \ 0 \ 0 \\ \hline \mathbf{1} \ 0 \ 0 \ 0 \ 0 \ 1 \ 0 \ 0 \ 0 \ 1 = R(5) < 0 \end{array}$$

$$R(5) < 0 \Rightarrow q(2) = 0; \ Q(1) = 10100$$

Step 6

$$P^*(6) = \mathbf{0} \ 1 \ 0 \ 1 \ 0 \ 0 \ 1 \ 1 \ 0 \ 0$$
$$R(6) = R(5) + P^*(6) = \mathbf{1} \ 0 \ 0 \ 0 \ 0 \ 1 \ 0 \ 0 \ 0 \ 1$$

$$\begin{array}{c} \mathbf{0} \ 1 \ 0 \ 1 \ 0 \ 0 \ 1 \ 1 \ 0 \ 0 \\ \hline \mathbf{1} \ 0 \ 1 \ 0 \ 1 \ 1 \ 1 \ 0 \ 1 = R(6) < 0 \end{array}$$

$$R(6) < 0 \Rightarrow q(1) = 0; \ Q(0) = 101000$$

Step 7

$$P^*(7) = \mathbf{0} \ 1 \ 0 \ 1 \ 0 \ 0 \ 0 \ 1 \ 1$$
$$R(6) + P^*(7) = \mathbf{1} \ 0 \ 1 \ 0 \ 1 \ 1 \ 1 \ 0 \ 1$$

$$\begin{array}{c} \mathbf{0} \ 1 \ 0 \ 1 \ 0 \ 0 \ 0 \ 1 \ 1 \\ \hline \mathbf{0} \ 0 \ 0 \ 0 \ 0 \ 0 \ 0 \ 0 \ 0 = R(7) = 0 \end{array}$$

$$R(7) = 0 \Rightarrow q(0) = 1; \ Q = 1010001,$$

the exact square root.

Comments 7.8 The square rooting methods developed in the preceding sections are classic and easy to implement. The main component involved in the step complexity is the signed sum. One digit is obtained at each step (digit recurrence), so, for a desired precision p, p n-digit adding stages are involved, combinationally or sequentially implemented. To speed-up the process, fast adders may be used. SRT type algorithms have been developed for square rooting, using the feature of carry-save redundant adders ([ERC2004]). On the other hand, convergence algorithms have been developed, such as Newton–Raphson, reviewed in the following section.

7.4.4   Convergence Method—Newton–Raphson

The priming function for square root computation could be

$$f(x) = x^2 - X, \tag{7.92}$$

which has a root at $x = X^{1/2}$.

To solve $f(x)=0$, the same equation used for inverse computation can be used:

$$x_{i+1} = x_i - f(x_i)/f'(x_i) \tag{7.93}$$

or

$$x_{i+1} = 1/2(x_i + X/x_i). \tag{7.94}$$

Formula (7.94) involves a division and an addition at each step, which makes Newton–Raphson's method apparently less attractive in this case. Nevertheless, computing the inverse square root $1/X^{1/2}$, then multiplying by the radicand X, leads to a more effective solution.

The priming function is now

$$f(x) = 1/x^2 - X, \tag{7.95}$$

with root $x=1/X^{1/2}$, and the equation of convergence is given by

$$x_{i+1} = (x_i/2).(3 - x.x_i^2), \tag{7.96}$$

which involves three multiplications, one subtraction (3's complement), and one division by 2 (shift in base 2). A final multiplication is needed: the inverse square root by the radicand.

7.5   BIBLIOGRAPHY

[CAO2001] J. Cao, B. W. Wei, and J. Cheng, High-performance architecture for elementary functions generation. In: *Proceedings of the 15th IEEE Symposium on Computer Arithmetic*, 2001, pp. 136–144.

[CHE1972] T. C. Chen, Automatic computation of exponentials, logarithms, ratios, and square roots. *IBM J. Res. Dev.*, **16**: 380–388 (1972).

[COD1980] W. J. Cody and W. Waite, *Software Manual for the Elementary Functions*, Prentice-Hall, Englewood, Cliffs, NJ, 1980.

[DEL1989] J.-M. Delosme, CORDIC algorithms, theory and extensions. *SPIE Adv. Algorithms Architecture Signal Processing IV* **1152**: 131–145 (1989).

[ERC1977] M. D. Ercegovac, A general hardware-oriented method for evaluation of functions and computations in a digital computer. *IEEE Trans. Comput.*, **26**(7): 667–680 (1977).

[ERC1987] M. D. Ercegovac and T. Lang, Fast cosine/sine implementation using on-line CORDIC. In: *Proceedings of the 21st Asilomar Conference on Signals, Systems, Computers*, 1987, pp. 222–226.

[ERC1994] M. D. Ercegovac and T. Lang, *Division and Square-Root: Digit-Recurrence Algorithms and Implementations*, Kluwer Academic Publishers, Norwell, MA, 1994.

[ERC1995] M. D. Ercegovac, FPGA implementation of polynomial evaluation algorithms. In: *Proceedings of SPIE Photonics East '95 Conference*, Vol. 2607, pp. 177–188, 1995.

[ERC2004] M. D. Ercegovac and T. Lang, *Digital Arithmetic*. Morgan Kaufmann, San Francisco, CA, 2004.

[GAR1959] H. L. Garner, The residue number system. *IRE Trans. Electron. Comput.*, **EC 8**: 140–147 (1959).

[HUA1983] C. Huang, A fully parallel mixed-radix conversion algorithm for residue number applications. *IEEE Trans. Comput.* **32**(4): 398–402 (1983).

[KOS1991] D. K. Kostopoulos, An algorithm for the computation of binary logarithms. *IEEE Trans. Comput.*, **40**(11): 1267–1270 (1991).

[MAR1990] P. W. Markstein, Computations of elementary functions on IBM RISC System/6000 processor. *IBM J. Res. Dev.*, 111–119 (1990).

[OBE1999] S. F. Oberman, Floating point division and square root algorithms and implementation in the AMD-K7 microprocessor. In: *Proceedings of the 14th IEEE Symposium Computer on Arithmetic*, 1999, pp. 106–115.

[PAL2000] V. Paliouras, K. Karagianni, and T. Stouraitis, A floating-point processor for fast and accurate sine/cosine evaluation. *IEEE Trans. Circuits Systems II: Analog and Digital Signal Processing*, **47** (5): pp. 441–451 (2000).

[SPE1965] W. H. Specker, A class of algorithms for ln (x), exp (x), sin (x), cos (x), $\tan^{-1}(x)$ and $\cot^{-1}(x)$. *IEEE Trans. Electron. Comput.*, **EC 14**: 85–86 (1965).

[SZA1967] N. S. Szabo and R.I. Tanaka, *Residue Arithmetic and Its Applications to Computer Technology*. McGraw-Hill, New York, 1967.

[TAN1991] P. K. Tang, Table look-up algorithms for elementary functions and their error analysis. In: *Proceedings of the 10th IEEE Symposium on Computer Arithmetic*, 1991, pp. 232–236.

[VOL1959] J. E. Volder, The CORDIC trigonometric computing technique. *IRE Trans. Electron. Comput.*, **EC 8**: 330–334 (1959).

[VOL2000] J. E. Volder, The birth of CORDIC. *J. VLSI Signal Processing Syst.*, **25**: 101–105 (2000).

8

FINITE FIELD OPERATIONS

Finite field operations are used as computation primitives for executing numerous cryptographic algorithms, especially those related with the use of public keys (*asymmetric cryptography*). Classical examples are ciphering deciphering, authentication, and digital signature protocols based on *RSA*-type or *elliptic curve* algorithms. Other classical applications of finite fields are *error correcting codes* and *residue number systems*. This chapter proposes algorithms allowing the execution of the main arithmetic operations (addition, subtraction, multiplication) in finite rings Z_m and polynomial rings $Z_p[x]/f(x)$. In the case of Z_m, an exponentiation algorithm based on the *Montgomery multiplication* concept is also described. If p is prime and $f(x)$ is an irreducible polynomial, then Z_p, $Z_p[x]/f(x)$, $GF(p)$, and $GF(p^n)$ are finite fields for which inversion algorithms are proposed.

8.1 OPERATIONS IN Z_m

Given a natural number $m > 1$, the set $Z_m = \{0, 1, \ldots, m - 1\}$ is a ring whose operations are defined modulo m (Chapter 2):

$$(x + y) \bmod m, (x - y) \bmod m, \text{ and } (x.y) \bmod m.$$

Synthesis of Arithmetic Circuits: FPGA, ASIC, and Embedded Systems
By Jean-Pierre Deschamps, Géry J. A. Bioul, and Gustavo D. Sutter
Copyright © 2006 John Wiley & Sons, Inc.

8.1.1 Addition

Given two natural numbers x and y belonging to the interval $0 \le x, y < m$, compute $z = (x + y) \bmod m$. Taking into account that

$$0 \le x + y < 2.m,$$

z must be equal to either $x + y$ or $x + y - m$. The corresponding algorithm is the following.

Algorithm 8.1   Modulo m Addition

```
z1:=x+y; z2:=z1 - m;
if z2>=0 then z:=z2; else z:=z1; end if;
```

Assume now that $B^{n-1} < m \le B^n$ and that x and y are n-digit base-B numbers. Consider three cases:

if $x + y < m$ then $x + y < B^n$, $(x + y) + (B^n - m) < B^n$, $z = x + y$;

if $m \le x + y < B^n$ then $(x + y) + (B^n - m) \ge B^n$, $z = x + y - m = ((x + y) + (B^n - m)) \bmod B^n$;

if $B^n \le x + y$ then $(x + y - B^n) + (B^n - m) = x + y - m = z$.

So Algorithm 8.1 can be substituted by the following one where all operands have n digits.

Algorithm 8.2   Base B Modulo m Addition

```
z1:=(x+y) mod B**n; c1:=(x+y)/B**n;
z2:=(z1+B**n - m) mod B**n; c2:=(z1+B**n - m)/B**n;
if c1=0 and c2=0 then z:=z1; else z:=z2; end if;
```

Example 8.1   Assume that $B = 10$, $n = 3$, $m = 750$, so that $B^n - m = 250$:

if $x = 247$ and $y = 391$ then $x + y = 638$, $((x + y) \bmod B^n) + (B^n - m) = 638 + 250 = 888$, so that $c_1 = c_2 = 0$ and $z = (x + y) \bmod B^n = 638$;

if $x = 247$ and $y = 597$ then $x + y = 844$, $((x + y) \bmod B^n) + (B^n - m) = 844 + 250 = 1094$, so that $c_1 = 0$, $c_2 = 1$, and $z = (((x + y) \bmod B^n) + (B^n - m)) \bmod B^n = 094$;

if $x = 247$ and $y = 912$ then $x + y = 1159$, $((x + y) \bmod B^n) + (B^n - m) = 159 + 250 = 409$, so that $c_1 = 1$, $c_2 = 0$, and $z = ((x + y) \bmod B^n) + (B^n - m)) \bmod B^n = 409$.

8.1.2 Subtraction

Given two natural numbers x and y belonging to the interval $0 \le x, y < m$, compute $z = (x - y) \bmod m$. Taking into account that

$$-m < x - y < m,$$

z must be equal to either $x - y$ or $x - y + m$. The corresponding algorithm is the following.

Algorithm 8.3   Modulo m Subtraction

```
z1:=x − y; z2:=z1+m;
if z1<0 then z:=z2; else z:=z1; end if;
```

If $B^{n-1} < m \le B^n$, and x and y are n-digit base-B numbers, consider two cases:

  if $-m < x - y < 0$ then $B^n - m < x + (B^n - y) < B^n$, $x + (B^n - y) + m > B^n$,
  $z = x - y + m = (x + (B^n - y) + m) \bmod B^n$;
  if $0 < x - y$ then $B^n < x + (B^n - y)$, $z = x - y = (x + (B^n - y)) \bmod B^n$.

Algorithm 8.3 can be substituted by the following algorithm where all operands have n digits.

Algorithm 8.4   Base B Modulo m Subtraction

```
z1:=(x+B**n − y) mod B**n; c1:=(x+B**n − y)/B**n;
z2:=(z1+m) mod B**n;
if c1=1 then z:=z1; else z:=z2; end if;
```

Example 8.2   Assume that $B = 10$, $n = 3$, $m = 750$:

  if $x = 247$ and $y = 391$ then $B^n - y = 609$, $x + (B^n - y) = 247 + 609 = 856$, so that $c_1 = 0$ and $z = (((x + (B^n - y)) \bmod B^n) + m) \bmod B^n = (856 + 750) \bmod 1000 = 606$;
  if $x = 391$ and $y = 247$ then $B^n - y = 753$, $x + (B^n - y) = 391 + 753 = 1144$, so that $c_1 = 1$ and $z = (x + (B^n - y)) \bmod B^n = 144$.

8.1.3 Multiplication

Given x and $y \in Z_m = \{0, 1, \dots, m - 1\}$, compute $z = x.y \bmod m$. Assume that x, y, and m are represented in base B and that $m < B^n$ (if $m = B^n$ the modulo m reduction is trivial).

8.1.3.1 Multiply and Reduce   The first algorithm consists of (1) multiplying x by y, obtaining a $2.n$-digit intermediate result p, and (2) reducing p modulo m. The following multiplication and division procedures must have been defined:

```
procedure multiply (x, y: in digit_vector (0..n-1); z: out
digit_vector (0..2*n-1));

procedure divide (x: in digit_vector (0..2*n-1); y: in
digit_vector (0..n-1); q: out digit_vector (0..n-1); r:
out digit_vector (0..n-1));
```

Given two natural numbers x and y, the first procedure generates the product $z = x.y$, and the second one the quotient q and the remainder r such that $x = q.y + r$, with $r < y$. For that purpose any one of the multiplication (Chapter 5) and division (Chapter 6) algorithms can be used. The following algorithm is based on the property:

$z = x.y \bmod m$ if, and only if, $z < m$ and there exists a natural number q such that $x.y = q.m + z$.

Algorithm 8.5   Base B Modulo m Multiplication, Multiply and Reduce

```
multiply (x, y, p);
divide (p, m, q, z);
```

8.1.3.2 Modified Shift-and-Add Algorithm   An alternative solution consists of using Algorithm 5.1 and reducing modulo m in every step.

Algorithm 8.6   Modulo m Shift-and-Add Algorithm

```
p(n):=0;
for i in 0..n-1 loop
  p(n-1-i):=(p(n-i)*B+x(n-1-i)*y) mod m;
end loop;
z:=p(0);
```

Assume now that $B^{n-1} < m \le B^n$ and that x and y are n-digit base-B numbers. Observe that

$$p(n-i).B + x(n-1-i).y \le (m-1).B + (B-1).(m-1)$$
$$= (2.B-1).(m-1) < (2.B-1).B^n \tag{8.1}$$

so that $p(n-i).B + x(n-1-i).y$ is an $(n+2)$-digit number and

$$p(n-i).B + x(n-1-i).y = q.m + r \tag{8.2}$$

where

$$q < 2.B - 1 \qquad\qquad (8.3)$$

is a 2-digit number. In order to execute Algorithm 8.6, two procedures must be defined: the first one computes $x.B + a.y$, where x and y are n-digit numbers and a is a digit:

```
procedure shift_and_add (x, y: in digit_vector (0..n-1); a: in
digit; z: out digit_vector(0..n+1));
```

the second one is a division procedure:

```
procedure divide (x: in digit_vector (0..n+1); y: in digit_
vector (0..n-1); q: out digit_vector (0..1); r: out
digit_vector (0..n-1));
```

Algorithm 8.6 can be substituted by the following algorithm where all operands have n digits:

Algorithm 8.7 Base-B Modulo m Shift-and-Add Algorithm

```
p(n):=0;
for i in 0..n-1 loop
   shift_and_add (p(n-i), y, x(n-1-i), z1);
   divide (z1, m, q, p(n-1-i));
end loop;
z:=p(0);
```

In base $B = 2$ the execution of the main operation of Algorithm 8.6, namely,

$$p(n - 1 - i) = (p(n - i).2 + x(n - 1 - i).y) \text{ modulo } m,$$

can be performed in a slightly different way. According to (8.2) and (8.3)

$$p(n - i).2 + x(n - 1 - i).y = q.m + r$$

where $q < 3$, so that q is either 0, 1, or 2.

Algorithm 8.8

```
p1:=p(n-i)*2; p2:=p1+x(n-i-1)*y-m;
if p2<0 then p3:=p2+m; p(n-1-i):=p3;
else
   p3:=p2-m;
   if p3<0 then p(n-1-i):=p2; else p(n-1-i):=p3; end if;
end if;
```

Algorithm 8.8 can be simplified. On the one hand $p2$ and $p3$ cannot be simultaneously negative:

$$p2 = 2.p(n - i) + x(n - i - 1).y - m,$$

so that $-m \le p2 < 2.m$; if $p2<0$ then $p3 = p2 + m \ge -m + m = 0$. On the other hand, instead of computing

$$p2 = p1 + x(n - i - 1).y - m,$$

the value of $k = m - y$ can be precalculated (outside the main loop) so that $p2$ is equal to either $p1 - m$ if $x(n - i - 1) = 0$ or $p1 - k = p1 - m + y$ if $x(n - i - 1) = 1$. The modified algorithm is the following.

Algorithm 8.9   Base-2 Modulo m Shift-and-Add Algorithm

```
p(n):=0; k:=m-y;
for i in 0..n-1 loop
   if x(n-i-1)=0 then w:=m; else w:=k; end if;
   p1:=p(n-i)*2; p2:=p1-w;
   if p2<0 then p3:=p2+m; else p3:=p2-m; end if;
   if p3<0 then p(n-1-i):=p2; else p(n-1-i):=p3; end if;
end loop;
z:=p(0);
```

8.1.3.3 Montgomery Multiplication   In some cases the use of the *Montgomery product* concept ([MON1985]) allows one to reduce the computation complexity. Only the binary case ($B=2$) will be studied. The corresponding algorithm is based on the fact that, given three n-bit natural numbers x, y, and m, such that

$$m \text{ odd}, x < m, \text{ and } y < m,$$

it is relatively easy to find a natural number $z < m$ such that

$$(z.2^n) \bmod m = x.y \bmod m. \tag{8.4}$$

As m is an odd number, the greatest common divisor of 2^n and m is 1, so that there exits a natural number, denoted 2^{-n}, such that $2^{-n}. 2^n = 1 \bmod m$, and the preceding relation can be written in the form

$$z = x.y.2^{-n} \bmod m. \tag{8.5}$$

Relation (8.5) defines the Montgomery product of x by y. The following algorithm computes z.

Algorithm 8.10 Montgomery Product

```
r(0):=0;
for i in 1..n loop
  a:=r(i-1)+x(i-1)*y;
  r(i):=(a+a(0)*m)/2;
end loop;
if r(n)<m then z:=r(n);
else z:=r(n)-m; end if;
```

It is based on the following lemmas.

Lemma 8.1

$$r(i).2^i \equiv (x(i-1).2^{i-1} + x(i-2).2^{i-2} + \cdots + x(0).2^0).y \bmod m, \quad \forall i > 0.$$

Proof First observe that if m is odd ($m(0)=1$) then at each step $a + a(0).m$ is even:

$$(a + a(0).m) \bmod 2 = (a(0) + a(0).m(0)) \bmod 2 = (a(0) + a(0)) \bmod 2 = 0.$$

Then the property is demonstrated by induction. At the first execution of the iteration,

$$a = r(0) + x(0).y = x(0).y;$$
$$r(1).2 = a + a(0).m = x(0).y + a(0).m \equiv x(0).y \bmod m.$$

At step number i,

$$a = r(i-1) + x(i-1).y,$$
$$r(i).2^i = a.2^{i-1} + a(0).2^{i-1}.m = r(i-1).2^{i-1} + x(i-1).2^{i-1}.y + a(0).2^{i-1}.m$$
$$\equiv (x(i-2).2^{i-2} + \cdots + x(0).2^0).y + x(i-1).2^{i-1}.y \bmod m$$
$$\equiv (x(i-1).2^{i-1} + \cdots + x(0).2^0).y \bmod m.$$

Lemma 8.2

$$r(i) < 2.m.$$

Proof The property is demonstrated by induction. At the first execution of the iteration,

$$a = r(0) + x(0).y = x(0).y \le y < m,$$
$$r(1) = (a + a(0).m)/2 \le (a + m)/2 < (m + m)/2 = m.$$

At step number i,

$$a = r(i-1) + x(i-1).y < 2.m + y < 3.m,$$
$$r(i) = (a + a(0).m)/2 < (3.m + m)/2 = 2.m.$$

A direct consequence of Lemmas 8.1 and 8.2 is that

$$r(n).2^n \equiv x.y \bmod m \quad \text{and} \quad r(n) < 2.m,$$

so that z is either $r(n)$ or $r(n) - m$.

Assume that the procedure `Montgomery_product` has been defined:

procedure   Montgomery_product   (x,   y,   m: in   bit_vector
(0..n − 1); z: out bit_vector (0..n − 1));

and that the value of

$$\exp\_2n = 2^{2.n} \bmod m \tag{8.6}$$

has been previously computed. Then $z = x.y \bmod m$ can be computed as follows:

$$z = (x.y.2^{-n}).2^{2.n}.2^{-n} \bmod m = (x.y.2^{-n}).(\exp\_2n).2^{-n} \bmod m.$$

Algorithm 8.11   Modular Product Based on the Montgomery Product

```
Montgomery_product (x, y, m, z1);
Montgomery_product (z1, exp_2n, m, z);
```

Example 8.3 $n = 8$, $m = 239$, $x = 217$, $y = 189$; in base 2, $x = 11011001$; $\exp\_2n = 2^{16} \bmod 239 = 50$.
    First compute the Montgomery product of x and y:

$$r(0) = 0,$$
$$a = r(0) + x(0).y = 189; \ r(1) = (189 + 239)/2 = 214;$$
$$a = r(1) + x(1).y = 214; \ r(2) = 214/2 = 107;$$
$$a = r(2) + x(2).y = 107; \ r(3) = (107 + 239)/2 = 173;$$
$$a = r(3) + x(3).y = 173 + 189 = 362; \ r(4) = 362/2 = 181;$$
$$a = r(4) + x(4).y = 181 + 189 = 370; \ r(5) = 370/2 = 185;$$
$$a = r(5) + x(5).y = 185; \ r(6) = (185 + 239)/2 = 212;$$
$$a = r(6) + x(6).y = 212 + 189 = 401; \ r(7) = (401 + 239)/2 = 320;$$

$a = r(7) + x(7).y = 320 + 189 = 509; \ r(8) = (509 + 239)/2 = 374;$

$z_1 = 374 - 239 = 135;$

in base 2 $z_1 = 10000111;$

then compute the Montgomery product of z_1 and exp_2n:

$r(0) = 0,$

$a = r(0) + x(0).y = 50; \ r(1) = 50/2 = 25;$

$a = r(1) + x(1).y = 25 + 50 = 75; \ r(2) = (75 + 239)/2 = 157;$

$a = r(2) + x(2).y = 157 + 50 = 207; \ r(3) = (207 + 239)/2 = 223;$

$a = r(3) + x(3).y = 223; \ r(4) = (223 + 239)/2 = 231;$

$a = r(4) + x(4).y = 231; \ r(5) = (231 + 239)/2 = 235;$

$a = r(5) + x(5).y = 235; \ r(6) = (235 + 239)/2 = 237;$

$a = r(6) + x(6).y = 237; \ r(7) = (237 + 239)/2 = 238;$

$a = r(7) + x(7).y = 238 + 50 = 288; \ r(8) = 288/2 = 144;$

$z = 144;$

conclusion: $217 \times 189 \bmod 239 = 144$.

In the case of multioperand modular products an elegant presentation—not always an effective one—is based on the definition of a mapping T from Z_m into Z_m:

$$T(x) = x.2^n \bmod m. \tag{8.7}$$

Use the following notation for representing the Montgomery product:

$$MP(x,y) = x.y.2^{-n} \bmod m. \tag{8.8}$$

Then the following properties are evident:

$$T(x) = MP(x, exp\_2n), \tag{8.9}$$

$$T^{-1}(x) = MP(x, 1), \tag{8.10}$$

$$MP(T(x), T(y)) = T(x.y \bmod m), \tag{8.11}$$

$\forall x$ and y in Z_m.

According to (8.11), the transformation T replaces the mod m product by the Montgomery product. The following algorithm computes the product

$$z = x_1.x_2. \cdots .x_k \bmod m.$$

Algorithm 8.12   Multioperand Modular Product Based on the Montgomery Product

```
for i in 1..k loop Montgomery_product(x(i), exp_2n, m,
y(i)); end loop;
p(1):=y(1);
for i in 2..k loop Montgomery_product(p(i-1), y(i), m,
p(i)); end loop;
z:=Montgomery_product(p(k), 1, m, z);
```

The preceding algorithm includes $2.k$ Montgomery products, instead of k modular products if a classical multioperand product algorithm were used. Generally, the shorter computation time of the Montgomery product does not compensate the multiplication by 2 of the number of primitive operations. This drawback disappears if many operands are known to be identical, as is the case if an exponential function such as x^k is computed (Section 8.1.4).

8.1.3.4 Specific Ring   In the preceding algorithms m is a parameter whose value is any natural number greater than 1. For some particular values of m, specific algorithms can be defined. As an example, if $m = B^k - c$ for some small c, the modulo m reduction is easier. Assume that x is a $2.n$-digit number (the product of two n-digit numbers) and $m = B^n - c$, with $c \ll B^n$. Then x can be decomposed in the form $x = x_1.B^n + x_0$, with x_1 and x_0 smaller than B^n, so that

$$x \bmod m = (x_1.B^n + x_0) \bmod m = (x_1.c + x_0) \bmod m,$$

where

$$x' = x_1.c + x_0 \ll x_1.B^n + x_0 = x.$$

So instead of reducing x modulo m, the first operation consists of computing x', and then reducing x'. If x' is still greater than B^n, the same transformation can be performed, that is, $x' = x'_1.B^n + x'_0$, $x'' = x'_1.c + x'_0$, and so on. Eventually a number z is obtained such that $z < B^n$ and $x \bmod m = z \bmod m$.

Algorithm 8.13   Modulo m Reduction

```
z:=x;
while z>=B**n loop
  z1:=z/B**n; z0:=z mod B**n; z:=z1*c+z0;
end loop;
z:=z mod m;
```

In base $B = 2$, with $m \geq 2^{n-1}$, the last instruction is replaced by

```
if z >=m then z:=z-m; end if;
```

Example 8.4

$B = 2$, $n = 8$, $m = 239$, $x = 217$, $y = 189$;

In order to compute $z = x.y \bmod 239$, first compute $p = x.y = 41013$; then reduce p mod $239 = 2^8 - 17$:

$$41013 = 160.256 + 53 \equiv 160.17 + 53 = 2773;$$
$$2773 = 10.256 + 213 \equiv 10.17 + 213 = 383;$$
$$383 = 1.256 + 127 \equiv 1.17 + 127 = 144;$$
$$Z = 144.$$

Even more specific algorithms can be used.

Example 8.5 Assume again that $B = 2$, $n = 8$, $m = 239 = 2^8 - 17$ and that x is a $2.n$-bit number. The computation of $x \bmod 239$ can be performed as follows:

decompose x in the form $x = x_2.2^{12} + x_1.2^8 + x_0$;
replace 2^{12} by $33 = 2^{12} \bmod 239$, and 2^8 by $17 = 2^8 \bmod 239$, so that $x' = x_2.33 + x_1.17 + x_0$;
x' is a 10-bit number that can be decomposed in the form $x' = x'_1.2^8 + x'_0$;
replace 2^8 by 17 so that $x'' = x'_1.17 + x'_0$;
x'' is a 9-bit number, smaller than $3.17 + 256 = 307$, so that $x \bmod 239$ is equal to either x'' or $x'' - 239$.

If $x = 41013$ then

$$41013 = 10.4096 + 0.256 + 53 \equiv 10.33 + 0.17 + 53 = 383;$$
$$383 = 1.256 + 127 \equiv 1.17 + 127 = 144.$$

8.1.4 Exponentiation

Given x and $y \in Z_m = \{0, 1, \ldots, m - 1\}$, compute $e = y^x \bmod m$. Assume that x, y, and m are represented in base 2 and that $m < 2^n$. Then

$$x = x(0) + 2.x(1) + \cdots + 2^{n-1}.x(n - 1),$$

and e can be written in the form (a so-called Horner scheme)

$$e = ((\cdots((1^2.y^{x(n-1)})^2.y^{x(n-2)})^2 \cdots)^2.y^{x(1)})^2.y^{x(0)} \bmod m.$$

The corresponding algorithm is the following.

Algorithm 8.14

```
e:=1;
for i in 1..n loop
  e:=(e*e) mod m;
  if x(n-i)=1 then e:=(e*y) mod m; end if;
end loop;
```

This algorithm includes a lot of mod m products. Nevertheless, all the operands are either 1, y, or a previously obtained value (e), so that an alternative solution is the use of the Montgomery product (Section 8.1.3.3, relations (8.7) to (8.11)). The computation is performed as follows:

1. Substitute the initial operands 1 and y by $T(1) = 2^n$ mod m and $T(y) = MP(y, exp\_2n)$.
2. Execute the main loop of Algorithm 8.14 substituting the mod m products by Montgomery products.
3. Compute $T^{-1}(e) = MP(e, 1)$.

Assume that $exp\_n = 2^n$ mod m and $exp\_2n = 2^{2n}$ mod m have been previously computed. The following algorithm computes $e = y^x$ mod m:

Algorithm 8.15

```
e_transformed:=exp_n;
Montgomery_product (y, exp_2n, m, y_transformed);
for i in 1..n loop
  Montgomery_product (e_transformed, e_transformed, m,
  e_transformed);
  if x(n-i)=1 then
  Montgomery_product (e_transformed, y_transformed, m,
  e_transformed); end if;
end loop;
Montgomery_product (e_transformed, 1, m, z);
```

8.2 OPERATIONS IN $GF(p)$

If p is a prime number, then Z_p is the Galois field $GF(p)$, and every nonzero element y of Z_p has an inverse y^{-1}. Unless p is small—in which case all inverses could have been previously computed and stored in a table—the computation of $z = x^{-1}$ mod p is based on the extended Euclidean algorithm (Chapter 2, Section 2.1.2), which allows the expression of the greatest common divider of two natural numbers x and y in the form

$$gcd(x, y) = b.x + c.y$$

where b and c are integers. Given x and p, the computation of $z = x^{-1} \bmod p$ is made up of a sequence of integer divisions:

$$
\begin{array}{ll}
r(0) = p & c(0) = 0 \\
r(1) = x & c(1) = 1 \\
r(0) = r(1).q(1) + r(2) & c(2) = c(0) - c(1).q(1) \\
r(1) = r(2).q(2) + r(3) & c(3) = c(1) - c(2).q(2) \\
r(2) = r(3).q(3) + r(4) & c(4) = c(2) - c(3).q(3) \\
\text{and so on.}
\end{array}
$$

It has been demonstrated (Chapter 2, Section 2.1.2) that $r(i) = b(i).p + c(i).x$, so that

$$r(i) \equiv c(i).x \bmod p.$$

Taking into account that

$$\cdots r(3) < r(2) < x < p,$$

and

$$\cdots gcd(r(2), r(3)) = gcd(r(1), r(2)) = gcd(r(0), r(1)) = gcd(p, x) = 1,$$

after some finite number of steps, a remainder $r(i+2)$ is obtained such that

$$r(i+2) = 0 \quad \text{and} \quad gcd(r(i+1), r(i+2)) = r(i+1) = 1;$$

so

$$1 \equiv c(i+1).x \bmod p$$

and

$$z = c(i+1) \bmod p.$$

The corresponding algorithm is the following.

Algorithm 8.16   Inversion in Z_p

```
r_i:=p; r_iplus1:=x; c_i:=0; c_iplus1:=1;
while r_iplus1>1 loop
  q:=r_i/r_iplus1; r_iplus2:=r_i mod r_iplus1;
  c_iplus2:=(c_i-q*c_iplus1) mod p;
  r_i:=r_iplus1; r_iplus1:=r_iplus2; c_i:=c_iplus1;
  c_iplus1:=c_iplus2;
end loop;
z:=c_iplus1;
```

As a matter of fact, it can be demonstrated that $-p/2 < c(i) < p/2$ so that, in the preceding algorithm, it is not necessary to perform the mod p reduction at each step. The reduction can be performed at the end of the computation, substituting the last instruction by

```
if c_iplus1<0 then z:=c_iplus1+p; else z:=c_iplus1; end if;
```

Example 8.6 Compute the inverse of 114 mod 239:

$$r(0) = 239, r(1) = 144, c(0) = 0, c(1) = 1$$
$$q(1) = 239/144 = 1, r(2) = 239 \bmod 144 = 95, c(2) = 0 - 11 = -1$$
$$q(2) = 144/95 = 1, r(3) = 144 \bmod 95 = 49, c(3) = 1 + 1.1 = 2$$
$$q(3) = 95/49 = 1, r(4) = 95 \bmod 49 = 46, c(4) = -1 - 2.1 = -3$$
$$q(4) = 49/46 = 1, r(5) = 49 \bmod 46 = 3, c(5) = 2 + 3.1 = 5$$
$$q(5) = 46/3 = 15, r(6) = 46 \bmod 3 = 1, c(6) = -3 - 5.15 = -78$$
$$z = -78 + 239 = 161.$$

8.3 OPERATIONS IN $Z_p[x]/f(x)$

Given a polynomial

$$f(x) = f_0 + f_1.x + f_2.x^2 + \cdots + f_{n-1}.x^{n-1} + f_n.x^n$$

of degree n ($f_n \neq 0$) whose coefficients belong to Z_p (p prime), the set $Z_p[x]$ /$f(x)$ of polynomials of degree less than n, modulo $f(x)$, is a finite ring (Chapter 2, Section 2.2.2).

8.3.1 Addition and Subtraction

Given two polynomials

$$a(x) = a_0 + a_1.x + a_2.x^2 + \cdots + a_{n-1}.x^{n-1} \quad \text{and}$$
$$b(x) = b_0 + b_1.x + b_2.x^2 + \cdots + b_{n-1}.x^{n-1},$$

the addition and the subtraction are defined as follows:

$$a(x) + b(x) = (a_0 + b_0) + (a_1 + b_1).x + (a_2 + b_2).x^2 + \cdots$$
$$+ (a_{n-1} + b_{n-1}).x^{n-1}, \tag{8.12}$$
$$a(x) - b(x) = (a_0 - b_0) + (a_1 - b_1).x + (a_2 - b_2).x^2 + \cdots$$
$$+ (a_{n-1} - b_{n-1}).x^{n-1}, \tag{8.13}$$

where $a_i + b_i$ and $a_i - b_i$ are computed modulo p. Assume that two procedures

```
procedure modular_addition (a, b: in coefficient; m: in
module; c: out coefficient);
procedure modular_subtraction (a, b: in coefficient; m: in
module; c: out coefficient);
```

have been defined. They compute $(a + b)$ mod m and $(a - b)$ mod m (see Sections 8.1.1 and 8.1.2). Then the addition and subtraction of polynomials are performed componentwise.

Algorithm 8.17   Addition of Polynomials

```
for i in 0..n-1 loop
  modular_addition (a(i), b(i), p, c(i));
end loop;
```

Algorithm 8.18   Subtraction of Polynomials

```
for i in 0..n-1 loop
  modular_subtraction (a(i), b(i), p, c(i));
end loop;
```

8.3.2   Multiplication

Given two polynomials

$$a(x) = a_0 + a_1.x + a_2.x^2 + \cdots + a_{n-1}.x^{n-1} \quad \text{and}$$
$$b(x) = b_0 + b_1.x + b_2.x^2 + \cdots + b_{n-1}.x^{n-1},$$

their product $z(x)=a(x).b(x)$ can be computed as follows:

$$
\begin{aligned}
z(x) &= a_0.b(x) + a_1.b(x).x + a_2.b(x).x^2 + \cdots + a_{n-1}.b(x).x^{n-1} \\
&= (\cdots((0.x + a_{n-1}.b(x)).x + a_{n-2}.b(x)).x + \cdots + a_1.b(x)).x \\
&\quad + a_0.b(x).
\end{aligned}
\tag{8.14}
$$

The corresponding formal algorithm is the following.

Algorithm 8.19

```
z:=zero;
for i in 1..n loop z:=z*x+a(n-i)*b; end loop;
```

The computation primitives necessary for executing Algorithm 8.19 are:

     the multiplication of a polynomial by x,
     the multiplication of a polynomial by a coefficient,
     the addition of polynomials.

The addition is performed componentwise (Algorithm 8.17). The multiplication of a polynomial by a coefficient is also computed componentwise. Assume that a procedure

```
procedure modular_product (a, b: in coefficient; m: in module;
c: out coefficient);
```

has been defined. It computes $c = a.b \bmod m$ (see Section 8.1.3). Then the following procedure computes the product of $a(x)$ by a coefficient b:

```
procedure by_coefficient (a: in polynomial; b: in coefficient;
p: in module; c: out polynomial)
is begin
  for i in 0..n-1 loop modular_product(a(i), b, p, c(i));
end loop;
end procedure;
```

It remains to generate a procedure for computing the multiplication of a polynomial $a(x)$ by x. First observe that

$$f_0 + f_1.x + f_2.x^2 + \cdots + f_{n-1}.x^{n-1} + f_n.x^n \equiv 0 \bmod f,$$

so that

$$x^n \equiv r_0 + r_1.x + r_2.x^2 + \cdots + r_{n-1}.x^{n-1} \bmod f, \tag{8.15}$$

where

$$r_i = -f_i/f_n \bmod p. \tag{8.16}$$

Compute now $a(x).x$:

$$(a_0 + a_1.x + a_2.x^2 + \cdots + a_{n-1}.x^{n-1}).x = a_0.x + a_1.x^2 + \cdots$$
$$+ a_{n-2}.x^{n-1} + a_{n-1}.x^n$$
$$\equiv a_0.x + a_1.x^2 + \cdots + a_{n-2}.x^{n-1} + (r_0 + r_1.x + r_2.x^2 + \cdots + r_{n-1}.x^{n-1})$$
$$= a_{n-1}.r_0 + (a_0 + a_{n-1}.r_1).x + (a_1 + a_{n-1}.r_2).x^2 + \cdots$$
$$+ (a_{n-2} + a_{n-1}.r_{n-1}).x^{n-1}.$$

The corresponding procedure is

```
procedure by_x (a: in polynomial; p: in module; b:
out polynomial) is
begin
  modular_product (a(n-1), r(0), p, b(0));
  for i in 1..n-1 loop
```

```
  modular_product (a(n-1), r(i), p, c);
  modular_addition (a(i-1), c, p, b(i));
end loop;
end procedure;
```

Thus Algorithm 8.19 is equivalent to the following one.

Algorithm 8.20   Multiplication of Polynomials, First Version

```
for i in 0..n-1 loop z(i):=0; end loop;
for i in 1..n loop
  by_x(z, p, z1);
  by_coefficient(b, a(n-i), p, z2);
  for j in 0..n-1 loop modular_addition(z1(j), z2(j),
  p, z(j)); end loop;
end loop;
```

The preceding algorithm can be decomposed at the coefficient level. The operations corresponding to the main loop are the following:

```
next_z(0)=(z(n-1).r(0)+b(0).a(n-i)) mod p,
next_z(1)=(z(0)+z(n-1).r(1)+b(1).a(n-i)) mod p,
next_z(2)=(z(1)+z(n-1).r(2)+b(2).a(n-i)) mod p,
...
next_z(n-1)=(z(n-2)+z(n-1).r(n-1)+b(n-1).a(n-i)) mod p,
z=next_z.
```

The complete algorithm is the following (the values of $r_i = -f_i/f_n$ mod p should have been previously computed).

Algorithm 8.21   Multiplication of Polynomials, Second Version

```
for i in 0..n-1 loop z(i):=0; end loop,
for i in 1..n loop
  modular_product (z(n-1), r(0), p, c(0)); -- c₀=zₙ₋₁.r₀ mod p
  modular_product (a(n-i), b(0), p, d(0)); -- d₀=aₙ₋ᵢ.b₀ mod p
  modular_addition (c(0), d(0), p, next_z(0));
                            -- next_z₀=zₙ₋₁.r₀+aₙ₋ᵢ.b₀ mod p
  for i in 1..n-1 loop
    modular_product (z(n-1), r(i), p, c(i)); -- cᵢ=zₙ₋₁.rᵢ mod p
    modular_product (a(n-i), b(i), p, d(i)); -- dᵢ=aₙ₋ᵢ.bᵢ mod p
    modular_addition (c(i), d(i), p, e(i)); -- eᵢ=zₙ₋₁.rᵢ+aₙ₋ᵢ.bᵢ mod p
    modular_addition (z(i-1), e(i), p, next_z(i));
                            -- next_zᵢ=zᵢ₋₁+zₙ₋₁.rᵢ+aₙ₋ᵢ.bᵢ mod p
  end loop;
  z:=next_z;
end loop;
```

Instead of addressing a new coefficient of a at each step $(a(n-1), a(n-2), \ldots, a(0))$, an alternative solution is to use a procedure

```
right_rotate procedure(a: inout polynomial)
```

that substitutes $a_0 + a_1.x + a_2.x^2 + \cdots + a_{n-1}.x^{n-1}$ by $a_{n-1} + a_0.x + a_1.x^2 + \cdots + a_{n-2}.x^{n-1}$.

Algorithm 8.22 Multiplication of Polynomials, Third Version

```
for i in 0..n-1 loop z(i):=0; end loop;
for i in 1..n loop
  modular_product (z(n-1), r(0), p, c(0));
  modular_product (a(n-1), b(0), p, d(0));
  modular_addition (c(0), d(0), p, next_z(0));
  for i in 1..n-1 loop
    modular_product (z(n-1), r(i), p, c(i));
    modular_product (a(n-1), b(i), p, d(i));
    modular_addition (c(i), d(i), p, e(i));
    modular_addition (z(i-1), e(i), p, next_z(i));
  end loop;
  z:=next_z;
  righ_rotate(a);
end loop;
```

8.4 OPERATIONS IN $GF(p^n)$

If f is an irreducible polynomial then $Z_p[x]/f(x)$ is the Galois field $GF(p^n)$, so that every nonzero polynomial $a(x)$ has a multiplicative inverse $a^{-1}(x)$. Given two polynomials

$$a = a_0 + a_1.x + a_2.x^2 + \cdots + a_{n-1}.x^{n-1} \quad \text{and}$$

$$f = f_0 + f_1.x + f_2.x^2 + \cdots + f_{n-1}.x^{n-1} + f_n.x^n,$$

a variant of the extended Euclidean algorithm (Chapter 2, Section 2.1.2) allows the expression of the greatest common divider of f and a in the form

$$gcd(f, a) = b.a + c.f.$$

In particular, if $gcd(f, a) = 1$ then $a^{-1}(x) = b(x) \bmod f$.

The following formal algorithm, in which $degree(a)$ returns the degree of a and $swap(a, b)$ interchanges a and b, computes $z(x) = a^{-1}(x)$.

Algorithm 8.23

```
u:=f; v:=a; c:=0; e:=1;
m:=degree(u); t:=degree(v);
if t=0 then result:=(v(0))⁻¹;
else
  while t>0 loop
    if m<t then swap(u,v); swap(c,e); swap(m,t);
    q:=u(m)*(v(t))⁻¹*xᵐ⁻ᵗ; r:=u-(v*q); cc:=c-(e*q);
    u:=v; v:=r; c:=e; e:=cc;
    m:=t; t:=deg(v);
  end loop;
z:=e*(v(0))⁻¹;
end if;
```

Example 8.7

$$p = 2, n = 4, f(x) = 1 + x + x^4, a(x) = 1 + x^2;$$

As $p = 2$, Algorithm 8.23 can be simplified; in particular, $u(m) = v^{-1}(t) = v(0) = 1$.

Compute $a^{-1}(x)$:

$u = 1 + x + x^4, v = 1 + x^2, c = 0, e = 1$

$m = 4, t = 2$

$q = x^2, r = 1 + x + x^4 - (1 + x^2).x^2 = 1 + x + x^2, cc = 0 - 1.x^2 = x^2$

$u = 1 + x^2, v = 1 + x + x^2, c = 1, e = x^2$

$m = 2, t = 2$

$q = 1, r = 1 + x^2 - (1 + x + x^2).1 = x, cc = 1 - x^2.1 = 1 + x^2$

$u = 1 + x + x^2, v = x, c = x^2, e = 1 + x^2$

$m = 2, t = 1$

$q = x, r = 1 + x + x^2 - x.x = 1 + x, cc = x^2 - (1 + x^2).x = x + x^2 + x^3$

$u = x, v = 1 + x, c = 1 + x^2, e = x + x^2 + x^3$

$m = 1, t = 1$

$q = 1, r = x - (1 + x).1 = 1, cc = 1 + x^2 - (x + x^2 + x^3).1 = 1 + x + x^3$

$u = 1 + x, v = 1, c = x + x^2 + x^3, e = 1 + x + x^3$

$m = 1, t = 0$

$z = 1 + x + x^3.$

Effectively,

$$(1 + x^2).(1 + x + x^3) = 1 + x + x^3 + x^2 + x^3 + x^5 = 1 + x + x^2 + x^5$$
$$= (1 + x + x^4).x + 1 = 1 \bmod (1 + x + x^4).$$

In order to execute Algorithm 8.23, the following procedures must be defined:

- the procedure

  procedure degree (a: **in** polynomial; deg: **out** natural);

  computes the degree *deg* of *a*;
- the procedure

  procedure invert (a: **in** coefficient; p: **in** module; b: **out** coefficient);

  computes $a^{-1} \bmod p$; Algorithm 8.16 could be used;
- the procedure by_coefficient has already been defined;
- the procedure

  procedure add (a, b: **in** polynomial; p: **in** module; c: **out** polynomial);

  computes the sum of two polynomials; Algorithm 8.17 could be used;
- the procedure

  procedure sub (a, b: **in** polynomial; p: **in** module; c: **out** polynomial);

  computes the difference of two polynomials; Algorithm 8.18 could be used;
- the procedure

  procedure shift (a: **in** polynomial; k: **in** natural; c: **out** polynomial);

  computes $c(x) = a(x).x^k$; it is equivalent to a *k*-position right-shift of the coefficients of *a*, with the *k* lower-degree coefficients set to 0.

The following algorithm is deduced from Algorithm 8.23 and from the previous procedure definitions (*zero* and *one* stand for the polynomials 0 and 1, respectively):

Algorithm 8.24   Inversion in $GF(p^n)$

```
u:=f(0..n-1); f_n:=f(n); v:=a; c:=zero; e:=one;
degree(v,t);
if t=0 then
  invert(v(0), p, result(0)); for i in 1..n-1
    loop result(i):=0; end loop;
```

```
else
   j:=n-t;                              --the initial value of m is deg(f)=n
   invert(v(t), p, inverted_v_t);  --(v(t))⁻¹
   k:=(f_n*inverted_v_t) mod p;     --f_n.(v(t))⁻¹
   by_coefficient(v, k, p, k_v);    --v.f_n.(v(t))⁻¹
   shift(k_v, j, shifted_v);        --v.f_n.(v(t))⁻¹.xⁿ⁻ᵗ=v.q
   sub(u, shifted_v, p, r);         --r=u - v.q
   by_coefficient(e, k, p, e_v);    --e.f_n.(v(t))⁻¹
   shift(e_v, j, shifted_e);        --e.f_n.(v(t))⁻¹.xⁿ⁻ᵗ=e.q
   sub(c, eq, p, cc);               --cc=c - e.q
   degree(r, deg_v);
   j:=t - deg_v;
   if j>=0 then u:=v; v:=r; c:=e; e:=cc; m:=t; t:=deg_v;
   else u:=r; c:=cc; m:=deg_v; end if;
   while t>0 loop
      j:=m-t;
      invert(v(t), p, inverted_v_t);  --(v(t))⁻¹
      k:=(u(m)*inverted_v_t) mod p;   --u(m).(v(t))⁻¹
      by_coefficient(v, k, p, k_v);   --v.u(m).(v(t))⁻¹
      shift(k_v, j, shifted_v);       --v.u(m).(v(t))⁻¹.xⁿ⁻ᵗ=v.q
      sub(u, shifted_v, p, r);        --r=u - v.q
      by_coefficient(e, k, p, e_v);   --e.u(m).(v(t))⁻¹
      shift(e_v, j, shifted_e);       --e.u(m).(v(t))⁻¹.xⁿ⁻ᵗ=e.q
      sub(c, eq, p, cc);              --cc=c - e.q
      degree(r, deg_v);
      j:=t - deg_v;
      if j>=0 then u:=v; v:=r; c:=e; e:=cc; m:=t; t:=deg_v;
      else u:=r; c:=cc; m:=deg_v; end if;
   end loop;
   invert(v(0), p, inverted_v_t);      --(v(0))⁻¹
   by_coefficient(e, inverted_v_t, p, result);   --e. (v(0))⁻¹
end if;
```

A different method, based on a modification of the Itoh–Tsujii algorithm ([ITO1988]), can be used if $f(x)$ is a binomial ([WOO2000], [BAI2001]). First observe that if $r = 1 + p + p^2 + \cdots + p^{n-1}$, then (Chapter 2, Section 2.2.4) $(a(x))^r$ is an element of Z_p (a 0-degree polynomial). As

$$(a(x))^{-1} = (a(x))^{r-1}/(a(x))^r,$$

the problem is reduced to the computation of exponential functions in $GF(p^n)$ and to the inversion in $GF(p)$.

The following formal algorithm computes $z(x) = a^{-1}(x)$.

Algorithm 8.25

```
b:=1;
for i in 1..n-1 loop
   d:=a**(p**i);           --d=aʳ⁽ⁱ⁾ where r(i)=pⁱ
   b:=b*d;                 --b=1.aʳ⁽¹⁾.....aʳ⁽ⁱ⁻¹⁾.aʳ⁽ⁱ⁾
end loop;                  --b=aʳ⁻¹
```

```
g:=b*a;                              --g=a^r
k:=(1/g(0)) mod p;                   --k=1/a^r
z:=b*k;                              --z=a^{r-1}/a^r=a^{-1}
```

In order to execute the preceding algorithm the following procedures must be defined:

- the procedure

  procedure multiply (a, b, f: **in** polynomial; p: **in** module; z: **out** polynomial);

  computes the product of a by b modulo f; Algorithm 8.20, 8.21, or 8.22 could be used;
- the procedures by_coefficient and invert have already been defined;
- the procedure

  procedure **exponentiation** (a, f: **in** polynomial; p: **in** module; i: **in** natural; b: **out** polynomial);

  computes $a^{r(i)}$ modulo f, where $r(i) = p^i$.

It remains to generate the preceding procedure. First recall (Chapter 2, Section 2.2.5) that if $\alpha, \beta, \ldots, \gamma$ are elements of $GF(p^n)$, then

$$(\alpha + \beta + \cdots + \gamma)^p = \alpha^p + \beta^p + \cdots + \gamma^p. \tag{8.17}$$

More generally, if $r(i) = p^i$, then

$$(\alpha + \beta + \cdots + \gamma)^{r(i)} = \alpha^{r(i)} + \beta^{r(i)} + \cdots + \gamma^{r(i)}.$$

Observe also that, given a coefficient a_k (an element of Z_p), then

$$a_k^p = a_k,$$

and, more generally,

$$a_k^{r(i)} = a_k. \tag{8.18}$$

From (8.17) and (8.18) the following relation is deduced:

$$(a(x))^{r(i)} = a_0 + a_1 x^{r(i)} + a_2 x^{2.r(i)} + \cdots + a_{n-1} x^{(n-1).r(i)}.$$

Assume now that f is a binomial

$$f(x) = x^n - c$$

and that n divides $p - 1$ ($p \bmod n = 1$), so that

$$p^i = q(i).n + 1.$$

Then

$$x^{k.r(i)} = x^{k.q(i).n}.x^k = (x^n)^{k.q(i)}.x^k = c^{k.q(i)}.x^k.$$

The values of

$$f_{ki} = c^{k.q(i)} \bmod p \qquad (8.19)$$

can be computed in advance (an algorithm for computing f_{ki} is given in Appendix 8.1), so that

$$(a(x))^{r(i)} = a_0 + f_{1i}.a_1.x + f_{2i}.a_2.x^2 + \cdots + f_{(n-1)i}.a_{n-1}.x^{n-1}. \qquad (8.20)$$

The corresponding exponentiation procedure is the following:

```
procedure exponentiation (a, f: in polynomial; p: in module;
 i: in natural; b: out polynomial) is
begin
  b(0):=a(0);
  for k in 1..n-1 loop modular_product (f(k,i), a(k), p, b(k));
end loop;
end procedure;
```

The complete inversion algorithm is deduced from Algorithm 8.25.

Algorithm 8.26   Inversion, Second Version

```
b:=one;
for i in 1..n-1 loop
   exponentiation (a, f, p, i, d);            --d=(a(x))^r(i)
   multiply (b, d, f, p, e);                  --e:=b.(a(x))^r(i)
end loop;                                     --e=(a(x))^r-1
multiply (e, a, f, p, g);                     --g=(a(x))^r
h:=g(0);
invert (h, p, k);                             --k=h^-1
by_coefficient (e, k, p, z);                  --z=(a(x))^r-1/(a(x))^r
```

In order to reduce the number of calls to the exponentiation procedure the following property can be used.

Property 8.1 If $s = 1 + p + p^2 + \cdots + p^k$ and $t = 1 + p + p^2 + \cdots + p^l$, where k is odd and $l = (k-1)/2$, then

$$(a(x))^s = (a(x))^t.((a(x))^{t.u},$$

where $u = p^{k-l}$.

Proof

$$t.u = (1 + p + p^2 + \cdots + p^l).p^{k-l} = p^{k-l} + p^{k-l+1} + \cdots + p^k,$$

where

$$k - l = k - (k-1)/2 = (k+1)/2 = l + 1,$$

so that

$$t + t.u = 1 + p + p^2 + \cdots + p^k.$$

If $k+1$ is a power of 2, that is, $k = 2^m - 1$, then $l = 2^{m-1} - 1$, and the same decomposition can be recursively applied. The following algorithm computes $z(x) = (a(x))^s$, where $s = 1 + p + p^2 + \cdots + p^k$ with $k = 2^m - 1$.

Algorithm 8.27

```
b(0):=a;
for j in 0..m-1 loop
   exponentiation (b(2*j), f, p, 2**j, b(2*j+1));
   multiply (b(2*j), b(2*j+1), f, p, b(2*(j+1)));
end loop;
z:=b(2*m);
```

Example 8.8 $k = 7, m = 3$; in the following computation scheme $r(i)$ stands for p^i, so that $r(i).r(j) = r(i+j)$.

$$b(0) = a,$$

$$b(1) = a^{r(1)},$$

$$b(2) = b(0).b(1) = a^{1+r(1)},$$

$$b(3) = (b(2))^{r(2)} = a^{r(2)+r(3)},$$

$$b(4) = b(3).b(4) = a^{1+r(1)+r(2)+r(3)},$$

$$b(5) = (b(4))^{r(4)} = a^{r(4)+r(5)+r(6)+r(7)},$$

$$b(6) = b(4).b(5) = a^{1+r(1)+r(2)+r(3)+r(4)+r(5)+r(6)+r(7)},$$

$$z = b(6).$$

Assume now that $n-1$ is a power of 2, that is,

$$n = 2^m + 1. \tag{8.21}$$

Then

$$r = 1 + p + p^2 + \cdots + p^{n-1} = 1 + p + p^2 + \cdots + p^{k+1}, \quad \text{where } k = 2^m - 1,$$

and

$$(a(x))^{r-1} = (a(x))^{s \cdot p}, \quad \text{where } s = 1 + p + p^2 + \cdots + p^k.$$

The preceding algorithm can be used for computing

$$b(2.m) = (a(x))^s.$$

It remains to compute

$$e = (b(2.m))^p = (a(x))^{s \cdot p} = (a(x))^{r-1}$$

and

$$g = e.a = (a(x))^r.$$

The complete inversion algorithm, when

$$f(x) = x^n - c, \ p \bmod n = 1, \ n = 2^m + 1, \tag{8.22}$$

is the following.

Algorithm 8.28   Inversion, Third Version

```
b(0):=a;
for j in 0..m-1 loop
   exponentiation (b(2*j), f, p, 2**j, b(2*j+1));
   multiply (b(2*j), b(2*j+1), f, p, b(2*(j+1)));
end loop;                              --b(2.m)=(a(x))^s
exponentiation (b(2*m), f, p, 1, e);   --e=(a(x))^s·p=(a(x))^r-1
multiply (e, a, f, p, g);              --g=(a(x))^r
h:=g(0);
invert (h, p, k);                      --k=h^-1
by_coefficient (e, k, p, z);           --z=(a(x))^r-1/(a(x))^r
```

Observe that the main iteration is executed m times instead of $n - 1 = 2^m$ times as in Algorithm 8.26.

Example 8.9   If $p = 239$ and $f(x) = x^{17} - 2$, then Algorithm 8.28 can be applied:

$$239 = 14..17 + 1, n = 17 = 2^4 + 1;$$

the coefficients f_{ki} can be computed with Algorithm A 8.1.

Another example is the binomial $x^6 - 2$ with

$$p = 42{,}798{,}677{,}629 = 2^{32} - 387;$$
$$42{,}798{,}677{,}629 = 7{,}133{,}112{,}938.6 + 1.$$

As $n - 1$ is not a power of 2, Algorithm 8.28 must be slightly modified.

8.5   BIBLIOGRAPHY

[BAI2001] D. V. Bailey and C. Paar, Efficient arithmetic in finite field extensions with application in elliptic curve cryptography. *J. Cryptol.*, **14**(3): 153–176 (2001).

[ITO1988] T. Itoh and S. Tsujii, A fast algorithm for computing multiplicative inverses in $GF(2^m)$ using normal bases. *Math. Computation* **44**(4): 519–521 (1985).

[MON1985] P. Montgomery, modular multiplication without trial division. *Math. Computation* **44**(4): 519–521 (1985).

[ROS1999] M. Rosing, *Elliptic Curve Cryptography*. Manning Publications, Greenwich, CT, 1999.

[WOO2000] A. D. Woodbury, Elliptic curve cryptography on smart cards without coprocessors. *IFIP CARDIS*, 71–92 (2000).

APPENDIX 8.1   COMPUTATION OF f_{ki}

First compute the value of $q(i)$ such that $p^i = q(i).n + 1$.

Lemma A8.1

$$\forall i > 1 : q(i) = p.q(i-1) + q(1). \tag{A8.1}$$

Proof   By induction,

$$p^i = p.p^{i-1} = p.(q(i-1).n + 1) = p.q(i-1).n + p = p.q(i-1).n + q(1).n + 1$$
$$= (p.q(i-1) + q(1)).n + 1,$$

so that

$$q(i) = p.q(i-1) + q(1).$$

Then compute $c^{q(i)} \bmod p$.

Lemma A8.2

$$c^{q(i)} \bmod p = c^{i.q(1)} \bmod p. \tag{A8.2}$$

Proof   By induction,

$$c^{q(i)} \bmod p = c^{p.q(i-1)+q(1)} \bmod p = (c^{q(i-1)})^p.c^{q(1)} \bmod p = c^{q(i-1)}.c^{q(1)} \bmod p$$
$$= c^{(i-1).q(1)}.c^{q(1)} \bmod p = c^{i.q(1)} \bmod p.$$

It remains to compute $f_{ki} = c^{k.q(i)}$.

Lemma A8.3

$$f_{ki} = b^{k.i} \bmod p, \quad \text{where } b = c^{q(1)} \bmod p. \tag{A8.3}$$

Proof   According to (A8.2),

$$c^{k.q(i)} \bmod p = c^{i.q(1).k} \bmod p = (c^{q(1)})^{k.i} \bmod p.$$

Example A8.1   (Complete Ada source code available.) Consider the following case:

$$c = 2, p = 239, n = 17.$$

First observe that $239 = 14.17+1$ so that

$$q(1) = 14 \text{ and } b = 2^{14} \bmod 239 = 132;$$

then compute

$$f_{11} = 132,$$
$$f_{12} = 132.132 \bmod 239 = 216,$$
$$f_{13} = 216.132 \bmod 239 = 71,$$
$$\cdots$$
$$f_{21} = (f_{11})^2 \bmod 239 = 132.132 \bmod 239 = 216,$$
$$f_{22} = (f_{12})^2 \bmod 239 = 216.216 \bmod 239 = 51,$$
$$f_{23} = (f_{13})^2 \bmod 239 = 71.71 \bmod 239 = 22,$$
$$\cdots$$

$$f_{31} = (f_{11})^3 \bmod 239 = 216.132 \bmod 239 = 71,$$

$$f_{32} = (f_{12})^3 \bmod 239 = 51.216 \bmod 239 = 22,$$

$$f_{33} = (f_{13})^3 \bmod 239 = 22.71 \bmod 239 = 128,$$

and so on.

The following Ada program computes all the coefficients f_{ki} (the complete source code is available).

Algorithm A8.1   Ada Program for Computing f_{ki}

```
procedure frobenius is
  type frobenius_matrix is array (0..n-1, 0..n-1) of
  coefficient;
  f: frobenius_matrix;
  q, qq: polynomial;
  quotient, power: coefficient;
  cr: character;
begin
  quotient:=p/n;
  for i in 1..n-1 loop
    q(i):=(p*q(i-1)+quotient) mod p;
  end loop;
  power:=(2**quotient) mod p;
  qq(0):=1;
  for i in 1..n-1 loop
    qq(i):=(power*qq(i-1)) mod p;
  end loop;
  for i in 1..n-1 loop
    f(0,i):=1;
    for k in 1..n-1 loop
      f(k,i):=(f(k-1, i)*qq(i)) mod p;
    end loop;
  end loop;
  for k in 1..n-1 loop
    for i in 1..n-1 loop
      put("f("); put(k); put(","); put(i); put(")=");
       put(f(k,i));
      new_line;
    end loop;
    get(cr);
  end loop;
end frobenius;
```

9

HARDWARE PLATFORMS

This chapter is devoted to the hardware platforms available to implement the algorithms described in the preceding chapters. In the first section, some generalities in electronic system design are presented. The hardware platforms are then classified as instruction-set processor, ASIC based, and reconfigurable hardware. Special emphasis is given to FPGA technologies.

9.1 DESIGN METHODS FOR ELECTRONIC SYSTEMS

With the passing of time, *integrated circuit* (IC) technology has provided a variety of implementation formats for system designers. The implementation format defines the technology to be used, how the switching elements are organized and how the system functionality will be materialized. The implementation format also affects the way systems are designed and sets the limits of the system complexity. Today the majority of IC systems are based on *complementary metal-oxide semiconductor* (CMOS) technology. In modern digital systems, CMOS switching elements are prominent in implementing basic Boolean functions such as AND, OR, and NOT. With respect to the organization of switching elements, *regularity* and *granularity* of elements are essential parameters. The regularity has a strong impact on the design effort, because the reusability of a fairly regular design can be very simple. The problem raised by the regularity is that the structure may limit the usability and the performances of the resource. The granularity expresses the level of functionality encapsulated into one design object. Examples of fine-grain,

Synthesis of Arithmetic Circuits: FPGA, ASIC, and Embedded Systems
By Jean-Pierre Deschamps, Géry J. A. Bioul, and Gustavo D. Sutter
Copyright © 2006 John Wiley & Sons, Inc.

medium-grain, and coarse-grain are logic gates, *arithmetic and logic units* (ALUs), and *intellectual property components* (processor, network interfaces, etc.), respectively. The granularity affects the number of required design objects and, thereby, the required design or integration effort.

Depending on how often the structure of the system can be changed, the three main approaches for implementing its functionality are *dedicated systems, reconfigurable systems*, and *programmable systems*. In a dedicated system, the structure is fixed at the design time, as in *application-specific integrated circuits* (ASICs). In programmable systems, the data path of the processor core, for example, is configured by every instruction fetched from memory during the decode-phase. The traditional microprocessor-based computer is the classical example. In reconfigurable systems, the structure of the system can be altered by changing the configuration data, as in field programmable gate arrays (FPGAs).

9.1.1 Basic Blocks of Integrated Systems

The basic building blocks for digital ICs are input, output, data path, memory, and control, as in a common computer (Figure 9.1). Additionally, a communication network is necessary to interconnect the blocks. The implementation format of each basic block can, at least theoretically, be any combination of the previous classes.

The **data path** consists of regular elements, as in reconfigurable arrays, or dedicated pipelined blocks, as in superscalar processors. The granularity of data path elements can vary from single gates to processor arrays in multiprocessor architectures. A typical data path consists of an interconnection of basic logic (AND, OR, etc.), arithmetic operators (adders, multipliers, shifters, complement), and registers to store the intermediate results.

The **memory** can be characterized by size, number of ports, latency, and bandwidth. Latency is the access delay of a randomly chosen data element and bandwidth is the data rate. Memory components are typically two-dimensional regular structures, and latency is inversely proportional to the size of memory. The bandwidth depends on the memory buses, the internal organization of the

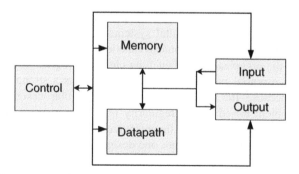

Figure 9.1 Components of a generic digital processor; arrows represent interconnection network.

memory, and the access logic. The memory is typically organized hierarchically. The faster and more expensive memories are near the data path; slower, bigger and cheaper memories are used to store less-frequently accessed data.

The main two memory classes are volatile memories, such as, for example, static random-access memory (SRAM) or dynamic random-access memory (DRAM), and nonvolatile memories, such as read-only memory, (ROM), and FLASH memory. SRAM is a fast memory, but typical implementations need six transistors per bit. DRAM is a dense memory, with only one transistor per bit, but the latency (i.e., access delay) is high. The FLASH memory also suffers from high latencies. The DRAM and FLASH memories are internally asynchronous and have different latencies for random and sequential accesses. For DRAM, in particular, a variety of solutions have been proposed for speeding up the overall performances, among them fast page mode accesses, synchronous interfaces, and intelligent control interfaces.

The **communication network** is another important component of an electronic system. Communication channels can be divided into dedicated channels (signals) and shared channels (buses and networks). The buses connect subsystems, and networks connect full systems according to the classical definition. The dedicated channels may be static point-to-point connections or dynamic switched connections. The buses can be further divided into data path memory and input/output (I/O) buses, parallel and serial buses, or synchronous and asynchronous buses, according to their purpose or physical implementation.

The **control module** determines what actions are carried out by the data path at any time, when and how memory, I/O modules, and data path are communicated or related. A controller is implemented as a finite state machine (FSM). The logic of a FSM can be implemented in different ways, with basic logic gates (AND, OR, NOT), using arrays of programmable logic devices (PLDs), or programming a memory (microprogramming). The way to implement FSM registers depends on the selected technology.

The **input/output modules** are used to connect the system to the outside world. These modules are most often slower than the other system parts. The throughput necessary in the I/O defines the communication network and the whole system organization.

9.1.2 Recurring Topics in Electronic Design

In the electronics industry, competition of one form or another leads to smaller, faster, cheaper, and better products and related manufacturing techniques. Gordon Moore's ([MOO1965]) insight, that the density of chips would double every 18 months, has proved incredibly accurate, and there is no end in sight. Nowadays, electronics industry leaders apply this principle to forecast three generations ahead. This competition and rapid growth create outstanding electronic design challenges.

9.1.2.1 Design Challenge: Optimizing Design Metrics The obvious design goal is to construct and implement within the desired functionality, but the key design challenge comes from the need for simultaneous optimizations with respect to numerous design metrics. The most common metrics are cost, size (physical space required by the system), performance, power consumption, flexibility, time-to-prototype (time needed to build a first working version of the system), time-to-market, maintainability, correctness, and safety. Competition generally exists between design metrics criteria; improving one may worsen others (Figure 9.2). Expertise with both software and hardware is needed to optimize design metrics. The designer must feel comfortable with various technologies in order to choose the best for a given application within given constraints. Some key concepts in electronic design are presented in what follows.

9.1.2.2 Cost in Integrated Circuits
When costs in electronic design are considered, one needs to worry about two main kinds of costs:

1. Cost of development, sometimes called *nonrecurring engineering* (NRE) cost.
2. Cost of manufacturing each copy of the system, called *unit cost* (UC).

The *total cost, TC*, is then readily calculated as

$$TC = NRE + UC.Q,$$

where Q stands for the quantity of units. The final cost, *FC*, per product (*per-product cost*) is then equal to

$$FC = TC/Q = NRE/Q + UC.$$

Trade-off strategies have to be implemented in relation to NRE and manufacturing costs. For example, according to Sperling ([SPE2003]), the NRE cost for the design of an ASIC, within the 100 nm, or less technology, can run around several million U.S. dollars; thus the manufactured quantities need to be important, that is, great enough to offset the impact of the NRE cost on the final cost. Figure 9.3 shows

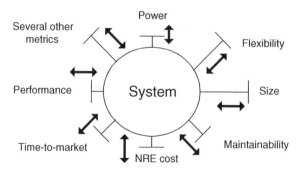

Figure 9.2   Design metrics competition.

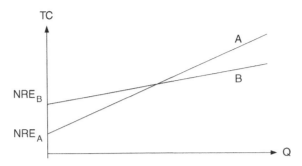

Figure 9.3 Cost evaluation as a function of manufactured quantities.

the evolution of the total costs with respect to manufactured quantities, marking low and high initial NRE costs (lines A and B, respectively).

9.1.2.3 Moore's Law In 1965, just four years after the first planar integrated circuit was discovered, Gordon Moore, cofounder of Intel™, made his famous observation about chip density growth. This is since referred to as 'Moore's law'. In his original paper ([MOO1965]), Moore observed an exponential growth in the number of transistors per integrated circuit and predicted that this trend would continue. In the following years, the pace slowed down a little bit, but data density has doubled approximately every 18 months, and this is the current definition of Moore's law. Figure 9.4 shows transistor density increment for Intel™ processors. In today's electronic world, these rules are also valid for memory capacities and computer systems performance.

9.1.2.4 Time-to-Market The time-to-market is the time required to develop a product up to the point it can be sold to customers. In today's electronics, it is

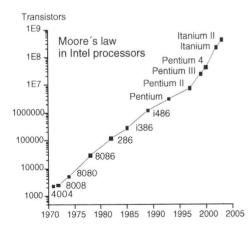

System	Year of Introduction	Transistors
4004	1971	2,250
8008	1972	2,500
8080	1974	5,000
8086	1978	29,000
286	1982	120,000
Intel386	1985	275,000
Intel486	1989	1,180,000
Pentium	1993	3,100,000
Pentium II	1997	7,500,000
Pentium III	1999	24,000,000
Pentium 4	2000	42,000,000
Itanium	2002	220,000,000
Itanium 2	2003	410,000,000

Figure 9.4 Moore's law for transistor capacity in Intel™ microprocessors.

one of the most important metrics. In most design projects, the average time-to-market constraint is about 6 to 12 months; extra delays could lead to unaffordable costs. An important related concept is the *market window*, that is, the period during which the product would have the highest sales. Figure 9.5 shows a typical market window and a simplified revenue model. In this model, the product life is equal to 2.P and the revenue's peak occurs at the half-life P; D represents the time delay.

This simplified model assumes that both market rise and market fall behaviors are linear; the market rise slope is the same for both on-time and delayed entries while the market fall assumes, for both cases, the same date for market zero. Time–revenue diagrams of market entries define triangles representing market penetration; triangle areas represent revenues. The percentage of revenue losses, materialized by the difference between the on-time and the delayed zone areas are then readily computed as $1 - (2.P - D)(P - D)/2.P^2$, where 2.$P$ is the full lifetime of the product and D the delay. For instance, a delay of 5 weeks for a product with a lifetime 2.P of one year (52 weeks) generates a loss of roughly 27%, but with a 3-month delay (13 weeks) the loss rises to 62.5%! This shows that delays are extremely expensive and therefore one of the most important driving forces in the IC industry; this motivates efforts toward new methodologies, design methods, and EDA tools.

9.1.2.5 Performance Metric The performance design metric is widely used to measure the "system quality," but it is also widely abused. The clock frequency and quantity of instructions per second are common criteria, but they are not always good enough as performance measures. For instance, in a digital camera, the user cares more about how fast it starts up or processes images, than about the internal clock frequency or instructions processing speed of the internal processor. More accurate and useful performance metrics are *latency* and *throughput*. Latency (response time) measures the time between the start and end of a task: in the camera

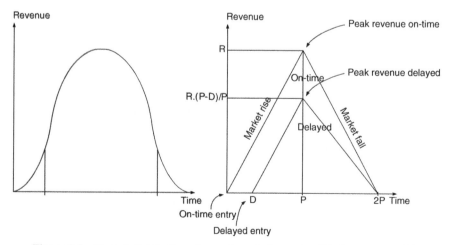

Figure 9.5   Typical distribution of a market window—simplified revenue model.

example, it could be the time to process an image, for example, 0.25 second (camera A) or 0.20 second (camera B). The throughput is defined as the quantity of tasks performed per second, for example, for the above cameras, 4 and 5 images per second, respectively. Observe that the throughput can be greater than the inverse of the latency thanks to possible concurrency or pipelining, for example, camera A could process 8 images per second, instead of 4, by capturing a new image while the previous image is being stored. Another useful metric is *speedup*, that is, comparing two performances: in the previous example, the throughput *speedup* of the pipelined processor of camera A over the one of camera B, is computed as A's performance/B's performance, that is, $8/5 = 1.6$.

9.1.2.6 The Power Dimension With the booming market of portable electronic devices, power consumption has turned out to be an important parameter in the design of integrated circuits. It allows avoiding expensive packaging: the chip life-time is increased, cooling is simplified, and battery-powered systems take advantage of increased autonomy and reduced weight.

There are two modes of power dissipation in integrated circuits: power generated during *static operation* or *dynamic operation. Static power* dissipation comes from currents flowing while no switching occurs. These include currents due to *pn*-junctions, static currents due to device biasing, and leakage currents. *Dynamic power* dissipation is a result of switching activities, whenever currents cause capacitances to be charged or discharged while performing logic operations. In CMOS devices the dissipated dynamic power P is proportional to the loading of capacitances C, the switching frequency F, and the square of supply voltage V:

$$P = k.C.F.V^2.$$

While power was becoming important in CMOS devices, designers have developed a number of tools and techniques to reduce the required power consumption. In CMOS devices, most of the power is used for voltage value switching on a wire; therefore most of the power reduction techniques try hard to ensure that a signal is not changed unless it really should be, then preventing other wasteful power sources. The power saving techniques range from simply turning off the processor/system when inactive—a technique used in almost all portable systems—to a careful power control of individual chip components. Observe, moreover, that power is very strongly related to the chip performance. A circuit can almost always be designed to require less energy for a task, if it is given more time to complete it. This has recently led to a set of techniques for dynamical control of the performances, to be kept as small as necessary to minimize the power used.

9.2 INSTRUCTION SET PROCESSORS

This section is devoted to architectures that execute a reduced set of instructions (*reduced instruction set computer*—RISC). Belonging to this category are

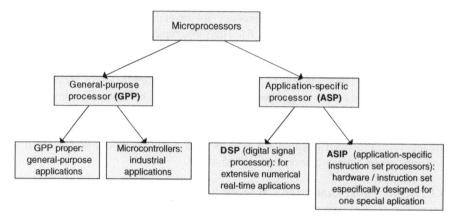

Figure 9.6 Classification of microprocessors according to level of specialization.

processors, microprocessors, *digital signal processors* (DSPs), *application specific instruction set processors* (ASIPs), and others. A first classification of microprocessors is presented in Figure 9.6; it is based on the levels of specialization.

The first microprocessors were sequential processors based on Von Neumann architecture. Initially, poor compilers and the lack of processor-memory bandwidth resulted in complex instruction sets: *complex instruction set computers* (CISCs). In the early 1980s, developments of RISC and VLSI technologies enabled the implementation of pipelined architectures, larger register banks, and more address space in a single-chip processor. In the 1990s the focus was on the exploitation of instruction-level parallelism, which eventually resulted in the modern *general-purpose processors* (GPPs). Superscalar processors and *very long instruction word* (VLIW) architectures are examples of dynamic and static parallelism.

An important dimension in processor architectures is the application orientation that has led to a variety of different types of instruction sets and organizations. Today, together with general-purpose processors (GPPs), *microcontrollers* share the feature of including memory and peripherals integrated into the same chip. Digital signal processors (DSPs) were invented to perform stream-based processing, such as filtering. The *Harvard architecture*,[1] involving advanced addressing, efficient interfaces, and powerful functional units such as fast multipliers and barrel shifters, provides superior performances in limited application spaces. *Multimedia processors* are targeted by applications where data parallelism can be exploited efficiently, for example, real-time compression and decompression of audio and video streams, and generation of computer graphics. Multimedia processors can be divided into microprocessors with multimedia instruction extensions and highly parallel DSPs. *Network processors* have an effective interconnection network between the processing elements, operating in parallel, and efficient instructions for packet

---

[1]The name comes from the *Harvard Mark* 1 relay-based computer, with stored instructions on punched tape and data in relay latches.

classification. *Reconfigurable data path processors* (RDPPs), have coarse-grain reconfigurable functional units such as ALUs or multipliers. Application-specific instruction set processors (ASIPs) are designed for a particular application set. The general ASIP idea is that the complete instruction set, together with the selection of the architecture template, is based on the application analysis. Application-specific processors are synthesized from the application description using a built-in architecture template. The strategy is to extract the computation resources from the application description and to synthesize the control that minimizes the resources within given performance constraints.

9.2.1 Microprocessors

The main classification starts from the differences between CISC and RISC processors. Actually, that is quite a philosophical classification, since most of today's processors are combinations of these models. Other criteria in the hardware classification of microprocessors are VLIW and superscalar architectures. Figure 9.7 describes the main microprocessor classes.

Another classification can be made according to the memory access. There are two fundamental memory access architectures: *Von Neumann* and *Harvard.* (Figure 9.8). In the first one, the memory is shared between instructions (program) and data; one data bus and one address bus only are used between processor and memory. Instructions and data have to be fetched in sequential order (known as the *Von Neumann bottleneck*); this limits the operation bandwidth. On the other hand, the Harvard architecture uses different memories for their instructions and data, requiring dedicated buses for each of them. Instructions and operands can therefore be fetched simultaneously. Different program and data bus widths are possible, allowing program and data memory to be better optimized with respect

Complex Instruction Set Computer (CISC)	Reduced Instruction Set Computer (RISC)
• Large number of complex addressing modes • Many versions of instructions for different operands • Different execution times for instructions • Few processor registers • Microprogrammed control logic	• One instruction per clock cycle • Memory accesses by dedicated load/store instructions • Few addressing modes • Hard-wired control logic
Very Long Instruction Word (VLIW)	Superscalar Processors
• Statically determined instruction-level parallelism (under compiler control) • Instructions are made of different machine operations whose execution is started in parallel • Many parallel functional units • Large register sets	• Subclasses of RISCs or CISCs • Multiple instruction pipelines for overlapping execution of instructions • Parallelism not necessarily exposed to the compiler

Figure 9.7 Types and characteristics of microprocessors, using the hardware structures as classification criteria.

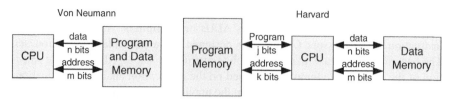

Figure 9.8   Von Neumann and Harvard architectures.

to the architectural requirements. A compromise between these two approaches is known as *modified Harvard architecture*, where programs and data are cached separately but are ultimately stored in one memory and connected over one bus.

The Von Neumann design is simpler than the Harvard one. Von Neumann's architecture has an efficient utilization of memory; it is the choice for most of the general-purpose processors. On the other hand, most DSPs and microcontrollers use Harvard architecture for streaming data; it allows greater and more predictable memory bandwidth.

9.2.2   Microcontrollers

Microcontrollers are single-chip computers; they are relatively slow and have very little memory, but cost less and are very easy to interface with real world devices. They are typically programmed in either *C* language (even subsets of *C*) or assembly languages. Microcontrollers are like single-chip computers; they are often embedded into other systems acting as processing/controlling units. For example, modern keyboards, microwave ovens, or washing machines use microcontrollers for decoding and controlling simple functions.

Microcontrollers usually adopt RISC architecture with a very small instruction set. A microcontroller virtually holds a complete system within it, with a CPU core, memory (ROM and RAM), and I/O circuits. Furthermore, a timer unit is available for operations based on time periods. A serial port is used for the data communication between devices or to a PC. Typically, small ROM-type memories are used to store the program codes. Another small RAM is used for data storage and stack management tasks. Traditionally, an 8- or 16-bit data path is used. Some microcontroller ports can be used to operate LEDs and relays, as well as logic circuit inputs. Recent high-end families of microcontrollers use a 32-bit data path, bigger memories, additional I/O capabilities such as A/D-D/A converters, faster standards of communications (e.g. CAN, USB, Ethernet), as well as connections with radio frequency or infrared circuits.

9.2.3   Embedded Processors Everywhere

Computing systems are everywhere. One first thinks about general-purpose computers (e.g., PCs, laptops, mainframes, servers), but another type of computing system, that is far more common, is the embedded computing system. These are computing systems embedded within electronic devices (radios, TV sets, phones, most home appliances, etc.). In fact, as a rough definition, practically any computing system other than a general-purpose computer. Billions of units are produced yearly, versus millions of general-purpose units only. They number perhaps close to one hundred per household or per high-end automobile.

The average new car has dozens of microprocessors inside. Embedded processors are present in every PC system: the keyboard and the mouse hold processors and there is a small CPU in each hard disk drive, floppy drive, CD-ROM/DVD drive, and so on. Except for graphics chips, most of these tiny helpers are 8-bit processors sourced by a number of companies. The very first IBM^{TM} PC/XT system already included about half a dozen different processor chips besides the 8088 CPU. The volume of 8-bit embedded chips is enormous and growing steadily. Today the estimated sale of these little processors is rounding three *billion* chips per year!

9.2.4 Digital Signal Processors

Digital signal processing deals with digital representations of signals; digital processors are used to analyze, modify, or extract information from signals: the *digital signal processor* (DSP) is an electronic system that processes digital signals. Internally, the signals are digitally represented as sequences of samples. Digital signals are obtained from physical signals via transducers (e.g., microphones) and *analog-to-digital converters* (ADCs); then digital signals can be converted back to physical analog signals using *digital-to-analog converters* (DACs) as shown in Figure 9.9.

Some of the most common application areas of DSPs are:

- *Image Processing:* pattern recognition, robotic vision, image enhancement, facsimile, satellite weather map, 3-D rendering, and animation.
- *Instrumentation and Control:* spectrum analysis, position and rate control, noise reduction, data compression, guidance, and GPS processing.
- *Audio and Video Processing:* speech recognition, speech synthesis, text to speech, digital audio, equalization, and machine vision.

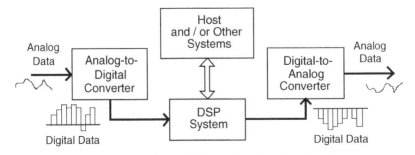

Figure 9.9 Typical DSP interaction with the real world.

- *Military:* secure communications, radar processing, sonar processing, and missile guidance.
- *Telecommunications:* echo cancellation, adaptive equalization, video conference, data communications, digital cellular telephony, pagers, wireless systems, and modems.
- *Biomedical:* patient monitoring, scanners, electronic brain mappers, ECG analysis, and X-ray storage and enhancement.

Typical algorithms implemented to carry out these tasks deal with finite (infinite) impulse response (FIR and IIR) filtering, frequency–time transformation, fast Fourier transform (FFT), and other correlation and convolution operations. The most important operations that a DSP needs to achieve are repetitive numerical computations with attention to numerical reliability (precision). Most of the tasks are real-time processes, and a high bandwidth memory is necessary, mostly through array accesses.

Though there are many DSPs, they are mostly designed with the same few basic operations in mind: they share the same set of basic characteristics. Most DSP operations require additions and multiplications. So DSPs usually involve hardware adders and multipliers, which can be used in parallel within a single instruction: multiply and accumulate (MAC) units.

The main differences between digital signal processors and general-purpose processors are related to the essence of problems they can respectively solve. Infinite streams of data, to be processed in real-time, are common applications for DSPs. Normally, DSPs have relatively small programs and data storage requirements; in addition, they run intensive arithmetic processes with a low amount of control and branching (in critical loops). Other remarkable DSP features are:

- They tend to be designed for just one program. Hence OS are much simpler; there is neither virtual memory or protection, nor typical OS facilities.
- Sometimes they run hard real-time applications. One must cope with anything that could happen in a time slot: all possible interruptions or exceptions must be accounted for, and their combined delays must be subtracted from the time interval.
- Algorithms are the most important and the binary compatibility is not an issue.
- A high amount of I/O with analog interfaces.
- Like other embedded processors, they are energy and cost efficient.

9.2.5 Application-Specific Instruction Set Processors

An application-specific instruction set processor (ASIP), alternatively referred to as a programmable platform, is a stored-memory CPU whose architecture is tailored for a particular set of applications. Programmability allows changes to the implementation so it can be used in several different products. Usually the ASIPs have high data path utilization. Application-specific architecture provides smaller silicon area and higher speed with respect to a general-purpose processor. Performance/cost enhancements are achieved using special-purpose registers and buses to complete the required computations, avoiding useless generality. Furthermore, special-purpose function units are implemented to perform long operations in fewer cycles. In addition, special-purpose control units allow execution of common combinations of instructions in fewer cycles. ASIPs mainly deal with audio/video processing, network processing, cellular telephony, wireless applications, and more. Another approach, related to ASIPs, is called reconfigurable instruction set processors (RISPs), where a processor core is coupled with reconfigurable logic and internal memory in order to extend functionalities and capabilities.

9.2.6 Programming Instruction Set Processors

The programming languages, used to program instruction set processors, depend on applications and specializations. The general-purpose processors use a wide variety of programming languages based on different programming paradigms, but when one moves to more specific fields, where microcontrollers, DSPs, or ASIPs are preferred, *assembly* language or C/C^{++} are most often used. Java is also present in embedded system design but it is still dominated by C/C^{++}.

The assembly language is a human-readable representation of a machine language that uses a reduced vocabulary of short words, such as, for example, MOV A(x),B(x). Some years ago, when CPU speed and storage space used to be measured in kilohertz and kilobytes, respectively, assembly language was the most cost-efficient way to implement programs; it is less used nowadays, as megas and gigas are more common prefixes, so that efficiency is getting less critical. Nevertheless, today's small microcontrollers are still programmed in assembly languages.

C offers smart compromises between the efficiency of coding in assembly language and the convenience and portability of writing in a well-structured, high-level language. By keeping many of its commands and syntax analogous to those common machine languages, and with several generations of optimizing compilers behind it, C makes it easy to write efficient codes without resigning readability. C^{++} is probably the most widely supported language today, and most commercial software products are written in C^{++}. The name reflects why. When it was introduced, it took all the benefits of the then-reigning development language C. Then it added the next set of features programmers were looking for: object-oriented programming. So, programmers didn't have to throw anything away and redo it: they just added these techniques to their repertoire, as needed. Free and commercial tools are available from various sources for just about any operating system (OS).

Most general-purpose processors are dominated by Windows, Linux, and Unix based OSs. In the embedded systems world, where real-time constraints, small memories, and other specific features appear, the range of possibilities is widening. In very small microcontrollers, no OS at all appears. For almost all OSs, the C/C^{++} is the de-facto programming language.

9.3   ASIC DESIGNS

Application-specific integrated circuits (ASICs) refer to those integrated circuits specifically built for preset tasks. Why use an ASIC solution instead of another off-the-shelf technology—programmable logic device (PLD, FPGA), or a microprocessor/microcontroller system? There are, indeed, many advantages in ASICs with respect to other solutions: increased speed, lower power consumption, lower cost (for mass production), better design security (difficult reverse engineering), better control of I/O characteristics, and more compact board design (less complex PCB, less inventory costs). However, there are important disadvantages: long turnaround time from silicon vendors (several weeks), expensive for low-volume production, very high NRE cost (high investment in CAD tools, workstations, and engineering manpower), and, finally, once committed to silicon the design cannot be changed.

Application-specific components can be classified into full-custom ASICs, semi custom ASICs, and field programmable ICs (Figure 9.10). This latter, sometimes referred to as programmable ASICs, will be analyzed in Section 9.4: programmable logic.

9.3.1   Full-Custom ASIC

In a full-custom ASIC all mask layers are customized (Figure 9.11). Full-custom designs offer the highest performance and the smallest die size, with the disadvantages of increased design time, higher complexity and costs, together with the highest risk of failure. This design option only makes sense when neither libraries nor IP cores are available, or when very high performances are required. Time

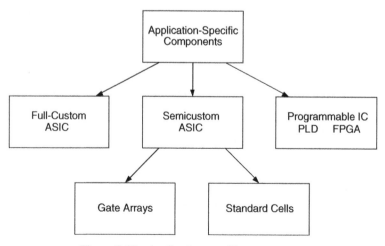

Figure 9.10   Application-specific components.

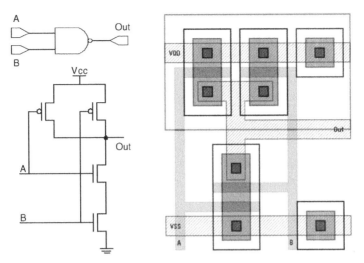

Figure 9.11 In a full-custom design every layer must be defined.

after time, fewer projects are really "full-custom," because of the very high cost and the prohibitively slow time-to-market. Most of the full-custom works are related to library cell generation or minor parts of a full design. Examples of full-custom IC specific parts are high-voltage (automobile, avionic), analog processing and analog/digital communication devices, sensors and transducers. Traditionally, microprocessors and memories were exclusively full-custom, but the industry is increasingly turning to semicustom ASIC techniques in these areas too.

9.3.2 Semicustom ASIC

In order to reduce the unaffordable cost of full custom in most projects, a wide variety of design approaches have been developed to shorten design time, cut down costs, and automate the processes. These approaches are commonly called semicustom. Semicustom designs are performed at logic (gate) level. As such, they lose some of the flexibility available from a full-custom fashion—that is the price paid for a much easier design technique. Semicustom solutions can be further categorized into *gate array* and *standard cell.*

9.3.2.1 Gate-Array ASIC *Gate arrays* (GAs) are basically composed of continuous arrays of *p*- and *n*-type transistors. The silicon vendor provides *master* or *base wafers*, to be then personalized according to the interconnection information supplied by the customer. Therefore, the designer supplies the personalized information that defines the connections between transistors in the gate array. Although a gate array standardizes the chip at the geometry level, user interaction is still typically carried out at logic level. The mapping, from transistors to gates, is performed through an ad hoc CAD tool. The gate array (also called masked gate array, or

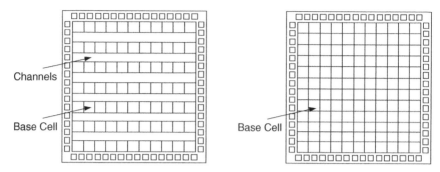

Figure 9.12 Gate array architectures: channeled and channelless gate arrays.

prediffused array) uses library components and macros that reduce the development time. Two main types of gate arrays can be mentioned: *channeled* and *channelless* (Figure 9.12). In a channeled gate array, the interconnections are drawn within predefined spaces (channels) between rows of logic cells. In a channelless gate array (channelfree gate array or sea-of-gates), there are no connection channels; the connections are drawn with the upper metal layers, that is, on the top of the logic cells. In both cases, only some mask layers (the upper ones) must be customized.

9.3.2.2 Standard-Cell-Based ASIC Standard cells are logic components (e.g., gates, multiplexers, adders, flip-flops) previously designed and stored in a library. A design is created using these library cells as inputs to a CAD system: logic schematic diagram or *hardware description language* (HDL) code description. Next, a further CAD tool automatically converts the design into a chip layout. Standard-cell designs are typically organized on the chip, as rows of constant height cells (Figure 9.13). Together with logic-level component cells, standard-cell systems typically offer-higher-level functions such as multipliers and memory arrays. This allows the use of predesigned (or automatically generated) high-level components to complete the design.

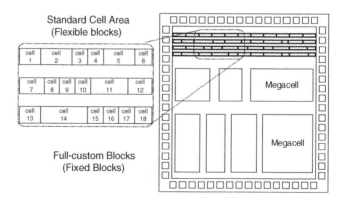

Figure 9.13 Typical standard cell layout.

9.3.3   Design Flow in ASIC

Figure 9.14 shows a typical semicustom ASIC design flow (excepted for the test vector generation). The steps in a traditional ASIC design flow (with a brief description) are:

- *Design Entry:* enters the design using either a hardware description language such as VHDL or Verilog (see Section 9.5) or a schematic entry.
- *Logic Synthesis:* from the HDL or schematic entry, extracts a netlist, that is, a description of the logic cells and their connections. The synthesis tool can infer a hardware implementation with the behavior as the HDL description.
- *System Partitioning:* divides a large system into ASIC-sized pieces.
- *Prelayout (Behavioral) Simulation:* checks the circuit working.
- *Floorplanning:* arranges the different blocks of the circuit on the chip.
- *Placement:* sets the cell locations in a block.
- *Routing:* creates the connections between cells and blocks.

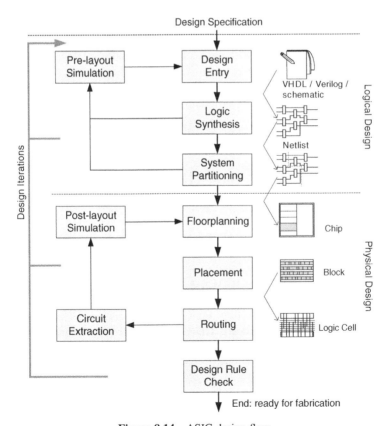

Figure 9.14   ASIC design flow.

- *Extraction (Back Annotation)*: determines the resistance and capacitance of the interconnections and calculates delays for simulation purposes.
- *Postlayout (Physical) Simulation:* checks the circuit working after including the delays created by interconnection loads.
- *Design Rule Check (DRC):* verifies that the circuit layout complies with the specifications of the design rules. DRC tools can range from a simple physical spacing check-up to complex tests.

9.4 PROGRAMMABLE LOGIC

Logic devices can be classified into two broad categories: fixed and programmable. Circuits in a fixed logic device are permanent: they perform one function or set of functions, and once manufactured, they cannot be changed. On the other hand, programmable logic devices (PLDs) are standard, off-the-shelf parts that can be modified at any time to perform any number of functions. A key benefit of using PLDs is that, during the design phase, designers can change the circuitry as often as they want until the design operates satisfactorily. PLDs are based on rewritable memory technology: to modify the design, the device only needs to be reprogrammed. Reusability is a further attractive feature of PLDs. Many types of programmable logic devices are currently available. The range of market products includes small devices capable of implementing a handful of logic equations up to huge FPGAs that can hold an entire processor core plus a number of peripherals. Besides this impressive diversity of sizes, numerous alternative architectures are offered to the designer. Within programmable logic devices, two major types deserve to be highlighted: the *complex programmable logic device* (CPLD) and *field programmable gate array* (FPGA). They are described below.

9.4.1 Programmable Logic Devices (PLDs)

At the low end of the spectrum stand the original programmable logic devices (PLDs). They were the first chips that could be used as hardware implementation of a flexible digital logic design. For instance, a couple of the 74xxx board parts could be removed and replaced by a single PLD. Other names also stand for this class of device: *programmable logic array* (PLA), *programmable array of logic* (PAL), and *generic array logic* (GAL). A PLD is made of a fully connected set of macrocells. These macrocells typically consist of some combinational logic (typically AND/OR gates and a flip-flop: Figure 9.15). A small Boolean equation can thus be built within each macrocell. This equation will convert the state of some binary inputs into a binary output and, if necessary, store that output in a flip-flop until the next clock edge. Obviously, the characteristics of the available logic gates and flip-flops are specific to each manufacturer and product family. But the general idea holds for any product. Hardware descriptions for these simple PLDs are generally either written in languages like ABEL or PALASM (the HDL equivalent of assembler) or drawn with the help of a schematic capture tool.

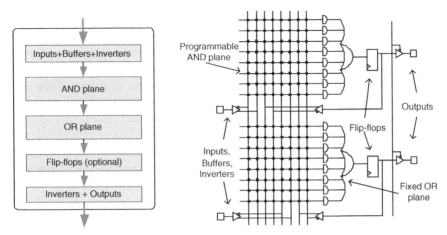

Figure 9.15   Typical PLD architecture.

As chip densities increased, PLD manufacturers naturally developed their products toward larger parts, called complex programmable logic devices (CPLDs). In a certain respect, CPLDs can be described as several PLDs (plus some programmable interconnection) in a single chip. The larger size of a CPLD allows implementing either more logic equations or more complicated designs.

Figure 9.16 contains a block diagram of a typical CPLD: within each logic block stands the equivalent of one PLD. Because CPLDs can hold larger designs than PLDs, their potential uses are quite wide-ranging. Sometimes they are used for simple applications, like address decoding, but more often they contain high-performance control-logic or complex finite state machines. At the high-end (in terms of numbers of gates), there is some overlapping with FPGAs in potential applications. Traditionally, CPLDs have been preferred over FPGAs whenever high-performance logic is required. Because of its less flexible internal architecture, delays through a CPLD are more predictable and usually shorter.

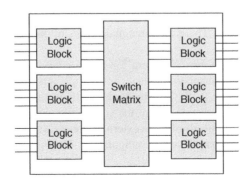

Figure 9.16   Internal structure of a theoretical CPLD.

9.4.2   Field Programmable Gate Array (FPGA)

Field programmable gate arrays (FPGAs) can be used to implement just about any hardware design. One common use of the FPGA is the prototyping of a piece of hardware that will eventually be implemented later into an ASIC. Nevertheless, FPGAs have been increasingly used as the final product platforms. Their use depends, for a given project, on the relative weights of desired performances, development, and production costs. See Section 9.1.2.2.

9.4.2.1 Why FPGA? A Short Historical Survey
By the early 1980s, most of the typical logic circuit systems were implemented within a small variety of standard *large scale integrated* (LSI) circuits: microprocessors, bus-I/O controllers, system timers, and so on. Nevertheless, every system still had the need for random "glue logic" to connect the large ICs, for example, generate global control signals and data formatting (serial to parallel, multiplexing, etc.). Custom ICs were often designed to replace the large amount of glue logic and consequently reduce system complexity and manufacturing cost, as well as improve performances. However, custom ICs are expensive to develop, while generating time-to-market (TTM) delays because of the prohibitive design time. Therefore the custom IC approach was only viable for products with very high volume (lowering the NRE cost impact), and not TTM sensitive. Coping with this problem, Xilinx™ (a startup company) introduced, in 1984,[2] the FPGA technology as an alternative to custom ICs for implementing glue logic. Thanks to *computer-aided design* (CAD) tools, FPGA circuits can be implemented in a relatively short amount of time: no physical layout process, no mask making, no IC manufacturing, lower NRE costs, and short TTM.

9.4.2.2 Basic FPGA Concepts
The basic FPGA architecture consists of a two-dimensional array of logic blocks and flip-flops with means for the user to configure (i) the function of each logic blocks, (ii) the inputs/outputs, and (iii) the interconnection between blocks (Figure 9.17). Families of FPGAs differ from each other by the physical means for implementing user programmability, arrangement of interconnection wires, and basic functionality of the logic blocks.

Programming Methods   There are three main types of programmability:

- *SRAM Based* (e.g., Xilinx™, Altera™): FPGA connections are achieved using pass-transistors, transmission gates, or multiplexers that are controlled by SRAM cells (Figure 9.18). This technology allows fast in-circuit reconfiguration. The major disadvantages are the size of the chip, required by the RAM technology, and the needs of some external source (usually external nonvolatile memory chips) to load the chip configuration. The FPGA can be programmed an unlimited number of times.

---

[2]The original idea has been published and patented by Sven E. Wahlstrom ([WAH1967]).

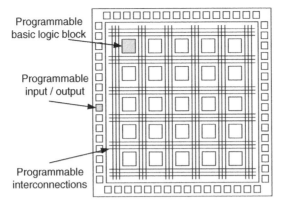

Figure 9.17 Basic architecture of FPGA: two-dimensional array of programmable logic cells, interconnections, and input/ouput.

- *Antifuse Technology* (e.g., Actel™, Quicklogic™): an antifuse remains in a high-impedance state until it is programmed into a low-impedance or "fused" state (Figure 9.18). This technology can be used only once on *one-time programmable* (OTP) devices; it is less expensive than the RAM technology.
- *EPROM/EEPROM Technology* (various PLDs): this method is the same as that used in EPROM/EEPROM memories. The configuration is stored within the device, that is, without external memory. Generally, in-circuit reprogramming is not possible.

Look-Up Tables The way logic functions are implemented in a FPGA is another key feature. Logic blocks that carry out logical functions are *look-up tables* (LUTs), implemented as memory, or multiplexer and memory. Figure 9.19 shows these alternatives, together with an example of memory contents for some basic operations. A $2^n \times 1$ ROM can implement any n-bit function. Typical sizes for n are 2, 3, 4, or 5.

In Figure 9.19a, an n-bit LUT is implemented as a $2^n \times 1$ memory; the input address selects one of 2^n memory locations. The memory locations (latches) are normally loaded with values from the user's configuration bit-stream. In Figure 9.19b,

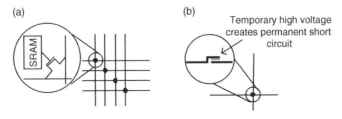

Figure 9.18 Programming methods: (a) SRAM connection and (b) antifuse.

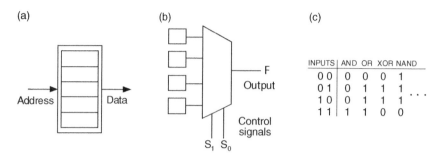

Figure 9.19 Look-up table implemented as (a) memory or (b) multiplexer and memory. (c) Memory contents example for different logic functions.

the multiplexer control inputs are the LUT inputs. The result is a general-purpose "logic gate." An n-LUT can implement any n-bit function.

An n-LUT is a direct implementation of a function *truth table*. Each latch location holds the value of the function corresponding to one input combination. An example of a 4-LUT is shown in Figure 9.20.

FPGA Logic Block A simplified FPGA logic block can be designed with a LUT, typically a 4-input LUT, implementing a combinational logic function, and a register that optionally stores the output of the logic generator (Figure 9.21).

9.4.3 Xilinx™ Specifics

This section is devoted to the description of the Xilinx Virtex family ([XIL2001]) and, in particular, the Spartan II ([XIL2004c]), a low cost version of Virtex. The Virtex II ([XIL2004a]) device family is a more recent and powerful architecture,

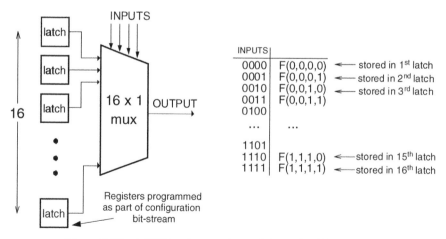

Figure 9.20 4-LUT implementation and the truth-table contents.

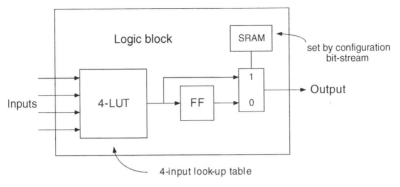

Figure 9.21   A basic FPGA logic block.

sharing most of the capabilities and basic concepts of Virtex. Spartan III ([XIL2004b]) is the low-cost version of Virtex II. Finally, Virtex II-Pro features additional hardwired Power-PC processors. For simplicity, minor details are omitted in the following.

All Xilinx FPGAs contain the same basic resources (Figure 9.22):

- Configurable logic blocks (CLBs), containing combinational logic and register resources.
- Input/output blocks (IOBs), interface between the FPGA and the outside world.
- Programmable interconnections (PIs).
- RAM blocks.
- Other resources: three-state buffers, global clock buffers, boundary scan logic, and so on.

Furthermore, Virtex II and Spartan III devices contain resources such as dedicated multipliers and a digital clock manager (DCM). The Virtex II-Pro also includes embedded Power-PC processors and full-duplex high-speed serial transceivers.

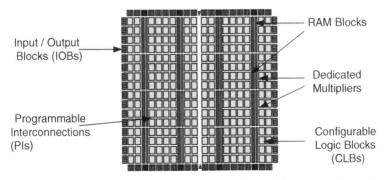

Figure 9.22   Example of distribution of CLBs, IOBs, PIs, RAM blocks, and multipliers in Virtex II.

9.4.3.1 Configurable Logic Blocks (CLBs) The basic building block of Xilinx CLBs is *the slice*. Virtex and Spartan II hold two slices in one CLB, while Virtex II and Spartan III hold four slices per CLB. Each slice contains two 4-input function generators (F/G), carry logic, and two storage elements. Each function generator output drives both the CLB output and the D-input of a flip-flop. Figure 9.23 shows a detailed view of a single Virtex slice. Besides the four basic function generators, the Virtex/Spartan II CLB contains logic that combines function generators to provide functions of five or six inputs. The look-up tables and storage elements of the CLB have the following characteristics:

- *Look-Up Tables (LUTs):* Xilinx function generators are implemented as 4-input look-up tables. Beyond operating as a function generator, each LUT can be programmed as a (16×1)-bit synchronous RAM. Furthermore, the two LUTs can be combined within a slice to create a (16×2)-bit or (32×1)-bit synchronous RAM, or a (16×1)-bit dual-port synchronous RAM. Finally, the LUT can also provide a 16-bit shift register, ideal for capturing high-speed data.
- *Storage Elements:* The storage elements in a slice can be configured either as edge-triggered D-type flip-flops or as level-sensitive latches. The D-inputs can be driven either by the function generators within the slice or directly from the slice inputs, bypassing the function generators. As well as clock and clock enable signals, each slice has synchronous set and reset signals.

9.4.3.2 Input/Output Blocks (IOBs) The Xilinx IOB includes inputs and outputs that support a wide variety of I/O signaling standards. The IOB storage elements act either as D-type flip-flops or as latches. For each flip-flop, the set/reset (SR) signals can be independently configured as synchronous set, synchronous reset, asynchronous preset, or asynchronous clear. Pull-up and pull-down resistors and an optional weak-keeper circuit can be attached to each pad. IOBs are programmable and can be categorized as follows:

- *Input Path:* A buffer in the IOB input path is routing the input signals either directly to internal logic or through an optional input flip-flop.
- *Output Path:* The output path includes a 3-state output buffer that drives the output signal onto the pad. The output signal can be routed to the buffer directly from the internal logic or through an optional IOB output flip-flop. The 3-state control of the output can also be routed directly from the internal logic or through a flip-flop that provides synchronous enable and disable signals.
- *Bidirectional Block:* This can be any combination of input and output configurations.

9.4.3.3 RAM Blocks Xilinx FPGA incorporates several large RAM memories (*block select RAM*). These memory blocks are organized in columns along the chip. The number of blocks, ranging from 8 up to more than 100, depends on the

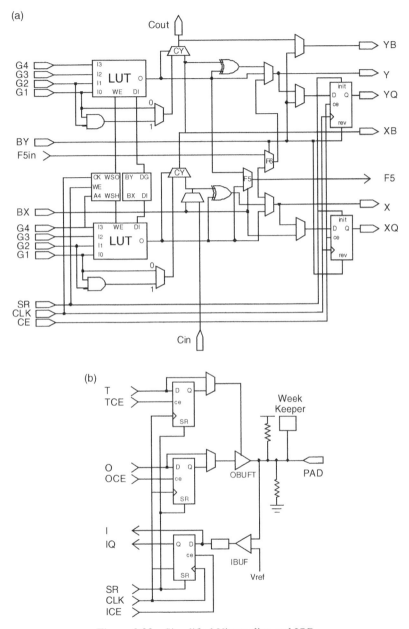

Figure 9.23 Simplified Virtex slice and IOB.

device size and family. In Virtex/Spartan II, each block is a fully synchronous dual-ported 4096-bit RAM, with independent control signals for each port. The data width of the two ports can be configured independently. In Virtex II/ Spartan III, each block provides 18-kbit storage.

9.4.3.4 Programmable Routing Adjacent to each CLB stands a *general routing matrix* (GRM). The GRM is a switch matrix through which resources are connected (Figure 9.24); the GRM is also the means by which the CLB gains access to the general-purpose routing. Horizontal and vertical routing resources for each row or column include:

- *Long Lines:* bidirectional wires that distribute signals across the device. Vertical and horizontal long lines span the full height and width of the device.
- *Hex Lines* route signals to every third or sixth block away in all four directions.
- *Double Lines:* route signals to every first or second block away in all four directions.
- *Direct Lines:* route signals to neighboring blocks—vertically, horizontally, and diagonally.
- *Fast Lines:* internal CLB local interconnections from LUT outputs to LUT inputs.

The routing performance factor of internal signals is the longest delay path that limits the speed of any worst-case design. Consequently, the Xilinx routing architecture and its place-and-route software were defined in a single optimization process.

Xilinx devices provide high-speed, low-skew *clock distribution.* Virtex provides four primary global nets that drive any clock pin; instead, Virtex II has 16 global clock lines—eight per quadrant.

9.4.3.5 Arithmetic Resources in Xilinx FPGAs Modern FPGAs have special circuitry to speed-up arithmetic operations. Therefore adders, counters, multipliers, and other common operators work much faster than the same operations built from LUTs and normal routing only.

Dedicated carry logic provides fast arithmetic carry capability for high-speed arithmetic functions. There is one carry chain per slice; the carry chain height is 2 bits per slice. The arithmetic logic includes one XOR gate that allows a 1-bit full adder to be implemented within the available LUT (see Section 11.1.10). In addition, a dedicated AND gate improves the efficiency of multiplier implementations (see Section 12.1.7).

The dedicated carry path can also be used to cascade function generators for implementing wide logic functions.

9.4.4 FPGA Generic Design Flow

The FPGA design flow has several points in common with the semicustom ASIC design flow. Figure 9.25 presents a simplified FPGA design flow. The successive process phases (blocks) of Figure 9.25 are described as follows:

- *Design Entry:* creation of design files using schematic editor or hardware description language (Verilog, VHDL, Abel).

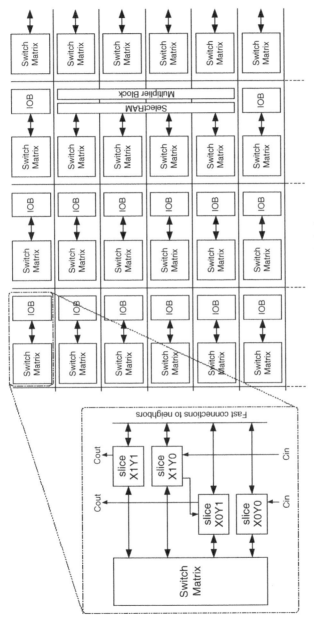

Figure 9.24 Routing in Virtex II devices.

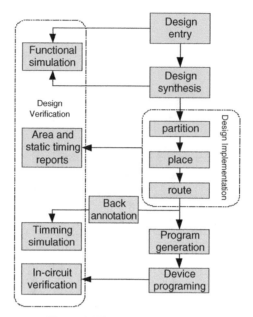

Figure 9.25   FPGA design flow.

- *Design Synthesis:* a process that starts from a high level of logic abstraction (typically Verilog or VHDL) and automatically creates a lower level of logic abstraction using a library of primitives.
- *Partition (or Mapping):* a process assigning to each logic element a specific physical element that actually implements the logic function in a configurable device.
- *Place:* maps logic into specific locations in the target FPGA chip.
- *Route:* connections of the mapped logic.
- *Program Generation:* a bit-stream file is generated to program the device.
- *Device Programming:* downloading the bit-stream to the FPGA.
- *Design Verification:* simulation is used to check functionalities. The simulation can be done at different levels. The functional or behavioral simulation does not take into account component or interconnection delays. The timing simulation uses back-annotated delay information extracted from the circuit. Other reports are generated to verify other implementation results, such as maximum frequency and delay and resource utilization.

The partition (or mapping), place, and route processes are commonly referred to as *design implementation.*

9.5 HARDWARE DESCRIPTION LANGUAGES (HDLs)

A hardware description language (HDL) is a computer language designed for formal description of electronic circuits. It can describe a circuit operation, its structure, and the input stimuli to verify the operation (using simulation). A HDL model is a text-based description of the temporal behavior and/or the structure of an electronic system. In contrast to a software programming language, the HDL syntax and semantics include explicit notations for expressing *time* and *concurrencies*, which are the primary attributes of hardware. Languages, whose only characteristics are to express circuit connectivity within a hierarchy of blocks, are properly classified as *netlist* languages. One of the most popular netlist formats and industry standards is EDIF, acronym for *Electronic Data Interchange Format* ([EIA2004]).

Traditional programming languages such as C/C^{++} (augmented with special constructions or class libraries) are sometimes used for describing electronic circuits. They do not include any capability for expressing time explicitly and, consequently, are not proper hardware description languages. Nevertheless, several products based on C/C^{++} have recently appeared: Handel-C ([CEL2004]), System-c ([SYS2004]), and other Java-like based such as JHDL ([JHD2004]) or Forge ([XIL2004d]).

Using a proper subset of nearly any hardware description or software programming language, software programs called *synthesizers* can infer hardware logic operations from the language statements and produce an equivalent netlist of generic hardware primitives to implement the specified behavior.

9.5.1 Today's and Tomorrow's HDLs

The two main players in this field are VHDL and Verilog. VHDL stands for VHSIC (very high speed integrated circuits) hardware description language. In the 1980s the U.S. Department of Defense and the IEEE sponsored the development of this hardware description language with the goal to develop very high-speed integrated circuits. It has now become one of the industry's standard languages used to describe digital systems. Around the same time another language, later called Verilog, with similarity to the C-language syntax, was developed. In 1989, Cadence Company acquired the license, and later, in 1990, opened Verilog to the public. Both VHDL and Verilog are powerful languages that allow describing and simulating complex digital systems. Verilog is popular within Silicon Valley companies; while VHDL is used more by governments, in Europe, in Japan, and in most of the universities worldwide. Most major CAD frameworks now support both languages.

Another recognized HDL is ABEL (advanced Boolean equation language); it has been specifically designed for programmable logic devices. ABEL is less powerful than the formerly mentioned languages and is less used in the industry. Growth in complexity and strict time-to-market requirements for new system designs demand faster and simpler ways to describe system behaviors. The C^{++} and Java extensions, to support hardware description, seem to have a future because of the possibility to describe both hardware and software. The biggest challenge in this field is to have powerful synthesizers that can recognize and extract hardware and software from a previous nonhardware-oriented code. Most examples in this book are presented in VHDL.

9.6   FURTHER READINGS

This chapter presented several topics related to the hardware platforms available to implement algorithms; further readings are recommended. In the field of ASIC design, outstanding references are [RAB2003] and [SMI1997]; a prominent Internet site is [SMI2004]. Furthermore, the web sites of electronic design automation (EDA) major companies (Cadence, Mentor, Synopsys, etc.) offer excellent sources of information.

The field of embedded systems design is a world in itself. The literature on design methodologies within different technologies has been booming in the last years. Some good surveys can be found in [VAH2002] and [WOL2001]. A good introduction to the FPGA is available on the Xilinx web page ([XIL2004]), offering detailed information on products, tools, and datasheets. Other manufacturers such as Altera, Atmel, and Quicklogic also feature very complete web pages. Finally, a number of annual symposia are dedicated to FPGA applications and technologies, among them the *IEEE Symposium on FPGAs for Custom Computing Machines* (FCCM), the *International Workshop on Field Programmable Logic and Applications* (FPL), and the *ACM/SIGDA International Symposium on Field-Programmable Gate Arrays* (FPGA).

About VHDL, [ASH1996], [MAZ1993], and [RUS2000] may be consulted, while numerous free-access tutorials can be found on the internet: [SPI2001], [ZHA2004], [GLA2004]. A number of free VHDL simulators, tutorials, examples, and useful references are also available from most CAD companies.

9.7   BIBLIOGRAPHY

[ASH1996] P. J. Ashenden, *The Designer's Guide to VHDL*. Morgan Kaufmann, San Francisco, CA, 1996.

[CEL2004] Celoxica Ltd, *Handel-C: Software-Compiled System Design* [online], [cited Sept. 2004], available from: http://www.celoxica.com/.

[EIA2004] Electronic Industries Alliance (EIA), *The Electronic Design Interchange Format (EDIF) Standard* [online], [cited Sept. 2004], available from: http//:www.edif.org.

[GLA2004] W. H. Glauert, *VHDL Tutorial* [online]. Universität Erlangen-Nürnberg [cited Sept. 2004], available from http://www.vhdl-online.de/.

[JHD2004] *JHDL Java Hardware Description Language Home Page* [online]. BYU (Brigham Young University) JHDL, Open Source FPGA CAD Tools [cited Sept. 2004], available from: http://www.jhdl.org.

[MAZ1993] S. Mazor and P. Langstraat, *A Guide to VHDL*. Kluwer Academic Publishers, Norwell, MA, 1993.

[MOO1965] G. E. Moore, Cramming more components onto integrated circuits. *Electronics* **38**(8), Apr. 19(1965).

[RAB2003] J. Rabaey, A. Chandrakasan, and B. Nikolic, *Digital Integrated Circuits*, 2nd ed. Prentice Hall, Englewood Cliffs, NJ, 2003.

[RUS2000] A. Rushton, *VHDL for Logic Synthesis*, 2nd ed. Wiley, Hoboken, NJ, 1998.

[SMI1997] M. J. S. Smith, *Application-Specific Integrated Circuits*. Addison-Wesley, Reading, MA, 1997.

[SMI2004] Michael Smith, *ASICs... the website* [online], [cited Sept. 2004], available from http://www-ee.eng.hawaii.edu/~msmith/ASICs/HTML/ASICs.htm.

[SPE2003] E. Sperling, FPGAs vs. ASICs [online]. *Electronic News*, Dec. 2003, available from http://www.reed-electronics.com/electronicnews/.

[SPI2001] J. Van der Spiegel, *VHDL Tutorial*. University of Pennsylvania, Dept. of Electrical Engineering, 2001. Available from http://www.seas.upenn.edu/~ee201/vhdl/vhdl_primer.html.

[SYS2004] The System-C community. *Everything You Want to Know About System-C* [online], [cited Sept. 2004], available from http://www.systemc.org.

[VAH2002] F. Vahid and T. Givargis, *Embedded System Design: A Unified Hardware/Software Introduction*. Wiley, Hoboken, NJ, 2002.

[WAH1967] S. E. Wahlstrom, Programmable logic arrays—cheaper by the millions. *Electronics* **40**(25): 90–95, (Dec. 1967).

[WOL2001] W. H. Wolf, *Computers as Components Principles of Embedded Computing System Design*. Morgan Kaufmann, XXX, 2001.

[XIL2001] Xilinx, Inc., *Virtex 2.5 V Field Programmable Gate Arrays*, DS003-1 (v2.5), April 2001. Available from http//:www.xilinx.com.

[XIL2004] Xilinx, Inc., *The Web Page for Programmable Logic* [online], [cited Sept. 2004], available from: http://www.xilinx.com.

[XIL2004a] Xilinx, Inc., *Virtex-II Platform FPGAs: Complete Data Sheet*, DS031 (v3.3), June 2004. Available from http//:www.xilinx.com.

[XIL2004b] Xilinx, Inc., *Spartan-III FPGA Family: Complete Data Sheet*, DS099 (v1.5), July 2004. Available from: http//: www.xilinx.com.

[XIL2004c] Xilinx, Inc., *Spartan-II 2.5V FPGA Family: Complete Data Sheet*, DS001 (v2.5), Sept. 2004. Available from http//:www.xilinx.com.

[XiL2004d] Xilinx, Inc., *Forge Registration Instructions* [online]. A Java-based design language/application that produces a synthesizable Verilog netlist [cited Sept. 2004], available from http://www.xilinx.com/ise/advanced/forge_get.htm.

[ZHA2004] W. Zhang, *VHDL Tutorial: Learn by Example*, [online]. University of California Riverside [cited Sept. 2004], available from http://www.cs.ucr.edu/content/esd/labs/tutorial/.

10

CIRCUIT SYNTHESIS: GENERAL PRINCIPLES

This chapter is a summary of digital system architecture. The general problem dealt with is the synthesis of a digital circuit implementing some given algorithm, in such a way that a set of conditions related to the costs and the delays are satisfied. The costs to be taken into account could be

- the number of cells in the case of an *application specific integrated circuit* (ASIC) or a *field programmable gate array* (FPGA),
- the number of integrated circuits if standard components are used.

There are other costs, among them those related to the circuit packages, such as

- the number of pins of the integrated circuits,
- the electric power consumption.

The most important timing conditions concern the data input and output operations, among others,

- the maximum delay between the input of a data set and the output of the result (*latency*),
- the maximum sample frequency (*throughput*).

Synthesis of Arithmetic Circuits: FPGA, ASIC, and Embedded Systems
By Jean-Pierre Deschamps, Géry J. A. Bioul, and Gustavo D. Sutter
Copyright © 2006 John Wiley & Sons, Inc.

10.1 RESOURCES

In order to synthesize a digital circuit the designer has to develop—or states the necessity to develop—computation, memory, and connection resources:

- The *computation resources* are deduced from the operations included in the algorithm; they are characterized by their functions, their computation times, and their costs.
- The *memory resources* are registers, banks of registers, random access or read-only memories, stacks, and queues, characterized by their minimum setup and hold times, maximum propagation time, and read and write cycles, among other features, as well as by their costs.
- The *connection resources* are multiplexers and tristate buffers used for controlling the transfer of data between computation resources and registers; they are characterized by their propagation times and costs.

Consider two examples

Example 10.1 (Combinational circuit; complete VHDL source code available.) Synthesize an n-bit adder based on Algorithm 4.2 with $B = 2$. The corresponding computation resources are:

- the carry-propagate function p defined by $p(a, b) = 1$ iff $a + b = 1$, that is, a 2-input XOR gate;
- the carry-generate function g defined by $g(a, b) = 1$ if $a + b > 1$, any value if $a + b = 1$, 0 if $a + b < 1$ (Comment 4.1(2)) so that g could be chosen equal to a, b or $a \cdot b$; let it be b;
- the 3-operand mod 2 sum: $mod\_sum(a, b, c) = (a + b + c) \bmod 2$, that is, a 3-input XOR gate equivalent to two 2-input ones.

The only connection resource type is a 2-to-1 multiplexer able to transfer either $q(i)$ or $g(i)$ to $q(i + 1)$.

If there is no restriction as regards the number of resources, a combinational (memoryless) circuit can be synthesized. An example is shown in Figure 10.1 (its corresponding FPGA implementation is described in Chapter 11).

Its costs and delays are equal to

$$C_{adder}(n) = n.(2.C_{XOR2} + C_{mux2-1}) \quad \text{and}$$
$$T_{adder}(n) = 2.T_{XOR2} + (n - 1).T_{mux2-1} \tag{10.1}$$

where C_{XOR2} and C_{MUX2-1} are, respectively, the costs of a 2-input XOR gate and of a 2-to-1 one-bit multiplexer, while T_{XOR2} and T_{mux2-1} are the corresponding propagation times.

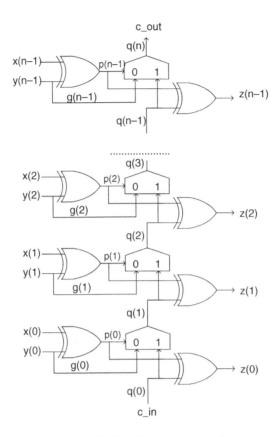

Figure 10.1 Combinational *n*-bit adder.

The corresponding VHDL model is the following one:

```
entity example10_1 is
port (
  x, y: in std_logic_vector(n-1 downto 0);
  c_in: in std_logic;
  z: out std_logic_vector(n-1 downto 0);
  c_out: out std_logic
);
end example10_1;
architecture circuit of example10_1 is
  signal p, g: std_logic_vector(n-1 downto 0);
  signal q: std_logic_vector(n downto 0);
begin
  q(0)<=_in;
  iterative_step: for i in 0 to n-1 generate
    p(i)<=x(i) xor y(i);
    g(i)<=y(i);
```

```
     with p(i) select q(i+1)<=q(i) when '1', g(i) when others;
     z(i)<=p(i) xor q(i);
   end generate;
   c_out<=q(n);
end circuit;
```

Example 10.2 (Complete VHDL source code available.) As a second example—
this one including memory resources—an m-operand n-bit adder is synthesized. It is
based on Algorithm 4.11 with $B = 2$. The only computation resource is an n-bit
adder. A memory resource, namely, an n-bit register, is necessary in order to
store the value of the variable *accumulator*. Two connection resources must be
used: the first one, an m-to-1 n-bit multiplexer, selects the second operand $x(j)$ as
a function of j; the other one, a 2-to-1 n-bit multiplexer loads either 0 or the
adder output within the register.

The corresponding circuit is made up of a data path (Figure 10.2) and a control
unit generating the signals op_select, clear, and load. In order to allow the connection of the circuit to some main processor, the control unit is also in charge of a
start/done communication protocol.

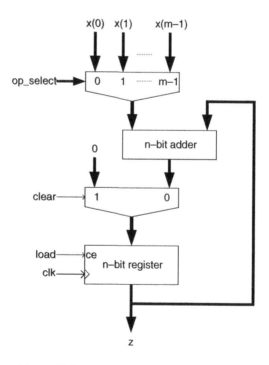

Figure 10.2   Data path of an m-operand adder.

The cost C and the delay T of the data path are equal to

$$C = n.(m - 1).C_{\text{mux2}-1} + C_{\text{adder}}(n) + n.C_{\text{FF}} \quad \text{and}$$

$$(\log_2 m).T_{\text{mux2}-1} + T_{\text{adder}}(n) + T_{\text{FF}}$$

(10.2)

where C_{FF} is the cost of a 1-bit register (e.g, a D flip-flop) and T_{FF} is the propagation time—it is assumed that the m-to-1 multiplexer is implemented by a tree-like $\log_2 m$ level circuit made up of 2-to-1 multiplexers.

The datapath VHDL model is the following one:

```
entity data_path is
port(
  operands: in operand_matrix;
  clk, clear, load: in std_logic;
  op_select: in std_logic_vector(logm-1 downto 0);
  z: out std_logic_vector(n-1 downto 0)
  );
end data_path;

architecture circuit of data_path is
  signal op_1, op_2, adder_out, reg_in, reg_out:
  std_logic_vector(n-1 downto 0);
begin
  op_1 <=operands(conv_integer(op_select));
  op_2<=reg_out;
  adder_out<=op_1+op_2;
  with clear select reg_in<=adder_out when '0',
  conv_std_logic_vector(0, n) when others;
  process(clk)
  begin
    if clk'event and clk='1' then
    if load='1' then reg_out<=reg_in; end if;
    end if;
  end process;
  z<=reg_out;
end circuit;
```

The control unit is a finite state machine whose VHDL model is the following one:

```
entity control_unit is
port(
  clk, start, reset: in std_logic;
  done, clear, load: out std_logic;
  op_select: out std_logic_vector(logm-1 downto 0)
  );
end control_unit;
```

```
architecture fsm of control_unit is
  subtype state is integer range-3 to m;
  signal current_state: state;
begin
  process(clk, reset)
  begin
    case current_state is
      when -3=>clear<='1'; load<='0';
      op_select<=conv_std_logic_vector(0, logm); done<='1';
      when -2=>clear<='1'; load<='0';
      op_select<=conv_std_logic_vector(0, logm); done<='1';
      when -1=>clear<='1'; load<='1';
      op_select<=conv_std_logic_vector(0, logm); done<='1';
      when 0 to m-1=>clear<='0'; load<='1';
      op_select<=conv_std_logic_vector(current_state, logm);
      done<='0';
      when m=>clear<='0'; load<='0';
      op_select<=conv_std_logic_vector(0, logm); done<='1';
    end case;
    if reset='1' then current_state<=-3;
    elsif clk'event and clk='1' then
      case current_state is
        when -3=>if start='0' then
        current_state<=current_state+1; end if;
        when -2=>if start='1' then
        current_state<=current_state+1; end if;
        when -1=>current_state<=current_state+1;
        when 0 to m-1=>current_state<=current_state+1;
        when m=>current_state<=-3;
      end case;
    end if;
  end process;
end fsm;
```

If a minimum number of state-encoding variables are used, and if the combinational cost is small compared with the state-register one, the cost of the control unit is roughly equal to $\log_2(m + 4).C_{FF}$.

The total cost of the m-operand adder is equal to

$$C(n, m) \cong n.(m - 1).C_{mux2-1} + C_{adder}(n) + (n + \log_2(m + 4)).C_{FF} \qquad (10.3)$$

and its computation time to

$$T(n, m) = m.T_{clk}, \quad \text{where } T_{clk} > (\log_2 m).T_{mux2-1} + T_{adder}(n) + T_{FF} \qquad (10.4)$$

10.2 PRECEDENCE RELATION AND SCHEDULING

The precedence relation and its graphical representation—a *data flow graph*—define which operations must be completed before starting a new one. Consider a first example:

Example 10.3   The precedence graph of Algorithm 4.2 (carry-chain adder), with $n = 4$, is shown in Figure 10.3.

If one chooses to execute the algorithm in just one cycle (parallel implementation), then the precedence graph is practically equivalent to the corresponding combinational circuit. As an example, the carry-chain adder of Figure 11.3 can be directly deduced from the precedence graph of Figure 10.3. If a sequential implementation is considered, the precedence graph allows scheduling the operations, that is, deciding in which cycle every operation is performed.

A sequential implementation is given in Example 10.4.

Example 10.4   (Complete VHDL source code available.) The data flow graph corresponding to Algorithm 4.10 (long-operand addition), with $n = 128$ and

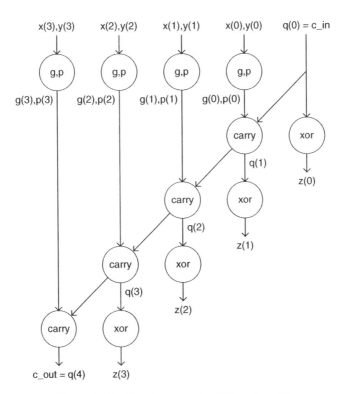

Figure 10.3   Precedence graph of Algorithm 4.2.

$s = 32$, is shown in Figure 10.4. A possible 4-cycle schedule is indicated. In every cycle a 32-digit addition is performed so that only one 32-digit adder is necessary. The scheduled precedence graph also gives information about the memory resources: every arc crossing a cycle dotted line corresponds to a variable computed during some cycle and to be used during one of the following ones. So some memory resource is necessary in order to store its value (except in the case of inputs assumed constant during the whole computation). A practical implementation of the data path, with $B = 2$, is shown in Figure 10.5. The memory resources are four 32-bit registers —with a clock enable input—storing the four parts of the result and a D flip-flop that stores the successive carries. The corresponding cost C and delay T are equal to

$$C = 192.C_{\text{mux2}-1} + C_{\text{adder}}(32) + 129.C_{\text{FF}} \tag{10.5}$$

$$T = 4.T_{\text{clk}}, \quad \text{where } T_{\text{clk}} > 2.T_{\text{mux2}-1} + T_{\text{adder}}(32) + T_{\text{FF}}. \tag{10.6}$$

The following VHDL model describes a generic version of the data path whose parameters are the number of bits n and the number of slices $m = n/s$:

```
entity data_path is
port (
  x, y: in long_operand;
  clk, reset, load_cy,: in std_logic;
```

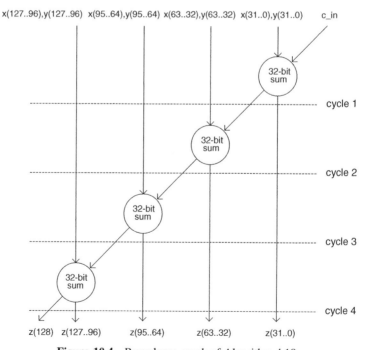

Figure 10.4 Precedence graph of Algorithm 4.10.

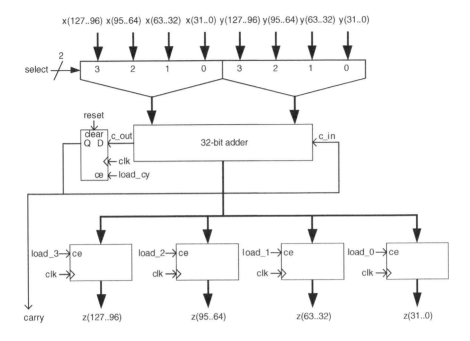

Figure 10.5 A 128-bit adder.

```
  word_select: std_logic_vector(logm-1 downto 0);
  load: std_logic_vector(m-1 downto 0);
  z: out long_operand;
  carry: out std_logic
  );
end data_path;

architecture circuit of data_path is
  signal op_1, op_2, adder_out: std_logic_vector(n downto 0);
  signal c_in, c_out: std_logic;
begin
  op_1<=('0'&x(conv_integer(word_select)));
  op_2<=('0'&y(conv_integer(word_select)));
  adder_out<=op_1+op_2+c_in;
  c_out<=adder_out(n);
  process(clk)
  begin
    if reset='1' then c_in<='0';
    elsif clk'event and clk='1' then
    if load_cy='1' then c_in<=c_out; end if;
    end if;
  end process;
```

```
  process(clk)
  begin
    if clk'event and clk='1' then
    for i in 0 to m-1 loop
    if load(i)='1' then z(i)<=adder_out(n-1 downto 0);
    end if; end loop;
    end if;
  end process;
  carry<=c_in;
end circuit;
```

An additional control unit must be designed in order to generate the control signals as well as controlling the start/done communication protocol. It can be modeled by a finite state machine:

```
entity control_unit is
port (
  clk, reset, start: in std_logic;
  done, load_cy: out std_logic;
  word_select: std_logic_vector(logm-1 downto 0);
  load: out std_logic_vector(m-1 downto 0)
  );
end control_unit;

architecture fsm of control_unit is
  subtype state is integer range -3 to m;
  signal current_state: state;
  begin
  process(clk, reset)
  begin
    case current_state is
      when -3=>load<=conv_std_logic_vector(0, m); load_cy<=
      '0'; word_select<=conv_std_logic_vector(0, logm);
      done<='1';
      when -2=>load<=conv_std_logic_vector(0, m); load_cy<=
      '0'; word_select<=conv_std_logic_vector(0, logm);
      done<='1';
      when -1=>load<=conv_std_logic_vector(0, m); load_cy<=
      '0'; word_select<=conv_std_logic_vector(0, logm);
      done<='1';
      when 0 to m-1=>load<=conv_std_logic_vector(2**
      (current_state), m); load_cy<='1'; word_select<=
      conv_std_logic_vector(current_state,
      logm); done<='0';
      when m=>load<=conv_std_logic_vector(0, m); load_cy<=
      '0'; word_select<=conv_std_logic_vector(0, logm);
      done<='1';
    end case;
    if reset='1' then current_state<=-3;
```

```
    elsif clk'event and clk='1' then
      case current_state is
        when -3=>if start='0' then current_state<=
        current_state+1; end if;
        when -2=>if start='1' then current_state<=
        current_state+1; end if;
        when -1=>current_state<=current_state+1;
        when 0 to m-1=>current_state<=current_state+1;
        when m=>current_state<=-3;
      end case;
    end if;
  end process;
end fsm;
```

If a minimum number of state-encoding variables are used, and if the combinational cost is small compared with the state-register one, the cost of the control unit is roughly equal to $\log_2(m + 4){\cdot}C_{FF}$. The total cost of the long-operand adder is equal to

$$C(n, m) \cong 2.(m - 1).(n/m).C_{mux2-1} + C_{adder}(n/m) + (n + 1 + \log_2 (m + 4)).C_{FF} \tag{10.7}$$

and its computation time to

$$T(n, m) = m.T_{clk}, \quad \text{where } T_{clk} > (\log_2 m).T_{mux2-1} + T_{adder}(n/m) + T_{FF}. \tag{10.8}$$

10.3 PIPELINE

A very useful implementation technique, especially for signal processing circuits, is pipelining.

Example 10.5   Consider again a 128-bit adder made up of four 32-bit adders. A parallel (combinational) implementation is described in Figure 10.6. The computation time (*latency*) of the circuit is roughly equal to 4.T, where T is the computation time of a 32-bit adder, so that the maximum sample rate of the input operands x and y is equal to $1/(4.T)$. The corresponding pipelined circuit is shown in Figure 10.7: a register is inserted between the computation resources assigned to successive cycles, in such a way that a new addition can be started as soon as the first cycle of the preceding addition has been completed, that is, every T seconds. In this way the latency is still equal to 4.T. Nevertheless, the sample rate is equal to $1/T$ instead of $1/(4.T)$.

Comments 10.1

1. The extra cost of the pipeline registers could appear to be prohibitive. Nevertheless, in many data processing systems there is a continuous flow of

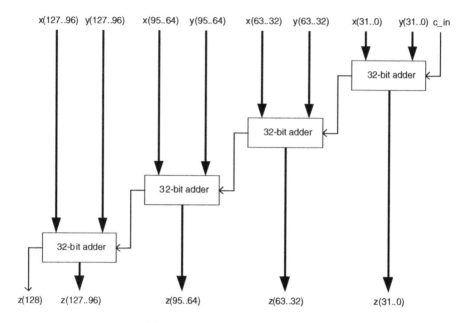

Figure 10.6   Parallel 128-bit adder.

new operands so that dynamic latches, instead of static ones, can be used, and an n-bit register practically reduces to n pass transistors (full-custom implementation). Latchless pipelining techniques have also been reported ([FLY1997]).

2. The insertion of pipeline registers also has a positive effect on the power consumption: the presence of synchronization barriers all along the circuit drastically reduces the number of generated spikes.

3. Most often, the basic cell of the field programmable gate arrays includes a flip-flop so that the insertion of pipeline registers does not necessarily increase the total cost, computed in terms of used basic cells. The pipeline registers could consist of flip-flops not used in the nonpipelined version.

10.4   SELF-TIMED CIRCUITS

Instead of synthesizing synchronous circuits, an alternative solution, especially in the case of large circuits, is *self-timing*. As a matter of fact, the synchronous approach has some pitfalls:

- It assumes that all clock events happen at the same time over the complete circuit; this is not the case in reality (*clock skew*).
- The simultaneous transition of all clock signals might generate noise problems.

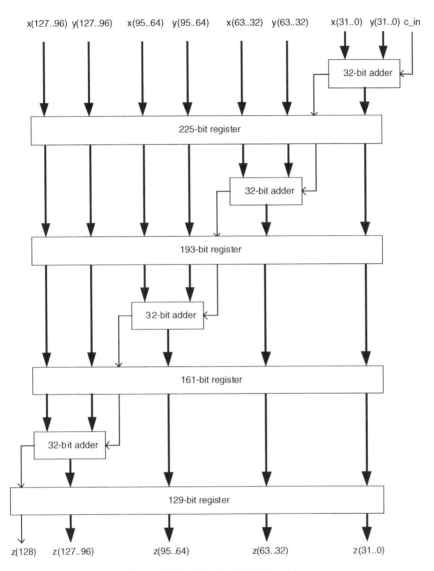

Figure 10.7   Pipelined 128-bit adder.

- The latency and throughput of the circuit are linked to the worst-case delay of the slowest element instead of the average case.

As a generic example, consider the pipelined circuit of Figure 10.8. To each block, for example number i, are associated a maximum delay $t_{max}(i)$ and an average one $t_{av}(i)$. The latency and throughput of the circuit of Figure 10.8 are equal to $n.T_{clk}$

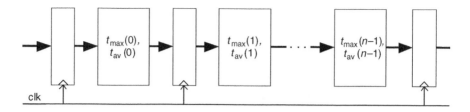

Figure 10.8   Generic pipelined circuit.

and $1/T_{\text{clk}}$, respectively where $T_{\text{clk}} > \max\{t_{\max}(0), t_{\max}(1), \ldots, t_{\max}(n-1)\}$, that is,

$$latency > n.\max\{t_{\max}(0), t_{\max}(1), \ldots, t_{\max}(n-1)\},$$
$$throughout < 1/\max\{t_{\max}(0), t_{\max}(1), \ldots, t_{\max}(n-1)\}. \tag{10.9}$$

A self-timed version of the same circuit is shown in Figure 10.9. The control is based on a `request/acknowledge` handshaking protocol:

- the `Req` input of block 0 is raised; if block 0 is free the data is registered (`en`), and the `Ack` signal is raised.
- The `start` signal of block 0 is raised; after some amount of time the `done` signal of block 0 goes high indicating the completion of the computation.
- A `Req` to block number 1 is issued; if block 1 is free, the output of block 0 is registered and the `Ack` signal is raised; and so on.

If the probability distribution of the internal data were uniform, inequalities (10.9) would be substituted by the following ones:

$$average\ latency > t_{\text{av}}(0) + t_{\text{av}}(1) + \cdots + t_{\text{av}}(n-1),$$
$$average\ throughput < 1/\max\{t_{\text{av}}(0) + t_{\text{av}}(1), \ldots, t_{\text{av}}(n-1)\}. \tag{10.10}$$

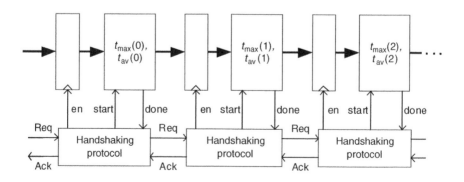

Figure 10.9   Self-timed pipelined circuit.

A detailed presentation of self-timed circuit design is given in [RAB2003]. As regards the synthesis of the functional blocks (the data path), the done signal must be generated. A systematic way of detecting the end of the computation consists of using a redundant encoding of the binary signals: every signal s is represented by a pair (s_1, s_0) according to the definition of Table 10.1.

The circuit will be designed in such a way that during the initialization (reset), and as long as the value of s has not yet been computed, $(s_1, s_0) = (0, 0)$. Once the value of s is known, $s_1 = s$ and $s_0 = not(s)$.

Assume that the circuit includes n signals $s_1, s_2, \ldots, s_n$. Every signal s_i is substituted by a pair (s_{i1}, s_{i0}). Then the done flag is computed as follows:

$$\text{done} = (s_{11} + s_{10}).(s_{21} + s_{20}) \cdot \ldots \cdot (s_{n1} + s_{n0}). \tag{10.11}$$

During the initialization (*reset*) and as long as at least one of the signals is in transition, the corresponding pair is equal to $(0, 0)$, so that done $= 0$. The done flag will be raised only when all signals have a stable value.

Example 10.6   (Complete VHDL source code available). In order to synthesize an n-bit adder to be used within a self-timed circuit, Algorithm 4.2 must be replaced by Algorithm 10.1. The carry-propagate (p) and carry-generate (g) functions have already been defined (Example 10.1). The carry-kill function k is the complement of the carry-generate function g. It can be defined as follows (see comment 4.1(2)): $k(a, b) = 1$ if $a + b < 1$, any value if $a + b = 1$, and 0 if $a + b > 1$, so that k can be chosen equal to (for example) not (b).

Algorithm 10.1   Self-Timed Adder

```
--computation of the carry-generate, carry-kill and carry-
propagate functions:
for i in 0..n-1 loop
  g(i):=g(x(i),y(i)); k(i):=k(x(i), y(i)); p(i):=p(x(i),
  y(i));
end loop;
--carry computation:
q(0):=c_in; qb(0):=not(c_in);
for i in 0..n-1 loop
  if start=1 then
```

TABLE 10.1

s	s_1	s_0
Reset or in transition	0	0
0	0	1
1	1	0

```
    if p(i)=1 then q(i+1):=q(i); qb(i+1):=qb(i); else
    q(i+1):=g(i); qb(i+1):=k(i);
      end if;
   else q(i+1):=0; qb(i+1):=0;
    end if;
end loop;
--completion detection
done:=(q(1) or qb(1)) and (q(2) or qb(2)) and...and (q(n) or
qb(n));
--sum computation
for i in 0..n-1 loop
  z(i):=(x(i)+y(i)+q(i)) mod 2;
end loop;
z(n):=q(n);
```

In order to generate a circuit similar to the one of Figure 10.1, Algorithm 10.1 is slightly modified.

Algorithm 10.2   Self-Timed Adder (Modified)

```
for i in 0..n-1 loop
  g(i):=start and g(x(i),y(i)); k(i):=start and k(x(i), y(i));
  p(i):=start and p(x(i),y(i));
end loop;
q(0):=c_in; qb(0):=not(c_in);
for i in 0..n-1 loop
  if p(i)=1 then q(i+1):=q(i); qb(i+1):=qb(i); else
  q(i+1):=g(i); qb(i+1):=k(i); end if;
end loop;
done:=(q(1) or qb(1)) and (q(2) or qb(2)) and...and
(q(n) or qb(n));
for i in 0..n-1 loop
  z(i):=(p(i)+q(i)) mod 2;
end loop;
z(n):=q(n);
```

The basic cell of the corresponding iterative circuit is shown in Figure 10.10.

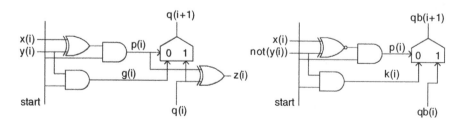

Figure 10.10   Iterative circuit basic cell.

It remains to compute the done condition:

$$\text{done} = (q(1) + qb(1)).(q(2) + qb(2)). \cdots .(q(n) + qb(n)). \tag{10.12}$$

The average computation time of an n-bit adder is proportional to $\log(n)$, so that the computation of the done condition should be performed by a circuit whose maximum delay is (at most) proportional to $\log(n)$. An example with $n = 64$ is given in Figure 10.11. If the delay of the 4-input OR-AND gates is equal to the delay of the 4-input NAND gates then the computation time is equal to $3.t_{gate}$, where t_{gate} is the common delay. More generally, the delay is equal to $(\log_4(n)).t_{gate}$.

The following VHDL model can be used to simulate the circuit (equation (10.12) is computed by a process corresponding to the behavior of the corresponding circuit block):

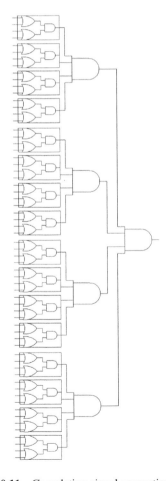

Figure 10.11 Completion signal generation ($n = 64$).

```
entity example10_6 is
port (
  x, y: in std_logic_vector(n-1 downto 0);
  c_in, start: in std_logic;
  z: out std_logic_vector(n-1 downto 0);
  c_out, done: out std_logic
);
end example10_6;

architecture circuit of example10_6 is
  signal p, g, k: std_logic_vector(n-1 downto 0);
  signal q, qb: std_logic_vector(n downto 0);
begin
  q(0)<=c_in; qb(0)<=not(c_in);
  iterative_step: for i in 0 to n-1 generate
    p(i)<=start and (x(i) xor y(i));
    g(i)<=start and y(i);
    k(i)<=start and not(y(i));
    with p(i) select q(i=1)<=q(i) when '1', g(i) when others;
    with p(i) select qb(i+1)<=qb(i) when '1', k(i) when
    others; z(i)<=p(i) xor q(i);
  end generate;
  process(q, qb)
    variable accumulator: std_logic;
  begin
    accumulator:=q(1) or qb(1);
    for i in 2 to n loop
      accumulator:=accumulator and (q(i) or qb(i));
    end loop;
    done<=accumulator;
  end process;
  c_out<=q(n);
end circuit;
```

10.5 BIBLIOGRAPHY

[DEM1994] G. De Micheli, *Synthesis and Optimization of Digital Circuits*, McGraw-Hill, New York, 1994.

[DES2002] J.-P. Deschamps, *Síntesis de circuitos digitales: un enfoque algorítmico [Synthesis of Arithmetic Circuits: An Algorithmic Approach]*, Thomson, Madrid, 2002 (In Spanish).

[FLY1997] M. Flynn, *Modern Research in Computer Arithmetic*. Stanford Architecture and Arithmetic Group, 1997.

[GAJ1994] D. D. Gajski, F. Vahid, S. Narayan, and J. Gong, *Specification and Design of Embedded Systems*. Prentice Hall, Englewood Cliffs, NJ, 1994.

[RAB2003] J. M. Rabaey, A. Chandrakasan, and B. Nikolic, *Digital Integrated Circuits: A Design Perspective*. Prentice Hall, Engle-wood Cliffs, NJ, 2003.

11

ADDERS AND SUBTRACTORS

Two-operand addition is a primitive operation included in practically all arithmetic algorithms. As a consequence, the efficiency of an arithmetic circuit strongly depends on the way the adders are implemented. A key point in two-operand adder implementation is the way the carries are computed. It has been seen in Chapter 4 that the computation time of a circuit based on the classical pencil and paper algorithm is proportional to the number n of digits of the operands. If this type of algorithm is used, it is important to reduce the multiplicative constant (delay per digit). Another option is to reduce the value of n, that is, to change the numeration system base. In order to get very fast adders, some of the logarithm-delay algorithms presented in Chapter 4 can be used. Another important topic, dealt with in this chapter, is the implementation of multioperand adders. The stored-carry form encoding defined in Chapter 4 is used to synthesize fast multioperand adders. Some ideas for implementing (relatively) low-cost and fast asynchronous adders are presented in the last section.

11.1 NATURAL NUMBERS

11.1.1 Basic Adder (Ripple-Carry Adder)

The structure of an n-digit *ripple-carry adder* is shown in Figure 11.1. The *full adder* (FA) cell calculates $q(i + 1)$ and $z(i)$ as a function of $x(i)$, $y(i)$, and $q(i)$, according to

Synthesis of Arithmetic Circuits: FPGA, ASIC, and Embedded Systems
By Jean-Pierre Deschamps, Géry J. A. Bioul, and Gustavo D. Sutter
Copyright © 2006 John Wiley & Sons, Inc.

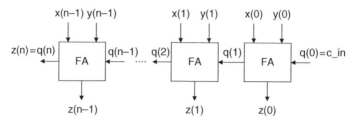

Figure 11.1   Basic adder.

the iteration body of Algorithm 4.1:

$$q(i+1) = 1 \text{ if } x(i) + y(i) + q(i) > B - 1, \quad q(i+1) = 0 \text{ otherwise;} \qquad (11.1)$$
$$z(i) = (x(i) + y(i) + q(i)) \bmod B.$$

Let C_{FA} and T_{FA} be the cost and the computation time of an FA cell. The cost and computation time of an n-digit basic adder are equal to

$$C_{\text{basic-adder}}(n) = n.C_{FA} \quad \text{and} \quad T_{\text{basic-adder}}(n) = n.T_{FA}. \qquad (11.2)$$

Examples 11.1

1. In base 2 the FA equations (11.1) are

$$q(i+1) = x(i).y(i) \vee x(i).q(i) \vee y(i).q(i), z(i) = x(i) \oplus y(i) \oplus q(i) \qquad (11.3)$$

   ($\vee$: *or* function, $\oplus$: *xor* function).

2. In base 10 the FA equations are

$$q(i+1) = 1 \text{ if } x(i) + y(i) + q(i) > 9, \quad q(i+1) = 0 \text{ otherwise;} \qquad (11.4)$$
$$z(i) = (x(i) + y(i) + q(i)) \bmod 10.$$

The decimal digits can be represented as 4-bit binary numbers (BCD—binary-coded decimal representation) so that a decimal full adder is a 9-input, 5-output binary circuit. It can be implemented using classical methods and tools of combinational logic synthesis. Another option is to decompose it further on. The following algorithm computes (11.4):

```
s:=x(i)+y(i)+q(i);
if s>9 then z(i):=(s+6) mod 16; q(i+1):=1; else z(i):=s;
q(i+1):=0; end if;
```

The corresponding circuit (Figure 11.2) is made up of a 4-bit binary adder that computes s (a 5-bit number), a 4-input 1-output combinational circuit that computes

$$q(i+1) = s_4 \vee s_3.(s_2 \vee s_1),$$

and another 4-bit binary adder that computes the BCD 4-bit expression of the sum

$$(z_3(i) \, z_2(i) \, z_1(i) \, z_0(i)) = (s_3 \, s_2 \, s_1 \, s_0) + (0 \; q(i+1) \; q(i+1) \; 0),$$

so that

$$z_0(i) = s_0,$$
$$z_1(i) = s_1 \oplus q(i+1), \; c_2 = s_1.q(i+1),$$
$$z_2(i) = s_2 \oplus q(i+1) \oplus c_2, \; c_3 = s_2.q(i+1) \vee s_2.c_2 \vee q(i+1).c_2,$$
$$z_3(i) = s_3 \oplus q(i+1).$$

Observe that in the preceding example z_1 and z_3 can be computed by *half-adder* (*HA*) cells whose equations are deduced from (11.3) substituting one of the operands, say, $y(i)$, by zero:

$$q(i+1) = x(i).q(i), \; z(i) = x(i) \oplus q(i).$$

More generally (base B), the half-adder equations are

$$q(i+1) = 1 \text{ if } x(i) + q(i) > B - 1, \; = 0 \text{ otherwise;} \qquad (11.5)$$
$$z(i) = (x(i) + q(i)) \bmod B.$$

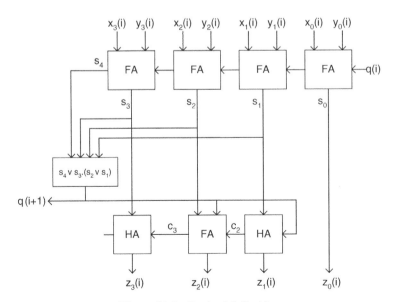

Figure 11.2   Decimal full adder.

11.1.2   Carry-Chain Adder

The structure of an n-digit adder with separate carry calculation is shown in Figure 11.3. It is based on Algorithm 4.2. The *G-P* (*generate–propagate*) cell calculates the *generate* and *propagate* functions (4.1), that is,

$$g(i) = 1 \text{ if } x(i) + y(i) > B - 1, \ = 0 \text{ otherwise,}$$
$$p(i) = 1 \text{ if } x(i) + y(i) = B - 1, \ = 0 \text{ otherwise.}$$

(11.6)

The *Cy.Ch.* (*carry-chain*) cell computes the next carry, that is,

$$q(i + 1) = q(i) \text{ if } p(i) = 1, \quad q(i + 1) = g(i) \text{ otherwise,}$$

so that $g(i)$ generates a carry, whatever happened upstream in the carry chain, and $p(i)$ propagates the carry from level $i - 1$. The *mod B sum* cell calculates

$$z(i) = (x(i) + y(i) + q(i)) \text{ mod } B.$$

Let C_{GP} and T_{GP}, $C_{Cy.Ch.}$ and $T_{Cy.Ch.}$, and C_{sum} and T_{sum} be the cost and the computation time of a G-P cell, Cy.Ch. cell, and mod B sum cell, respectively. The cost and computation time of an n-digit carry-chain adder are equal to

$$C_{\text{carry-chain-adder}}(n) = n.(C_{GP} + C_{Cy.Ch.} + C_{sum}),$$
$$T_{\text{carry-chain-adder}}(n) = T_{GP} + (n - 1).T_{Cy.Ch.} + T_{sum}.$$

(11.7)

As regards the computation time, the critical path is shaded in Figure 11.3. It has been assumed that $T_{sum} > T_{Cy.Ch.}$.

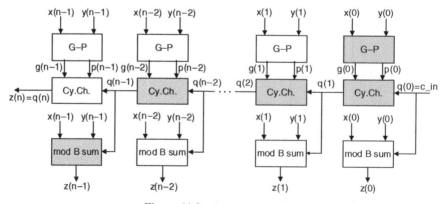

Figure 11.3   Carry-chain adder.

Another interesting time is the delay $T_{carry}(n)$ from $q(0)$ to $q(n)$ assuming that all propagate and generate functions have already been calculated:

$$T_{carry}(n) = n.T_{Cy.Ch.}.$$ (11.8)

Comments 11.1

1. The carry-chain cells are binary circuits while the generate–propagate and the mod B sum cells are B-ary ones.

2. The carry-chain cell is functionally equivalent to a 2-to-1 binary multiplexer (Figure 11.4a), so that, according to (11.7), the *per-digit delay* of a carry-chain adder is equal to the delay T_{mux2-1} of a 2-to-1 binary multiplexer, whatever the base B. Furthermore (Comment 4.1.(2)) the definition of the *generate* function can be relaxed:

$$g(i) = 1 \text{ if } x(i) + y(i) > B - 1, \ g(i) = 0 \text{ if } x(i) + y(i) < B - 1, \ g(i)$$
$$= 0 \text{ or } 1 \text{ (don't care) otherwise.}$$

   If $B = 2$ and the carry-chain cell of Figure 11.4a is used, then $p(i) = x(i) \oplus y(i)$ and $g(i)$ can be chosen equal to for example, $y(i)$. The corresponding n-bit adder is shown in Figure 10.1. Its cost and computation time are given by (10.1).

3. Another way to synthesize the carry-chain cell—the *Manchester carry chain*—is shown in Figure 11.4b. First observe that the complement of the carries is computed. It works as follows: the output node (the complement of $q(i + 1)$) is precharged when the synchronization signal clk is equal to 0; when clk $= 1$ the output node is discharged if either $p(i) = 1$ and the preceding node (the complement of $q(i)$) has been discharged, or if $g(i) = 1$. In

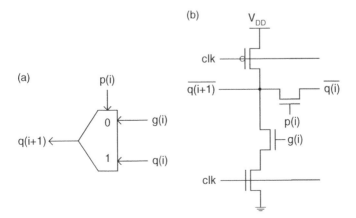

Figure 11.4 Carry-chain cells.

order that it works properly $g(i)$ and $p(i)$ should not be equal to 1 at the same time so that the definition of g cannot be relaxed as in the preceding case.

Example 11.2 (Complete VHDL source code available). Generate a generic n-digit base-B carry-chain adder:

```
entity example11_2 is
port (
  x, y: in digit_vector(n-1 downto 0);
  c_in: in std_logic;
  z: out digit_vector(n-1 downto 0);
  c_out: out std_logic
);
end example11_2;

architecture circuit of example11_2 is
  signal p, g: std_logic_vector(n-1 downto 0);
  signal q: std_logic_vector(n downto 0);
begin
  q(0)<=c_in;
  iterative_step: for i in 0 to n-1 generate
    p(i)<='1' when x(i)+y(i)=B-1 else '0';
    g(i)<='1' when x(i)+y(i)>B-1 else'0';
    with p(i) select q(i+1)<=q(i) when '1', g(i) when others;
    z(i)<=(x(i)+y(i)+conv_integer(q(i))) mod B;
  end generate;
  c_out<=q(n);
end circuit;
```

11.1.3 Carry-Skip Adder

Consider a group of s successive cells within a carry chain (Figure 11.5). If all propagate functions $p(i.s)$, $p(i.s + 1)$, ..., $p(i.s + s - 1)$ within the group are equal to 1 then $q(i.s + s) = q(i.s)$, and the carry $q(i.s)$ is said to be propagated through the group. In the contrary case, there is at least one cell, say, number $i.s + j$, such that $p(i.s + j) = 0$ so that $q(i.s + j + 1) = g(i.s + j)$. Assume that cell number $i.s + j$ is the last one such that $p(i.s + j) = 0$ and $p(i.s + j + 1) = \cdots$

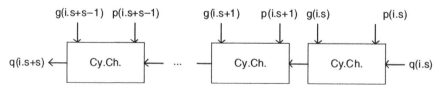

Figure 11.5 An s-bit carry chain.

$= p(i.s + s - 1) = 1$. Then $q(i.s + s) = g(i.s + j)$, and the carry $q(i.s + s)$ is said to be locally computed within the group.

An n-bit *carry-skip* carry chain is made up of n/s s-bit carry chains interconnected through 2-to-1 multiplexers (Figures 11.6 and 11.7). Besides the generate and propagate functions, the generalized propagate functions $p(i.s + s - 1:i.s) = p(i.s + s - 1) \cdots p(i.s + 1).p(i.s)$ must also be computed.

The structure of a carry-skip adder is shown in Figure 11.8. It is made up of n G-P cells, n Cy.Ch. cells, n mod B sum cells, n/s 2-to-1 multiplexers, and n/s s-input AND gates (or any equivalent circuit) for computing $p(i.s + s - 1:i.s)$. Its cost and computation time are equal to

$$
\begin{aligned}
C_{\text{carry-skip-adder}}(n,s) &= n.(C_{\text{GP}} + C_{\text{Cy.Ch.}} + C_{\text{sum}}) \\
&\quad + (n/s).(C_{\text{mux2-1}} + C_{\text{AND}(s)}), \quad (11.9) \\
T_{\text{carry-skip-adder}}(n,s) &= T_{\text{GP}} + s.T_{\text{Cy.Ch.}} + (n/s - 1).T_{\text{mux2-1}} \\
&\quad + (s - 1).T_{\text{Cy.Ch.}} + T_{\text{sum}}.
\end{aligned}
$$

The critical path of the carry chain is shaded in Figure 11.7. It has been assumed that $s.T_{\text{Cy.Ch.}} > T_{\text{AND}(s)}$ and $T_{\text{sum}} > T_{\text{Cy.Ch.}} + T_{\text{mux2-1}}$, so that in the first group $p(s - 1:0)$ is computed in parallel with the multiplexer inputs, and in the last group the critical path is from $q(n - s)$ to $z(n - 1)$ through $s - 1$ Cy.Ch. cells and one mod B sum cell.

As before another interesting time is the delay $T_{\text{carry}}(n,s)$ from $q(0)$ to $q(n)$ assuming that all propagate, generate, and generalized propagate functions have already been calculated:

$$
T_{\text{carry}}(n,s) = s.T_{\text{Cy.Ch.}} + (n/s).T_{\text{mux2-1}}. \quad (11.10)
$$

Comments 11.2

1. For great values of n and s the computation time (11.9) is roughly equal to $(n/s).T_{\text{mux2-1}} + 2.s.T_{\text{Cy.Ch.}}$. It must be compared with (11.7), that is, (approximately), $n.T_{\text{Cy.Ch.}}$. The computation time reduction is due to the fact that the locally generated carries are calculated in parallel.

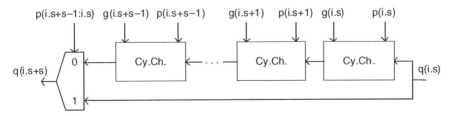

Figure 11.6 Carry chain: s-bit group.

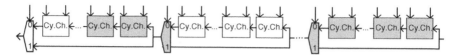

Figure 11.7 Carry-skip carry chain.

2. The s rightmost Cy.Ch. cells of Figure 11.7 belong to the critical path, so that the first multiplexer should be deleted, unless the corresponding adder is used as a building block for larger adders.

Multilevel carry-skip adders can be defined. For example, each s-bit carry chain in Figure 11.7 could in turn be substituted by s/t t-bit carry chains. The corresponding delay $T_{carry}(n, s, t)$ from $q(0)$ to $q(n)$ assuming that all propagate, generate, and generalized propagate functions have already been calculated is

$$T_{carry}(n, s, t) = T_{carry}(s, t) + (n/s).T_{mux2\text{-}1},$$

so that, according to (11.10),

$$T_{carry}(n, s, t) = t.T_{Cy.Ch.} + (s/t + n/s).T_{mux2\text{-}1}. \qquad (11.11)$$

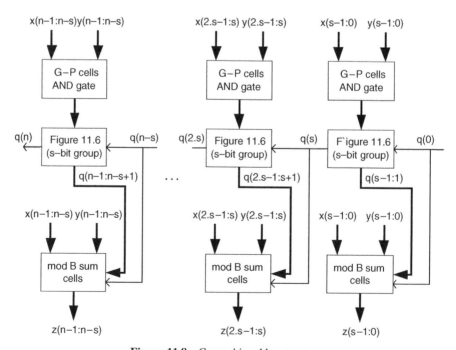

Figure 11.8 Carry-skip adder structure.

If three levels are used then

$$T_{carry}(n, s, t, u) = u.T_{Cy.Ch.} + (t/u + s/t + n/s).T_{mux2-1}, \qquad (11.12)$$

and so on. In particular, if $n = s^m$,

$$T_{carry}(n, s^{m-1}, s^{m-2}, \ldots, s) = s.T_{Cy.Ch.} + (s^2/s + s^3/s^2 + \cdots + s^m/s^{m-1}).T_{mux2-1}$$
$$= s.T_{Cy.Ch.} + (m - 1).s.T_{mux2-1},$$

$$(11.13)$$

that is,

$$T_{carry}(n, s^{m-1}, s^{m-2}, \ldots, s) = s.T_{Cy.Ch.} + (\log_s n - 1).s.T_{mux2-1}. \qquad (11.14)$$

If the computation time of every generalized propagation function is assumed to be shorter than the propagation time of the corresponding (partial) carry chain, then the delay of the complete adder is equal to (11.14) plus the delay of a G-P cell and of a mod B sum cell, that is, a logarithmic delay.

Example 11.3 (Complete VHDL source code available). Generate a generic n-digit base-B carry-skip adder:

```
entity carry_skip is
port(
  x, y: in digit_vector(s-1 downto 0);
  c_in: in std_logic;
  c_out: out std_logic_vector(s downto 1)
);
end carry_skip;

architecture circuit of carry_skip is
  signal p, g: std_logic_vector(s-1 downto 0);
  signal generalized_p: std_logic;
  signal q: std_logic_vector(s downto 0);
begin
  q(0)<=c_in;
  iterative_step: for i in 0 to s-1 generate
    p(i)<='1' when x(i)+y(i)=B-1 else '0';
    g(i)<='1' when x(i)+y(i)>B-1 else'0';
    with p(i) select q(i+1)<=q(i) when '1', g(i) when others;
  end generate;
  process(p)
    variable accumulator: std_logic;
  begin
    accumulator:=p(0);
    for i in 1 to s-1 loop accumulator:=accumulator and p(i);
```

```
      end loop;
      generalized_p<=accumulator;
    end process;
    with generalized_p select c_out(s)<=c_in when '1', q(s) when
    others;
    carries: for i in 1 to s-1 generate c_out(i)<=q(i);
    end generate;
end circuit;

entity example11_3 is
port(
  x, y: in digit_vector(n-1 downto 0);
  c_in: in std_logic;
  z: out digit_vector(n-1 downto 0);
  c_out: out std_logic
);
end example11_3;

architecture circuit of example11_3 is
  component carry_skip...end component;
  signal q: std_logic_vector(n downto 0);
begin
  q(0)<=c_in;
  ext_iteration: for i in 0 to n_div_s-1 generate
    csa_carry_chain: carry_skip port map(x(i*s+s-1 downto
    *s), y(i*s+s-1 downto i*s), q(i*s), q(i*s+s downto
    i*s+1));
    int_iteration: for j in 0 to s-1 generate
      z(i*s+j)<=(x(i*s+j)+y(i*s+j)+conv_integer(q(i*s+j)))
      mod B;
    end generate;
  end generate;
  c_out<=q(n);
end circuit;
```

11.1.4 Optimization of Carry-Skip Adders

Assume that in the carry-skip adder of Figure 11.7 the last carry of group number j, namely, $q(j.s + s - 1)$, has been generated or killed (Example 10.6) within the first cell of group number i and propagated toward group number j. Then the computation time of $q(j.s + s - 1)$, assuming that all the propagate, generate, and generalized propagate functions have already been calculated, is equal to

$$t(i, j) = (2.s - 1).T_{Cy.Ch.} + (j - i).T_{mux2-1}.$$

In particular (worst case), if $j = n/s - 1$ and $i = 0$, then

$$t(0, n/s - 1) = (2.s - 1).T_{Cy.Ch.} + (n/s - 1).T_{mux2-1}. \tag{11.15}$$

The theoretical minimum value of $t(0, n/s - 1)$ is obtained when $2.s.T_{\text{Cy.Ch.}} = n/s.T_{\text{mux2-1}}$, that is,

$$s = (n.\alpha/2)^{1/2} \tag{11.16}$$

where

$$\alpha = T_{\text{mux2-1}}/T_{\text{Cy.Ch.}} \cdot \tag{11.17}$$

The corresponding value of (11.15) is approximately

$$t(0, n/s - 1) \cong 4.s.T_{\text{Cy.Ch.}} = (8.n.T_{\text{Cy.Ch.}}.T_{\text{mux2-1}})^{1/2}. \tag{11.18}$$

A better solution is obtained if the groups are allowed to have different sizes. Assume that the group sizes are $s_0, s_1, \ldots, s_{k-1}$, where $s_0 + s_1 + \cdots + s_{k-1} = n$. Then the computation of the last carry of group number j, assuming that it has been generated or killed within the first cell of group number i, and propagated toward group number j, is equal to

$$t(i, j) = (s_i + s_j - 1).T_{\text{Cy.Ch.}} + (j - i).T_{\text{mux2-1}}. \tag{11.19}$$

The corresponding optimization problem can be stated as follows:

define $k, s_0, s_1, \ldots, s_{k-1}$, such that the cost function

$$c(k, s_0, s_1, \ldots, s_{k-1}) = \max_{i,j} \{(s_i + s_j).T_{\text{Cy.Ch.}} + (j - i).T_{\text{mux2-1}}\}$$

be minimum and

$$s_0 + s_1 + \cdots + s_{k-1} = n. \tag{11.20}$$

Obviously it's a complex problem. Nevertheless, an interesting solution is deduced from the following observation: if the difference $j - i$ is big (it means that group i is one of the first ones and group j one of the last ones), then s_i and s_j should be small; conversely, if the difference $j - i$ is small (it could mean that groups i and j are close to the center), then s_i and s_j are allowed to have bigger values. Thus choose

$$s_0 = s_{k-1}, s_1 = s_{k-2}, \ldots, s_{k/2-1} = s_{k/2} \tag{11.21}$$

(assuming that k is even). Furthermore, in order that the values of the computation times $t(0, k - 1), t(1, k - 2), \ldots, t(k/2 - 1, k/2)$ be equal, then

$$2.s_1.T_{\text{Cy.Ch.}} + (k - 3).T_{\text{mux2-1}} = 2.s_0.T_{\text{Cy.Ch.}} + (k - 1).T_{\text{mux2-1}},$$
$$2.s_2.T_{\text{Cy.Ch.}} + (k - 5).T_{\text{mux2-1}} = 2.s_0.T_{\text{Cy.Ch.}} + (k - 1).T_{\text{mux2-1}},$$
$$\cdots$$
$$2.s_{k/2-1}.T_{\text{Cy.Ch.}} + (k - (k - 1)).T_{\text{mux2-1}} = 2.s_0.T_{\text{Cy.Ch.}} + (k - 1).T_{\text{mux2-1}},$$

that is,

$$s_1 = s_0 + \alpha, \; s_2 = s_0 + 2.\alpha, \ldots, \; s_{k/2-1} = s_0 + (k/2 - 1).\alpha, \qquad (11.22)$$

where α is defined by (11.17). The value of s_0 is deduced from (11.20), (11.21), and (11.22):

$$n = 2.(s_0 + (s_0 + \alpha) + (s_0 + 2.\alpha) + \cdots + (s_0 + (k/2 - 1).\alpha))$$

$$= k.s_0 + (k/2 - 1).(k/2).\alpha,$$

and

$$s_0 = n/k - \tfrac{1}{2}.(k/2 - 1).\alpha. \qquad (11.23)$$

The corresponding value of $t(0,k - 1)$ is approximately equal to

$$t(0, k - 1) \cong 2.s_0.T_{\text{Cy.Ch.}} + (k - 1).T_{\text{mux2-1}}$$
$$= 2.(n/k - \tfrac{1}{2}.(k/2 - 1).\alpha).T_{\text{Cy.Ch.}} + (k - 1).T_{\text{mux2-1}}. \qquad (11.24)$$

In order to minimize (11.24), the value of k is chosen in such a way that

$$2.(n/k).T_{\text{Cy.Ch.}} = k.T_{\text{mux2-1}} - (k/2).\alpha.T_{\text{Cy.Ch.}},$$

that is, using (11.17)

$$2.(n/k) = (k/2).\alpha, \qquad (11.25)$$

so that

$$k = (4.n/\alpha)^{1/2}. \qquad (11.26)$$

Substituting k by (11.26) in (11.23) yields

$$s_0 = \alpha/2. \qquad (11.27)$$

According to (11.19), (11.21), and (11.17) the value of the carry computation time is

$$t(0,k - 1) = (2.s_0 - 1).T_{\text{Cy.Ch.}} + (k - 1).T_{\text{mux2-1}},$$

and, for k great enough,

$$t(0,k - 1) \cong k.T_{\text{mux2-1}} = (4.n.T_{\text{Cy.Ch.}}.T_{\text{mux2-1}})^{1/2}. \qquad (11.28)$$

Thus the variable size approach allows the reduction of the delay by a factor $2^{1/2}$ (compare (11.18) and (11.28)).

Comment 11.3   Generally, $T_{\text{mux2-1}}$ and $T_{\text{Cy.Ch.}}$ have the same order of magnitude, so that $\alpha \cong 1$; s_0 should be chosen equal to 1 and k to approximately $((1 + 4.n)^{1/2} - 1)$ in order that (11.23) be satisfied with $s_0 = \alpha = 1$.

Example 11.4   With $n = 64$ and $s_0 = \alpha = 1$, the number of groups is equal to $((1 + 4.64)^{1/2} - 1) \cong 15$. A possible choice is

$$s_0 = 1, s_1 = 2, s_2 = 3, s_3 = 4, s_4 = 5, s_5 = 6, s_6 = 7, s_7 = 8,$$
$$s_8 = 7, s_9 = 6, s_{10} = 5, s_{11} = 4, s_{12} = 3, s_{13} = 2, s_{14} = 1.$$

Calculating some carry computation times (11.19):

$$t(0, 14) = T_{\text{Cy.Ch.}} + 14.T_{\text{mux2-1}},$$
$$t(7, 8) = 14.T_{\text{Cy.Ch.}} + T_{\text{mux2-1}},$$
$$t(4, 13) = 6.T_{\text{Cy.Ch.}} + 9.T_{\text{mux2-1}},$$
$$\cdots$$

The fixed group-size approach, with $\alpha = 1$, gives the following results:

$$s = (n.\alpha/2)^{1/2} \cong 6 \quad \text{and} \quad k = n/s \cong 10.$$

Choose, for example, ten groups, six of 6 digits and four of 7 digits. Then (11.15)

$$t(0, 9) = 11.T_{\text{Cy.Ch.}} + 9.T_{\text{mux2-1}},$$

which is significantly greater than the above values with heterogeneous size blocks.

11.1.5   Base-B^s Adder

In a base-B^s adder, every slice of Figure 11.3 is replaced by the circuit of Figure 11.9. The combinational circuit computes the generalized generate and propagate functions (Definitions 4.1)

$$G = g(i.s + s - 1:i.s) \quad \text{and} \quad P = p(i.s + s - 1:i.s). \tag{11.29}$$

Its cost and computer time are equal to

$$C_{\text{base-}B^{**}s}(n,s) = n.(C_{\text{GP}} + C_{\text{Cy.Ch.}} + C_{\text{sum}}) + (n/s).C_{\text{circ.comb.}},$$
$$T_{\text{base-}B^{**}s}(n,s) = T_{\text{GP}} + T_{\text{circ.comb.}} + (n/s + s - 2).T_{\text{Cy.Ch.}} + T_{\text{sum}}. \tag{11.30}$$

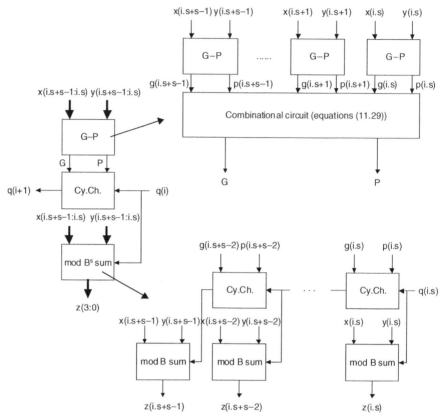

Figure 11.9 Base-B^s adder cell.

The computation of G and P can be performed with the dot operation (Definitions 4.1):

$$(G,P) = (g(i.s + s - 1:i.s), (p(i.s + s - 1:i.s))$$
$$= (g(i.s + s - 1), p(i.s + s - 1)) \cdot (g(i.s + s - 2), p(i.s + s - 2))$$
$$\bullet \cdots \bullet (g(i.s), p(i.s)).$$

Let C_{dot} and T_{dot} be the cost and computation time of a dot operator. As the dot operation is associative, the computation can be executed by a ($\log_2 s$)-level tree of ($s - 1$) dot operators, so that

$$C_{circ.comb.} = (s - 1).C_{dot} \quad \text{and} \quad T_{circ.comb.} = (\log_2 s).T_{dot}$$

and, using those values in (11.30),

$$C_{\text{base-}B^{**}s}(n,s) = n.(C_{GP} + (1 - 1/s).C_{dot} + C_{Cy.Ch.} + C_{sum}),$$
$$T_{\text{base-}B^{**}s}(n,s) = T_{GP} + (\log_2 \ s).T_{dot} + (n/s + s - 2).T_{Cy.Ch.} + T_{sum}. \tag{11.31}$$

11.1.6 Carry-Select Adder

Once again the adder is decomposed into n/s groups of s adder steps. For every group a *conditional carry chain* is generated (Figure 11.10). It computes the functions $q_0(i.s + j)$ and $q_1(i.s + j)$ according to the canonical expansion

$$q(i.s + j) = not(q(i.s)).q_0(i.s + j) + q(i.s).q_1(i.s + j), \qquad (11.32)$$

that is, the carry is equal to $q_0(i.s + j)$ when $q(i.s) = 0$, and to $q_1(i.s + j)$ when $q(i.s) = 1$.

An s-digit carry-select adder cell is shown in Figure 11.11. It is made up of an s-bit conditional carry chain, s 2-to-1 binary multiplexers that select the actual carry value (either $q_0(i.s + j)$ or $q_1(i.s + j)$) as a function of $q(i.s)$, as well as the G-P and mod B sum cells.

The complete adder structure is shown in Figure 11.12. Its cost and computation time (see the shaded blocks) are equal to

$$C_{\text{carry-select}}(n,s) = n.(C_{\text{GP}} + 2.C_{\text{Cy.Ch.}} + C_{\text{mux2-1}} + C_{\text{sum}}) - s.(C_{\text{Cy.Ch.}} + C_{\text{mux2-1}}),$$
$$T_{\text{carry-select}}(n, s) = T_{\text{GP}} + s.T_{\text{Cy.Ch.}} + (n/s - 1).T_{\text{mux}} + T_{\text{sum}}. \qquad (11.33)$$

Multilevel carry-select adders can be defined. With two s-bit conditional carry chains (Figure 11.10), as well as two s-bit 2-to-1 multiplexers, a $(2.s)$-bit conditional carry chain can be generated (Figure 11.13). Assume that

$$q(s + j) = not(q(s)).q_0'(s + j) + q(s).q_1'(s + j).$$

Then replace in the preceding equation $q(s)$ by $not(q(0)).q_0(s) + q(0).q_1(s)$:

$$
\begin{aligned}
q(s + j) =\ & [not(q(0)).not(q_0(s)) + q(0).not(q_1(s))].q_0'(s + j) \\
& + [not(q(0)).q_0(s) + q(0).q_1(s)].q_1'(s + j) \\
=\ & not(q(0)).[not(q_0(s)).q_0'(s + j) \\
& + q_0(s).q_1'(s + j)] + q(0).[not(q_1(s)).q_0'(s + j) + q_1(s).q_1'(s + j)];
\end{aligned}
$$

thus

$$
\begin{aligned}
q_0(s + j) &= not(q_0(s)).q_0'(s + j) + q_0(s).q_1'(s + j), \\
q_1(s + j) &= not(q_1(s)).q_0'(s + j) + q_1(s).q_1'(s + j).
\end{aligned}
$$

Let $T_{\text{cond.carry}}(s)$ be the computation time of a conditional carry chain. Then, according to Figure 11.13,

$$T_{\text{cond.carry}}(2.s) = T_{\text{cond.carry}}(s) + T_{\text{mux2-1}}. \qquad (11.34)$$

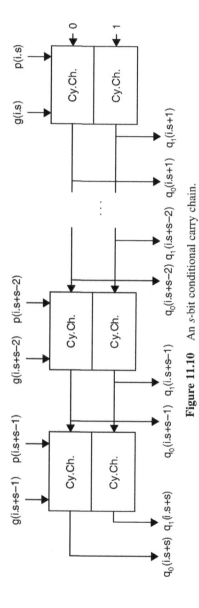

Figure 11.10 An *s*-bit conditional carry chain.

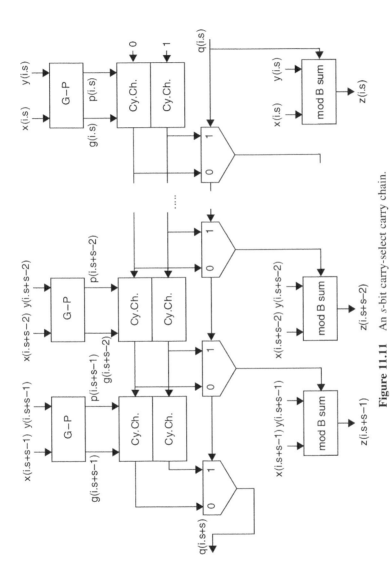

Figure 11.11 An *s*-bit carry-select carry chain.

305

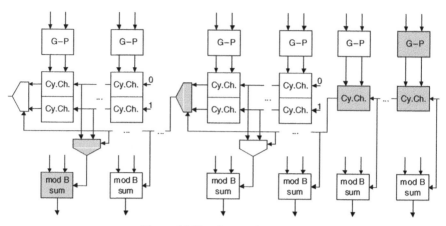

Figure 11.12   Carry-select adder.

A recursive use of the decomposition of Figure 11.13 allows one to generate a $(2^p.s)$-bit conditional carry chain whose computation time is equal to

$$T_{\text{cond.carry}}(2^p.s) = T_{\text{cond.carry}}(s) + p.T_{\text{mux2-1}},$$

that is,

$$T_{\text{cond.carry}}(n) = T_{\text{cond.carry}}(s) + \log_2(n/s).T_{\text{mux2-1}}. \tag{11.35}$$

The computation time of the corresponding n-digit adder is equal to (11.35) plus the delay of one G-P cell and of one mod B sum cell—a logarithmic behavior.

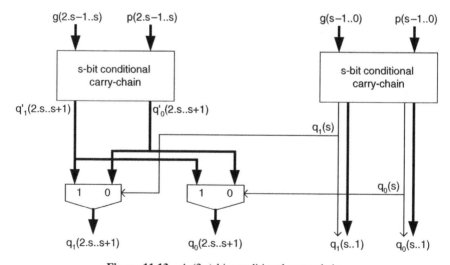

Figure 11.13   A $(2.s)$-bit conditional carry chain.

Example 11.5 (Complete VHDL source code available). Generate a generic *n*-digit base-*B* carry-select adder:

```
entity carry_select is
port (
  x, y: in digit_vector(s-1 downto 0);
  c_in: in std_logic;
  c_out: out std_logic_vector(s downto 1)
);
end carry_select;

architecture circuit of carry_select is
  signal p, g: std_logic_vector(s-1 downto 0);
  signal q0, q1: std_logic_vector(s downto 0);
begin
  q0(0)<='0'; q1(0)<='1';
  iterative_step: for i in 0 to s-1 generate
    p(i)<='1' when x(i)+y(i)=B-1 else '0';
    g(i)<='1' when x(i)+y(i)>B-1 else'0';
    with p(i) select q0(i+1)<=q0(i) when '1', g(i) when
    others;
    with p(i) select q1(i+1)<=q1(i) when '1', g(i) when
    others;
    with c_in select c_out(i+1)<=q0(i+1) when '0', q1(i+1)
    when others;
  end generate;
end circuit;

entity example11_5 is
port (
  x, y: in digit_vector(n-1 downto 0);
  c_in: in std_logic;
  z: out digit_vector(n-1 downto 0);
  c_out: out std_logic
);
end example11_5;

architecture circuit of example11_5 is
  component carry_select ...
  signal q: std_logic_vector(n downto 0);
begin
  <substitute the carry_skip component by the carry_select
  one in example11_3>
end circuit;
```

11.1.7 Optimization of Carry-Select Adders

In the carry-select adder of Figure 11.12, assuming that all the generate and propagate functions have been previously computed, the computation time of the carries of group number j is equal to

$$t(j) = s.T_{\text{Cy.Ch.}} + j.T_{\text{mux2-1}}. \tag{11.36}$$

In particular (worst case),

$$t(n/s - 1) = s.T_{\text{Cy.Ch.}} + (n/s - 1).T_{\text{mux2-1}}. \tag{11.37}$$

The minimum value of (11.37) is obtained when $s.T_{\text{Cy.Ch.}} = n/s.T_{\text{mux2-1}}$, that is, when

$$s = (n.\alpha)^{1/2} \tag{11.38}$$

where α is defined by (11.17). The corresponding value of (11.37) is approximately equal to

$$t(n/s - 1) \cong 2.s.T_{\text{Cy.Ch.}} = (4.n.T_{\text{Cy.Ch.}}.T_{\text{mux2-1}})^{1/2}. \tag{11.39}$$

As for carry-skip adders, a better solution is obtained if the groups are allowed to have different sizes, say, $s_0, s_1, \ldots, s_{k-1}$, where $s_0 + s_1 + \cdots + s_{k-1} = n$. Every multiplexer receives two types of input signals: the locally generated carries (q_0 and q_1) and the output carry of the preceding group whose value controls the selection of either q_0 or q_1. If all groups have the same size, then the locally generated carries of the latest groups are available much sooner than the control signal propagated all along the previous groups. In order that both types of signals arrive more or less at the same time, the latest blocks should be longer than the first ones. For group number j the computation time of the control signal is

$$t_{\text{control}}(j) = s_0.T_{\text{Cy.Ch.}} + (j - 1).T_{\text{mux2-1}}, \tag{11.40}$$

and the computation time of the local carries is

$$t_{\text{carries}}(j) = s_j.T_{\text{Cy.Ch.}}. \tag{11.41}$$

Choosing s_j in such a way that

$$s_0.T_{\text{Cy.Ch.}} + j.T_{\text{mux2-1}} = s_j.T_{\text{Cy.Ch.}}$$

that is,

$$s_j = s_0 + j.\alpha, \tag{11.42}$$

where α is defined by (11.17), the computation time of $q(j)$ is equal to

$$t(j) = \max \{ s_0.T_{\text{Cy.Ch.}} + j.T_{\text{mux2-1}}, s_j.T_{\text{Cy.Ch.}} + T_{\text{mux2-1}} \}$$

$$= s_j.T_{\text{Cy.Ch.}} + T_{\text{mux2-1}}. \tag{11.43}$$

Compute the value of s_0 such that both (11.42) and (11.20) are satisfied:

$$n = s_0 + (s_0 + \alpha) + (s_0 + 2.\alpha) + \cdots + (s_0 + (k-1).\alpha)$$
$$= k.s_0 + (k-1).(k/2).\alpha,$$

and

$$s_0 = n/k - \frac{1}{2}.(k-1).\alpha. \qquad (11.44)$$

According to (11.42), (11.43), and (11.44),

$$t(k-1) = (n/k - \frac{1}{2}.(k-1).\alpha + (k-1).\alpha).T_{\text{Cy.Ch.}} + T_{\text{mux2-1}} \qquad (11.45)$$

whose minimum value is obtained when

$$n/k = \frac{1}{2}.k.\alpha,$$

that is,

$$k = (2.n/\alpha)^{1/2}. \qquad (11.46)$$

Substituting k by the preceding value in (11.44) yields

$$s_0 = \alpha/2. \qquad (11.47)$$

According to (11.42), (11.43), (11.46), and (11.47),

$$t(k-1) = s_{k-1}.T_{\text{Cy.Ch.}} + T_{\text{mux2-1}} = (k.\alpha - \alpha/2).T_{\text{Cy.Ch.}} + T_{\text{mux2-1}}$$
$$\cong (2.n.T_{\text{mux2-1}}.T_{\text{Cy.Ch.}})^{1/2}. \qquad (11.48)$$

The delay has been reduced by a factor of $2^{1/2}$ (compare (11.38) with (11.47)).

Comment 11.4   Generally, $T_{\text{mux2-1}}$ and $T_{\text{Cy.Ch.}}$ have the same order of magnitude, so that $\alpha \cong 1$; s_0 should be chosen equal to 1 and k to approximately $((\frac{1}{4} + 2.n)^{1/2} - \frac{1}{2})$ in order that (11.44) be satisfied with $s_0 = \alpha = 1$.

Example 11.6 With $n = 64$ and $s_0 = \alpha = 1$, the number of groups is equal to $((\frac{1}{4} + 2.64)^{1/2} - 1/2) \cong 11$. A possible choice is

$$s_0 = 1, \; s_1 = 2, \; s_2 = 3, \; s_3 = 4, \; s_4 = 5, \; s_5 = 6,$$
$$s_6 = 7, \; s_7 = 8, \; s_8 = 9, \; s_9 = 9, \; s_{10} = 10.$$

According to (11.43),

$$t(10) = 10.T_{\text{Cy.Ch}} + T_{\text{mux2-1}}.$$

The fixed group-size approach, with $\alpha = 1$, gives the following results:

$$s = (n.\alpha)^{1/2}8 \quad \text{and} \quad k = n/s = 8.$$

Thus (11.39)

$$t(10) = 16.T_{\text{Cy.Ch.}}.$$

11.1.8   Carry-Lookahead Adders (CLAs)

Figure 11.14 shows an n-bit carry-lookahead carry chain, based on the recursive algorithm 4.8. It is made up of two types of components:

the *dot* component implements equations (4.5);
the *carry* component implements equations (4.6), that is,

$$q(i.s + j) = g(i.s + j - 1:i.s) \lor p(i.s + j - 1:i.s).q(i.s), \quad \forall j = 1, 2, \ldots, s - 1.$$

Let $C_{\text{dot}}(s)$ and $T_{\text{dot}}(s)$ be the cost and delay of an s-bit *dot* component, and C_{carry} and T_{carry} the cost and delay of a circuit computing the switching function $f = x \lor y.z$. The cost $C_{\text{cla}}(n)$ and the computation time $T_{\text{cla}}(n)$ of the n-bit carry-lookahead carry chain are equal to

$$C_{\text{cla}}(n) = (n/s).(C_{\text{dot}}(s) + (s - 1).C_{\text{carry}}) + C_{\text{cla}}(n/s),$$
$$T_{\text{cla}}(n) = T_{\text{dot}}(s) + T_{\text{cla}}(n/s) + T_{\text{carry}}. \tag{11.49}$$

A recursive use of equations (11.49) demonstrates that

$$C_{\text{cla}}(n) = [(n/s) + (n/s^2) + \cdots + (n/s^k)].(C_{\text{dot}}(s) + (s - 1).C_{\text{carry}}) + C_{\text{cla}}(n/s^k),$$
$$T_{\text{cla}}(n) = k.(T_{\text{dot}}(s) + T_{\text{carry}}) + T_{\text{cla}}(n/s^k). \tag{11.50}$$

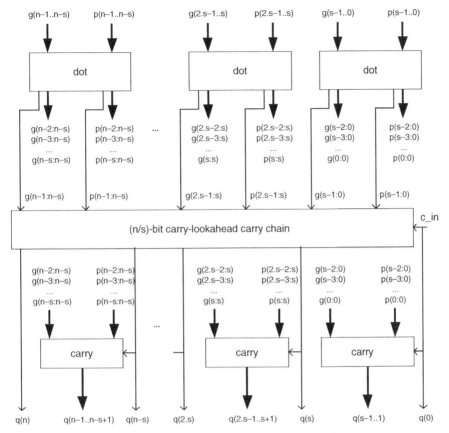

Figure 11.14 n-bit carry-lookahead carry chain.

In particular, if $n = s^k$ then, as $C_{cla}(1) = C_{carry}$ and $T_{cla}(1) = T_{carry}$,

$$C_{cla}(n) = [(n - 1)/(s - 1)].C_{dot}(s) + n.C_{carry},$$

$$T_{cla}(n) = (\log_s n).T_{dot}(s) + (1 + \log_s n)T_{carry}. \qquad (11.51)$$

The computation time of the corresponding n-digit adder is equal to (11.51) plus the delay of a G-P cell and of a mod B sum cell: that is, a logarithmic computation time.

Example 11.7   (Complete VHDL source code available.) Generate a generic two-level n-digit base-B carry-lookahead adder. The following model is based on Figure 11.14. In order to obtain a three-level carry-lookahead adder, the behavioral description of the (n/s)-bit carry chain should be replaced by a structural description based on Figure 11.14 with n substituted by n/s, and so on.

```
entity cla_carry_chain is
port (
  g, p: in std_logic_vector(n-1 downto 0);
  c_in: in std_logic;
  c_out: out std_logic_vector(n downto 0)
);
end cla_carry_chain;

architecture circuit of cla_carry_chain is
  component dot...end component;
  component carry...end component;
  signal q: std_logic_vector(n downto 0);
  signal gg, pp: std_logic_vector(n-1 downto 0);
  signal ggg, ppp, generalized_ggg, generalized_ppp:
  std_logic_vector(n_div_s-1 downto 0);
  signal qqq: std_logic_vector(n_div_s downto 1);
begin
  dot_iteration: for i in 0 to n_div_s-1 generate
    dot_instantiation: dot port map(g(i*s+s-1 downto i*s),
    p(i*s+s-1 downto i*s), gg(i*s+s-1 downto i*s),pp(i*s+s-1
    downto i*s));
  end generate;
  input_connections: for i in 0 to n_div_s-1 generate
    ggg(i)<=gg(i*s+s-1); ppp(i)<=pp(i*s+s-1);
  end generate;
  -------------------------------------------------
  --behavioral description of an (n_div_s)-bit cla carry
  --chain:
  generalized_ggg(0)<=ggg(0); generalized_ppp(0)<=ppp(0);
  cla_carry_chain_description: for i in 1 to n_div_s-1 generate
    qqq(i)<=generalized_ggg(i-1) or (generalized_ppp(i-1) and
    c_in);
    dot_operation(ggg(i), ppp(i),
    generalized_ggg(i-1), generalized_ppp(i-1),
    generalized_ggg(i), generalized_ppp(i));
  end generate;
  qqq(n_div_s)<=generalized_ggg(n_div_s-1) or
  (generalized_ppp(n_div_s-1) and c_in);
  -------------------------------------------------
  output_connections: for i in 1 to s generate
    q(i*s)<=qqq(i);
  end generate;
  q(0)<=c_in;
  carry_iteration: for i in 0 to s-1 generate
    carry_instantiation: carry port map(gg(i*s+s-2 downto i*s),
```

```
  pp(i*s+s-2 downto i*s), q(i*s), q(i*s+s-1 downto i*s+1));
 end generate;
 output_carries_iteration: for i in 0 to s**2 generate
  c_out(i)<=q(i);
 end generate;
end circuit;

entity example11_7 is
port (
 x, y: in digit_vector(n-1 downto 0);
 c_in: in std_logic;
 z: out digit_vector(n-1 downto 0);
 c_out: out std_logic
);
end example11_7;

architecture circuit of example11_7 is
 component cla_carry_chain...end component;
 signal p, g: std_logic_vector(n-1 downto 0);
 signal q: std_logic_vector(n downto 0);
begin
 iterative_step: for i in 0 to n-1 generate
  p(i)<='1' when x(i)+y(i)=B-1 else '0';
  g(i)<='1' when x(i)+y(i)>B-1 else'0';
  z(i)<=(x(i)+y(i)+conv_integer(q(i))) mod B;
 end generate;
 cla_carry_chain_instantiation: cla_carry_chain port map(g, p,
 c_in, q);
 c_out<=q(n);
end circuit;
```

Definition 11.1 A *carry-lookahead generator* (Figure 11.15a) is a $(2.s+1)$-input $(s+1)$-output combinational circuit equivalent to the circuit of Figure 11.15b. It computes the following switching functions:

$$g(s-1{:}0) = g(s-1) \vee g(s-2).p(s-1) \vee g(s-3).p(s-2).p(s-1) \vee \cdots$$

$$\vee g(0).p(1).p(2).\cdots.p(s-1),$$

$$p(s-1{:}0) = p(0).p(1).\cdots.p(s-2).p(s-1),$$

$$q(1) = g(0) \vee c\_in.p(0),$$

$$q(2) = g(1{:}0) + c\_in.p(1{:}0) = g(1) \vee g(0).p(1) \vee c\_in.p(0).p(1),$$

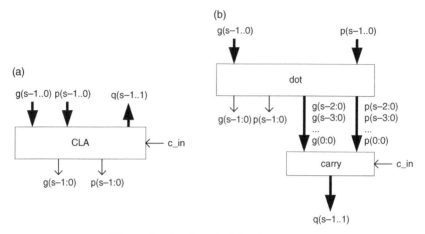

Figure 11.15   Carry-lookahead generator.

$$q(3) = g(2{:}0) + c\_in.p(2{:}0) = g(2) \vee g(1).p(2) \vee g(0).p(1).p(2)$$
$$\vee\ c\_in.p(0).p(1).p(2),$$

$$\cdots$$

$$q(s-1) = g(s-2{:}0) + c\_in.p(s-2{:}0) = g(s-2) \vee g(s-3).p(s-2)$$
$$\vee\ g(s-4).p(s-3).p(s-2) \vee \cdots \vee g(0).p(1).p(2).\cdots.p(s-2)$$
$$\vee\ c\_in.p(0).p(1).\cdots.p(s-3).p(s-2).$$

An n-bit carry-lookahead carry chain made up of carry-lookahead generators is shown in Figure 11.16 (just another way to draw the circuit of Figure 11.14).

Example 11.8   Let $n = 16$, $s = 4$, that is, $n/s = 4$. The CLA equations can be written

$$q(0) = c\_in,$$
$$q(1) = g(0) \vee c\_in.p(0),$$
$$q(2) = g(1) \vee g(0).p(1) \vee c\_in.p(0).p(1),$$
$$q(3) = g(2) \vee g(1).p(2) \vee g(0).p(1).p(2) \vee c\_in.p(0).p(1).p(2),$$
$$q(4) = g(3{:}0) \vee c\_in.p(3{:}0),$$
$$q(5) = g(4) \vee q(4).p(4),$$
$$q(6) = g(5) \vee g(4).p(5) \vee q(4).p(4).p(5),$$
$$q(7) = g(6) \vee g(5).p(6) \vee g(4).p(5).p(6) \vee q(4).p(4).p(5).p(6),$$
$$q(8) = g(7{:}4) \vee g(3{:}0).p(7{:}4) \vee c\_in.p(3{:}0).p(7{:}4),$$
$$q(9) = g(8) \vee q(8).p(8),$$

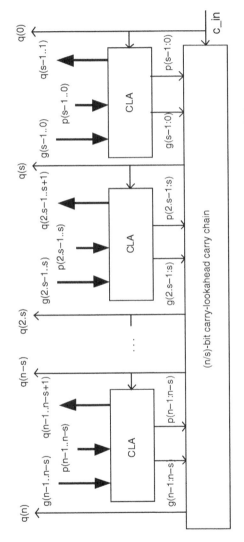

Figure 11.16 An *n*-bit carry-lookahead carry chain (second version).

$q(10) = g(9) \lor g(8).p(9) \lor q(8).p(8).p(9),$

$q(11) = g(10) \lor g(9).p(10) \lor g(8).p(9).p(10) \lor q(8).p(8).p(9).p(10),$

$q(12) = g(11{:}8) \lor g(7{:}4).p(11{:}8) \lor g(3{:}0).p(7{:}4).p(11{:}8)$

$\qquad\qquad \lor c\_in.p(3{:}0).p(7{:}4).p(11{:}8),$

$q(13) = g(12) \lor q(12).p(12),$

$q(14) = g(13) \lor g(12).p(13) \lor q(12).p(12).p(13),$

$q(15) = g(14) \lor g(13).p(14) \lor g(12).p(13).p(14) \lor q(12).p(12).p(13).p(14),$

$q(16) = g(15{:}12) \lor g(11{:}8).p(15{:}12) \lor g(7{:}4).p(11{:}8).p(15{:}12) \lor$

$\qquad\qquad g(3{:}0).p(7{:}4).p(11{:}8).p(15{:}12) \lor c\_in.p(3{:}0).p(7{:}4).p(11{:}8).p(15{:}12).$

The complete circuit is shown in Figure 11.17.

Definition 11.2 An *augmented full adder* (*AFA*), whose symbol is shown in Figure 11.18, calculates $g(i)$, $p(i)$, and $z(i)$ as a function of $x(i)$, $y(i)$, and $q(i)$:

$$g(i) = 1 \text{ if } x(i) + y(i) > B - 1, \ = 0 \text{ otherwise,}$$

$$p(i) = 1 \text{ if } x(i) + y(i) = B - 1, = 0 \text{ otherwise,}$$

$$z(i) = (x(i) + y(i) + q(i)) \bmod B.$$

Carry-lookahead generators and augmented full adders are the building blocks for synthesizing carry-lookahead adders.

Example 11.9 The circuit of Figure 11.19 is an s^2-digit carry-lookahead adder.

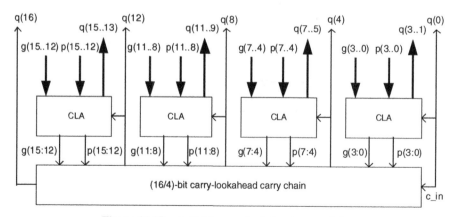

Figure 11.17 A 16-bit carry-lookahead carry-chain.

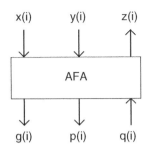

Figure 11.18 Augmented full adder.

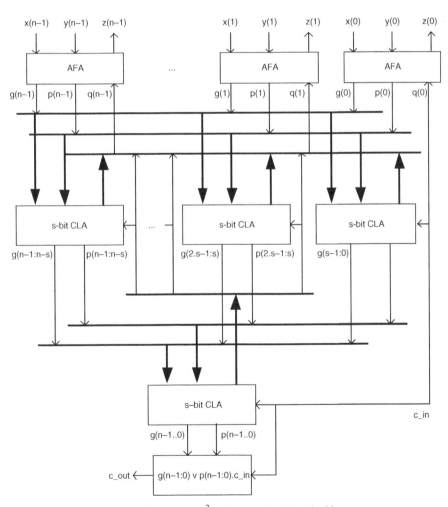

Figure 11.19 An s^2-digit carry-lookahead adder.

11.1.9 Prefix Adders

An n-bit *prefix carry chain*, based on the recursive algorithm 4.5 (BRE1982), is shown in figure 11.20. It is made up of components implementing the associative dot operation (Definitions 4.1), so that (4.5)

$$(g(i), p(i)) \bullet (g(i-1), p(i-1)) \bullet (g(i-2), p(i-2)) \bullet \cdots \bullet (g(i-k), p(i-k))$$
$$= (g(i{:}i-k), p(i{:}i-k)).$$

An n-bit prefix carry chain generates all the generalized *generate* and *propagate* functions of type $g(i{:}0)$ and $p(i{:}0)$, respectively. They have been underlined in Figure 11.20.

Let C_{dot} and T_{dot} be the cost and delay of the dot operator. According to Figure 11.20, the cost and computation time of an n-bit prefix carry chain (a *Brent–Kung* carry chain) are equal to

$$C_{Brent-Kung}(n) = C_{Brent-Kung}(n/2) + (n-1).C_{dot},$$
$$T_{Brent-Kung}(n) = T_{Brent-Kung}(n/2) + 2.T_{dot}. \tag{11.52}$$

A recursive use of equations (11.52) yields

$$C_{Brent-Kung}(n) = C_{Brent-Kung}(n/2^k) + (2^k - 1).n/2^{k-1}.C_{dot} - k.C_{dot},$$
$$T_{Brent-Kung}(n) = T_{Brent-Kung}(n/2^k) + 2.k.T_{dot}. \tag{11.53}$$

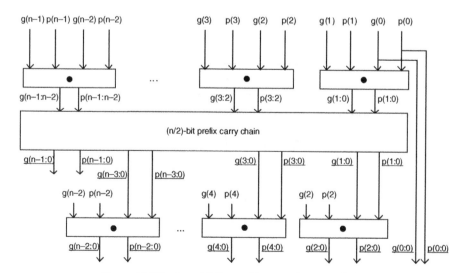

Figure 11.20 An n-bit Brent–Kung prefix carry chain.

In particular, if $n = 2^{k+1}$ then

$$C_{\text{Brent-Kung}}(n) = C_{\text{Brent-Kung}}(2) + (2.n - 4).C_{\text{dot}} - (\log_2 n - 1).C_{\text{dot}}$$
$$= (2.n - 2 - \log_2 n).C_{\text{dot}}, \tag{11.54}$$
$$T_{\text{Brent-Kung}}(n) = (2.\log_2 n - 1).T_{\text{dot}}.$$

Comment 11.5 According to (11.54), $T_{\text{Brent-Kung}}(4) = 3.T_{\text{dot}}$. It corresponds to the following scheduled algorithm:

1. $(g(3:2),\ p(3:2)):= (g(3),\ p(3)) \bullet (g(2),\ p(2));\ (g(1:0),\ p(1:0)):= (g(1), p(1)) \bullet (g(0), p(0));$
2. $(g(3:0), p(3:0)):= (g(3:2), p(3:2)) \bullet (g(1:0), p(1:0));$
3. $(g(2:0), p(2:0)):= (g(2), p(2)) \bullet (g(1:0), p(1:0));$

Nevertheless, the operations of cycles 2 and 3 could be executed in parallel, so that $T_{\text{Brent-Kung}}(4) = 2.T_{\text{dot}}$. More generally,

$$T_{\text{Brent-Kung}}(n) = (2.\log_2 n - 2).T_{\text{dot}}. \tag{11.55}$$

In order to get the carries, an additional *carry* component (Figure 11.15b) should be added for computing

$$q(i) = g(i - 1:0) + p(i - 1:0).c\_in, \quad \forall i = 1, 2, \ldots, n - 1.$$

Example 11.10 (Complete VHDL source code available.) Generate a generic n-digit Brent–Kung base-B prefix adder. The following model is based on Figure 11.20.

```
entity prefix_2 is...
...
entity prefix_n_div_2 is...

entity prefix_n is
port (
  g, p: in std_logic_vector(n-1 downto 0);
  gg, pp: out std_logic_vector(n-1 downto 0)
);
end prefix_n;

architecture circuit of prefix_n is
  component prefix_n_div_2...end component;
  signal a, b, aa, bb: std_logic_vector(n_div_2-1 downto 0);
begin
```

```
  first_iteration: for i in 0 to n_div_2-1 generate
    dot_operation(g(2*i+1), p(2*i+1), g(2*i), p(2*i), a(i),
    b(i));
  end generate;
  component_instantiation: prefix_n_div_2 port map (a, b, aa,
  bb);
  second_iteration: for i in 1 to n_div_2-1 generate
    dot_operation(g(2*i), p(2*i), aa(i-1), bb(i-1), gg(2*i),
    pp(2*i));
  end generate;
  gg(0)<=g(0); pp(0)<=p(0);
  third_iteration: for i in 0 to n_div_2-1 generate
    gg(2*i+1)<=a(i); pp(2*i+1)<=b(i);
  end generate;
end circuit;

entity example11_10 is
port (
  x, y: in digit_vector(n-1 downto 0);
  c_in: in std_logic;
  z: out digit_vector(n-1 downto 0);
  c_out: out std_logic
);
end example11_10;


architecture circuit of example11_10 is
  component prefix_n...end component;
  signal p, g, pp, gg: std_logic_vector(n-1 downto 0);
  signal q: std_logic_vector(n downto 0);
begin
  q(0)<=c_in;
  iterative_step: for i in 0 to n-1 generate
    p(i)<='1' when x(i)+y(i)=B-1 else '0';
    g(i)<='1' when x(i)+y(i)>B-1 else'0';
    z(i)<=(x(i)+y(i)+conv_integer(q(i))) mod B;
  end generate;
  component_instantiation: prefix_n port map(g, p, gg, pp);
  carry_computation: for i in 1 to n generate
    q(i)<=gg(i-1) or (pp(i-1) and q(i-1));
  end generate;
  c_out<=q(n);
end circuit;
```

Another n-bit *prefix carry chain*, based on the recursive algorithm 4.6 ([LAD1980]), is shown in Figure 11.21.

Let C_{dot} and T_{dot} be the cost and delay of the dot operator. According to Figure 11.21 the cost and computation time of an n-bit prefix carry chain (a

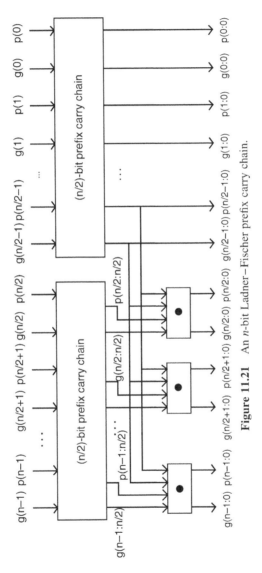

Figure 11.21 An n-bit Ladner–Fischer prefix carry chain.

Ladner–Fischer carry chain) are equal to

$$C_{\text{Ladner–Fischer}}(n) = 2.C_{\text{Ladner–Fischer}}(n/2) + (n/2).C_{\text{dot}},$$
$$T_{\text{Ladner–Fischer}}(n) = T_{\text{Ladner–Fischer}}(n/2) + T_{\text{dot}}. \tag{11.56}$$

A recursive use of equations (11.56) yields

$$C_{\text{Ladner–Fischer}}(n) = 2^k.C_{\text{Ladner–Fischer}}(n/2^k) + k.(n/2).C_{\text{dot}},$$
$$T_{\text{Ladner–Fischer}}(n) = T_{\text{Ladner–Fischer}}(n/2^k) + k.T_{\text{dot}}. \tag{11.57}$$

In particular, if $n = 2^{k+1}$ then

$$C_{\text{Ladner–Fischer}}(n) = (n/2).C_{\text{Ladner–Fischer}}(2) + (\log_2 n - 1).(n/2).C_{\text{dot}}$$
$$= (\log_2 n).(n/2).C_{\text{dot}}, \tag{11.58}$$
$$T_{\text{Ladner–Fischer}}(n) = T_{\text{Ladner–Fischer}}(2) + (\log_2 n - 1).T_{\text{dot}} = (\log_2 n).T_{\text{dot}}$$

Comments 11.6

1. Several types of adders have been analyzed. As regards their cost and computation time, they can be classified into three groups:
   - The ripple-carry and the carry-chain adders have $O(n)$ cost and computation time, and the same occurs with the slice-based structures (carry-skip, base-B^s, carry-select) if s (the slice size) is a previously defined constant.
   - In the case of the carry-skip and carry-select adders, the value of s can be optimized (Sections 11.1.4 and 11.1.7). Furthermore, the groups can have different sizes. Then the cost is $O(n)$ and the computation time $O(n^{1/2})$.
   - The carry-lookahead and the prefix adders have $O(n)$ cost (or even $O(n.\log(n))$ and $O(\log(n))$ computation time.
2. Other parallel-prefix structures have been proposed, including [KOG1973], [HAN1987], and [SUG1990], all of them characterized by a logarithmic computation time.

11.1.10 FPGA Implementation of Adders

11.1.10.1 Carry-Chain Adders FPGAs generally contain dedicated computation resources for generating fast adders. As an example, the Virtex-family programmable arrays (Chapter 9) include logic gates and multiplexers that, along with the general-purpose look-up tables, allow one to build effective carry-chain adders. The basic adder cell (Figure 11.22) allows implementation of the ripple adder of Figure 10.1. The carry chain is made up of multiplexers belonging to adjacent configurable blocks. The look-up table is used for implementing the exclusive-or function:

$$p(i) = x(i) \; xor \; y(i).$$

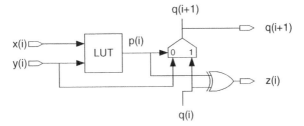

Figure 11.22 Basic cell.

The computation time $T_{adder}(n)$ of an n-bit adder is equal to

$$T_{adder}(n) = T_{LUT} + (n - 1).T_{mux\text{-}cy} + T_{XOR2} \qquad (11.59)$$

where t_{LUT} is the computation delay of a general-purpose look-up table, $t_{mux\text{-}cy}$ the delay of a dedicated multiplexer along with the delay of the connection to the next adjacent block, and t_{XOR2} the delay of the 2-input XOR gate. It has been assumed that $t_{XOR2} > t_{mux\text{-}cy}$. The delay from $q(0)$ to $q(n)$, assuming that all $p(i)$ functions have already been calculated, is

$$T_{carry}(n) = n.T_{mux\text{-}cy}. \qquad (11.60)$$

Every slice of the Virtex family includes two cells so that the cost $C_{adder}(n)$ of an n-bit adder is equal to

$$C_{adder}(n) = n/2 \text{ slices}. \qquad (11.61)$$

The computation time t_{LUT} of a look-up table, as well as the average propagation time $t_{connection}$ of a general-purpose connection, are much longer than $t_{mux\text{-}cy}$ (the delay of a dedicated multiplexer plus the connection to the next adjacent block). As a consequence, techniques such as the carry-lookahead or prefix adders, based on more complex computation resources (carry-lookahead generator, dot operation) and connection schemes, are generally inefficient. However, the carry-skip and carry-select techniques can be used. As an example, the FPGA implementation of carry-skip adders is analyzed in the next section.

11.1.10.2 Carry-Skip Adders   A key point in order to get fast circuits is the use of dedicated carry-logic multiplexers for any iterative subcircuit belonging to one of the critical paths. In the case of a carry-skip carry chain, both the circuits computing $p(i.s + s\text{-}1\text{:}i.s)$ in every group (Figure 11.6) and the carry-skip multiplexers (shaded in Figure 11.7) can be implemented with carry-logic multiplexers. The computation of $p(i.s + s - 1\text{:}i.s)$ is performed by the circuit of Figure 11.23. The look-up tables

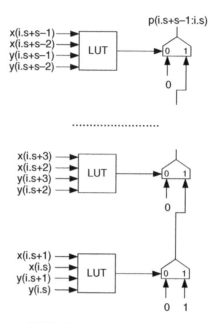

Figure 11.23 Computation of $p(i.s + s - 1 : i.s)$.

compute

$$p(i.s).p(i.s + 1) = (x(i.s) \; xor \; y(i.s)).(x(i.s + 1) \; xor \; y(i.s + 1)),$$
$$p(i.s + 2).p(i.s + 3) = (x(i.s + 2) \; xor \; y(i.s + 2)).(x(i.s + 3) \; xor \; y(i.s + 3)),$$
and so on.

The corresponding cost and delay are equal to

$$C_p = s/4 \text{ slices}, \quad T_p = T_{\text{LUT}} + (s/2).T_{\text{mux-cy}}. \tag{11.62}$$

The multiplexers that select the output carry of every group—the so-called carry-skip multiplexers—belong to the critical path of the circuit, so that they must be implemented with dedicated carry multiplexers, as shown in Figure 11.24, where $qq(i.s)$ is the carry locally generated by group number $i - 1$. Observe that the connection of $p(i.s + s - 1 : i.s)$ to the internal multiplexer must be done through the look-up table. The corresponding cost and propagation time are, respectively, equal to

$$C_{\text{skip}} = (n/s)/2 \text{ slices}, \quad T_{\text{skip}} = T_{\text{LUT}} + (n/s).T_{\text{mux-cy}}. \tag{11.63}$$

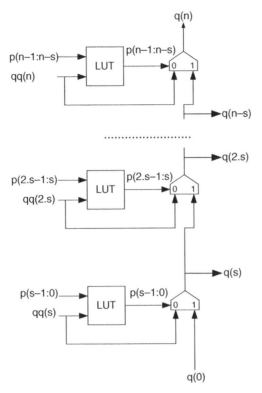

Figure 11.24 Carry-skip multiplexers.

The implementation of an s-bit group of the adder is shown in figure 11.25. Its cost and delay are equal to

$$C_{\text{group}} = s/2 \text{ slices}, \quad T_{\text{group}} = T_{\text{LUT}} + s.T_{\text{mux-cy}}. \tag{11.64}$$

The complete circuit architecture is shown in Figure 11.26. The cost and computation time of an n-bit s-group carry-skip adder are equal to

$$
\begin{aligned}
C_{\text{carry-skip-adder}}(n,s) &= (n/s).(C_{\text{group}} + C_p) + C_{\text{skip}} = 0.75.n + 0.5.(n/s), \\
T_{\text{carry-skip-adder}} &= \max \{T_{\text{group}} + T_{\text{connection}}, T_p + T_{\text{connection}} + T_{\text{LUT}}\} \\
&\quad + (n/s + s - 2).T_{\text{muc-cy}} + T_{\text{connection}} + T_{\text{XOR2}} \\
&= T_{\text{LUT}} + \max \{(2.s + n/s - 2).T_{\text{mux-cy}.}, T_{\text{LUT}} \\
&\quad + (1.5.s + n/s - 2).T_{\text{mux-cy}.}\} + 2.T_{\text{connection}} + T_{\text{XOR2}}.
\end{aligned} \tag{11.65}
$$

As mentioned earlier (Comment 11.2(2)), the first carry-skip multiplexer is not necessary unless the adder is used as a building block for generating larger adders.

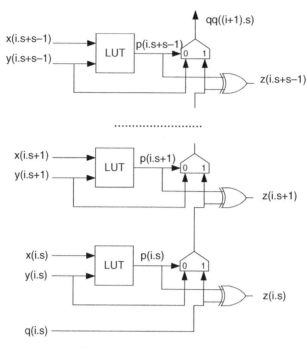

Figure 11.25   An s-bit group.

According to (11.65) the shortest theoretical delay is obtained when $2.s + n/s$ or $1,5.s + n/s$ is minimum, that is, when

$$s \cong (n/2)^{1/2}, n/s \cong (2.n)^{1/2}, \quad \text{or} \quad s \cong (n/1,5)^{1/2}, n/s \cong (1,5.n)^{1/2}. \quad (11.66)$$

11.1.10.3 Experimental Results Several adders have been implemented [BIO2003] within a Spartan II-family FPGA. The synthesis was performed with the Xilinx Synthesis Technology (XST) and the physical implementation with the Xilinx ISE (Integrated System Environment) Version 5.1. In order to take advantage of the resources, the design was implemented instantiating the low-level FPGA components and using relative placement & routing (RPR).

The results are summarized in Table 11.1. The first column ($s = n$) gives the delay of a traditional adder. The last column gives the frequency increase of the fastest adder with respect to the traditional one. Additionally, Table 11.2 gives the area expressed in terms of FPGA slices.

Thanks to the use of the dedicated carry-logic circuitry for all blocks included within the critical path, the frequency increase for long-operand adders is substantial: more than 500% for a 1024-bit adder.

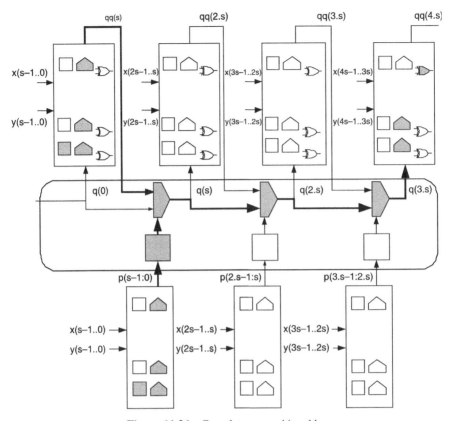

Figure 11.26 Complete carry-skip adder.

11.1.11 Long-Operand Adders

In the case of long-operand additions, the n-digit operands can be split down into s-digit groups and the addition computed according to Algorithm 4.10. The complete circuit is made up of an s-digit adder (procedure `natural_addition` of algorithm

TABLE 11.1 Experimental Results: Delay

	Delay in ns				Frequency
n	$s = n$	$s = 8$	$s = 16$	$s = 32$	Increase
64	14 ns	13 ns	12 ns	—	13%
96	16 ns	14 ns	13 ns	—	21%
128	23 ns	14 ns	14 ns	—	63%
256	38 ns	—	16 ns	17 ns	141%
512	77 ns	—	20 ns	20 ns	296%
1024	159 ns	—	28 ns	25 ns	531%

TABLE 11.2   Experimental Results: Area

	Area in Slices				Area Overhead		
n	$s = n$	$s = 8$	$s = 16$	$s = 32$	$s = 8$	$s = 16$	$s = 32$
64	32	47	41	—	47%	28%	—
96	48	73	66	—	52%	38%	—
128	64	99	91	—	55%	42%	—
256	128	—	191	179	—	49%	40%
512	256	—	391	375	—	53%	46%
1024	512	—	791	767	—	54%	50%

4.10), connection resources giving access to the s-digit groups, a D-flip-flop that stores the carries (q in Algorithm 4.10), and a control unit whose kernel is an (n/s)-state counter. An example was seen in Chapter 10 (Example 10.4), where the access to the successive groups was through (n/s)-to-1 s-bit multiplexers.

11.1.12   Multioperand Adders

11.1.12.1 Sequential Multioperand Adders   In order to compute $z = x^{(0)} + x^{(1)} + \cdots + x^{(m-1)}$, where every $x^{(i)}$ is a natural number, Algorithm 4.11 can be used. The corresponding sequential circuit is made up of an n-digit adder, an n-digit register, and a control unit; furthermore, some kind of connection resource (equivalent to an m-to-1 n-digit multiplexer) must be used in order to enter the m operands.

Example 11.11   (Complete VHDL code available.) Generate a generic m-operand n-digit adder. The circuit is shown in Figure 11.27.
    The corresponding VHDL description ($B = 2$) is the following one:

```
entity example11_11 is
port (
  x: in operands;
  clk, start, reset: in std_logic;
  done: out std_logic;
  z: inout std_logic_vector(n-1 downto 0)
);
end example11_11;

architecture circuit of example11_11 is
  signal adder_out, op_1: std_logic_vector(n-1 downto 0);
  signal operand_select: std_logic_vector(logm-1 downto 0);
  signal load, clear: std_logic;
  subtype state is integer range -3 to m;
  signal current_state: state;
```

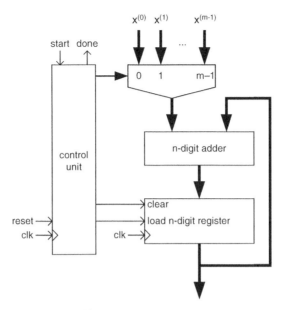

Figure 11.27   Sequential multioperand adder.

```
begin
  --data path:
  op_1<=x(conv_integer(operand_select));
  adder_out<=op_1+z;
  process(clk)
  begin
    if clear='1' then z<=conv_std_logic_vector(0,n);
    elsif clk'event and clk='1' then
      if load='1' then z<=adder_out; end if;
    end if;
  end process;
  --control unit
  process(clk, reset)
  begin
    case current_state is
      when -3=>load<='0'; clear<='0';
      operand_select<=conv_std_logic_vector(0, logm);
      done<='1';
      when -2=>load<='0'; clear<='0';
      operand_select<=conv_std_logic_vector(0, logm);
      done<='1';
      when -1=>load<='0'; clear<='1';
      operand_select<=conv_std_logic_vector(0, logm);
      done<='1';
      when 0 to m-1=>load<='1'; clear<='0';
```

```
        operand_select<=conv_std_logic_vector(current_state,
        logm); done<='0';
        when m=>load<='0'; clear<='0';
        operand_select<=conv_std_logic_vector(0, logm);
        done<='1';
      end case;
      if reset='1' then current_state<=-3;
      elsif clk'event and clk='1' then
        case current_state is
          when -3=>if start='0' then current_state <=
          current_state+1; end if;
          when -2=>if start='1' then current_state <=
          current_state+1; end if;
          when -1=>current_state<=current_state+1;
          when 0 to m-1=>current_state<=current_state+1;
          when m=>current_state<=-3;
        end case;
      end if;
    end process;
end circuit;
```

The cost and computation time of the preceding m-operand n-digit sequential adder are equal to

$$C_{\text{sequential}}(m, n) = C_{\text{adder}}(n) + C_{\text{register}}(n) + C_{\text{control}}(m) + C_{\text{multiplexer}}(m, n),$$
$$T_{\text{sequential}}(m, n) = m.(T_{\text{adder}}(n) + T_{\text{Pmax}} + T_{\text{SUmin}}) \cong m.T_{\text{adder}}(n), \qquad (11.67)$$

where $C_{\text{adder}}(n)$ is the cost of an n-digit adder, $C_{\text{register}}(n)$ the cost of an n-digit register, $C_{\text{control}}(m)$ the cost of the control unit, approximately proportional to $\log_2 m$ (the number of internal state variables of an m-state machine), $C_{\text{multiplexer}}(m, n)$ the cost of an m-to-1 n-digit multiplexer, $T_{\text{adder}}(n)$ the computation time of an n-digit adder, T_{Pmax} the maximum propagation time of a D-flip-flop, and T_{SUmin} its minimum set up time.

In the case of a long-multioperand adder, the n-digit register is substituted by an (n/s)-word s-digit register bank, the 2-operand n-digit adder by a 2-operand s-digit adder, and the control unit is an $m.(n/s)$-state counter.

In conclusion, the computation time of an m-operand n-digit sequential adder is approximately proportional to $m.\log_2 n$ if a fast adder is used.

11.1.12.2 Combinational Multioperand Adders   The combinational circuit that corresponds to Algorithm 4.11 is an iterative circuit made up of m-1 2-operand n-digit adders. If every 2-operand n-digit adder is a simple ripple-carry adder, then the complete circuit is a two-dimensional array made up of $(m - 1).n$ full adders (Figure 11.28).

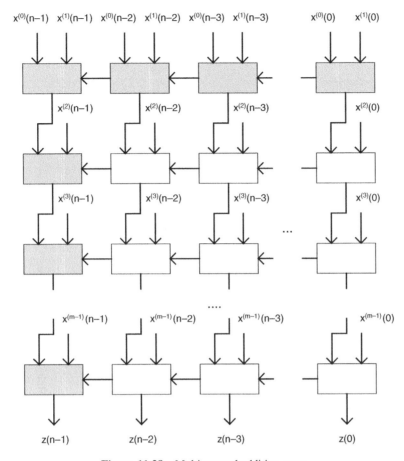

Figure 11.28 Multioperand addition array.

The corresponding cost and computation time—one of the critical paths has been shaded—are equal to

$$C_{\text{adder-array}}(m, n) = (m - 1).n.C_{\text{FA}},$$
$$T_{\text{adder-array}}(m, n) = (m + n - 2).T_{\text{FA}}. \tag{11.68}$$

As regards the computation time, a better solution is a binary tree of 2-operand n-digit adders instead of an iterative circuit. An example in which every 2-operand n-digit adder is a simple ripple-carry adder is shown in Figure 11.29 (with $n = 3$ and $m = 8$). The depth of the tree is equal to $\log_2 m$. Its cost and computation time—see the shaded critical path—are equal to

$$C_{\text{adder-tree}}(m, n) = (m - 1).n.C_{\text{FA}},$$
$$T_{\text{adder-tree}}(m, n) = (n + \log_2 m - 1).T_{\text{FA}}. \tag{11.69}$$

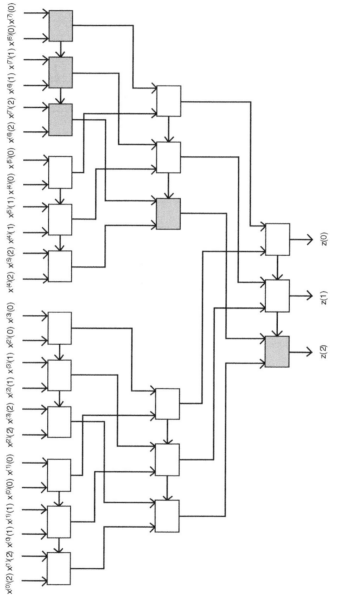

Figure 11.29 Multioperand addition tree.

Comments 11.7

1. In both the array (Figure 11.28) and the tree (Figure 11.29) circuits, the ripple-carry adder can be substituted by a faster one. Then the computation time would be proportional to the product of the number of computation steps (the number of lines of the array or the depth of the tree) by $\log_2 n$, that is, $(m-1).\log_2 n$ or $\log_2 m.\log_2 n$. Observe that if $m > (n + \log_2 n)/(\log_2 n - 1)$, then $(m-1).\log_2 n$ is greater than $m + n$, and if $\log_2 m > n/(\log_2 n - 1)$ then $\log_2 m.\log_2 n$ is greater than $\log_2 m + n$, so that, for certain values of m and n, the use of fast 2-operand n-digit adders could generate a slower multioperand adder than the one obtained with simple carry-ripple ones.

2. All operands as well as the result were assumed to be n-digit base-B numbers. If all the operands belong to the same range, and the result is known to be an n-digit number whatever the value of the operands, then the operands can be represented with $(n - k)$ digits where $k \cong \log_B m$. The previously described circuits should be pruned and the cost evaluation modified.

11.1.12.3 Carry-Save Adders   The stored-carry encoding (Algorithm 4.13) consists of representing the result of a 3-operand n-digit addition under the form of two n-digit numbers:

$$w + x + y = u + v. \tag{11.70}$$

According to Algorithm 4.13, the computation of u and v can be performed by an n-cell iterative circuit (Figure 11.30) whose basic cell, a 3-operand 1-digit adder, implements the two following functions:

$$\begin{aligned} q(i + 1) &= (w(i) + x(i) + y(i))/B, \\ z(i) &= (w(i) + x(i) + y(i)) \bmod B. \end{aligned} \tag{11.71}$$

In the binary case ($B = 2$), the 3-operand 1-bit adder is a full adder. This is not so in the nonbinary case ($B \geq 3$), since the maximum value of $(w(i) + x(i) + y(i))/B$, that

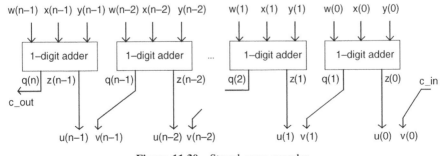

Figure 11.30   Stored-carry encoder.

is, $3.(B-1)/B = 3 - (3/B)$, could be equal to 0, 1, or 2 so the carry could be equal to 2.

An m-operand carry-save array (Algorithm 4.14) is shown in Figure 11.31. The result is assumed to be the sum of two n-digit numbers u and v, and the same comment as before (Comment 11.7(2)) can be made. In order to get the actual (non-encoded) result, an additional 2-operand n-digit adder is necessary for computing $u + v$ (last instruction of Algorithm 4.14). The corresponding cost and computation time (without the final addition) are equal to

$$C_{\text{carry-save-array}}(m, n) = (m-2).n.C_{1\text{-digit-adder}},$$
$$T_{\text{carry-save-array}}(m, n) = (m-2).T_{1\text{-digit-adder}}. \tag{11.72}$$

With the additional 2-operand n-digit adder, the cost and computation time are equal to

$$C_{\text{carry-save-adder}}(m, n) = (m-2).n.C_{1\text{-digit-adder}} + C_{\text{adder}}(n)$$
$$T_{\text{carry-save-adder}}(m, n) = (m-2).T_{1\text{-digit-adder}} + T_{\text{adder}}(n). \tag{11.73}$$

Comments 11.8

1. If one of the operands of the stored-carry encoder of Figure 11.30, say, y, has all its digits equal to 0 or 1, then the 1-digit adders can be substituted by full

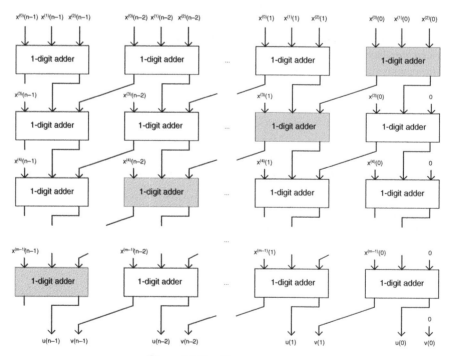

Figure 11.31 Carry-save array.

adders and v also has all its digits equal to 0 or 1. Thus, if one of the inputs of the carry-save array of Figure 11.31 has all its digits equal to 0 or 1, then all the 1-digit adders can be substituted by full adders.

2. The computation time and cost of the carry-save array of Figure 11.31 are practically the same as the ones of a simple combinational (carry-propagate) adder: compare (11.73) with (11.68) assuming that $T_{\text{1-digit-adder}} = T_{\text{FA}}$, $C_{\text{1-digit-adder}} = C_{\text{FA}}$ and that the 2-operand adder is a ripple-carry one so that $T_{\text{adder}}(n) = n.T_{\text{FA}}$ and $C_{\text{adder}}(n) = n.C_{\text{FA}}$. The conclusion will be different if a sequential implementation is considered.

A sequential implementation of the carry-save adder is shown in Figure 11.32. Initially u and v are equal to 0 so that (Comment 11.8(1)) the n-digit stored-carry encoder is made up of full adders and v has all its digits equal to 0 or 1. The computation time is equal to

$$T_{\text{sequential}}(m, n) \cong m.T_{\text{FA}} + T_{\text{adder}}(n). \tag{11.74}$$

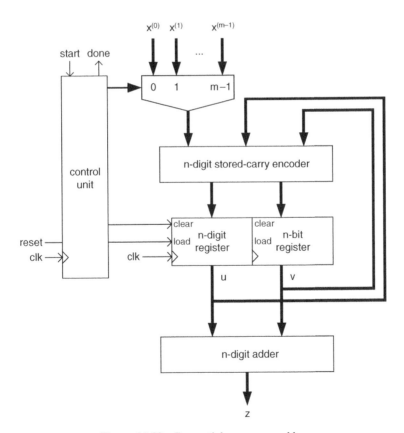

Figure 11.32 Sequential carry-save adder.

Compare now (11.74) with (11.67). If ripple-carry 2-operand adders are used, the computation times are approximately equal to

$m.n.T_{FA}$ (Figure 11.27, sequential carry-propagate adder),

$(m + n).T_{FA}$ (Figure 11.32, sequential carry-save adder).

The carry-save adder is much faster than the carry-propagate one.

Example 11.12 (Complete VHDL code available.) Generate a generic m-operand n-digit sequential carry-save adder (the 2-operand n-digit adder summing up u and v is not included).

```
entity example11_12 is
port (
  x: in operands;
  clk, start, reset: in std_logic;
  done: out std_logic;
  u: inout digit_vector(n-1 downto 0);
  v: inout std_logic_vector(n-1 downto 0)
);
end example11_12;

architecture circuit of example11_12 is
  signal op_1, reg_in_u: digit_vector(n-1 downto 0);
  signal reg_in_v: std_logic_vector(n-1 downto 0);
  signal operand_select: std_logic_vector(logm-1 downto 0);
  signal load, clear: std_logic;
  subtype state is integer range -3 to m;
  signal current_state: state;
begin
  --data path:
  op_1<=x(conv_integer(operand_select));
  reg_in_v(0)<='0';
  encoder: for i in 0 to n-2 generate
    reg_in_v(i+1)<='0' when op_1(i)+u(i)+conv_integer(v(i))
    <B else '1';
    reg_in_u(i)<=(op_1(i)+u(i)+conv_integer(v(i))) mod B;
  end generate;
  reg_in_u(n-1)<=(op_1(n-1)+u(n-1)+conv_integer(v(n-1)))
  mod B;
  process(clk)
  begin
    if clear='1' then u<=zero; v<=(others=>'0');
    elsif clk'event and clk='1' then
    if load='1' then u<=reg_in_u; v<=reg_in_v; end if;
    end if;
```

```
  end process;
  --control unit:
  <see example11_11>
end circuit;
```

11.1.12.4 Parallel Counters   The stored-carry encoder of Figure 11.30 is made up
of 3-operand 1-digit adders, each of them computing (11.71)

$$q(i + 1) = (w(i) + x(i) + y(i))/B,$$
$$z(i) = (w(i) + x(i) + y(i)) \bmod B.$$

In other words, it reduces the sum of three digits to the (weighted) sum of two digits:

$$w(i) + x(i) + y(i) = q(i + 1).B + z(i). \tag{11.75}$$

This type of computation resource is also called a *parallel (3,2)-counter* as it counts
the total number of units among $w(i)$, $x(i)$, and $y(i)$ and expresses the result as a
2-digit number. More generally, the following computation resource is defined.

Definition 11.3   A base-B (p,k)-*counter* is a p-input k-output circuit whose
behavior is defined by the following equation

$$x_0 + x_1 + \cdots + x_{p-1} = y_0 + y_1.B + \cdots + y_{k-1}.B^{k-1}, \tag{11.76}$$

where all x_i and y_j are B-ary digits. The output vector y represents the total number of
units among the p components of the input vector x.

As a matter of fact, a base-B (p,k)-counter is just a base-B p-operand 1-digit
adder. The maximum value of the first member of (11.76) is $p.(B - 1)$ so that the
following relation must be satisfied:

$$p.(B - 1) \le B^k - 1. \tag{11.77}$$

Thus the minimum value of k is given by the following relation:

$$k \ge \log_B (1 + p.(B - 1)). \tag{11.78}$$

If $k = \log_B(1 + p.(B - 1))$ then the full capacity of the counter is used. Whatever the
base B, this occurs if $p = B + 1$ and $k = 2$:

$$p.(B - 1) = (B + 1).(B - 1) = B^2 - 1 = B^k - 1.$$

More generally, a full capacity base-B counter can be generated for all values of
p such that $p.(B - 1)$ can be expressed in the form $B^k - 1$. If $B = 2$ the preceding

rule amounts to $p = B^k - 1$, for example, $p = 3$ and $k = 2$, $p = 7$ and $k = 3$, $p = 15$ and $k = 4$, and so on.

By connecting n (p, k)-counters in parallel, a (p, k)-stored-carry encoder is obtained. An example is given in Figure 11.33 with $B = 2$, $p = 7$, and $k = 3$. The behavior of the circuit is defined by the following equation:

$$x^{(0)} + x^{(1)} + \cdots + x^{(6)} + 2.c_{in\_1} + c_{in\_0a} + c_{in\_0b}$$

$$= u + v + w + (2.c_{out\_1} + c_{out\_0a} + c_{out\_0b}).2^n \qquad (11.79)$$

Observe that a (3,2)-stored-carry encoder is what was called a stored-carry encoder in the preceding section (Figure 11.30).

The (p,k)-stored-carry encoder can in turn be used as a building block for generating carry-save adders. As an example, the circuit of Figure 11.34 is a binary carry-save tree that computes the sum of 31 numbers $x^{(i)}$ and expresses the result in the following form:

$$x^{(0)} + x^{(1)} + \cdots + x^{(30)} = y + z + w.$$

In order to complete the adder, a (3,2)-stored-carry encoder (Figure 11.30) would substitute the sum $y + z + w$ by the sum of two numbers, say, $u + v$. Then it remains to compute $u + v$ with a 2-operand adder. As a matter of fact, every (7,3)-stored-carry encoder is made up of (7,3)-counters, that is, 7-operand 1-bit adders (Figure 11.33). Each of them can be synthesized with full adders (Figure 11.35). So

$$C_{(7,3)\text{-stored-carry-encoder}} = 4.n.C_{FA},$$
$$T_{(7,3)\text{-stored-carry-encoder}} = 3.T_{FA}.$$

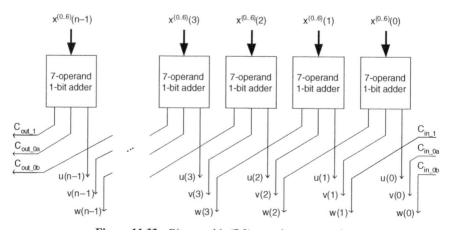

Figure 11.33   Binary n-bit (7,3)-stored-carry encoder.

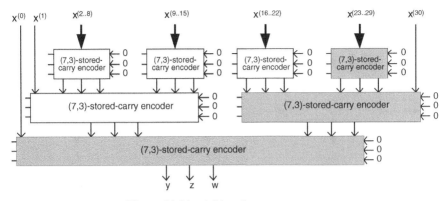

Figure 11.34 A 31-to-3 carry-save tree.

Then, according to Figures 11.34 and 11.35,

$$C_{\text{31-to-3-carry-save-tree}} = 7.C_{\text{(7,3)-stored-carry-encoder}} = 28.n.C_{\text{FA}},$$

$$T_{\text{31-to-3-carry-save-tree}} = 3.T_{\text{(7,3)-stored-carry-encoder}} = 9.T_{\text{FA}},$$

$$C_{\text{adder}}(31, n) = C_{\text{31-to-3-carry-save-tree}} + C_{\text{(3,2)-stored-carry-encoder}}$$
$$+ C_{\text{adder}}(n) = 29.n.C_{\text{FA}} + C_{\text{adder}}(n),$$

$$T_{\text{adder}}(31, n) = T_{\text{31-to-3-carry-save-tree}} + T_{\text{(3,2)-stored-carry-encoder}}$$
$$+ T_{\text{adder}}(n) = 10.T_{\text{FA}} + T_{\text{adder}}(n). \tag{11.80}$$

The same binary (7,3)-counter can be used in a different way. As an example, Figure 11.36 is a binary 5-operand ripple-carry adder.

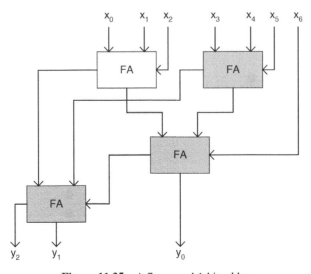

Figure 11.35 A 7-operand 1-bit adder.

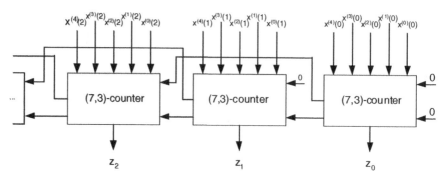

Figure 11.36   Binary 5-operand ripple-carry adder.

An easy method for understanding the working of many arithmetic circuits is the *dot notation*. As an example, the function of a (7,3)-counter (namely, to reduce the sum of seven digits to three digits) is shown in Figure 11.37a, and that of the corresponding (7,3)-stored-carry encoder (namely, to substitute the sum of seven numbers by the sum of three numbers) in Figure 11.37b.

This type of notation facilitates the understanding of more complex counters. As an example, the function of a (5,5;4)-counter (Figure 11.38a) is to reduce the sum of five 2-digit numbers to four digits. So it is defined by the following relation:

$$x^{(0)}(i) + \cdots + x^{(4)}(i) + B.(x^{(0)}(i+1) + \cdots + x^{(4)}(i+1)) = u(i) + B.u(i+1)$$

$$+ B^2.v(i+2) + B^3.v(i+3).$$

By connecting in parallel n such (5,5;4)-counters (Figure 11.38b), a (5,2)-stored-carry-encoder is generated. Its function is to substitute the sum of five numbers by the sum of two numbers.

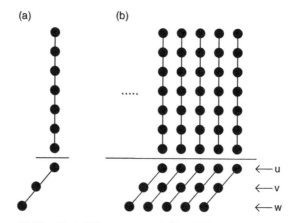

Figure 11.37   (a) A (7,3) counter and (b) a (7,3) stored-carry encoder.

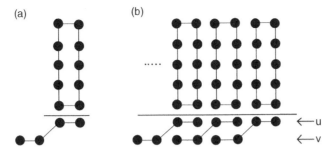

Figure 11.38 (a) A (5,5;4) counter and (b) a (5,2) stored-carry encoder.

A (5,5;4)-counter, implemented with (3,2)-counters, is shown in Figure 11.39, and a (5,2)-stored-carry encoder in Figure 11.40.

According to Figures 11.39 and 11.40, the cost and computation time of the (5,2)-stored-carry encoder are equal to

$$C_{(5,2)\text{-stored-carry-encoder}} = 6.C_{(3,2)\text{-counter}}.(n/2) \cong 3.n.C_{FA},$$
$$T_{(5,2)\text{-stored-carry-encoder}} = 3.T_{(3,2)\text{-counter}} \cong 3.T_{FA}.$$

The (5,2)-stored-carry encoder can be used for generating carry-save trees. For example, a 26-to-2 carry-save tree is shown in Figure 11.41. It expresses the sum

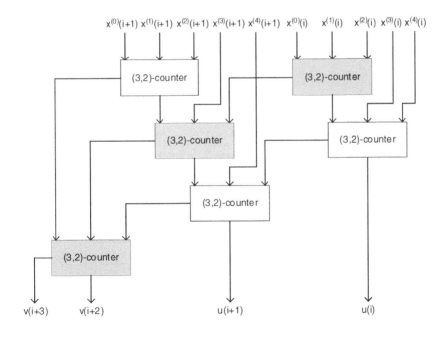

Figure 11.39 A (5,5;4) counter.

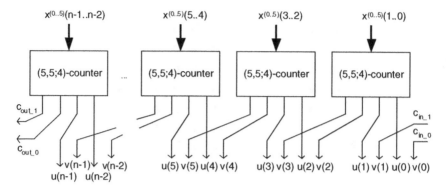

Figure 11.40   A (5,2) stored-carry encoder.

of 26 numbers $x^{(0)}, x^{(1)}, \ldots, x^{(25)}$, under the form $y + z$. With an additional 2-operand adder, a 26-operand adder is generated. Its cost and computation time are equal to

$$C(26, n) = 8.C_{(5,2)\text{-stored-carry-encoder}} + C_{\text{adder}}(n) \cong 24.n.C_{\text{FA}} + C_{\text{adder}}(n),$$

$$T(26, n) = 3.T_{(5,2)\text{-stored-carry-encoder}} + T_{\text{adder}}(n) \cong 9.T_{\text{FA}} + T_{\text{adder}}(n). \qquad (11.81)$$

The (5,5;4)-counter is a particular case of the following type of computation resource:

Definition 11.4   A $(p_{r-1}, p_{r-2}, \ldots, p_0; k)$-*counter* is a $(p_{r-1} + p_{r-2} + \cdots + p_0)$-input k-output combinational circuit whose behavior is defined by the following equation:

$$(\Sigma_{0 \le i \le p(0)} x_{i0}) + B.(\Sigma_{0 \le i \le p(1)} x_{i1}) + \cdots + B^{r-1}.(\Sigma_{0 \le i \le p(r-1)} x_{ir-1})$$

$$= y_0 + y_1.B + \cdots + y_{k-1}.B^{k-1},$$

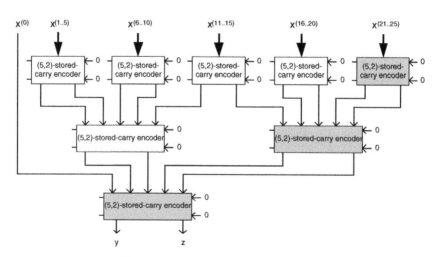

Figure 11.41   A 26-to-2 carry-save tree.

(where $p(j)$ stands for p_j). The output vector y represents the number of units within the input vector x_0, plus B times the number of units within the input vector x_1, plus B^2 times the number of units within the input vector x_2, and so on.

Example 11.13   (Complete VHDL code available.) Generate the VHDL model of a binary 31-to-3 carry-save tree (Figure 11.34):

```
entity seven_to_three is
port (
  x_0, x_1, x_2, x_3, x_4, x_5, x_6: in std_logic;
  y_2, y_1, y_0: out std_logic
);
end seven_to_three;

architecture circuit of seven_to_three is
  component full_adder...end component;
  signal a, b, c, d, e: std_logic;
begin
  fa_1: full_adder port map(x_3, x_4, x_5, b, a);
  fa_2: full_adder port map(x_0, x_1, x_2, d, c);
  fa_3: full_adder port map(c, a, x_6, e, y_0);
  fa_4: full_adder port map(d, b, e, y_2, y_1);
end circuit;

entity stored_carry_encoder is
port (
  x_0, x_1, x_2, x_3, x_4, x_5, x_6:
  in std_logic_vector(n-1 downto 0);
  u, v, w: out std_logic_vector(n-1 downto 0)
);
end stored_carry_encoder;

architecture circuit of stored_carry_encoder is
  component seven_to_three...end component;
  signal v_n, w_n, w_nn: std_logic;
begin
  v(0)<='0'; w(1)<='0'; w(0)<='0';
  main_loop: for i in 0 to n-3 generate
    iterative_step: seven_to_three port map (x_0(i),x_1(i),
    x_2(i),x_3(i),x_4(i),x_5(i),x_6(i),
    w(i+2), v(i+1), u(i)));
  end generate;
  second_last_step: seven_to_three port map
  (x_0(n-2),x_1(n-2),x_2(n-2),x_3(n-2),x_4(n-2),
  x_5(n-2),x_6(n-2), w_n, v(n-1), u(n-2));
  last_step: seven_to_three port map
  (x_0(n-1),x_1(n-1),x_2(n-1),x_3(n-1),x_4(n-1),
  x_5(n-1),x_6(n-1), w_nn, v_n, u(n-1));
```

```
end circuit;

entity carry_save_tree is
port (
  x_0, x_1,..., x_30: in std_logic_vector(n-1 downto 0);
  y, z, w: out std_logic_vector(n-1 downto 0)
);
end carry_save_tree;

architecture circuit of carry_save_tree is
  component stored_carry_encoder...end component;
  signal a1, a2,..., a12: std_logic_vector(n-1 downto 0);
  signal b1, b2, b3, b4, b5, b6: std_logic_vector(n-1
  downto 0);
begin
  encoder_1: stored_carry_encoder port map(x_2, x_3, x_4, x_5,
  x_6, x_7, x_8, a1, a2, a3);
  encoder_2: stored_carry_encoder port map(x_9, x_10, x_11,
  x_12, x_13, x_14, x_15, a4, a5, a6);
  encoder_3: stored_carry_encoder port map(x_16, x_17, x_18,
  x_19, x_20, x_21, x_22, a7, a8, a9);
  encoder_4: stored_carry_encoder port map(x_23, x_24, x_25,
  x_26, x_27, x_28, x_29, a10, a11, a12);
  encoder_5: stored_carry_encoder port map(x_1, a1, a2, a3, a4,
  a5, a6, b1, b2, b3);
  encoder_6: stored_carry_encoder port map(a7, a8, a9, a10,
  a11, a12, x_30,b4, b5, b6);
  encoder_7: stored_carry_encoder port map(x_0, b1, b2, b3, b4,
  b5, b6, y, z, w);
end circuit;
```

11.1.13 Subtractors and Adder-Subtractors

Given two n-digit natural numbers x and y, and an input borrow b_in, the difference $z=x-y-b\_in$ could be a negative number. So, in the case of natural numbers, the subtractors must have a status output (a flag), indicating that the result of the subtraction is not a natural number. A first option consists in implementing Algorithm 4.17. The corresponding circuit—a *ripple-carry subtractor*—is made up of n *full-subtractor* (*FS*) cells (Figure 11.42) whose behavior is the following:

$$q(i+1) = 1 \text{ if } x(i) - y(i) - q(i) < 0, \quad = 0 \text{ otherwise;}$$
$$z(i) = (x(i) - y(i) - q(i)) \bmod B. \tag{11.82}$$

Another option is to use a simplified version of the Algorithm 4.21.

Algorithm 11.1 Natural Number Subtraction

```
for i in 0...n loop y'(i):=B-1-y(i); end loop;
c_in:=1-b_in;
```

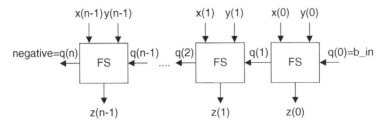

x(n-1) y(n-1)   x(1)   y(1)   x(0)   y(0)

negative=q(n)   q(n-1)   q(2)   q(1)   q(0)=b_in
                FS   ....   FS   FS

z(n-1)   z(1)   z(0)

Figure 11.42   Ripple-carry subtractor.

```
natural_addition(n, x, y', c_in, z, c_out);
negative:=1-c_out;
```

The advantage of the second method is that any type of adder can be used, so that all the adder implementations presented in the preceding sections can be considered. Furthermore, it's easy to design an adder/subtractor based on Algorithm 11.1 (Figure 11.43a):

- if $control = 0$ the circuit computes the $(n + 1)$-digit number $z = x + y + d\_in$, where $z(n) = d\_out$;
- if $control = 1$ and if $x - y - d\_in$ is nonnegative, the circuit computes the n-digit number $z = x - y - d\_in$ and the d_out output flag is put to 0; if $x - y - d\_in$ is negative, the d_out output flag is raised.

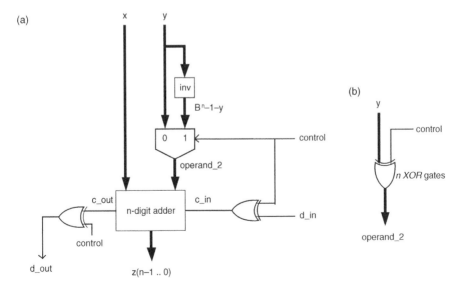

Figure 11.43   Adder-subtractor.

In the preceding circuit the combinational block *inv* is made up of n identical subcircuits that compute $B - 1 - y(i)$ for every digit $y(i)$ of y. If $B = 2$ then every subcircuit is an inverter. Furthermore, the n inverters and the multiplexer could be replaced by n XOR gates (Figure 11.43b).

Example 11.14   (Complete VHDL code available.) Generate the VHDL model of an adder-subtractor (Figure 11.43).

```
entity example11_14 is
port (
  x, y: in digit_vector(n-1 downto 0);
  control, d_in: in std_logic;
  z: out digit_vector(n-1 downto 0);
  d_out: out std_logic
);
end example11_14;
architecture circuit of example11_14 is
  signal minus_y, operand_2: digit_vector(n-1 downto 0);
  signal carries: std_logic_vector(n downto 0);
begin
  invert: for i in 0 to n-1 generate minus_y(i)<=B-1-y(i);
  end generate;
  with control select
  operand_2<=y when '0', minus_y when others;
  carries(0)<=control xor d_in;
  adder: for i in 0 to n-1 generate
    iterative_step: z(i)<=(x(i)+operand_2(i)+
    conv_integer(carries(i))) mod B;
    carries(i+1)<='0' when x(i)+operand_2(i)+
    conv_integer(carries(i))<B else '1';
  end generate;
  d_out<=carries(n) xor control;
end circuit;
```

11.1.14   Termination Detection

Self-timed circuits (Section 10.4) constitute an attractive option to build reliable and time-effective circuits. An example of their implementation has been seen in Chapter 10 (Example 10.6). In this section a slightly different approach is proposed: instead of computing the actual done condition a simpler condition is computed; nevertheless in most cases it will be equivalent to the done one ([BIO2003]).

For that purpose an n-bit adder is decomposed into n/s s-bit groups, and the propagation conditions $p(i.s + s - 1{:}i.s)$ are computed as in the case of an s-bit carry-skip chain (Figure 11.6). Assume now that all the propagation conditions $p(i.s + s - 1{:}i.s)$, $i = 0, \ldots, n/s - 1$, are equal to 0. Then all the carries must have been generated or killed within the group or its predecessor. As a consequence, sum completion for the n-bit adder is guaranteed after a time delay less than

$$T_{\text{completion}}(s) = T_{\text{adder}}(2.s), \tag{11.83}$$

where $T_{\text{adder}}(2.s)$ is the computation time of a partial adder made up of two successive groups.

The probability α of all propagation conditions $p(i.s + s - 1:i.s)$ being equal to 0 is

$$\alpha = (1 - (\tfrac{1}{2})^s)^{n/s}. \tag{11.84}$$

Observe that if s is great enough, then

$$\alpha \cong 1 - (n/s).(\tfrac{1}{2})^s, \tag{11.85}$$

in such a way that if $n/s \ll 2^s$, then $\alpha \cong 1$. Some particular values are given in Table 11.3.

Define the stat-done flag as follows:

$$\text{stat-done} = not(\,p(s - 1:0) \vee p(2.s - 1:s) \vee \cdots \vee p(n.s - 1:(n - 1).s)).$$

An example of how to use the stat-done flag is shown in Figure 11.44. The circuit of Figure 11.44a is assumed to be part of a signal processing system. It is made up of an n-bit adder that generates the stat-done flag and an output register. The clock period must be greater than both $T_{\text{completion}}$ (11.83) and the computation time of stat-done. The adder works as follows (Figure 11.44.b):

- if stat-done is equal to 1, the addition is performed within one clock cycle;
- if stat-done is equal to 0, a wait instruction is executed; the delay value is defined by the maximum computation time (a value that can be previously computed); according to Table 11.3 the value of s can be chosen in such a way that the probability of stat-done being equal to 0 is very small.

The minimum clock period is equal to

$$T_{\min}(n, s) = \max\{T_{\text{completion}}(s), T_{\text{stat-done}}(n, s)\} \tag{11.86}$$

TABLE 11.3 Probability α of All $p(i.s + s - 1:i.s)$ Being Equal to 0

	$n/s = 1$	$n/s = 2$	$n/s = 4$	$n/s = 8$	$n/s = 16$	$n/s = 32$	$n/s = 64$
$s = 8$	0.9960	0.9922	0.9844	0.9691	0.9392	0.8822	0.7784
$s = 12$	0.9997	0.9995	0.9990	0.9980	0.9961	0.9922	0.9844
$s = 16$	0.9999	0.9999	0.9999	0.9998	0.9997	0.9995	0.9990

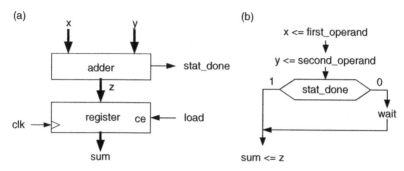

Figure 11.44   Statistical approach.

where $T_{\text{completion}}(s)$ is the computation time of a partial adder made up of two successive groups, and $T_{\text{stat-done}}(n, s)$ is the computation time of the stat-done flag.

The average computation time is equal to

$$T_{\text{average}}(n, s) = \alpha.T_{\min}(n, s) + (1 - \alpha).T_{\text{adder}}(n) = T_{\min}(n, s)$$
$$+ (1 - \alpha).(T_{\text{adder}}(n) - T_{\min}(n, s)). \qquad (11.87)$$

Using (11.85),

$$T_{\text{average}}(n, s) \cong T_{\min}(n, s) + (n/s).(1/2)^s.(T_{\text{adder}}(n) - T_{\min}(n, s)). \qquad (11.88)$$

For great values of n, the value of s can be chosen in such a way that

$$T_{\text{average}}(n, s) \cong T_{\min}(n, s). \qquad (11.89)$$

11.1.15   FPGA Implementation of the Termination Detection

The computation of the stat-done flag is performed as follows:

- Computation of $p(s - 1:0), p(2.s - 1: s), \ldots, p(n.s - 1:(n - 1).s)$. An FPGA implementation is shown in Figure 11.23; the corresponding cost C_p and delay T_p are equal to (11.62)

$$C_p = s/4 \text{ slices}, \quad T_p = T_{\text{LUT}} + (s/2).T_{\text{mux-cy}}.$$

- The computation of stat-done $= not(p(s - 1:0) \vee p(2.s - 1: s) \vee \cdots \vee p(n.s - 1:(n - 1).s))$ is performed with the circuit of Figure 11.45; the

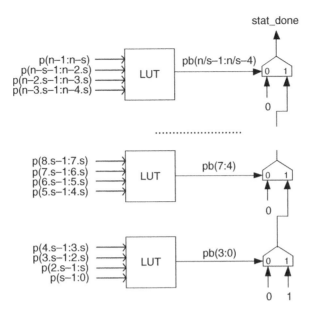

Figure 11.45 *stat-done* flag generation.

look-up tables compute

$$pb(3:0) = not(p(4.s - 1:3.s)).not(p(3.s - 1:2.s))$$
$$.not(p(2.s - 1:s)).not(p(s - 1:0)),$$
$$pb(7:4) = not(\dot{p}(8.s - 1:7.s)).not(p(7.s - 1:6.s))$$
$$.not(p(6.s - 1:5.s)).not(p(5 \cdot s - 1:4.s)),$$

and so on

Its cost C_{pb} and computation time T_{pb} are equal to

$$C_{pb} = n/(8.s) \text{ slices}, \quad T_{pb} = T_{\text{LUT}} + (n/(4.s)).T_{\text{mux-cy}}. \tag{11.90}$$

The cost $C_{\text{stat-done}}$ and computation time $T_{\text{stat-done}}$ of the stat-done flag are equal to

$$C_{\text{stat-done}} = C_p + C_{pb} = s/4 + n/(8.s) \text{ slices},$$
$$T_{\text{stat-done}} = T_p + T_{pb} = 2.T_{\text{LUT}} + ((s/2) + (n/(4.s))).T_{\text{mux-cy}} + T_{\text{connection}}. \tag{11.91}$$

The previous values (11.62), (11.90), and (11.91) are correct as long as both $s/4$ and $n/(8.s)$ are smaller than the number of rows of the selected circuit matrix. Assume now that $s/2$ is also smaller than the number of rows. Then every s-bit group of the adder (Figure 11.25) can be placed within one column, so that $T_{completion}$, namely the computation time of a $2 \cdot s$-digit adder, is equal to (see relation (11.59))

$$T_{completion} = T_{LUT} + (2.s - 1).T_{mux\text{-}cy} + T_{XOR2} + T_{connection}. \tag{11.92}$$

According to (11.86), (11.91), and (11.92), the minimum clock period T_{clk} of the system of Figure 11.44 is equal to

$$\begin{aligned} T_{min}(n, s) &= \max \{T_{completion}(s), T_{stat\text{-}done}(n, s)\} \\ &= T_{LUT} + T_{connection} + \max \{T_{LUT} + ((s/2) + (n/(4.s))).T_{mux\text{-}cy}, \\ &\quad (2.s - 1).T_{mux\text{-}cy} + T_{XOR2}\}. \end{aligned} \tag{11.93}$$

Example 11.15   Several adders have been implemented within a Spartan II FPGA. Tools and conditions are similar to the ones used in Section 11.1.10.3. The main results are summarized in Table 11.4.

11.2   INTEGERS

11.2.1   *B*'s Complement Adders and Subtractors

The B's complement adder of Figure 11.46a is deduced from Algorithm 4.18. It consists of an $(n + 1)$-digit adder and two instances of the combinational circuit *ext* (*digit extension*) whose function is to represent x and y with an additional digit (sign digit):

$$ext(a) = B - 1 \text{ if } a \geq B/2, \quad ext(a) = 0 \text{ if } a < B/2.$$

Another circuit is shown in Figure 11.46b. Instead of generating an $(n + 1)$-digit output, this second adder generates an n-digit output and an overflow flag (Equation 4.17) is raised if the result cannot be expressed with n digits.

TABLE 11.4   Experimental Results

n	s	$T_{adder}(n)$	$T_{stat\text{-}done}$	$T_{completion}$	T_{min}
256	8	38 ns	8 ns	6 ns	8 ns
256	16	38 ns	8 ns	7 ns	8 ns
256	32	38 ns	8 ns	13 ns	13 ns
512	8	77 ns	9 ns	6 ns	9 ns
512	16	77 ns	9 ns	7 ns	9 ns
512	32	77 ns	9 ns	13 ns	13 ns
1024	16	159 ns	10 ns	7 ns	10 ns
1024	32	159 ns	10 ns	13 ns	13 ns

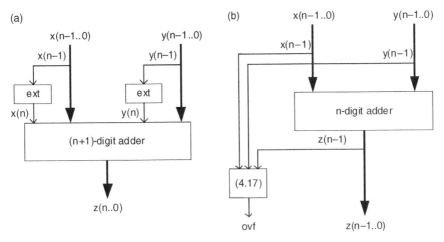

Figure 11.46 B's complement adders.

In order to synthesize a subtractor, or an adder-subtractor, another type of combinational circuit, namely, *inv*, is necessary. Given an n-digit number a, it computes

$$inv(a) = (B - 1 - a(n-1), B - 1 - a(n-2), \ldots, B - 1 - a(0)).$$

Two versions of a B's complement adder-subtractor are shown in Figure 11.47. The first one generates the exact $(n + 1)$-digit result. The other one generates an n-digit result and an overflow flag according to the relations (4.17) and (4.20). In both circuits the control signal defines the operation: addition (control $= 0$) or subtraction (control $= 1$).

Comments 11.9

1. If the reduced B's representation is used—in particular, if $B = 2$—the digit extension just consists of repeating the most significant bit.
2. If $B = 2$, the circuit *inv* is made up of n inverters. Furthermore the n inverters and the multiplexer could be replaced by n XOR gates (as in Figure 11.43b).

Example 11.16 (Complete VHDL code available.) Generate the VHDL model of a B's complement adder-subtractor (Figure 11.47a):

```
entity example11_16 is
port (
  x, y: in digit_vector(n-1 downto 0);
  control, d_in: in std_logic;
  z: out digit_vector(n downto 0)
);
end example11_16;
```

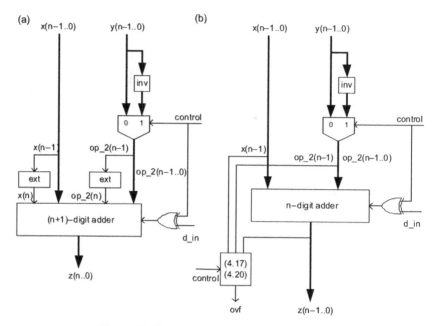

Figure 11.47 *B*'s complement adder-subtractors.

```
architecture circuit of example11_16 is
  signal minus_y, operand_2: digit_vector(n-1 downto 0);
  signal carries: std_logic_vector(n downto 0);
  signal x_n, operand_2_n: digit;
begin
  invert: for i in 0 to n-1 generate minus_y(i)<=B-1-y(i);
  end generate;
  with control select operand_2<=y when '0',
  minus_y when others;
  x_n<=0 when x(n-1)<B/2 else B-1;
  operand_2_n<=0 when operand_2(n-1)<B/2 else B-1;
  carries(0)<=control xor d_in;
  adder: for i in 0 to n-1 generate
    iterative_step:
    z(i)<=(x(i)+operand_2(i)+conv_integer(carries(i))) mod B;
    carries(i+1)<='0' when x(i)+operand_2(i)+conv_integer
    (carries(i))<B else '1';
  end generate;
  z(n)<=(x_n+operand_2_n+conv_integer(carries(n))) mod B;
end circuit;
```

11.2.2 Excess-*E* Adders and Subtractors

The circuit of Figure 11.48a, where E' stands for the $(n + 1)$-digit representation of $B^{n+1} - 1 - E$, is an excess-*E* adder based on Algorithm 4.22. The *pos* (*positive*)

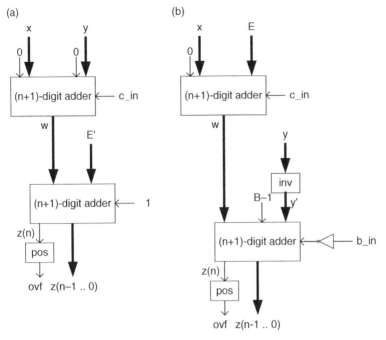

Figure 11.48 Excess-*E* adder and subtractor.

circuit detects whether $z(n)$ is greater than 0, or not—if $B = 2$, the *pos* circuit is a simple connection. An excess-*E* subtractor, based on Algorithm 4.23, is shown in Figure 11.48b. As before, the *inv* circuit computes the $(B - 1)$'s complement y' of y.

Example 11.17 (Complete VHDL code available.) Generate the VHDL model of an excess-*E* adder and subtractor.

```
--excess-E adder
entity example11_17 is
port (
  x, y: in digit_vector(n-1 downto 0);
  c_in: in std_logic;
  z: out digit_vector(n-1 downto 0);
  ovf: out std_logic
);
end example11_17;
architecture circuit of example11_17 is
  signal w: digit_vector(n downto 0);
  signal carries_1, carries_2: std_logic_vector(n downto 0);
  signal z_n: digit;
```

```
begin
  --first adder:
  carries_1(0)<=c_in;
  adder_1: for i in 0 to n-1 generate
    iterative_step: w(i)<=(x(i)+y(i) +
    conv_integer(carries_1(i))) mod B;
    carries_1(i+1)<='0' when x(i)+y(i)+conv_integer
    (carries_1(i))<B else '1';
  end generate;
  last_step_1: w(n)<=conv_integer(carries_1(n));
  --second adder:
  carries_2(0)<='1';
  adder_2: for i in 0 to n-1 generate
    iterative_step: z(i)<=(w(i)+minus_excess(i)
    +conv_integer(carries_2(i))) mod B;
    carries_2(i+1)<='0' when w(i)+minus_excess(i)+
    conv_integer(carries_2(i))<B else '1';
  end generate;
  last_step_2:z_n<=(w(n)+minus_excess(n)+conv_integer
  (carries_2(n))) mod B;
  ovf<='1' when z_n>0 else '0';
end circuit;

--excess-E subtractor
entity example11_17bis is
port (
  x, y: in digit_vector(n-1 downto 0);
  b_in: in std_logic;
  z: out digit_vector(n-1 downto 0);
  ovf: out std_logic
);
end example11_17bis;

architecture circuit of example11_17bis is
  signal w: digit_vector(n downto 0);
  signal carries_1, carries_2: std_logic_vector(n downto 0);
  signal z_n: digit;
begin
  --first adder:
  carries_1(0)<='0';
  adder_1: for i in 0 to n-1 generate
    iterative_step: w(i)<=(x(i)+excess(i)+
    conv_integer(carries_1(i))) mod B;
    carries_1(i+1)<='0' when x(i)+excess(i)+
    conv_integer(carries_1(i))<B else '1';
  end generate;
  last_step_1: w(n)<=(excess(n)+conv_integer(carries_1(n)))
  mod B;
  --second adder:
```

```
carries_2(0)<=not(b_in);
adder_2: for i in 0 to n-1 generate
  iterative_step: z(i)<=(w(i)+
  (B-1-y(i))+
  conv_integer(carries_2(i))) mod B;
  carries_2(i+1)<='0' when w(i)+(B-1-y(i))+
  conv_integer(carries_2(i))<B else '1';
end generate;
last_step_2: z_n<=(w(n)+(B-1)+conv_integer(carries_2(n)))
mod B;
ovf<='1' when z_n>0 else '0';
end circuit;
```

11.2.3 Sign-Magnitude Adders and Subtractors

The circuit of Figure 11.49 implements Algorithm 4.26. The combinational circuit *comp* (*comparator*) detects whether $a(n-1)$ is greater than or equal to $B/2$; if $B = 2$

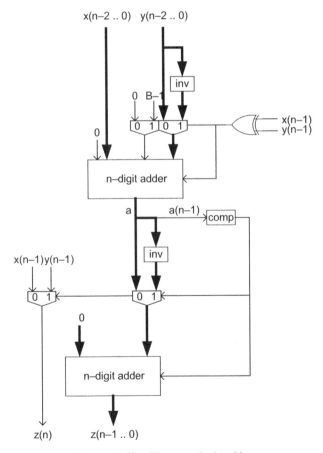

Figure 11.49 Sign-magnitude adder.

the *comp* circuit is a simple connection. As before, the *inv* circuit computes the $(B-1)$'s complement of y or a. The control signal defines the operation: addition (control $= 0$) or subtraction (control $= 1$).

As the sign-change operation amounts to inverting the sign bit, the synthesis of a subtractor, or of an adder-subtractor, is straightforward: it's just a matter of substituting $y(n-1)$ by *not* $(y(n-1))$, or by $(y(n-1)$ *xor control*$)$.

Example 11.18   (Complete VHDL code available.) Generate the VHDL model of a sign-magnitude adder (Figure 11.49):

```
entity example11_18 is
port (
  x, y: in digit_vector(n-2 downto 0);
  sign_x, sign_y: in std_logic;
  z: out digit_vector(n-1 downto 0);
  sign_z: out std_logic
);
end example11_18;

architecture circuit of example11_18 is
  signal minus_y: digit_vector(n-2 downto 0);
  signal operand_2, a, operand_2bis: digit_vector(n-1
  downto 0);
  signal minus_a: digit_vector(n-1 downto 0);
  signal carries_1, carries_2: std_logic_vector(n-1 downto 0);
begin
  invert_y: for i in 0 to n-2 generate minus_y(i)<=B-1-y(i);
end generate;
  carries_1(0)<=sign_x xor sign_y;
  with carries_1(0) select operand_2(n-2 downto 0)<=y when '0',
  minus_y when others;
  with carries_1(0) select operand_2(n-1)<=0 when '0',
  B-1 when others;
  adder_1: for i in 0 to n-2 generate
    iterative_step: a(i)<=(x(i)+operand_2(i)+
    conv_integer(carries_1(i))) mod B;
    carries_1(i+1)<='0' when x(i)+operand_2(i)+
    conv_integer(carries_1(i))<B else '1';
  end generate;
  a(n-1)<=(operand_2(n-1)+conv_integer(carries_1(n-1)))
  mod B;
  invert_a: for i in 0 to n-1 generate minus_a(i)<=B-1-a(i);
  end generate;
  carries_2(0)<='0' when a(n-1)<B/2 else '1';
  with carries_2(0) select operand_2bis<=a when '0',
  minus_a when others;
  with carries_2(0) select sign_z<=sign_x when '0',
  sign_y when others;
```

```
  adder_2: for i in 0 to n-2 generate
    iterative_step:
    z(i)<=(operand_2bis(i)+conv_integer(carries_2(i))) mod B;
    carries_2(i+1)<='0' when operand_2bis(i)+
    conv_integer(carries_2(i))<B else '1';
  end generate;
  z(n-1)<=(operand_2bis(n-1)+conv_integer(carries_2(n-1)))
  mod B;
end circuit;
```

11.3 BIBLIOGRAPHY

[BIO2003] G. Bioul, J.-P. Deschamps, and G. Sutter, Efficient FPGA implementation of carry-skip adders. In: *Proceedings of the 3rd Reconf. Computing and Applications (JCRA 03)*, UAM, Madrid, Sept. 2003, pp. 81–90.

[BRE1982] R. Brent and H. T. Kung, A regular layout for parallel adders. *IEEE Trans. Comput.*, **C-31**(3): 260–264 (1982).

[HAN1987] T. Han and D. A. Carlson, Fast area-efficient VLSI adders. In: *Proceedings of the 8th Symposium on Computer Arithmetic*, 1987, pp. 49–56.

[KOG1973] P. M. Kogge and H. S. Stone, A parallel algorithm for the efficient solution of a general class of recurrence equations. *IEEE Trans. Comput.*, **C-22**(8), 786–793 (1973).

[LAD1980] R. E. Ladner and M. J. Fischer, Parallel prefix computation. *J. ACM* **27**: 831–838 (1980).

[SUG1990] B. Sugla and D. Carlson, Extreme area-time tradeoffs in VLSI. *IEEE Trans. Comput.*, **39**(2), 251–257 (1990).

12

MULTIPLIERS

According to speed/cost requirements, the technology at hand, and a number of other circumstantial criteria, such as expandability, user-configurable features, copy protection, or power consumption, a great quantity of theoretical and practical multiplier implementations have been proposed in the literature. This chapter presents classic multipliers in base B with emphasis on base 2. In particular, attention is paid to *multiplication array* multipliers and adding tree reduction techniques. Based on the extended-Booth representation, the Per Gelosia multiplier is described as a particular multiplication array for signed-digit numbers. Some typical FPGA implementations are presented.

As a matter of fact, combinational multipliers are inherently faster, although generally less cost effective, than their corresponding (same algorithm) sequential implementation. The cost criterion is to be taken in a general theoretical context of hardware consumption, not directly related to the money price; it is well known that the price is more related to the batch size of production than to the gate cost itself. As mentioned in Chapter 9, PLA integrated circuit (IC) technology is a good example of inexpensive mass production; FPGA, for its reusability feature, may be considered cheap whenever it is used for special (low-quantity requirement) circuit design or simply for prototype design.

Synthesis of Arithmetic Circuits: FPGA, ASIC, and Embedded Systems
By Jean-Pierre Deschamps, Géry J. A. Bioul, and Gustavo D. Sutter
Copyright © 2006 John Wiley & Sons, Inc.

12.1 NATURAL NUMBERS

12.1.1 Basic Multiplier

According to the Hörner expansion presented in Chapter 5, formula (5.6),

$$Z/B^n = B^{-1}.(x_{n-1}.Y + B^{-1}.(x_{n-2}.Y + \cdots + B^{-1}.(x_1.Y + B^{-1}.(x_0.Y + 0))\cdots)),$$

is easily mapped into a combinational circuit to materialize Algorithm 5.2 (shift and add 2). A basic space iteration of the shift and add multiplier in base B is shown in Figure 12.1. The function Z implemented by this n-digit $\times$ m-digit multiplier is

$$Z = X.Y + D.$$

where $D = P(0)$ is an m-digit number.

Whenever $B > 2$, the size of the result Z is $m + n$; moreover, $(m + 1)$-digit adders are needed, because $x_i.Y$ may exceed $B^m - 1$. Otherwise, in the binary case, the size of the result is limited to $m + n - 1$, and m-digit adders meet the requirement. After each addition step, a digit result appears as the rightmost digit of the shifted sum. According to the case at hand, inverting the role of *multiplicand* and *multiplicator* may appear useful. The effects of this permutation are that the products $y_i.X$ are n-digit products (instead of m for $x_i.Y$), while the n $(m + 1)$-digit adders are switched for m $(n + 1)$-digit ones. Obviously the size of the result doesn't change.

The hardware cost of this circuit is high because of the n (resp. m) adders involved. The time is roughly equal to n $(m + 1)$-digit (resp. m $(n + 1)$-digit) adders. As will be observed later, if ripple-carry adders are used, this implementation reduces to the ripple-carry multiplier (Section 12.1.3.1).

Example 12.1 (Complete VHDL source code available.) Generate a generic n-digit by m-digit base-B basic multiplier. The first multiplier step called is mult_by_1_digit:

```
entity mult_by_1_digit is
Port (
  A: in digit_vector(M-1 downto 0);
  B: in digit_vector(M-1 downto 0);
  x_i: in digit;
  P: out digit_vector(M downto 0)
  );
end mult_by_1_digit;

architecture Behavioral of mult_by_1_digit is
begin
  process(B, A, x_i)
  variable carry: digit_vector(M downto 0);
  begin
```

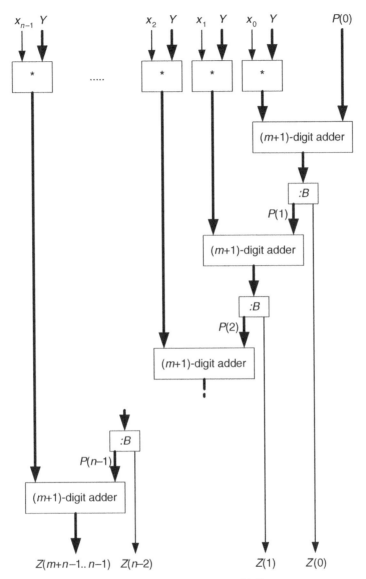

Figure 12.1 Basic base-B multiplier.

```
carry(0):=0;
for i in 0 to M-1 loop
  P(i)<=(B(i)*X_i+A(i)+carry(i)) mod BASE;
  carry(i+1):=(B(i)*X_i+A(i)+carry(i))/BASE;
end loop;
P(M)<=carry(M);
```

```
  end process;
end Behavioral;
```

The multiplier structure of Figure 12.1 is:

```
entity basic_base_B_mult is
port (
  X: in digit_vector(N-1 downto 0);
  Y: in digit_vector(M-1 downto 0);
  P: out digit_vector(N+M-1 downto 0)
  );
end basic_base_B_mult;

architecture simple_arch of basic_base_B_mult is
  type connections is array (0 to N) of digit_vector(M downto 0);
  signal wires: connections;
begin
 wires(0)<=(others=>0);
 iterac: for i in 0 to N-1 generate
   mult: mult_by_1_digit port map (wires(i)(M downto 1),
       Y, X(i), wires(i+1));
  p(i)<=wires(i+1)(0);
 end generate;
 p(M+N-1 downto N)<=wires(N)(M downto 1);
end simple_arch;
```

Example 12.2   (Complete VHDL source code available.) Generate a generic *n*-bit by *m*-bit base-2 basic multiplier. The first multiplier step called is mult_by_1_bit:

```
entity mult_by_1_bit is
Port (
  A: in std_logic_vector (M-1 downto 0);
  B: in std_logic_vector (M-1 downto 0);
  X_i: in std_logic;
  S: out std_logic_vector (M downto 0)
  );
end mult_by_1_bit;

architecture Behavioral of mult_by_1_bit is
begin
  add_mux: process(x_i,A,B)
  begin
    if x_i='1' then
      S<=('0' & A)+B;
    else
```

```
      S<=('0' & A);
    end if;
  end process;
end Behavioral;
```

The multiplier structure is:

```
entity basic_base2_mult is
port (
  X: in std_logic_vector (N-1 downto 0);
  Y: in std_logic_vector (M-1 downto 0);
  P: out std_logic_vector (N+M-1 downto 0)
  );
end basic_base2_mult;

architecture simple_arch of basic_base2_mult is
type connect is array (0 to N) of std_logic_vector (M downto 0);
signal wires: connect;
begin
  wires(0)<=(others=>'0');
  iterac: for i in 0 to N-1 generate
    mult: mult_by_1_bit port map (wires(i)(M downto 1), Y, X(i),
    wires(i+1));
    p(i)<=wires(i+1)(0);
  end generate;
  p(M+N-1 downto N)<=wires(N)(M downto 1);
end simple_arch;
```

12.1.2 Sequential Multipliers

Shift and add Algorithms 5.1 and 5.2 are actually more suited for time iteration, that is, using the same adder recursively. As an example, a sequential multiplier derived from Algorithm 5.2 is shown in Figure 12.2. Initially, the n-digit shift register contains X. If the m-digit register is preset to $P(0) = D$ then $Z = X.Y + D$ after n clock cycles.

12.1.3 Cellular Multiplier Arrays

Most combinational multipliers belong to the class of multiplication arrays. An essential characteristic of multiplication arrays is that they rest on computation primitives that are independent of the data size. The multiplication process consists of two main phases: in the first phase, the digit-by-digit products $x_i y_j$ are computed; in the second phase, the addition phase, those products are added. These phases are not necessarily successive. According to the type of implementation some mix can happen between making products and adding them. This occurs typically when a

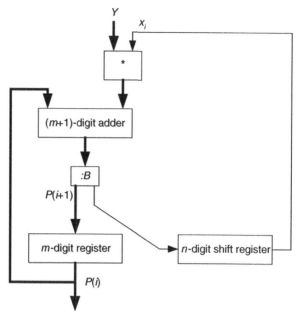

Figure 12.2   Sequential base-B shift and add multiplier.

reduced set of cells is used sequentially. Most multiplication arrays start from the basic pencil and paper scheme described in Chapter 5, Figure 5.1. Actually, in the literature, the cell arrays are generally presented according to this scheme, obviously not related to the *place and route* process result in the physical circuits. Most often, partial products are represented by simple dots, whose coordinates (i, j) in the scheme stand for the actual indices of the digit product being represented.

Example 12.3   (Complete VHDL source code available.) Generate a generic n-digit by m-digit base-B basic sequential multiplier. The basic cell `mult_by_1_digit` is similar as in Example 12.1. The circuit of Figure 12.2 including the state machine is:

```
entity basic_base_B_mult_seq is
port (
  clk: in std_logic;
  ini: in std_logic;
  X: in digit_vector(N-1 downto 0);
  Y: in digit_vector(M-1 downto 0);
  done: out std_logic;
  P: out digit_vector(N+M-1 downto 0)
  );
end basic_base_B_mult_seq;

architecture simple_seq_arch of basic_base_B_mult_seq is
  signal reg_X: digit_vector(N-1 downto 0);
```

```
  signal reg_Y, reg_P: digit_vector(M-1 downto 0);
  signal n_reg_P: digit_vector(M downto 0);
  signal counter: integer range 0 to N+1;
  signal work: std_logic;
begin
  state_mach: process (clk, work, ini)
  begin
    if clk'event and clk='0' then
      if ini='1' then
        work<='1'; counter<=0;
        reg_P<=(others=>0);
        reg_X<=X; reg_Y<=Y;
      elsif work='1' then
        counter<=counter+1;
        reg_P<=n_reg_P(M downto 1);
        reg_X<=n_reg_P(0) & reg_X(N-1 downto 1);
        if (counter=N) then
          P<=reg_P & reg_X;
          work<='0';
        end if;
      end if;
    end if;
  end process;
  mult: mult_by_1_digit port map (reg_P,reg_Y,reg_X(0),n_
  reg_P);
  done<=not work;
end simple_seq_arch;
```

12.1.3.1 Ripple-Carry Multiplier

The space iteration of Algorithm 5.4 (cellular ripple-carry algorithm) is materialized by the combinational circuit displayed in Figure 12.4. The basic cell (Figure 12.3) computes

$$c_{i(j+1)} = (p_{i(i+j)} + x_i.y_j + c_{ij})/B$$

and

$$p_{(i+1)(i+j)} = sum(i, j) = (p_{i(i+j)} + x_i.y_j + c_{ij}) \bmod B.$$

The implementation of the basic cell depends on the cost/speed trade-off to be considered by the designer. A full-custom high-speed circuit option would suggest, for binary-coded digits (e.g., high-radix or binary-coded decimal—BCD), a look-up table procedure or a 3-level combinational implementation. In the case of BCD digits, the problem at hand is that of a simultaneous synthesis of eight 16-variable

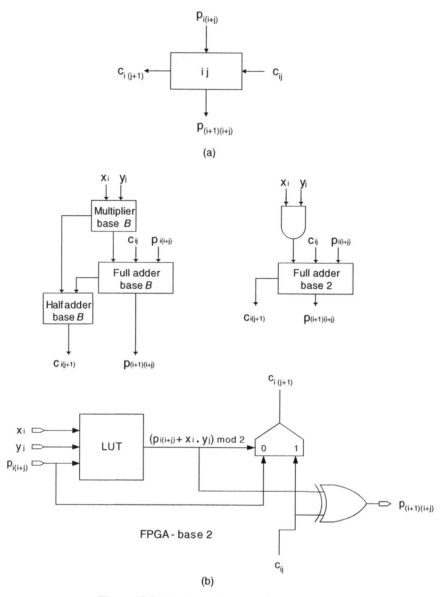

Figure 12.3 Basic cell: (a) symbol and (b) details.

functions. Thanks to the symmetry, this problem is affordable but the hardware cost could be prohibitive compared to the one suggested by Figure 12.3b with standard adders and multipliers or using FPGA. The circuit of Figure 12.4 displays the ripple-carry array for an n-digit $\times$ m-digit multiplier. It is the direct mapping of the precedence graph presented in Chapter 5, Figure 5.2.

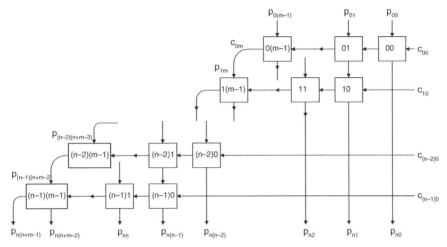

Figure 12.4 Ripple-carry multiplier.

Comment 12.1 In base 2, the circuit of Figure 12.3b is reduced to an AND (carry-free) gate for the binary product and a base-2 full adder. Although this could suggest a cell implementation with one additional gate delay, classic synthesis techniques readily provide 3-level implementations at a reasonable cost.

Example 12.4 (Complete VHDL source code available.) Generate a generic *n*-bit by *m*-bit base-2 ripple-carry multiplier. The first multiplier cell (Figure 12.3b) is:

```
entity basic_mul_cell is
Port (
  x_i, y_j: in std_logic;
  cin, pin: in std_logic;
  cout, pout: out std_logic;
  );
end basic_mul_cell;

architecture behavioral of basic_mul_cell is
  signal int_p: std_logic;
begin
  int_and<=x_i and y_j;
  cout<=(cin and pin) or (cin and int_p) or (pin and int_p);
  pout<=cin xor int_p xor pin;
end behavioral;
```

The multiplier structure (Figure 12.4) is:

```
entity ripple_carry_mult is
Port (
```

```
   X: in std_logic_vector(N-1 downto 0);
   Y: in std_logic_vector(M-1 downto 0);
   P: out std_logic_vector(M+N-1 downto 0));
end ripple_carry_mult;

architecture behavioral of ripple_carry_mult is

component basic_mul_cell

type connect is array (0 to N) of std_logic_vector
(M downto 0);
signal cin, pin, cout, pout: connect;
begin
  init: for i in 0 to N-1 generate cin(i)(0)<='0'; end generate;
  pin(0)<=(others=>'0');
  ext_loop: for i in 0 to N-1 generate
   int_loop: for j in 0 to M-1 generate
     cell: basic_mul_cell port map(X(i), Y(j),
         cin(i)(j), pin(i)(j), cout(i)(j), pout(i)(j));
     cin(i)(j+1)<=cout(i)(j);
     j0: if j=0 generate p(i)<=pout(i)(j); end generate;
     jn: if j>0 generate pin(i+1)(j-1)<=pout(i)(j); end generate;
   end generate;
   pin(i+1)(M-1)<=cin(i)(M);
  end generate;
  P(M+N-1 downto N)<=pin(N)(M-1 downto 0);
end behavioral;
```

12.1.3.2 Carry-Save Multiplier   The space iteration of Algorithm 5.5 is materialized by the combinational circuit displayed in Figure 12.5. The basic cell now computes

$$c_{(i+1)j} = (p_{i(i+j)} + x_i.y_j + c_{ij})/B$$

and

$$p_{(i+1)(i+j)} = sum(i,j) = (p_{i(i+j)} + x_i.y_j + c_{ij}) \bmod B.$$

This is a straightforward application of the carry-save technique of Chapter 11 (Figure 11.40). A carry-save multiplier, with $m=n$, is shown in Figure 12.5.

The basic cell is the same as that of Figure 12.3a, with a single difference: the carry output ($c_{i(j+1)}$ in the ripple-carry array) is now indexed as $c_{(i+1)j}$. This means that this carry is now connected as input to cell $(i+1,j)$, instead of cell $(i,j+1)$ for the ripple-carry. This reindexing technique corresponds to new connection assignments as it appears in Figure 12.5. Basically the array is similar to the ripple-carry one with respect to the number of cells but an additional n-bit adder

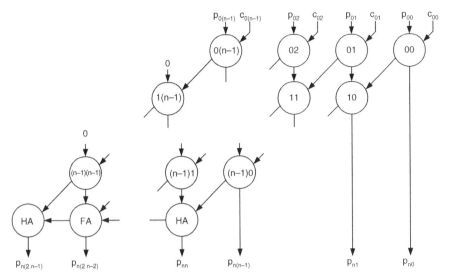

Figure 12.5 Carry-save multiplier.

is necessary at the bottom of the array. Observe that, due to the maximum length of the result, the leftmost half-adder has carry necessarily zeroed; this module may accordingly be reduced to a single XOR gate. As will be shown in the following section, the time saving of the carry-save array with respect to the ripple-carry one is asymptotically 33.3%, while the cost increase remains negligible (n).

Example 12.5 (Complete VHDL source code available.) Generate a generic n-bit by m-bit base-2 carry-save multiplier. The multiplier cell (Figure 12.3b) is the same as in Example 12.4. The multiplier structure (Figure 12.5) is:

```
entity carry_save_mult is
Port (
  X: in std_logic_vector(N-1 downto 0);
  Y: in std_logic_vector(M-1 downto 0);
  P: out std_logic_vector(M+N-1 downto 0));
end carry_save_mult;

architecture behavioral of carry_save_mult is

type connect is array (0 to N) of std_logic_vector (M downto 0);
signal cin, pin, cout, pout: connect;
begin
  pin(0)<=(others=>'0'); cin(0)<=(others => '0');
  ext_loop: for i in 0 to N-1 generate
    int_loop: for j in 0 to M-1 generate
      cell: basic_mul_cell port map(X(i), Y(j),
```

```
    cin(i)(j), pin(i)(j), cout(i)(j), pout(i)(j));
   cin(i+1)(j)<=cout(i)(j);
   j0:if j=0 generate p(i)<= pout(i)(j); end generate;
   jn:if j>0 generate pin(i+1)(j-1)<=pout(i)(j); end generate;
  end generate;
  pin(i+1)(M-1)<='0';
 end generate;
 P(M+N-1 downto N)<= pin(N)(M-1 downto 0)+cin(N)(M-1 downto 0);
end behavioral;
```

12.1.3.3 *Figures of Merit* Assuming that T_2 and C_2 are the respective time and gate complexity of the standard cells displayed in Figure 12.3, the ripple-carry (n-digit $\times$ n-digit) multiplier has overall figures given by:

$$T_{RCM} = (3.n - 2).T_2 \tag{12.1}$$

$$C_{RCM} = n^2.C_2 \tag{12.2}$$

while the carry-save implementation scheme gives

$$T_{CSM} = 2.n.T_2 \tag{12.3}$$

$$C_{CSM} = n.(n + 1).C_2 \tag{12.4}$$

(the same cost C_2 is assumed for the half and full adder cells).

Moreover, if the adding stage is implemented through a fast adding technique, formula (12.3) can be improved.

12.1.4 Multipliers Based on Dissymmetric $B^r \times B^s$ Cells

This section is a generalization of multiplier arrays to dissymmetric multiplication cells. First, a particular case is treated with x_i and y_j as 2-digit and 4-digit base-B numbers, respectively; in base 2 it corresponds to radix-16 by radix-4 multiplication. The elementary unit computes a 2-digit carry

$$c_{ij}\text{out} = (p_{ij}\text{in} + x_i.y_j + c_{ij}\text{in})/B^4 \quad \text{(integer division)} \tag{12.5}$$

and a 4-digit sum

$$p_{ij}\text{out} = sum(i, j) = (p_{ij}\text{in} + x_i.y_j + c_{ij}\text{in}) \bmod B^4. \tag{12.6}$$

A possible implementation is shown in Figure 12.6, where GHA and GFA are generalized half-adder (two 2-digit operands), and full adder (three 2-digit operands) respectively. Figure 12.7 illustrates a typical array for a 12-digit $\times$ 6-digit

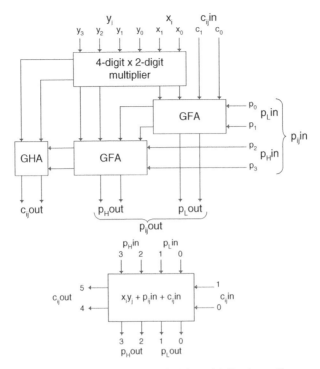

Figure 12.6 A 4-digit × 2-digit multiplication cell.

ripple-carry multiplication with additive operands; for clarity, inputs related to x_i and y_j have been omitted.

In Figure 12.7 the digits of the additive operands C and D are displayed at the top and right inputs of the array. The construction is self-explanatory and can readily be expanded to $4m$-digit by $2n$-digit arrays. Figure 12.7 shows that the concepts of carry and sum, as defined at formulas (12.5) and (12.6), are somewhat artificial; as a matter of fact, one could have defined a 4-bit carry and a 2-bit sum. Each file of cells behaves as a ripple-carry adder producing $x_i.Y + c_i + P_i\text{in}$, shifted 2 positions to the left with respect to the preceding file.

A carry-save array can be derived but the interconnection structure is somewhat more irregular than that of base-2 multiplications, as it appears in Figure 12.8a. As in the preceding array, inputs corresponding to x_i and y_j have been omitted; cell inputs and outputs have been labeled according to the exponent of the corresponding power of B weights. The cell inputs and outputs can be, expressed respectively, as

$$(z_2.B^4 + z_1.B^2 + z_0) + (p_H\text{in}.B^2 + p_L\text{in}) + c_L\text{in}$$

and

$$c_H\text{out}.B^4 + p_H\text{out}.B^2 + p_L\text{out},$$

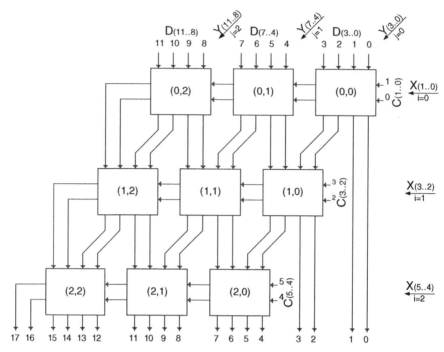

Figure 12.7 Multiplication ripple-carry array using 4-digit by 2-digit cells-X.Y + C + D.

where coefficients of powers of B are 2-digit numbers. So with regard to cell coordinates (i, j), the input and output labels are set, respectively, as

$$p_H\text{in} = 3 + 2i + 4j, 2 + 2i + 4j; \quad p_L\text{in} = 1 + 2i + 4j,$$
$$0 + 2i + 4j; \quad c_L\text{in} = 1 + 2i + 4j, 0 + 2i + 4j$$

and

$$p_H\text{out} = 3 + 2i + 4j, 2 + 2i + 4j; \quad p_L\text{out} = 1 + 2i + 4j,$$
$$0 + 2i + 4j; \quad c_H\text{out} = 5 + 2i + 4j, 4 + 2i + 4j.$$

To make the drawing simpler, inputs and outputs have been reorganized according to the cell presented in Figure 12.8b. The overall circuit is shown in Figure 12.8c; it is strictly equivalent to the one presented in Figure 12.8a.

Observe that a ripple-carry adder has been selected for adding the two numbers provided after the carry-save reduction. This alternative is arbitrary; the choice of the adder type is left to the designer.

Application of carry-save reduction to arbitrary m-digit by n-digit cells is manageable but circuits are less regular.

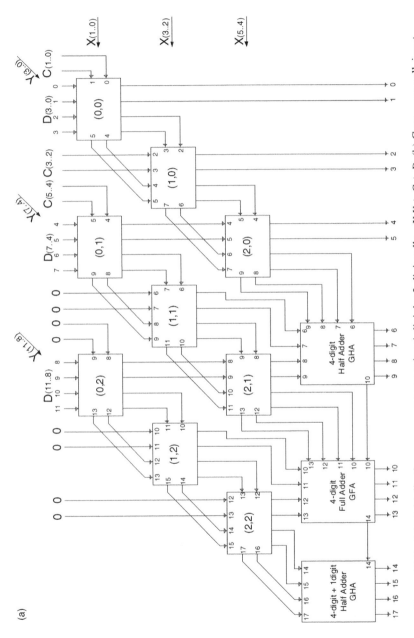

Figure 12.8 (a) Multiplication carry-save array using 4-digit by 2-digit cells—$X.Y + C + D$. (b) Carry-save cell, input–output settings. (c) Multiplication carry-save array using 4-digit by 2-digit cells—$X.Y + C + D$.

373

(b)

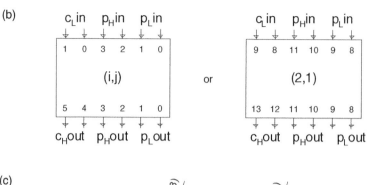

(c)

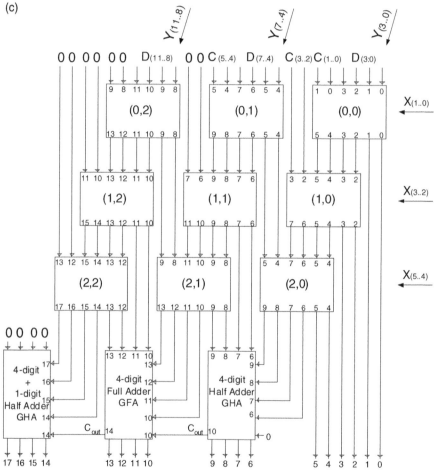

Figure 12.8 (*Continued.*)

Example 12.6 (Complete VHDL source code available.) Generate a generic *n*-bit by *m*-bit base-2 ripple-carry multiplier using a 4 by 2 digits multiplier cell. The basic 4 × 2 bits multiplier cell (Figure 12.6) is:

```
entity mul_4x2_cell is
Port (
  x_i: in std_logic_vector(1 downto 0);
  y_j: in std_logic_vector(3 downto 0);
  cin: in std_logic_vector(1 downto 0);
  din: in std_logic_vector(3 downto 0);
  cout: out std_logic_vector(1 downto 0);
  dout: out std_logic_vector(3 downto 0));
end mul_4x2_cell;

architecture behavioral of mul_4x2_cell is
signal int_prod, int_result: std_logic_vector(5 downto 0);
begin
  int_prod<=x_i*y_j;
  int_result<=int_prod+cin+din;
  dout<=int_result(3 downto 0);
  cout<=int_result(5 downto 4);
end behavioral;
```

The multiplier structure (Figure 12.7) is:

```
package mypackage is
  constant HORZ_CELL: natural:=2;
  constant VERT_CELL: natural:=5;
  constant N: natural:=VERT_CELL*2;
  constant M: natural:=HORZ_CELL*4;
end mypackage;

entity ripple_carry_4x2_mult is
Port (
  X: in std_logic_vector(N-1 downto 0);
  Y: in std_logic_vector(M-1 downto 0);
  P: out std_logic_vector(M+N-1 downto 0));
end ripple_carry_4x2_mult;

architecture behavioral of ripple_carry_4x2_mult is
  type connect_x2 is array (0 to VERT_CELL, 0 to HORZ_CELL) of
  std_logic_vector (1 downto 0); signal cin, cout: connect_x2;
  type connect_x4 is array (0 to VERT_CELL, 0 to HORZ_CELL) of
  std_logic_vector (3 downto 0); signal din, dout: connect_x4;
begin
```

```
  iniH: for i in 0 to HORZ_CELL-1 generate
    din(0, i)<="0000"; end generate;
  iniV: for i in 0 to VERT_CELL-1 generate
    cin(i, 0)<="00"; end generate;

  ext_loop: for i in 0 to VERT_CELL-1 generate
    int_loop: for j in 0 to HORZ_CELL-1 generate
      cell: mul_4x2_cell port map(
          X((i+1)*2-1 downto i*2),Y((j+1)*4-1 downto j*4),
          cin(i,j), din(i,j), cout(i,j), dout(i,j));
      cin(i,j+1)<=cout(i,j);
      j_0: if j=0 generate
      P((i+1)*2-1 downto i*2)<=dout(i,j)(1 downto 0);
      din(i+1,j)(1 downto 0)<=dout(i,j)(3 downto 2);
      end generate;

    jn0: if j>0 generate
      din(i+1,j-1)(3 downto 2)<=dout(i,j)(1 downto 0);
      din(i+1,j)(1 downto 0)<=dout(i,j)(3 downto 2);
      end generate;
    end generate;
    din(i+1,HORZ_CELL-1)(3 downto 2)<=cin(i,HORZ_CELL);
  end generate;

  outp_loop: for i in 0 to HORZ_CELL-1 generate
    P((i+1)*4+N-1 downto i*4+N)<=din(VERT_CELL,i);
  end generate;
end behavioral;
```

Example 12.7 (Complete VHDL source code available.) Generate a generic *n*-bit by *m*-bit sequential multiplier for signed operands. The basic multiplier cell is:

```
entity mult_by_1_bit is
Port (
  A: in std_logic_vector(M-1 downto 0);
  B: in std_logic_vector(M-1 downto 0);
  op: in std_logic; --(add/sub)=1 or nothing
  a_s: in std_logic; --add or subtract
  P: out std_logic_vector(M downto 0)
  );
end mult_by_1_bit;

architecture behavioral of mult_by_1_bit is
begin
  process(B, A, op, a_s)
  begin
```

```
    if op='1' then
      if a_s='0' then P<=(A(M-1) & A)+(B(M-1) & B);
      else P<=(A(M-1) & A)-(B(M-1) & B);
      end if;
    else P<=(A(M-1) & A);
    end if;
  end process;
end behavioral;
```

The multiplier including the state machine is:

```
entity signed_mult_seq is
port (
  clk: in std_logic;
  ini: in std_logic;
  X: in std_logic_vector(N-1 downto 0);
  Y: in std_logic_vector(M-1 downto 0);
  done: out std_logic;
  P: out std_logic_vector(N+M-1 downto 0)
  );
end signed_mult_seq;

architecture simple_seq_arch of signed_mult_seq is
  signal reg_X: std_logic_vector(N-1 downto 0);
  signal reg_Y, reg_P: std_logic_vector(M-1 downto 0);
  signal n_reg_P: std_logic_vector(M downto 0);
  signal counter: integer range 0 to N;
  signal work, add_sub: std_logic;
begin
  state_mach: process (clk, work, ini)
  begin
    if clk'event and clk='0' then
      if ini='1' then
        work<='1'; add_sub<='0';
        counter<=0;
        reg_X<=X; reg_Y<=Y; reg_P<=(others => '0');
      elsif work='1' then
        counter<=counter+1;
        reg_P<=n_reg_P(M downto 1);
        reg_X<=n_reg_P(0) & reg_X(N-1 downto 1);
        if (counter=N-2) then
          add_sub<='1';
        end if;
        if (counter=N-1) then
          P<=n_reg_P & reg_X(N-1 downto 1);
          work<='0';
```

```
      end if;
    end if;
  end if;
end process;
done<=not work;
mult: mult_by_1_bit port map (reg_P, reg_Y,
            reg_X(0), add_sub, n_reg_P);
end simple_seq_arch;
```

12.1.5 Multipliers Based on Multioperand Adders

A straightforward way to translate relation (5.4) to a multiplication circuit consists of (i) generating all (shifted) products $x_{n-1}.Y.B^{n-1}$, $x_{n-2}.Y.B^{n-2}, \ldots, x_2.Y.B^2$, $x_1.Y.B, x_0.Y$ and (ii) adding them. The corresponding circuit structure is shown in Figure 12.9.

The multioperand adder can be synthesized according to any one of the methods proposed in Section 11.1.12: carry-save array, carry-save tree (Wallace/Dadda tree) ([WAL1964], [DAD1965]), (p, k)-counter-based adders, and ripple-carry multioperand adders.

Examples 12.8

1. An 8-bit × 7-bit multiplier using carry-save tree.

Multioperand adding techniques, as described in Chapter 11, are used to perform a 2-stage reduction tree followed by a 2-operand sum. Parallel counters up to (7,3)

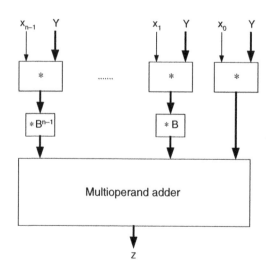

Figure 12.9 Multiplier structure.

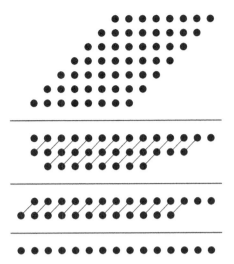

Figure 12.10 An 8-bit × 7-bit multiplier: dot diagram.

are used at the first stage, in such a way that the second stage provides two operands using full adders and half-adders only. Figure 12.10 shows the dot diagram while Figure 12.11 displays the carry-save tree according to Section 11.1.12.4. The detailed circuit is shown in Figure 12.12.

2. 8-bit × 7-bit multiplier using one-stage carry-save tree with (2, 3; 3), (1, 5; 3), (6, 3), (7, 3) counters and a 3-operand ripple-carry adder made up of parallel counters.

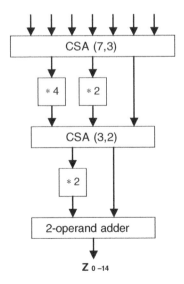

Figure 12.11 An 8-bit × 7-bit multiplier: 7-operand carry-save tree.

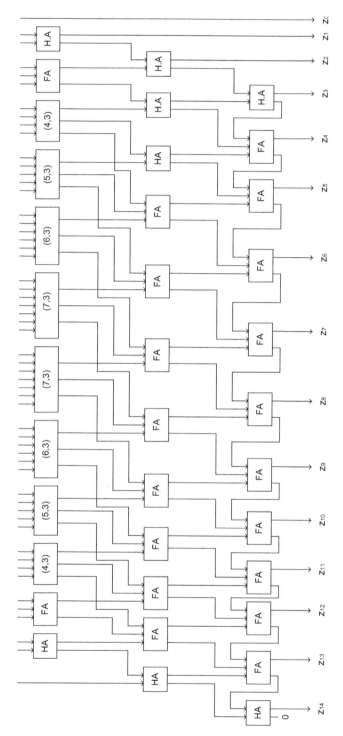

Figure 12.12 An 8-bit × 7-bit multiplier circuit.

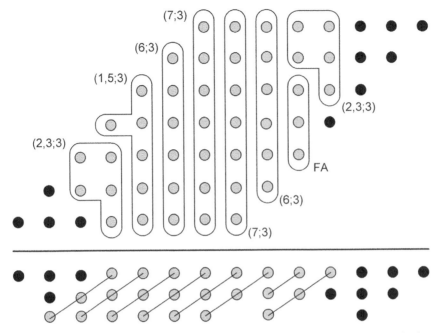

Figure 12.13 An 8-bit × 7-bit multiplier dot diagram. First stage is a carry-save reduction.

In this example, the carry-save reduction stage is carried out by counters, easy to synthesize as 7-input (at most) Boolean functions or LUTs. The dot diagram reduction process is illustrated in Figure 12.13. As in the preceding example, the three operands of the second stage could readily be reduced to two, by half adders and full adders. As an alternative a ripple-carry adder made up from (5, 3)-counters is presented in Figures 12.14 (dot diagram) and 12.15 (circuit).

3. When inexpensive fast counters are available, an m-bit by 31-bit multiplier can be designed as a 31-to-5 reduction stage—using (31, 5) counters—followed

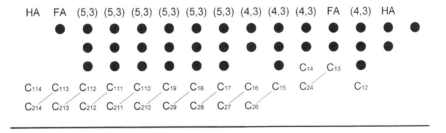

Figure 12.14 An 8-bit × 7-bit multiplier dot diagram. Second stage is a ripple-carry-counter adder.

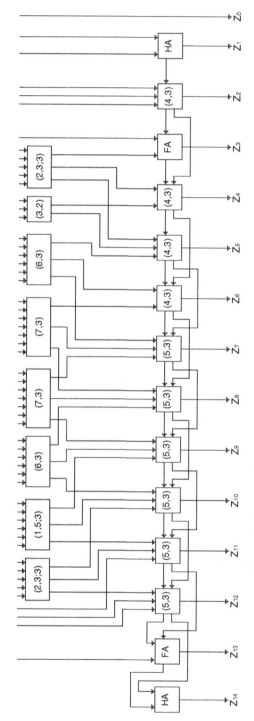

Figure 12.15 An 8-bit × 7-bit multiplier circuit with ripple-carry (5, 3)-counter adder.

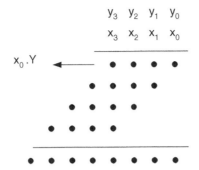

Figure 12.16 Binary products dot scheme.

by a 5-operand ripple-carry adder such as the one displayed in Figure 11.45 of the preceding chapter.

12.1.6 Per Gelosia Multiplication Arrays

12.1.6.1 Introduction Whenever base B is greater than 2, partial products appear less straightforward than what is involved in binary products (straight AND operation). Moreover, base-B elementary products generate carries. This basic difference between respective multiplication processes is illustrated in Figures 12.16 and 12.17, where dot schemes are presented without loss of generality for 4-digit multiplication.

Figure 12.16 displays the classical shift and add scheme, where each line i stands for the binary expression of $x_i.Y$. The problem is reduced to a multioperand sum

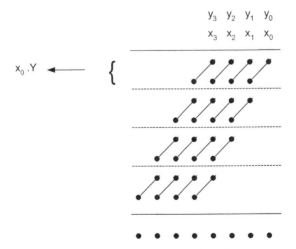

Figure 12.17 Base-B products dot scheme.

process. In Figure 12.17, the dot diagram suggests that $x_i.Y$ is not explicitly completed, $x_i.Y$ is presented as a subdiagram, where each $x_i.y_j$ appears as a shifted double dot dealing with carries generated through partial products generation. Per Gelosia technique doesn't compute $x_i.Y$, allowing a parallel treatment of all partial products $x_i.y_j$. Each 2-digit partial product is then part of a multioperand base-B adding scheme ([DAV1977]).

12.1.6.2 Adding Tree for Base-B Partial Products

The adding stage dot diagram of Figure 12.17, generalized to n-bit $\times$ n-bit B-ary multiplication, is represented in Figure 12.18. The maximum depth of the tree displayed in Figure 12.18 is $2.n - 1$; therefore, (m, k) B-ary counters can be fruitfully used to reduce the tree. An example is given hereafter for $n = 4$.

Example 12.9 We present a reduction tree for 4-digit $\times$ 4-digit multiplication in base 6. (3, 2), (5, 2), and (7, 2) counters are needed to proceed to a 2-operand reduction in one stage. Then a B-ary ripple-carry adder can be used (Figure 12.19).

Observe that, as far as $2.n - 1$ doesn't exceed $B + 1$, $(m, 2)$-counters can be used to get the two summands within one reduction stage (sufficient condition). Whenever n increases with respect to B, a k-operand reduction can be made with (m, k)-counters, with $k > 2$, and/or more reduction stages.

The 4-digit x 4-digit multiplication circuit is shown in Figure 12.20, where partial products $x_i.y_j$ are quoted as (P_{ij1}, P_{ij0}). With this indexing rule, each column L of the adding tree is characterized by:

$$i + j + k = L, \quad \text{with } k \in \{0, 1\},$$

L being the rank of the column.

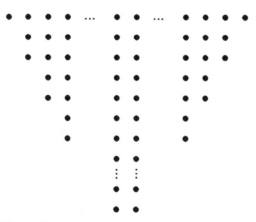

Figure 12.18 Adding stage dot diagram for n-digit base-B multiplication.

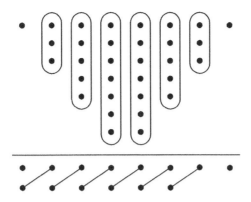

Figure 12.19 (4×4)-digit dot scheme.

Comment 12.2 According to Section 11.1.12 of Chapter 11, multioperand addition (m n-bit operands), using a carry-save reduction tree followed by a fast (logarithmic delay) adder, has an overall computation time given by $T(m,n) = O(\log m.n)$.

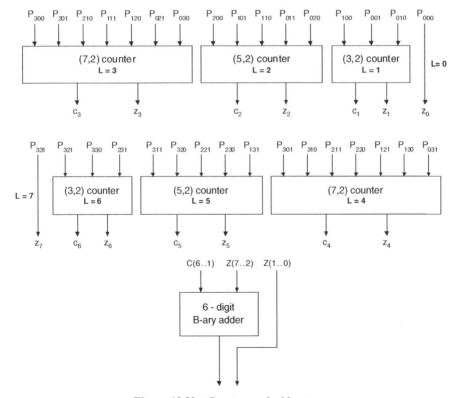

Figure 12.20 Counters and adder stage.

Considering that the partial products can be performed in parallel, this order of magnitude can be assumed for implementations of *m-digit by n-digit* multiplication based on this technique.

12.1.7 FPGA Implementation of Multipliers

In order to take advantage of the Virtex-family slice structure, relation (5.4) can be slightly modified as follows ($B = 2$, n even):

$$Z = (2.x_{n-1} + x_{n-2}).Y.2^{n-2} + \cdots + (2.x_3 + x_2).Y.2^2 + (2.x_1 + x_0).Y. \quad (12.7)$$

Every term of the preceding sum can be computed with an $(m + 1)$-cell iterative circuit whose basic *ij*-cell (computation of $(2.x_{i+1} + x_i).Y$) is shown in Figure 12.21.
The look-up table computes

$$x_i.y_j \ xor \ x_{i+1}.y_{j-1},$$

so that

$$p_i = x_i.y_j \ xor \ x_{i+1}.y_{j-1} \ xor \ c_{in} \ and$$
$$c_{out} = x_i.y_j.x_{i+1}.y_{j-1} \lor x_i.y_j.c_{in} \lor x_{i+1}.y_{j-1}.c_{in},$$

where $c_{in} = c_j$ and $c_{out} = c_{j+1}$.
Thus $(m + 1).n/2$ *ij*-cells are needed: $i = 2k$, $k \in [0,(n - 2)/2]$; $j \in [0, m]$; $y_{-1} = y_m = 0$.
It remains to compute the sum (12.7) of the so obtained (and previously shifted) $n/2$ terms. Figure 12.22 displays the complete circuit for $n = 6$, $m = 4$. As before, inputs x and y are not represented. Observe that the iterative line i actually computes

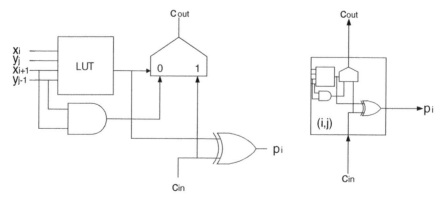

Figure 12.21   Slice configuration.

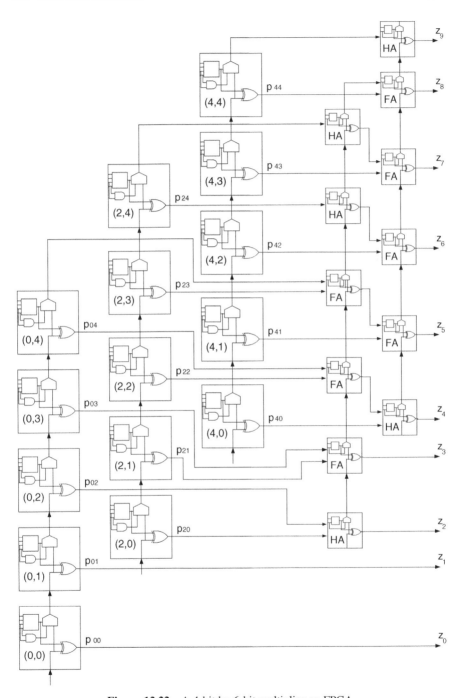

Figure 12.22 A 4-bit by 6-bit multiplier on FPGA.

a 2-operand sum. The look-up tables recursively compute the values of the propagate function as defined in Chapter 4. Figure 12.22 corresponds to ripple-carry adders implementation; nevertheless, each option remains open to the designer. The carry-skip technique (Chapter 11, Section 11.1.10.2) is particularly well suited for high-speed adders on FPGA ([BIO2003]). In Figure 12.22, a carry-save reduction tree, using full adders, provides two operands processed by a ripple-carry adder. This last adding step could also be left open to the designer.

The cost in terms of (i, j)-cells of this implementation is given by

$$C_{mn} = n.(m + 1)/2.$$

The cost of the carry-save reduction, with full adders, is asymptotically equal to

$$C_{\text{carry-save}} = O(n.m)$$

while the cost of the 2-operand adder is linear with $n + m$.

The delays mainly depend on the type of adders and elementary counter cells selected for carry-save reduction. This delay may be expressed as

$$T_{\text{mult}} = T_{\text{extended-adder}}(m) + T_{\text{carry-save}} + T_{\text{adder}},$$

where $T_{\text{extended-adder}}(m)$ stands for the time delay involved in the iterative circuit computing $(2.x_{i+1} + x_i).Y$, T_{adder} stands for the delay of the final adding stage, and $T_{\text{carry-save}}$ represents the carry-save reduction tree delay.

Observe that $T_{\text{extended-adder}}(m)$ depends only on m, because functions $(2.x_{i+1} + x_i).Y$ are computed in parallel. Whenever $n < m$, the number of operands to be added decreases but the operand length increases, so that the final adding stage will be accordingly longer.

12.2   INTEGERS

12.2.1   *B*'s Complement Multipliers

A straightforward implementation of Algorithm 5.7 (Section 5.3.1.1: mod B^{n+m} B's complement multiplication) consists of extending the representation of X and Y to $n + m$ digits and computing the $n + m$ less significant digits of $R(Z) = R(X).R(Y)$. For that purpose any natural-number multiplier can be used. As an example, assume that a carry-save multiplier is used (Figure 12.5). The adder, ripple-carry, or whatever is no longer necessary as the $n + m$ most significant digits must be truncated. So the array is limited to the rightmost $n + m$ columns of the array represented in Figure 12.5. The cost and computation time of the obtained array are

$$C(n, m) = ((m + n).(m + n + 1)/2).C_2, \quad T(n, m) = (m + n).T_2,$$

where C_2 and T_2 are the cost and propagation time of the cell of Figure 12.3.

If $m = n$, then $C(n) = n.(2.n + 1).C_2$, and $T(n, m) = 2.n.T_2$. The computation time is the same as in the case of the natural numbers (12.3) but the cost is almost twice the cost $C_{CSM} = n.(n + 1).C_2$ given by formula (12.4).

Another option is to implement Algorithm 5.8. A circuit similar to the ripple-carry multiplier of Figure 12.4 can be used. Every cell of the last row (Figure 12.3 with $i = n - 1$) must be replaced by a different one whose behavior is defined by the following rules:

```
if xₙ₋₁=0 then
   for j in 0...m-1 loop
   Pn(n-1+j)=P(n-1)(n-1+j);  C(n-1)(j+1)=0;
   endloop;
else
   C(n-1)1=(C(n-1)j+P(n-1)(n-1+j)+B-y₀)/B;
   Pn(n-1)=(C(n-1)j+P(n-1)(n-1+j)+B-y₀)mod B;
for j in 1...m-1 loop
   C(n-1)(j+1)=(C(n-1)j+P(n-1)(n-1+j)+B-1-yj)/B;
   Pn(n-1+j)=(C(n-1)j+P(n-1)(n-1+j)+B-1-yj)mod B; endloop;
```

The cost and the computation time are (practically) the same as in the case of a ripple-carry multiplier for natural numbers.

A third option is a straightforward implementation of relation (5.10) of Section 5.2.1.3:

$$Z = x'_{n-1}.y'_{m-1}.B^{n+m-2} + x'_{n-1}.Y_0.B^{n-1} + y'_{m-1}.X_0.B^{m-1} + X_0.Y_0.$$

The block diagram, shown in Figure 12.23, has to be interpreted in the following manner.

- The product $X_0.Y_0$ is an $(n + m - 2)$-bit positive number with sign extension.

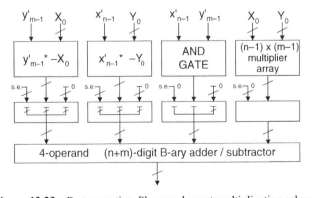

Figure 12.23 Postcorrection B's complement multiplication scheme.

- $x'_{n-1}, y'_{m-1} \in \{0,1\}$ stand for sign digits $x_{n-1}, y_{m-1} \in \{0, B-1\}$ (see Chapter 5, Section 5.2.1.3).
- If $x'_{n-1} = 0$ (resp. $y'_{m-1} = 0$), correction $x'_{n-1}.Y_0.B^{n-1}$ (resp. $y'_{m-1}.X_0.B^{m-1}$) vanishes.
- Whenever x'_{n-1} (resp. y'_{m-1}) is nonzero, the correcting term is built up by changing the sign of $(0, Y_0)$ left-shifted by $(n-1)$ positions with sign extension (resp. $(0, X_0)$ left-shifted by $(m-1)$ positions with sign extension).
- Whenever both x'_{n-1} and y'_{m-1} are nonzero, a left-shifted 1 $((n+m-2)$ positions plus sign extension) is added.

The above-mentioned operations are a lot simpler in 2's complement; moreover thanks to the simplifications suggested by Baugh and Wooley ([BAU1973]), the circuit of Figure 12.4 or 12.5 can be used with the following modifications:

- In the basic cell (Figure 12.3b), the and products $x_i.y_{m-1}$ $(i = 0,1, \ldots, n-2)$ and $x_{n-1}.y_j$ $(j = 0,1, \ldots, m-2)$ have to be complemented, which is readily achieved by replacing the AND by a NAND gate in the corresponding cells. Observe that $x_{n-1}.y_{m-1}$ remains unchanged.
- A bit 1 has to be added at levels $m-1$ and $n-1$ (a single 1 at level n if $m=n$).
- A bit 1 (sign) has to be added at level $m+n-1$ (at level $2n-1$ if $m=n$).

The cost and the computation time are (practically) the same as in the case of the corresponding multiplier array for natural numbers (Figure 12.4 or 12.5).

12.2.2   Booth Multipliers

12.2.2.1   Booth-1 Multiplier
One first considers a Booth-1 representation for the n-bit binary 2's complement number X. In this case $r = 1$ means $k = n$ and $B = 2$. According to formula (5.12) of Chapter 5, Booth-1 coding is defined as

$$x'_{-1} = 0; \; x'_i = -x_i + x_{i-1}, \quad i = 0, \ldots, n-1.$$

The circuit corresponding to Algorithm 5.10 (Section 5.2.3.1) is an iterative one made up of n steps (Figure 12.24a). Step number i generates the value of $P(i+1) = p_{i+1} z_i \cdots z_0$ as a function of $P(i) = p_i z_{i-1} \cdots z_0$, Y, x_{i-1}, and x_i according to the following recurrence formulas:

$$p_{i+1} = (p_i + (x_{i-1} - x_i).Y)/2; \quad z_i = (p_i + (x_{i-1} - x_i).Y) \bmod 2,$$
$$i = 0, \ldots, n-1; p_0 = 0. \tag{12.8}$$

The final result is $p_n z_{n-1} \cdots z_0$. Observe that all p_i are m-bit numbers and that the circuit corresponding to equation (12.8) behaves like an adder/subtractor controlled by x_{i-1} and x_i; in particular, the Boolean function $S = not\, x_{i-1} \wedge x_i$ may be used to control the sign of Y (negative if $S = 1$) while the function $P = x_{i-1} \oplus x_i$ controls if

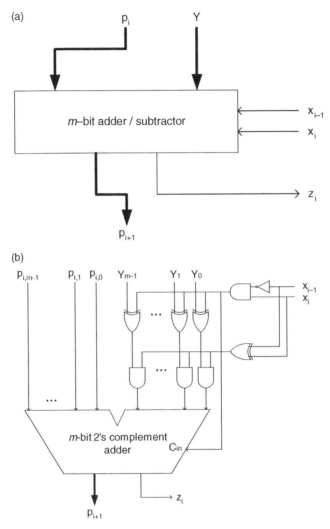

Figure 12.24   (a) Booth-1 multiplier step. (b) Booth-1 multiplier step implementation.

$\pm Y$ is added ($P = 1$) or not ($P = 0$). Figure 12.24b depicts a possible implementation of the m-bit adder/subtractor assuming both p_i and Y expressed in 2's complement.

Example 12.10   Let $X = 101011$ and $Y = 01101$ ($n = 6$, $m = 5$), both given in 2's complement representation The circuit of Figure 12.24b (p_0 is assumed 0) will compute recursively:

$$z_0 = (00000 - 01101) \bmod 2 = 1 \qquad x_{-1} - x_0 = -1$$
$$p_1 = (00000 - 01101)/2 \quad = 11001$$
$$z_1 = (11001 + 0) \bmod 2 \quad = 1 \qquad x_0 - x_1 = 0$$

$$p_2 = (11001 + 0)/2 \qquad\qquad = 11100$$
$$z_2 = (11100 + 01101) \bmod 2 = 1 \qquad\qquad x_1 - x_2 = +1$$
$$p_3 = (11100 + 01101)/2 \qquad = 00100$$
$$z_3 = (00100 - 01101) \bmod 2 = 1 \qquad\qquad x_2 - x_3 = -1$$
$$p_4 = (00100 - 01101)/2 \qquad = 11011$$
$$z_4 = (11011 + 01101) \bmod 2 = 0 \qquad\qquad x_3 - x_4 = +1$$
$$p_5 = (11011 + 01101)/2 \qquad = 00100$$
$$z_5 = (00100 - 01101) \bmod 2 = 1 \qquad\qquad x_4 - x_5 = -1$$
$$p_6 = (00100 - 01101)/2 \qquad = 11011$$
$$\mathbf{Z = 101011 \times 01101} \qquad = \mathbf{11011101111.}$$

12.2.2.2 Booth-2 Multiplier Let us focus now on the Booth-2 representation. In this case, $r = 2$, $k = n/2$, and $B = 2$. According to formula (5.12) of Chapter 5, Booth-2 coding is defined as

$$x'_{-1} = 0; \; x'_i = -2.x_{2.i+1} + x_{2.i} + x_{2.i-1}, \quad i = 0, \ldots, n/2 - 1,$$

thus $x'_i \in \{-2, -1, 0, 1, 2\}$.

The circuit corresponding to Algorithm 5.10 (Section 5.2.3.1) is now made up of $k = n/2$ iterative steps (Figure 12.25a). Step number i generates the value of $P(i + 1) = p_{i+1} z_{2.i+1} \cdots z_0$ as a function of $P(i) = p_i z_{2.i-1} \cdots z_0$, Y, $x_{2.i-1}$, $x_{2.i}$, and $x_{2.i+1}$:

$$p_{i+1} = (p_i + (x_{2.i-1} + x_{2.i} - 2.x_{2.i+1}).Y)/4,$$
$$z_{2.i+1} z_{2.i} = (p_i + (x_{2.i-1} + x_{2.i} - 2.x_{2.i+1}).Y) \bmod 4. \qquad (12.9)$$

Two bit results $z_{2.i+1}$ and $z_{2.i}$ are generated at each step. The final result is $p_k z_{2.k-1} \cdots z_0$. Observe that all p_i are still m-bit numbers but the circuit corresponding to equation (12.9) must now be able to compute $p_i + Y$, $p_i - Y$, $p_i + 2.Y$, and $p_i - 2.Y$. The corresponding behavior is that of an adder/subtractor/left-shifter controlled by $x_{2.i-1}$, $x_{2.i}$ and $x_{2.i+1}$, as shown in Figure 12.25a. The control functions may be expressed as

$$S = x_{2.i+1}$$
$$P = not[(x_{2.i-1} \wedge x_{2.i} \wedge x_{2.i+1}) \vee not(x_{2.i-1} \vee x_{2.i} \vee x_{2.i+1})] \qquad (12.10)$$
$$SL = not(x_{2.i} \oplus x_{2.i-1})$$

S controls the sign ($S = 1$ for negative), $P = 1$ enables the sum ($P = 0$ for adding zero), and SL controls a one-position left-shift of Y ($SL = 1$ for shift). Shift registers are common in most microprocessors, but for high-speed shift, more specific circuits may be designed (Figure 12.25b). A possible combinational implementation of the control circuit is presented in Figure 12.25c. In FPGA implementations, high-speed look-up tables are used to implement those control functions. Figure 12.25d presents a detailed combinational implementation of the Booth-2 multiplier step.

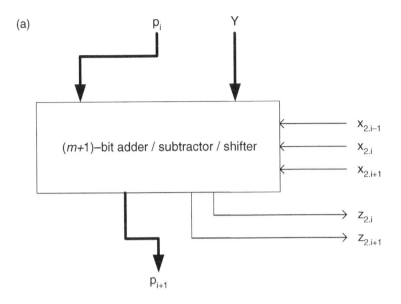

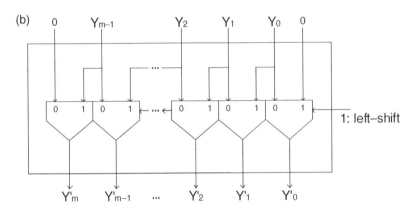

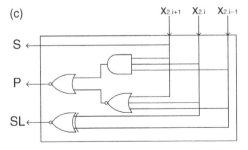

Figure 12.25 (a) Booth-2 multiplier step. (b) A 1-bit left-shifter. (c) Control circuit for Booth-2 multiplier. (d) Booth-2 multiplier step implementation.

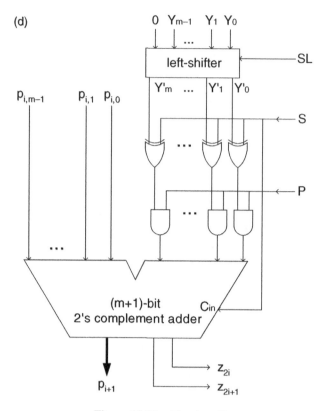

Figure 12.25 (*Continued.*)

The most prominent feature of the Booth-2 algorithm with respect to Booth-1 rests on the number of steps: $n/2$ instead of n. Matching up Figures 12.24a–12.24b to Figures 12.25a–12.25d highlights the fact that the hardware complexity is quite similar for both implementations. Booth-2 needs additional resources: a shifter, an $(m+1)$-bit adder instead of an m-bit adder, and a slightly more sophisticated control circuit.

Example 12.11 Let $X = 10101101$ and $Y = 01101$ ($n = 8$, $m = 5$), both given in 2's complement representation. The circuit of Figures 12.25a–12.25d (p_0 is assumed 0) will compute recursively:

$$
\begin{array}{lll}
z_1, z_0 = (000000 & +001101) \bmod 4 = 0, 1 & x_{-1} + x_0 - 2.x_1 = +1 \\
p_1 \ \ \ = (000000 & +001101)/4 & = 000011 \\
z_3, z_2 = (000011 & -001101) \bmod 4 = 1, 0 & x_1 + x_2 - 2.x_3 = -1 \\
p_2 \ \ \ = (000011 & -001101)/4 & = 111101 \\
z_5, z_4 = (111101 & -001101) \bmod 4 = 0, 0 & x_3 + x_4 - 2.x_5 = -1 \\
p_3 \ \ \ = (111101 & -001101)/4 & = 111100
\end{array}
$$

$$z_7, z_6 = (111100 \quad - 01101) \bmod 4 \quad = 1,1 \qquad\qquad x_5 + x_6 - 2.x_7 = -1$$
$$p_4 \quad = (111100 \quad - 01101)/4 \quad = 1011$$
$$\mathbf{Z} \quad = \mathbf{10101101 \times 01101} \quad = \mathbf{101111001001}$$

Note: For clarity, all successive values of p_i, but the last one, have been extended to 6 bits.

Comment 12.3 Higher-level Booth-r multipliers ($r > 2$) appear more intricate because of the complexity of the x_i'. For example, Booth-3 coding would generate x_i' digits in the range $[-4, +4]$. Besides adding, shifting and sign change, implementation of multiplication by 3 is also needed. A possible approach might consist of a previous computation of the suitable multiples of Y, to be then stored in a fast-access memory.

Example 12.12 (Complete VHDL source code available.) Generate a generic n-bit by m-bit Booth-1 multiplier for signed operands. The basic multiplier cell (Figure 12.24) is:

```
entity booth_1_cell is
Port (
  P: in std_logic_vector(M-1 downto 0);
  Y: in std_logic_vector(M-1 downto 0);
  x_i: in std_logic_vector(1 downto 0);
  S: out std_logic_vector(M downto 0)
  );
end booth_1_cell;

architecture behavioral of booth_1_cell is
begin
  the_mux: process(x_i,P, Y)
  begin
    case x_i is
      when "00"=>S<=(P(M-1) & P);
      when "01"=>S<=(P(M-1) & P)+(Y(M-1) & Y);
      when "10"=>S<=(P(M-1) & P)-(Y(M-1) & Y);
      when "11"=>S<=(P(M-1) & P);
      when others=>NULL;
    end case;
  end process;
end behavioral;
```

The complete Booth-1 multiplier is:

```
entity booth_1 is
port (
  X: in std_logic_vector (N-1 downto 0);
  Y: in std_logic_vector (M-1 downto 0);
```

```
  P: out std_logic_vector (N+M-1 downto 0)
  );
end booth_1;

architecture simple_arch of booth_1 is
  type connect is array (0 to N) of std_logic_vector (M downto 0);
  signal wires: connect;
  signal eX: std_logic_vector (N downto 0);
begin
  eX(N downto 1)<=X; eX(0)<='0';
  wires(0)<=(others=>'0');
  iterac: for i in 0 to N-1 generate
    mult: booth_1_cell port map (wires(i)(M downto 1), Y,
        eX(i+1 downto i), wires(i+1));
    p(i)<=wires(i+1)(0);
end generate;
P(M+N-1 downto N)<=wires(N)(M downto 1);
end simple_arch;
```

Example 12.13   (Complete VHDL source code available.) Generate a generic n-bit by m-bit Booth-2 multiplier for signed operands. The basic multiplier cell (Figure 12.25) is:

```
entity booth_2_cell is
Port (
  P: in std_logic_vector(M-1 downto 0);
  Y: in std_logic_vector(M-1 downto 0);
  X_i: in std_logic_vector(2 downto 0);
  Z: out std_logic_vector(1 downto 0);
  P_n: out std_logic_vector(M-1 downto 0)
  );
end booth_2_cell;

architecture behavioral of booth_2_cell is
signal long_P, long_Y, long_Y_2: std_logic_vector(M+1 downto
0);
signal S: std_logic_vector(M+1 downto 0);
begin
  long_P<=P(M-1) & P(M-1) & P;
  long_Y<=Y(M-1) & Y(M-1) & Y;
  long_Y_2<=Y(M-1) & Y & '0';
  the_mux: process(x_i, long_P, long_Y, long_Y_2)
  begin
    case x_i is
      when "000"=>S<=long_P;
      when "001"=>S<=long_P+ long_Y;
      when "010"=>S<=long_P+ long_Y;
      when "011"=>S<=long_P+long_Y_2;
```

```
      when "100"=>S<=long_P-long_Y_2;
      when "101"=>S<=long_P-long_Y;
      when "110"=>S<=long_P-long_Y;
      when "111"=>S<=long_P;
    end case;
  end process;
  P_n<=S(M+1 downto 2); Z<=S(1 downto 0);
end behavioral;
```

The complete Booth-2 multiplier is:

```
entity booth_2 is
port (
  X: in std_logic_vector (N-1 downto 0);
  Y: in std_logic_vector (M-1 downto 0);
  P: out std_logic_vector (N+M-1 downto 0)
  );
end booth_2;

architecture simple_arch of booth_2 is
type con is array (0 to N/2+1) of std_logic_vector (M-1 downto
0);
Signal wires: con
Signal eX: std_logic_vector (N+1 downto 0);
begin
  eX<=X(N-1) & X & '0';
  wires(0)<=(others=>'0');
  iter: for i in 0 to (N+1)/2-1 generate
    mult: booth_2_cell port map (wires(i), Y,
      eX(2*i+2 downto 2*i), p(2*i+1 downto 2*i),wires(i+1));
  end generate;
  p(M+N-1 downto N+(N mod 2))<=
                    wires((N+1)/2)(M-(N mod 2)-1 downto 0);
end simple_arch;
```

12.2.2.3 Signed-Digit Multiplier

12.2.2.3 Signed-Digit Multiplier Another type of multiplier can be deduced from Algorithm 5.12, in Section 5.2.3.2. It consists of a multiplier array processing Booth-coded signed digits. It is made up of an n/r by m/r array of signed-digit cells. The concept is introduced revisiting the example treated in Chapter 5: the 12-bit by 12-bit multiplier ($m = n = 12$) with Booth-3 ($r = 3$) digit coding. The general structure is shown in Figure 5.7. Assuming that the operands X and Y are initially given by their 2's complement representations, the method involves a preliminary coding according to formula (5.12) (see Algorithm 5.13: Booth_encode) and a final decoding process (Algorithm 5.14: Booth_decode). The Booth encoder and Booth decoder cells ($r = 3$) are shown in Figures 12.26 and 12.27, respectively. In this case, the bidimensional array uses cells computing functions $G(a, b, c, d)$

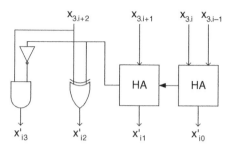

Figure 12.26 Booth-3 encoder cell.

and $H(a, b, c, d)$ such that $a.b + c + d = G.2^3 + H$; a, b, c and d being elements of the set $E = \{-4, -3, \ldots, +3, +4\}$. If 2's complement codification is selected for the elements of E, the Booth encoding of X (resp. Y) can be achieved with the circuit of Figure 12.26. The input–output relation is given by

$$-x_{3.i+2}.2^2 + x_{3.i+1}.2 + x_{3.i} + x_{3.i-1} = -x'_{i3}.2^3 + x'_{i2}.2^2 + x'_{i1}.2 + x'_{i01} \quad (12.11)$$

According to formula (5.14), for $r = 3$, cell outputs G and H are in the range $[-4, +3]$; actually as a, b, c, and d belong to the set E, the number (G, H) is in the following range:

$$(-3, 0) \leq (G, H) \leq (3, 0); \text{ in decimal } -24 \leq G.2^3 + H \leq 24$$

In the same way, the adding array (Figure 5.7) produces signed digits in the range $[-4, +3]$.

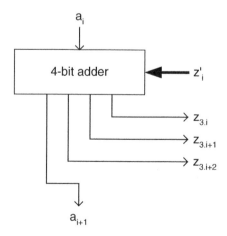

Figure 12.27 Booth-3 decoder cell.

A first option for the decoding of the multiplier output Z is an iterative circuit made up of $(m + n)/3$ steps (Algorithm 5.14a). Step number i generates the value of $A(i + 1) = a_{i+1} z_{3.i+2} \cdots z_0$ as a function of $A(i) = a_i z_{3.i-1} \cdots z_0$ and $Z(i)$:

$$a_{i+1} = (a_i + Z(i))/8; \quad z_{3.i+2} z_{3.i+1} z_{3.i} = (a_i + Z(i)) \bmod 8. \tag{12.12}$$

Algorithm 5.14b of Chapter 5 provides an alternative decoding circuit as described in Figure 12.28. Step i computes r new bits of the aimed 2's complement expression of $Z(i)$, as a function of a carry (from the preceding step) and the r corresponding bits (signed digit z_i) of the signed-digit expression of $Z(i)$. This function actually subtracts modulo 2^r from z_i' (z_i with its sign bit turned positive), the carry generated at the preceding step; the carry c_{i+1} is also generated. This carry c_{i+1} is the result of the potential positive adjustment of the first bit of $(z_i' - c_i)$. The circuit of Figure 12.28 is the materialization of the Boolean functions implementing the bitwise subtraction modulo-8:

$$\begin{aligned}
z_{3i}'' &= z_{3i} \oplus c_i, \\
z_{3i+1}'' &= z_{3i}.z_{3i+1} \vee z_{3i+1}.notc_i \vee c_i.notz_{3i}.notz_{3i+1}, \\
z_{3i+2}'' &= z_{3i+2} \oplus c_i.notz_{3i}.notz_{3i+1}, \\
c_{i+1} &= z_{3i+2} \vee c_i.notz_{3i}.notz_{3i+1}.
\end{aligned} \tag{12.13}$$

Figure 12.29 displays the generalized Booth-r decoder cell based on the same principle. The proposed cell may obviously be used sequentially or integrated in a full combinational iterative circuit.

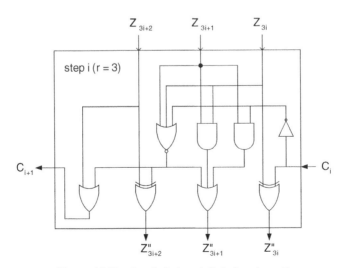

Figure 12.28 Booth-3 signed-digit decoder cell.

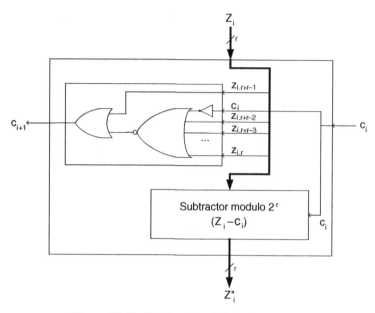

Figure 12.29 Booth-r signed-digit decoder cell.

Example 12.14 (Complete VHDL source code available.) Generate a generic n-bit by m-bit Booth-3 multiplier for signed operands. The basic multiplier cell (Figure 12.27) is:

```
entity booth_3_cell is
Port (
  P: in std_logic_vector(M-1 downto 0);
  Y: in std_logic_vector(M-1 downto 0);
  x_i: in std_logic_vector(3 downto 0);
  Z: out std_logic_vector(2 downto 0);
  P_n: out std_logic_vector(M-1 downto 0)
  );
end booth_3_cell;

architecture behavioral of booth_3_cell is
  signal s,1_P,1_Y,1_Y2,1_Y3,1_Y4: std_logic_vector(M+2 downto
  0);
begin
  1_P<=P(M-1) & P(M-1) & P(M-1) & P;
  1_Y<=Y(M-1)&Y(M-1)&Y(M-1)&Y;  1_Y2<=Y(M-1)&Y(M-1)&Y&'0';
  1_Y3<=1_Y+ 1_Y_2;  1_Y4<=Y(M-1) & Y & "00";
  the_mux: process(x_i,1_P, 1_Y, 1_Y2, 1_Y3, 1_Y4)
  begin
    case x_i is
      when "0000"|"1111"=>s<=1_P;
      when "0001"|"0010"=>s<=1_P+ 1_Y;
```

```
  when "0011"|"0100"=>s<=1_P+1_Y2;
  when "0101"|"0110"=>s<=1_P+1_Y3;
  when "0111"        =>s<=1_P+1_Y4;
  when "1000"        =>s<=1_P-1_Y4;
  when "1001"|"1010"=>s<=1_P-1_Y3;
  when "1011"|"1100"=>s<=1_P-1_Y2;
  when "1101"|"1110"=>s<=1_P-1_Y;
    end case;
  end process;
  P_n<=s(M+2 downto 3); Z<=s(2 downto 0);
end behavioral;
```

The complete Booth-3 multiplier is:

```
entity booth_3 is
port (
  X: in std_logic_vector (N-1 downto 0);
  Y: in std_logic_vector (M-1 downto 0);
  P: out std_logic_vector (N+M-1 downto 0)
  );
end booth_3;

architecture simple_arch of booth_3 is
  type connections is array (0 to (N+2)/3+1) of std_logic_vector
  (M-1 downto 0); signal wires: connections;
  signal eX: std_logic_vector (N+2 downto 0);
  constant bitsC: integer:=(3-(N mod 3)) mod 3;
begin
  eX<=X(N-1) & X(N-1) & X & '0'; wires(0)<=(others => '0');
  iter: for i in 0 to (N+2)/3-1 generate
    mult: booth_3_cell port map (wires(i), Y,
          eX(3*i+3 downto 3*i),p(3*i+2 downto 3*i),wires(i+1));
    end generate;
    P(M+N-1 downto N+bitsC)<=wires((N+2)/3)(M-bitsC-1 downto 0);
end simple_arch;
```

Comment 12.4   The Per Gelosia adding technique can be applied. As an example, if $m = n = 12$ and $r = 3$, X and Y are represented with 4 digits belonging to $E = \{-4, -3, \ldots, 4\}$. The product of two digits belongs to the interval $[-16, 16]$ and can be represented with two 3-bit 2's complement digits, thus within the range $[(-4, -4), (3, 3)]$ in decimal $[-36, 27]$; as before, the digits of p_{ij} are constrained to belong to the range $[-4, 3]$, which is sufficient to represent 2-digit numbers in the required range. A possible structure for the adding stage of the multiplier is described in what follows.

All the products $p_{ij} = x_i.y_j$ are computed; x_i and y_j can be encoded with 4 bits (binary Booth encoding within E), while the 2-digit number p_{ij} needs 6 bits only; this corresponds to a cell with 8 inputs and 6 outputs. Moreover, it is possible to design an 8-input cell performing the coding together with the product as 6 Boolean functions of 8 variables: each 4-bit Booth-coded input is replaced by the 3 bits from the corresponding slice plus the 1 bit, from the next rightmost slice, involved in the Booth-coding operation. Standard minimization techniques can be used to provide a three-level NAND-gate circuit ([DAV1977]); alternatively, ROM (LUT) or PLA can be used.

The weighted sum of all p_{ij} can be performed according to the computation scheme of Figure 12.19, where each dot stands for a 3-bit 2's complement number; the sum of seven (or less) elements of $\{-4, -3, \ldots, 3\}$ belongs to the interval $[-28, 21]$ and can be represented by two digits within $[(-3, -4), (3, -3)]$. So, $(7, 2)$, $(5, 2)$, and $(3, 2)$ counters can be synthesized to handle the overall sum. The scheme of Figure 12.20 is applicable but counters are now signed-digit counters; therefore, the design is somewhat different but with complexity comparable to that of base-8 counters. A $(7, 2)$ signed-digit counter made up of signed-digit adders (Figure 12.30a) is presented in Figure 12.30b. Another alternative for the reducing stage is a carry-save array made up of signed-digit full adders. The signed digit full adder (SDFA: Figure 12.31) is similar to the adder of

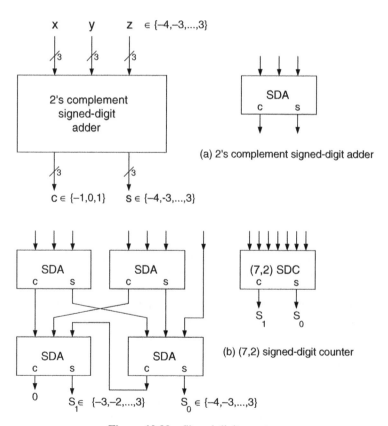

Figure 12.30   Signed-digit counter.

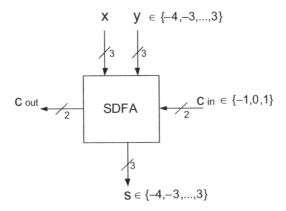

Figure 12.31 Signed-digit full adder.

Figure 12.30a but is somewhat simpler in the sense that only 2-bit carries within $\{-1,0,1\}$ are generated; this reduces the cell complexity to 8-input/5-output, instead of 9-input/6-output.

In the same way that Booth encoding can be skipped, thanks to a special design of the multiplying cells, the output adders (SDOFA: Figure 12.32) from the final adding stage (e.g., ripple-carry adder) can be designed to cope with the Booth decoding process too. Actually, each adder of the chain has to add, with carry, signed digits in the range $[-4,3]$. In order to avoid negative digits as a result (except for the last carry out), the adder cell is redesigned to generate a 3-bit positive digit sum S and a signed carry-out C in the range $[-2,1]$ accepting a carry-in in the same range $[-2,1]$. It is straightforward to note that the result of the above-mentioned sum remains in the (decimal) range $[-10, 7]$ and can be expressed in a

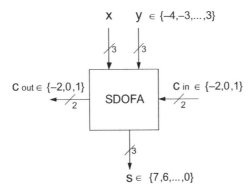

Figure 12.32 Signed-digit output full adder.

unique way in the form

$$(-C_1.2 + C_0).2^3 + (S_2.2^2 + S_1.2^1 + S_0), \quad C_i, S_j \in \{0, 1\} \qquad (12.14)$$

where C_i and S_j are the binary components of the signed carry-out C and the (positive) sum S, respectively. The inputs of the corresponding signed-digit output full adder (SDOFA) cell are similar to those of the SDFA of Figure 12.31 but the 3-bit output is now a positive 3-bit number while the carry-in and the carry-out are 2-bit signed digits in the range $[-2,1]$. The functions to be implemented are thus accordingly different. The last generated carry holds the bit-sign of the result.

12.2.3 FPGA Implementation of the Booth-1 Multiplier

An FPGA implementation of the circuit of Figure 12.24a is shown in Figure 12.33. The LUT functions are the following ones:

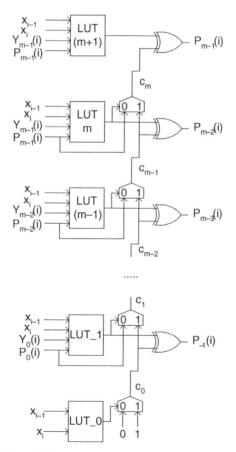

Figure 12.33 FPGA implementation of a Booth-1 multiplier step.

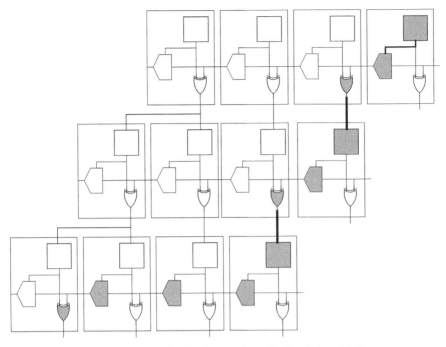

Figure 12.34   FPGA implementation of a Booth-1 multiplier.

if $x_{i-1} = 1$ and $x_i = 0$ then LUT_0 generates 0 and the other ones generate
  Y_j xor P_j (addition);

if $x_{i-1} = 0$ and $x_i = 1$ then LUT_0 generates 1 and the other ones generate
  not(Y_j) xor P_j (subtraction);

if $x_{i-1} = x_i$ then LUT_0 generates 0 and the other ones generate P_j, so that
  $c_{j+1} = not(P_j).P_j + P_j.c_j = 0$ ($c_0 = 0$) and $P_{j-1}(i+1) = 0$ xor $P_j(i) = P_j(i)$.

The general structure of the whole circuit (without LUT_0), along with one of the
critical paths, is shown in Figure 12.34. The cost is equal to

$$C(n, m) = n.(m+2)/2 \text{ slices},$$

and the computation time to

$$T(n, m) = (n+m).T_{\text{mux-cy}} + n.(T_{\text{LUT}} + T_{\text{XOR}}) + (n-1).T_{\text{connection}}.$$

Observe that if m is not much greater than n then

$$T(n, m) \approx n.(T_{\text{LUT}} + T_{\text{XOR}}) + (n-1).T_{\text{connection}},$$

so that the computation time is similar to that of a carry-save array.

12.3 BIBLIOGRAPHY

[BAU1973] C. R. Baugh and B. A. Wooley, A two's complement parallel array multiplication algorithm. *IEEE Trans. Comput.* **C-22**: 1045–1047 (1973).

[BIO2003] G. Bioul, J.-P. Deschamps, and G. Sutter. Spartan-II/Virtex Implementation of High Speed Adders. URJC Madrid, Technical Report, Feb. 2003.

[BOO1951] A. D. Booth, A signed binary multiplication technique. *Q. J. Mech. Appl. Math.* **4**: 236–240 (1951).

[DAD1965] L. Dadda, Some schemes for parallel multipliers. *Alta Frequenza* **34**: 349–356 (1965).

[DAV1977] M. Davio and G. Bioul, Fast parallel multiplication. *Philips Res. Rpts.* **32**: 44–70 (1977).

[WAL1964] C. S. Wallace, A suggestion for fast multipliers. *IEEE Trans. Electron. Comput.* **EC-13**: 14–17 (1964).

13

DIVIDERS

Chapter 6 shows that division is somewhat more intricate than the other three basic arithmetic operations. In the earliest computer applications, division was most often implemented as an assembly language program using the other arithmetic operations as primitives. Such was the case for multiplication too in elementary pioneer processors. The progress of technology together with increasing user needs have motivated designer efforts toward faster implementations of most arithmetic functions in general, and division in particular. This chapter presents implementations of the two most important classes of division algorithms, namely, digit recurrence (one digit at a time) and convergence types.

13.1 NATURAL NUMBERS

Let Y be an n-bit positive number and X a natural number belonging to the range $0 \leq X < Y$, so that it can also be represented as an n-bit number. The circuit corresponding to the basic division algorithm 6.1 (with $q(i)$ substituted by $q(p - i)$ in order that the least significant bit of q be $q(0)$) is an iterative circuit made up of p cells, which implement the division_step procedure (Figure 13.1). The divider structure is shown in Figure 13.2 (combinational and sequential implementations). In the binary case the division_step block (base-2 division step, Algorithm 6.2) consists of an $(n + 1)$-bit subtractor and an n-bit 2-to-1 multiplexer (Figure 13.1). The corresponding cost $C(n, p)$ and computation

Synthesis of Arithmetic Circuits: FPGA, ASIC, and Embedded Systems
By Jean-Pierre Deschamps, Géry J. A. Bioul, and Gustavo D. Sutter
Copyright © 2006 John Wiley & Sons, Inc.

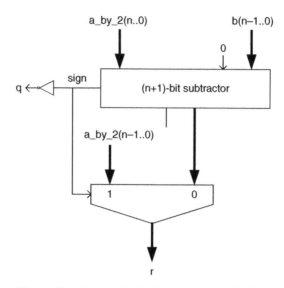

Figure 13.1 Basic cell of a binary restoring divider.

time $T(n, p)$ are equal to

$$C(n, p) = p.(C_{subtractor}(n + 1) + n.C_{mux}) \tag{13.1}$$

and

$$T(n, p) = p.(T_{subtractor}(n + 1) + T_{mux}). \tag{13.2}$$

Example 13.1 (Complete VHDL source code available.) Generate a generic n-bits base-2 restoring divider. The division step of Figure 13.1 is:

```
entity restoring_cell is
port (
  a_by_2: in STD_LOGIC_VECTOR (N downto 0);
  b: in STD_LOGIC_VECTOR (N-1 downto 0);
  q: out STD_LOGIC;
  r: out STD_LOGIC_VECTOR (N-1 downto 0)
);
end restoring_cell;

architecture cel_arch of restoring_cell is
signal subst: STD_LOGIC_VECTOR (N downto 0);
begin
  subst <= a_by_2 - b;
  multiplexer: process (a_by_2,b,subst)
```

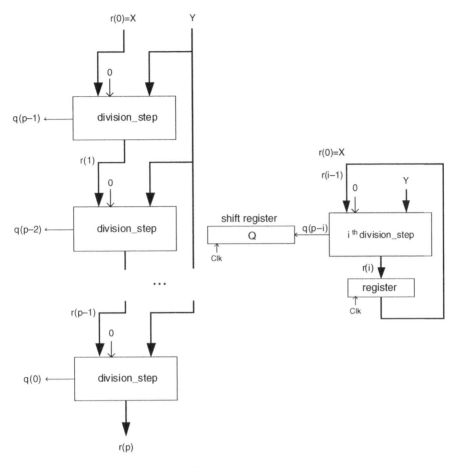

Figure 13.2 Divider structure.

```
  begin
    if subst(N)='1' then r<=a_by_2(N-1 downto 0);
    else r<=subst(N-1 downto 0);
    end if;
  end process;
  q<=not subst(N);
end cel_arch;
```

The divider structure of Figure 13.2 is:

```
entity div_rest_frac is
port (
  X: in STD_LOGIC_VECTOR (N-1 downto 0);
  Y: in STD_LOGIC_VECTOR (N-1 downto 0);
```

```
  Q: out STD_LOGIC_VECTOR (P-1 downto 0);
  R: out STD_LOGIC_VECTOR (N-1 downto 0)
);
end div_rest_frac;

architecture div_arch of div_rest_frac is
type connect is array (0 to P-1) of STD_LOGIC_VECTOR (N downto
0);
Signal rem_in, rem_out: connect;
begin
  rem_in(0)<=X&'0';
  divisor: for i in 0 to P-1 generate
    rest_cell: restoring_cell port map (rem_in(i),
              Y, Q(P-i-1), rem_out(i)(N-1 downto 0));
    rem_in(i+1)<=rem_out(i)(N-1 downto 0)&'0';
  end generate;

  R<=rem_out(P-1)(N-1 downto 0);
end div_arch;
```

Given an m-bit natural X and an n-bit positive integer Y, that is,

$$0 \le X < 2^m \quad \text{and} \quad 1 \le Y < 2^n,$$

synthesize an *integer divider*, that is, a circuit generating two natural numbers Q and R such that

$$X = Q.Y + R, \quad \text{with} \quad R < Y.$$

For that purpose (Section 6.1) Y must be substituted by $Y' = 2^m.Y$. Then the division of X by Y' is computed with an accuracy of m bits, so that

$$2^m.X = Q.Y' + R', \quad \text{with } R' < Y',$$

and

$$X = Q.Y'.2^{-m} + R'.2^{-m} = Q.Y + R'.2^{-m}, \quad \text{with } R = R'/2^m < Y.$$

A better option (Comment 6.1) is to substitute X by $X' = X/2$ and Y by $Y' = 2^{m-1}.Y$. The final remainder R' is divided by 2^{m-1}. The corresponding circuit is shown in Figure 13.3.

Observe that, at each step, the difference between an $(m + n)$-bit number (twice the previous remainder) and $Y.2^{m-1}$ is computed (Figure 13.4), so that the number of bits of the internal subtractor is $n + 1$ (not $n + m$) as the $m - 1$ least significant bits of $2.r(i)$ are just propagated to the cell output.

Example 13.2 (Complete VHDL source code available.) Generate a generic m-bit by n-bit base-2 integer divider. The basic cell is the same as before (Figure 13.1).

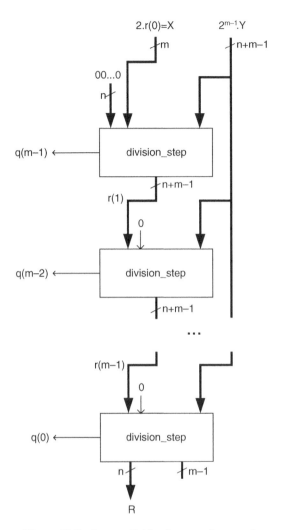

Figure 13.3 Integer divider for natural operands.

The VHDL model is based on Figure 13.3, taking into account the observation of Figure 13.4.

```
entity div_rest_nat is
port (
  X: in STD_LOGIC_VECTOR (M-1 downto 0);
  Y: in STD_LOGIC_VECTOR (N-1 downto 0);
  Q: out STD_LOGIC_VECTOR (M-1 downto 0);
  R: out STD_LOGIC_VECTOR (N-1 downto 0)
);
end div_rest_nat;
```

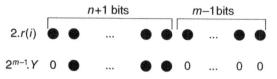

Figure 13.4   Basic operation.

```
architecture div_arch of div_rest_nat is
type connect is array (0 to M-1) of STD_LOGIC_VECTOR (N
downto 0);
signal wires_in, wires_out: connect;
signal zeros: STD_LOGIC_VECTOR (N-1 downto 0);
begin
  zeros<=(others=>'0');
  wires_in(0)<=zeros&X(M-1);
  divisor: for i in 0 to M-1 generate
    rest_cell: restoring_cell port map (wires_in(i),
              Y, Q(M-i-1), wires_out(i)(N-1 downto 0));
  end generate;
  wires_conections: for i in 0 to M-2 generate
    wires_in(i+1)<=wires_out(i)(N-1 downto 0)&X(M-i-2);
  end generate;
  R<=wires_out(M-1)(N-1 downto 0);
end div_arch;
```

In non binary cases the `division_step` block is more complex. The base-B division step described in Algorithm 6.3 consists of checking increasing values of the quotient-digit q up to the minimum value verifying

$$B.a < q.b, \tag{13.3}$$

where $B.a$ and b are the shifted remainder and the divisor, respectively; if q_{min} is the minimum value of q verifying (13.3), then $q_{min}-1$ is the asserted quotient-digit. This method is easy but quite ineffective. The restoring base-B division step described by Algorithm 6.4 is faster because only one value of q has to be checked at every step. The corresponding circuit is shown in Figure 13.5. It includes a 5-input 2-output look-up table, an $(n \times 1)$-digit multiplier, an $(n + 1)$-digit subtractor, an n-digit adder, and $(n + 1)$ 2-to-1 multiplexers. As far as B is greater than 2, the n-digit adder may not be replaced by a direct connection of a to the inputs 1 of the $(n + 1)$ 2-to-1 multiplexers; in this case the restoring process doesn't actually restore the previous remainder, but consists of adding one divisor to a negative new remainder. The adding stage could be nevertheless eliminated if two subtractions are executed in parallel ($-q_t.b$ and $-(q_t - 1).b$); then the multiplexers would select the correct remainder according to the sign of the result. This would provide a saving in cycle time, without significantly increasing the hardware cost.

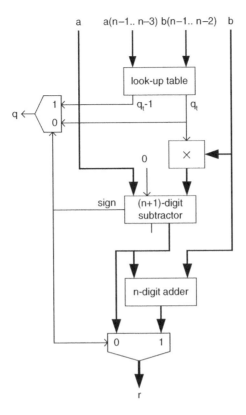

Figure 13.5 Basic cell of a nonbinary divider.

The cost and the computation time of the corresponding divider are, respectively,

$$C(n, p) = p.(C_{\mathrm{LUT}}(5, 2) + C_{\mathrm{multiplier}}(n \times 1) + C_{\mathrm{subtractor}}(n + 1)$$
$$+ C_{\mathrm{adder}}(n) + (n + 1).C_{\mathrm{mux}}) \tag{13.4}$$

and

$$T(n, p) = p.(T_{\mathrm{LUT}}(5, 2) + T_{\mathrm{multiplier}}(n \times 1) + T_{\mathrm{subtractor}}(n + 1)$$
$$+ T_{\mathrm{adder}}(n) + T_{\mathrm{mux}}). \tag{13.5}$$

Example 13.3 (Complete VHDL source code available.) Generate a generic *n*-digits base-B restoring divider. The division step of Figure 13.5 is:

```
entity rest_baseB_step is
port (
  a: in digit_vector(N-1 downto 0);
  b: in digit_vector(N-1 downto 0);
```

```
  q: out digit;
  r: out digit_vector(N-1 downto 0)
);
end rest_baseB_step;

architecture behavioral of rest_baseB_step is
signal at: digit_vector(2 downto 0);
signal bt: digit_vector(1 downto 0);
signal qt, qt_1: digit;
signal q_x_b, a_x_B, re, r_plus_b: digit_vector(N downto 0);
signal sign: bit;
begin
  at<=a(N-1 downto N-3);
  bt<=b(N-1 downto N-2);
  a_x_B(N downto 1)<=a; a_x_B(0)<=0;
  LUT: look_up_table port map(at, bt, qt, qt_1);
  mult: base_b_mult port map(qt, B, q_x_b);
  subt: base_b_subt port map(a_x_B, q_x_b, re, sign);
  adder: base_b_adder port map(re(N-1 downto 0), B, r_plus_b);
  multiplexers: process (sign,qt_1, qt, r_plus_b,re)
  begin
    if sign='1' then
      q<=qt_1; r<=r_plus_b(N-1 downto 0);
    else q<=qt; r<=re(N-1 downto 0);
    end if;
  end process;
end behavioral;
```

The divider structure is:

```
entity div_rest_baseB is
port (
  A: in digit_vector(N-1 downto 0);
  B: in digit_vector(N-1 downto 0);
  Q: out digit_vector(P-1 downto 0);
  R: out digit_vector(N-1 downto 0)
);
end div_rest_baseB;

architecture div_arch of div_rest_baseB is
type connections is array (0 to P) of digit_vector(N-1
downto 0);
Signal wires: connections;
begin
  wires(0)<=A;
  divisor: for i in 0 to P-1 generate
  rest_step: rest_baseB_step port map (wires(i),
    B, Q(P-i-1), wires(i+1));
  end generate;
```

```
   R<=wires(P)(N-1 downto 0);
end div_arch;
```

13.2 INTEGERS

13.2.1 Base-2 Nonrestoring Divider

Let Y be an n-bit positive number and X an integer belonging to the range $-Y \leq X <$ Y, so that it can be represented as an $(n + 1)$-bit 2's complement number. The circuit corresponding to the nonrestoring algorithm 6.6 (with $q(i)$ substituted by $q(p - i)$ in order that the least significant bit of q be $q(0)$) is shown in Figures 13.6 (basic cell), 13.7 (divider structure, combinational and sequential implementations), and 13.8 (correction circuit).

The cost and computation time of the corresponding divider are

$$C(n, p) = p.C_{\text{adder/subtractor}}(n + 1) + 2.C_{\text{mux}}$$
$$+ C_{\text{adder}}(p + 1) + C_{\text{adder}}(n + 1), \tag{13.6}$$

and

$$T(n, p) = p.T_{\text{adder/subtractor}}(n + 1) + T_{\text{mux}}$$
$$+ \max (T_{\text{adder}}(p + 1), T_{\text{adder}}(n + 1)). \tag{13.7}$$

Examples 13.4 (Complete VHDL source code available.) Generate a VHDL model of a generic base-2 nonrestoring divider (Figures 13.6, 13.7, and 13.8):

```
entity nonr_cell is
port (
  a: in STD_LOGIC_VECTOR (N-1 downto 0);
```

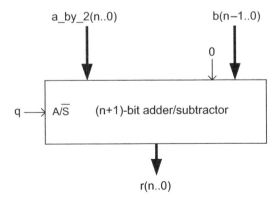

Figure 13.6 Nonrestoring divider: basic cell.

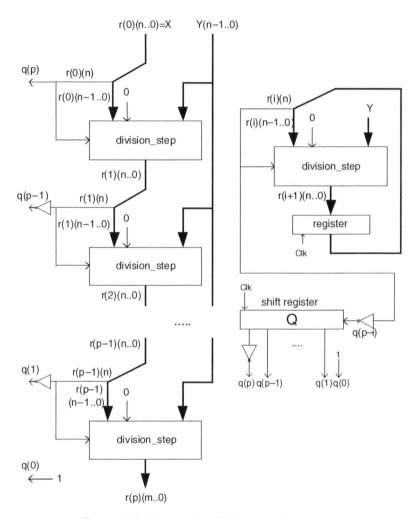

Figure 13.7 Nonrestoring divider: general structure.

```
  b: in STD_LOGIC_VECTOR (N-1 downto 0);
  q: in STD_LOGIC;
  r: out STD_LOGIC_VECTOR (N downto 0)
);
end nonr_cell;

architecture nr_cel_arch of nonr_cell is
signal a_by_2: STD_LOGIC_VECTOR (N downto 0);
begin
a_by_2<=a(N-1 downto 0)&'0';
adder_subtracter: process (a_by_2,b,q)
```

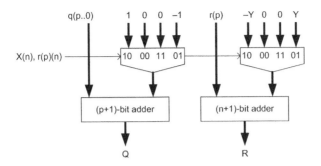

Figure 13.8 Nonrestoring divider: correction circuit.

```
begin
  if q='1' then r<=a_by_2+b;
  else r<=a_by_2 - b;
  end if;
end process;
end nr_cel_arch;
```

The final correction circuit of Figure 13.8 is:

```
entity correction_cell is
port (
  Q: in std_logic_vector(P downto 0);
  R: in std_logic_vector(N downto 0);
  Y: in std_logic_vector(N-1 downto 0);
  x_n: in std_logic;
  r_n: in std_logic;
  adj_Q: out std_logic_vector(P downto 0);
  adj_R: out std_logic_vector(N downto 0)
);
end correction_cell;

architecture correction_arch of correction_cell is
signal selector: std_logic_vector(1 downto 0);
begin
  selector<=x_n&r_n;
  correction: process (selector,Q,R,Y)
  begin
    case selector is
      when "10"=>adj_Q<=Q+1; adj_R<=R - Y;
      when "01"=>adj_Q<=Q-1; adj_R<=R+Y;
      when others=>adj_Q<=Q; adj_R<=R;
    end case;
  end process;
end correction_arch;
```

The divider structure is:

```
entity div_nr is
port (
  X: in STD_LOGIC_VECTOR (N downto 0);
  Y: in STD_LOGIC_VECTOR (N-1 downto 0);
  Q: out STD_LOGIC_VECTOR (P downto 0);
  R: out STD_LOGIC_VECTOR (N downto 0)
);
end div_nr;

architecture div_arch of div_nr is
type connect is array (0 to P) of STD_LOGIC_VECTOR (N downto 0);
Signal wires: connect;
Signal QQ: STD_LOGIC_VECTOR (P downto 0);
begin
 wires(0)<=X;
 divisor: for i in 0 to P-1 generate
 nr_step: nonr_cell port map (wires(i)(N-1 downto 0),
 Y, wires(i)(N), wires(i+1));
 end generate;

 QQ(P)<=wires(0)(N);
 Quotient: for i in 1 to P-1 generate
  QQ(P-i)<=not wires(i)(N);
 end generate;
 QQ(0)<='1';
 final_adjust: correction_cell port map (QQ(P downto 0),
 wires(P), Y, X(N), wires(P)(N), Q, R);
end div_arch;
```

Given two n-bit *normalized numbers*, that is, two natural numbers X and Y such that

$$2^{n-1} \le X < 2^n \text{ and } 2^{n-1} \le Y < 2^n,$$

synthesize a circuit generating the quotient Q and the remainder R with an accuracy of p bits, that is,

$$2^p.X = Q.Y + R.$$

In order that the dividend be smaller than the divider, X is substituted by $X/2 < 2^{n-1} \le Y$. According to Comment 6.3, it's only a matter of defining $2.r(0) = X$ (instead of $2.X$) and performing the division with an accuracy of $p + 1$ bits. The corresponding iterative circuit is shown in Figure 13.9. The final correction circuit is simpler: X is positive so that, in Figure 13.8, $X(n) = 0$.

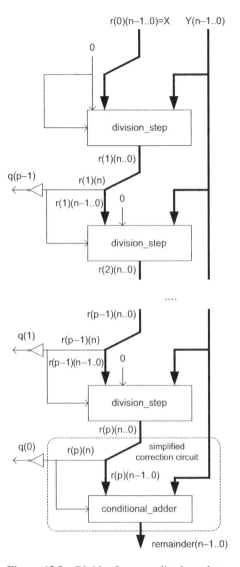

Figure 13.9   Divider for normalized numbers.

Example 13.5   (Complete VHDL source code available.) Generate the VHDL model of a divider for normalized base-2 numbers (Figure 13.9 with the basic cell of Figure 13.6):

```
entity nr_cell is
port (
  a_by_2: in STD_LOGIC_VECTOR (N downto 0);
  b: in STD_LOGIC_VECTOR (N-1 downto 0);
```

```
  q: in STD_LOGIC;
  r: out STD_LOGIC_VECTOR (N downto 0)
);
end nr_cell;

architecture nr_cel_arch of nr_cell is
begin
adder_subtracter: process (q,a_by_2,b)
begin
  if q='1' then r<=a_by_2+b;
  else r<=a_by_2 - b;
  end if;
end process;
end nr_cel_arch;
```

The correction cell of Figure 13.8 (if necessary) is reduced to a conditional adder, and the last quotient bit ($q(0)$) is the negative of the last-remainder's most-significant-bit.

```
entity cond_adder is
port (
  a: in STD_LOGIC_VECTOR (N-1 downto 0);
  b: in STD_LOGIC_VECTOR (N-1 downto 0);
  sel: in STD_LOGIC;
  r: out STD_LOGIC_VECTOR (N-1 downto 0)
);
end cond_adder;
architecture cond_adder_arch of cond_adder is
begin
  conditional_adder: process (sel,a,b)
  begin
    if sel='1' then r<=a+b;
    else r<=a;
    end if;
end process;
```

The divider structure of Figure 13.9 is:

```
entity div_nr_norm is
port (
  X: in STD_LOGIC_VECTOR (N-1 downto 0);
  Y: in STD_LOGIC_VECTOR (N-1 downto 0);
  Q: out STD_LOGIC_VECTOR (P-1 downto 0);
  R: out STD_LOGIC_VECTOR (N-1 downto 0)
);
end div_nr_norm;
```

```
architecture div_arch of div_nr_norm is
type connect is array (0 to P+1) of STD_LOGIC_VECTOR (N
downto 0); Signal r_in, r_out: connect;
Signal op: STD_LOGIC_VECTOR (P+1 downto 0);
begin
  wires_in(0)<='0'&X; op(0)<='0';
  divisor: for i in 0 to P generate
    nr_step: nr_cell port map (r_in(i), b=>Y, op(i), r_out(i));
    Q(P-i)<=not wires_out(i)(N); op(i+1)<=r_out(i)(N);
    r_in(i+1)<=r_out(i)(N-1 downto 0)&'0';
  end generate;
  rem_adj: cond_adder port map (r_out(P)(N-1 downto 0),
  Y, r_out(P)(N), R);
end div_arch;
```

13.2.2 Base-*B* Nonrestoring Divider

Figure 13.10 depicts the basic cell corresponding to the non restoring base-*B* division step of Algorithm 6.10. It includes an 8-input 2-output look-up table (LUT), a 2-digit by *n*-digit multiplier, and an $(n + 2)$-digit subtractor. The LUT inputs are *rt*, entered as the five leftmost digits $B.r(i)(n + 1..n - 3)$ of the shifted remainder, and *Yt*, entered as the three leftmost digits $Y(n - 1..n - 3)$ of the divisor; the 2-digit output corresponds to the result *qt* of the integer division $rt/Yt + 1$, namely the selected 2-digit quotient.

The LUT size and cost can become prohibitive for increasing values of *B*, so a fast 5-digit by 3-digit base-*B* divider can be designed as an alternative. The

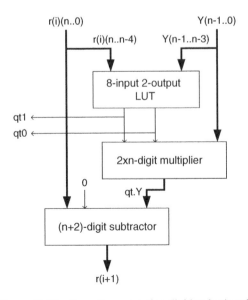

Figure 13.10 Base-*B* nonrestoring divider: basic cell.

multiplier output is the product $qt.Y$ of the selected quotient by the divisor. According to the resources at hand those products may be precalculated then stored for fast retrieval. The subtractor computes the new remainder as the difference $B.r(i) - qt.Y$. Finally, a carry-save computation of $r(i+1)$ could be implemented but conditions over the quotient-digit set have to be relaxed. The divider structure is shown in Figure 13.11.

The cost and computation time of the nonrestoring divider basic cell of Figure 13.10 are given by

$$C_{\text{division\_step}}(n) = C_{\text{LUT}}(8, 2) + C_{\text{multiplier}}(2 \times n) + C_{\text{subtractor}}(n + 2), \qquad (13.8)$$

and

$$T_{\text{division\_step}}(n) = T_{\text{LUT}}(8, 2) + T_{\text{multiplier}}(2 \times n) + T_{\text{subtractor}}(n + 2), \qquad (13.9)$$

or

$$T_{\text{division\_step}}(n) = T_{\text{LUT}}(8, 2) + T_{\text{multiplier}}(2 \times n) + T_{\text{full adder}} + T_{\text{sign-bit}}, \qquad (13.10)$$

if carry-save is implemented.

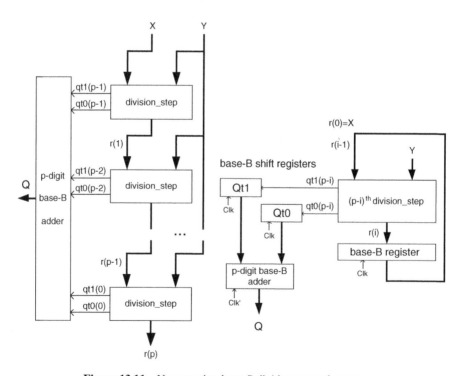

Figure 13.11 Nonrestoring base-B divider: general structure.

The cost and computation time of the non restoring base-B divider of Figure 13.11 are given by

$$C(n, p) = p.C_{\text{division\_step}}(n) + C_{\text{adder}}(p), \qquad (13.11)$$

and

$$T(n, p) = p.T_{\text{division\_step}}(n) + T_{\text{adder}}(p). \qquad (13.12)$$

Example 13.6 (Complete VHDL source code available.) Generate a generic n-digit base-B nonrestoring divider. The division step of Figure 13.10 is:

```
entity nr_baseB_step is
port (
  a: in digit_vector(N downto 0);
  b: in digit_vector(N-1 downto 0);
  q1, q0: out digit;
  r: out digit_vector(N downto 0)
);
end nr_baseB_step;

architecture behavioral of nr_baseB_step is
signal rt: digit_vector(4 downto 0);
signal yt: digit_vector(2 downto 0);
signal qt1,qt0: digit;
signal a_x_BASE, q_x_b, remainder: digit_vector(N downto 0);

begin
  rt<=a(N downto N-4);
  yt<=b(N-1 downto N-3);
  a_x_BASE(N downto 1)<=a(N-1 downto 0); a_x_BASE(0)<=0;
  LookUpTable:look_up_table port map (rt, yt, qt1, qt0);
  mult: base_b_2_x_n_mult port map (qt1&qt0, B, q_x_b);
  subtractor: base_b_subt port map (a_x_BASE, q_x_b, remainder);
  q1<=qt1; q0<=qt0;
  r<=remainder;
end behavioral;
```

The divider structure is:

```
entity div_nr_baseB is
port (
  A: in digit_vector(N-1 downto 0);
  B: in digit_vector(N-1 downto 0);
  Q: out digit_vector(P-1 downto 0);
  R: out digit_vector(N-1 downto 0)
);
end div_nr_baseB;
```

```
architecture div_arch of div_nr_baseB is
type connections is array (0 to P) of digit_vector(N downto 0);
Signal wires: connections;
signal Q_1, Q_0: digit_vector(P downto 0);
begin
  wires(0)<=O&A;
  divisor: for i in 0 to P-1 generate
    rest_step: nr_baseB_step port map (wires(i), B,
    Q_1(P-i), Q_0(P-i-1), wires(i+1));
  end generate;
  correction: rem_adjust port map (wires(P), B, Q_1(0), R);
  final_adder: base_B_adder port map (Q_1(P-1 downto 0),
            Q_0(P-1 downto 0), Q);
end div_arch;
```

13.2.3 SRT Dividers

13.2.3.1 SRT-2 Divider   The SRT-2 divider with full computation of the remainder will be presented first. Let Y be an n-bit normalized number (Example 6.4.2) and X an n-bit 2's complement integer, that is,

$$2^{n-1} \leq Y < 2^n \quad \text{and} \quad -2^{n-1} \leq X < 2^{n-1}.$$

Then Algorithm 6.7 (with $q(i)$ substituted by $q(p - i)$ in order that the least significant bit of q be $q(0)$) can be applied. At each step the following operation is performed:

$$r(i + 1) = 2.r(i) - q(p - i - 1).Y,$$

where $r(i + 1)$ and $r(i)$ are n-bit 2's complement numbers, and Y an n-bit natural and $q(p - i - 1)$ a signed bit (-1, 0 or 1) whose value is defined (Table 6.1) as a function of $w(n)$ and $w(n - 1)$, that is, $r(i)(n - 1)$ and $r(i)(n - 2)$.

The basic cell is shown in Figure 13.12. If en $= 0$, then $r =$ a_by_2; if en $= 1$ then $r =$ a_by_2 $\pm$ b where the operation is selected by op (0: add; 1: subtract). The divider structure is shown in Figure 13.13. The combinational circuit implements Table 13.1. Observe that op(p-i-1) can be chosen equal to q_pos(p-i-1), so that it is a 2-input 3-input combinational circuit. An additional (not represented) $(p + 1)$-bit subtractor generates $Q =$ q_pos-q_neg. Furthermore, a correction circuit, similar to that of Figure 13.8, is necessary if the condition $sign(R) = sign(X)$ must hold.

The cost and computation time of the non restoring divider basic cell of Figure 13.12 are given by

$$C_{\text{division\_step}}(n) = n.C_{\text{AND2}} + C_{\text{adder/subtractor}}(n) \tag{13.13}$$

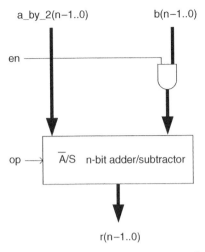

Figure 13.12 SRT-2 divider: basic cell.

and

$$T_{\text{division\_step}}(n) = T_{\text{AND2}} + T_{\text{adder/subtractor}}(n). \tag{13.14}$$

The cost and computation time of a nonrestoring SRT-2 divider (Figure 13.13 with an additional $(p + 1)$-bit subtractor for computing `q_pos-q_neg`), without the correction circuit, are given by

$$C(n, p) = p.C_{\text{division\_step}}(n) + p.C_{\text{LUT}}(2, 3) + C_{\text{subtractor}}(p + 1) \tag{13.15}$$

and

$$T(n, p) = p.T_{\text{division\_step}}(n) + p.T_{\text{LUT}}(2, 3) + T_{\text{subtractor}}(p + 1), \tag{13.16}$$

where $C_{\text{LUT}}(2, 3)$ and $T_{\text{LUT}}(2, 3)$ are the cost and computation time of a 2-input 3-output combinational circuit (e.g., a look-up table).

Comment 13.1 The SRT-2 algorithm is very similar to the base-2 nonrestoring algorithm. The latter is based on the diagram of Figure 6.4b while the former is based on the diagram of Figure 6.3b. At the circuit level, the similarity is quite evident: compare Figures 13.7 (nonrestoring) and 13.13 (SRT-2), taking into account that in the first case X is an $(n + 1)$-bit 2's complement number and in the second case an n-bit 2's complement number. The difference is that the SRT-2 method needs a 2-input table in order to select the next quotient bit value, while the nonrestoring algorithm only needs a 1-bit table (an inverter). Furthermore, the SRT-2 circuit needs an additional output subtractor, or some kind of on-the-fly conversion circuit. As regards the cost and computation time, the nonrestoring method should always

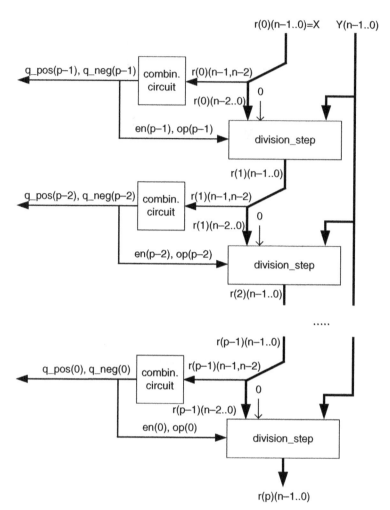

Figure 13.13   SRT-2 divider: general structure.

TABLE 13.1   Selection of `q_pos(p-i-1)`, `q_neg(p-i-1)`, `en(p-i-1)`, **and** `op(p-i-1)`

$r(i)(n-1)$	$r(i)(n-2)$	q_pos (p-i-1)	q_neg (p-i-1)	en(p-i-1)	op(p-i-1)
0	0	0	0	0	—
0	1	1	0	1	1
1	0	0	1	1	0
1	1	0	0	0	—

be better than (or equivalent to) the SRT-2 method. The real advantage of the base-2 SRT algorithm is when stored-carry encoding is used (next section).

Example 13.7 (Complete VHDL source code available.) Generate a generic n-bits base-2 SRT divider. The division step of Figure 13.12 is:

```vhdl
entity srt_cell is
port (
  r_by_2: in STD_LOGIC_VECTOR (N-1 downto 0);
  y: in STD_LOGIC_VECTOR (N-1 downto 0);
  en, op: in STD_LOGIC;
  r_n: out STD_LOGIC_VECTOR (N-1 downto 0)
);
end srt_cell;
architecture behavioral of srt_cell is
begin
  cell: process (en,op,y,r_by_2)
  begin
    if en='1' then
      if op='1' then r_n<=r_by_2-y;
      else r_n<=r_by_2+y;
      end if;
    else r_n<=r_by_2;
    end if;
  end process;
end behavioral;
```

The combinational circuit of Table 13.1 is:

```vhdl
entity comb_circ is
port (
  r: in STD_LOGIC_VECTOR (1 downto 0);
  q_pos, q_neg: out STD_LOGIC;
  en, op: out STD_LOGIC);
end comb_circ;

architecture behavioral of comb_circ is
begin
  combinational: process (r)
  begin
    case r is
      when "00"=>q_pos<='0'; q_neg<='0'; en<='0'; op<='-';
      when "01"=>q_pos<='1'; q_neg<='0'; en<='1'; op<='1';
      when "10"=>q_pos<='0'; q_neg<='1'; en<='1'; op<='0';
      when "11"=>q_pos<='0'; q_neg<='0'; en<='0'; op<='-';
      when others=>NULL;
    end case;
  end process;
end behavioral;
```

The divider structure of Figure 13.13 is:

```
entity SRT_radix2 is
port (
  X: in STD_LOGIC_VECTOR (N-1 downto 0);
  Y: in STD_LOGIC_VECTOR (N-1 downto 0);
  Q: out STD_LOGIC_VECTOR (P downto 0);
  R: out STD_LOGIC_VECTOR (N-1 downto 0)
);
end SRT_radix2;

architecture srt_arch of SRT_radix2 is
type connect is array (0 to P) of STD_LOGIC_VECTOR (N downto 0);
signal r_in, r_out: connect;
signal QQ, Q_pos, Q_neg, en, op: STD_LOGIC_VECTOR (P downto 0);
begin
 r_in(0)<=X&'0';
 divisor: for i in 0 to P-1 generate
  comb: comb_circ port map(r_in(i)(N downto N-1),
      Q_pos(P-i-1), Q_neg(P-i-1), en(P-i-1), op(P-i-1));
  div_step: srt_cell port map (r_in(i)(N-1 downto 0), Y,
          en(P-i-1), op(P-i-1),r_out(i)(N-1 downto 0));
  r_in(i+1)<=r_out(i)(N-1 downto 0)&'0';
 end generate;
 QQ<=('0'&Q_pos) - ('0'&Q_neg);
 final_adjust: correction_cell port map (QQ, r_in(P)(N downto 1),
Y, x_n=>X(N-1), r_in(P)(N), Q, R);
end srt_arch;
```

13.2.3.2 SRT-2 Divider with Carry-Save Computation of the Remainder

Let Y be an n-bit normalized number and X an integer belonging to the range $-Y \leq X < Y$, so that it can be expressed as an $(n + 1)$-bit 2's complement integer. Then Algorithm 6.8 can be applied. At each step the following operation is performed (recall that s' and c' stand for $s/2$ and $c/2$):

$$(s'(i+1) + c'(i+1)) = 2.s'(i) + 2.c'(i)) - q(p - i - 1).Y,$$

where $s'(i + 1)$, $c'(i + 1)$, $s'(i)$, and $c'(i)$ are $(n + 2)$-bit 2's complement numbers, Y is an n-bit natural number, and $q(p - i - 1)$ is a signed bit $(-1, 0$ or $1)$ whose value is defined (Figure 6.5 and Table 6.2) as a function of $s(i)(n + 2..n - 1)$ and $c(i)(n + 2..n - 1)$, that is, $s'(i)(n + 1..n - 2)$ and $c'(i)(n + 1..n - 2)$.

The basic cell is shown in Figure 13.14. The carry-save adder is a set of full adders working in parallel (Chapter 11), so that its computation time does not depend on the operand size. If $en = 0$, then $(s'(i + 1) + c'(i + 1)) = 2.s'(i) + 2. c'(i))$; if $en = 1$ then $(s'(i + 1) + c'(i + 1)) = 2.s'(i) + 2.c'(i)) \pm Y$, where the operation is selected by op (0: add; 1: subtract). The divider structure is shown in

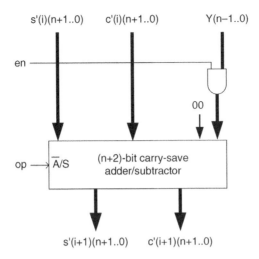

Figure 13.14 SRT-2 carry-save divider: basic cell.

Figure 13.16. The combinational circuit (Figure 13.15) implements the circuit of Table 13.2. An additional (not represented) p-bit subtractor generates $Q = $ q_pos $-$ q_neg. Another additional (not represented) $(n + 1)$-bit adder generates $r(p) = s'(p) + c'(p)$, that is, the decoded value of the final remainder. Furthermore, a correction circuit, similar to that of Figure 13.6, is necessary if the condition $sign(R) = sign(X)$ is to hold.

The cost and computation time of the carry-save basic cell of Figure 13.14 are given by

$$C_{\text{division\_step}}(n) = n.C_{\text{AND2}} + C_{\text{carry-save adder}}(n + 2) \tag{13.17}$$

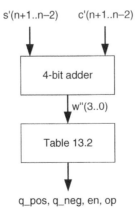

Figure 13.15 Selection of q_pos, q_neg, en, op.

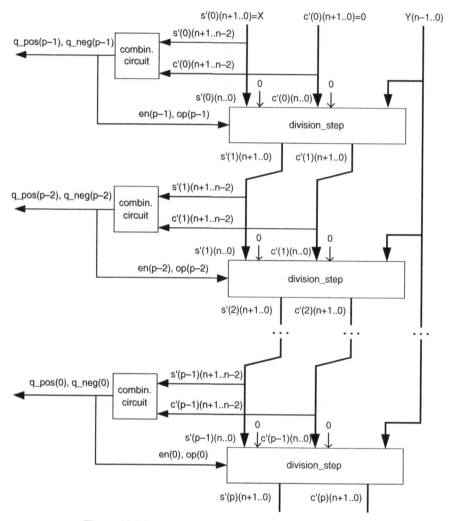

Figure 13.16 SRT-2 carry-save divider: general structure.

and

$$T_{\text{division\_step}} = T_{\text{AND2}} + T_{\text{carry-save adder}}(1). \qquad (13.18)$$

Observe that (Table 13.2) the control signals *op* and *en* can be chosen equal to

$$op = q\_neg \quad \text{and} \quad en = q\_pos \vee q\_neg, \qquad (13.19)$$

so that the circuit of Figure 13.15 can be synthesized with a 4-bit adder, a 4-input 2-output look-up table, and some additional logic gates. According to Comment

TABLE 13.2   Definitions

w''	q_pos	q_neg	en	op
0000	1	0	1	0
0001	1	0	1	0
0010	1	0	1	0
0011	1	0	1	0
0100 to 1010	—	—	—	—
1011	0	1	1	1
1100	0	1	1	1
1101	0	1	1	1
1110	0	1	1	1
1111	0	0	0	—

6.4, in some cases a 3-bit adder can be used. So, the cost and computation time of the circuit of Figure 13.15 are approximately equal to

$$C_{\text{comb.circuit}} = C_{\text{adder}}(4) + C_{\text{LUT}}(4, 2) \tag{13.20}$$

and

$$T_{\text{comb.circuit}} = T_{\text{adder}}(4) + T_{\text{LUT}}(4, 2). \tag{13.21}$$

The cost and computation time of a carry-save SRT-2 divider, that is, Figure 13.16 with an additional $(p + 1)$-bit subtractor for computing q_pos − q_neg, and an additional $(n + 1)$-bit adder for computing $r(p) = s'(p) + c'(p)$), without the correction circuit, are given by

$$C(n, p) = p.C_{\text{division\_step}}(n) + p.C_{\text{comb.circuit}} + C_{\text{subtractor}}(p + 1)$$
$$+ C_{\text{adder}}(n + 1), \tag{13.22}$$

and

$$T(n, p) = p.T_{\text{division\_step}} + p.T_{\text{comb.circuit}}$$
$$+ \max\left(T_{\text{subtractor}}(p + 1), T_{\text{adder}}(n + 1)\right). \tag{13.23}$$

Thus, the computation time is a linear (not quadratic) function of p and n.

Example 13.8   (Complete VHDL source code available.) Generate a generic n-bit base-2 SRT divider with carry-save remainder. The correction cell is similar to that of Figure 13.8. The division step of Figure 13.14 is:

```
entity srt_cs_cell is
port (
   y: in std_logic_vector (N-1 downto 0);
```

```
  s, c: in std_logic_vector (N+1 downto 0);
  en, op: in std_logic;
  next_s, next_c: out std_logic_vector (N+1 downto 0)
);
end srt_cs_cell;

architecture behavioral of srt_cs_cell is
signal op_y, sum: std_logic_vector (N+1 downto 0);
signal cy: std_logic_vector (N+2 downto 0);
begin
  process (en, op, y)
  begin
    if en='0' then op_y<=(others=>'0');
    else if op='1' then op_y<="00"&y;
         else op_y<=("11"&not(y))+1;
         end if;
    end if;
  end process;
  adder: for i in 0 to N+1 generate
    sum(i)<=op_y(i) xor c(i) xor s(i);
     cy(i+1)<=(op_y(i)and c(i))or(op_y(i)and s(i))or(s(i)and
     c(i));
  end generate;
  next_s<=sum;
  next_c<=cy(N+1 downto 1)&'0';
end behavioral;
```

The combinational circuit of Figure 13.16 is:

```
entity srt_cs_selection is
port (
  s, c: in std_logic_vector (3 downto 0);
  en, op: out std_logic;
  q_pos, q_neg: out std_logic
);
end srt_cs_selection;

architecture behavioral of srt_cs_selection is
signal t: std_logic_vector (3 downto 0);
begin
  t<=s+c;
  process (t)
  begin
    case t is
      when "0000"|"0001"|"0010"|"0011"=>
        q_pos<='1'; q_neg<='0'; en<='1'; op<='0';
      when "1011"|"1100"|"1101"|"1110"=>
        q_pos<='0'; q_neg<='1'; en<='1'; op<='1';
```

```
      when others=>--"1111" and 0100 to 1010
        q_pos<='0'; q_neg<='0'; en<='0'; op<='-';
    end case;
  end process;
end behavioral;
```

The divider structure is:

```
entity srt_carry_save_r2 is
port (
  X: in std_logic_vector (N downto 0);
  Y: in std_logic_vector (N-1 downto 0);
  Q: out std_logic_vector (P downto 0);
  R: out std_logic_vector (N downto 0)
);
end srt_carry_save_r2;

architecture rtl of srt_carry_save_r2 is
type rems is array (0 to P) of std_logic_vector (N+2 downto 0);
signal c, s, c_o, s_o: rems;
signal q_pos, q_neg, en, op: std_logic_vector (P-1 downto 0);
signal qq: std_logic_vector (P downto 0);
signal rr: std_logic_vector (N+1 downto 0);
begin
  s(0)<=X(N)&X&'0'; c(0)<=(others=>'0');
  gen_p: for i in 0 to P-1 generate
    selc: srt_cs_selection port map(s(i)(N+2 downto N-1),
            c(i)(N+2 downto N-1), en(P-i-1), op(P-i-1),
            q_pos(P-i-1), q_neg(P-i-1));
    cell: srt_cs_cell port map (Y, c(i)(N+1 downto 0),
            s(i)(N+1 downto 0), en(P-i-1), op(P-i-1),
            c_o(i)(N+1 downto 0), s_o(i)(N+1 downto 0));
    c(i+1)<=c_o(i)(N+1 downto 0)&'0';
    s(i+1)<=s_o(i)(N+1 downto 0)&'0';
  end generate;
  rr<=c(P)+s(P); qq<=('0'&q_pos) - q_neg;
  final_adjust: correction_cell port map (QQ, rr(N+1 downto 1),
            Y, X(N), rr(N+1), Q, R);
end rtl;
```

Comment 13.2 The circuit of Figure 13.15 is an 8-input 2-output combinational circuit (en and op are assumed to be generated by equations (13.19)). As this circuit belongs to the critical path ($p.T_{\text{comb.circuit}}$ in (13.22)), instead of a 4-bit adder and a 4-input 2-output combinational circuit (Figure 13.15), alternative options are an 8-input 2-output optimized logic circuit, generated with Boolean minimization techniques, or an 8-input 2-output look-up table. If the most significant bits

$s'(n + 1)$ and $c'(n + 1)$ are not used (Comment 6.4), the combinational circuit has 6 inputs and 2 outputs.

13.2.3.3 FPGA Implementation of the Carry-Save SRT-2 Divider A FPGA implementation of the binary SRT divider cell of Figure 13.16 is shown in Figure 13.17.

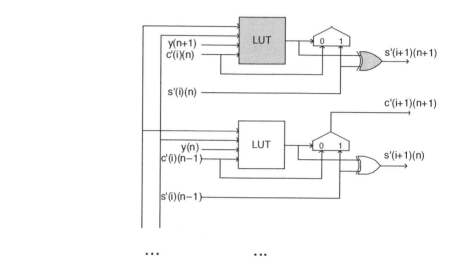

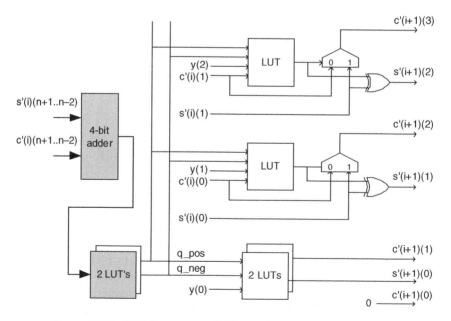

Figure 13.17 SRT-2 carry-save divider: basic cell FPGA implementation.

Assuming a ripple-carry FPGA implementation for the 4-bit adder, the cost and computation time of the cell of Figure 13.17 are given by

$$C_{cell} = (n/2 + 5) \text{ slices,} \tag{13.24}$$

$$T_{cell} = 3.T_{LUT} + 3.T_{mux\text{-}cy} + 3.T_{connection} + 2.T_{XOR2}. \tag{13.25}$$

The critical path is shaded in Figure 13.17. The cost and computation time of the binary SRT divider (Figure 13.16) are then

$$C(n, p) = p.C_{cell} + (p + n + 2)/2 \text{ slices,} \tag{13.26}$$

$$T(n, p) = p.(T_{cell} + T_{connection}) + T_{LUT} + \max{(p + 1, n + 1)}.T_{mux\text{-}cy}. \tag{13.27}$$

Observe that T_{cell} is a constant value so that the computation time is a linear function of p (and n if $n > p$).

13.2.4 SRT-4 Divider

A possible implementation of the basic cell for the SRT-4 divider is displayed in Figure 13.18. It corresponds to Algorithm 6.9. One assumes that $2.Y$ and $3.Y$ are

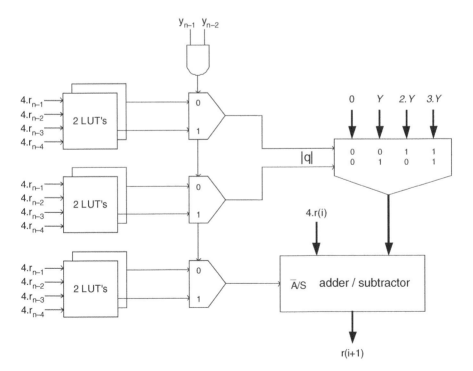

Figure 13.18 SRT-4 divider: basic cell.

readily available from some preliminary multiples generation and storage procedure. This cell is made up of an adder/subtractor whose inputs 0, Y, 2.Y, or 3.Y are selected through a 4-to-1 multiplexer. The q-selection circuit controls this multiplexer together with the $\bar{A}/S$ input of the adder/subtractor. This circuit takes advantage of the fact that the Y coordinates of the border line staircase threshold are identified by the Boolean equation $y_{n-1}.y_{n-2} = 1$, as shown in the $P–D$ diagram of Figure 6.10.

The main part of the circuit may be implemented by ROMs or look-up tables: a programmable logic array has been used in the q-selection hardware of the Pentium SRT-4 division implementation. The general structure of the SRT-4 divider is shown in Figure 13.19. A correction circuit similar to that of Figure 13.8 is still needed. Moreover, an output circuit converts the final quotient Q into a nonredundant 2's complement form. If Q is given in the form of a signed 2's complement 3-bit digit vector, a circuit implementing the Booth_decode Algorithm 5.14a or 5.14b may be used. To speed up the decoding operation, on-the-fly conversion algorithms have been presented in the literature ([ERC1987], [ERC1992], [ERC1994], [MON1994]).

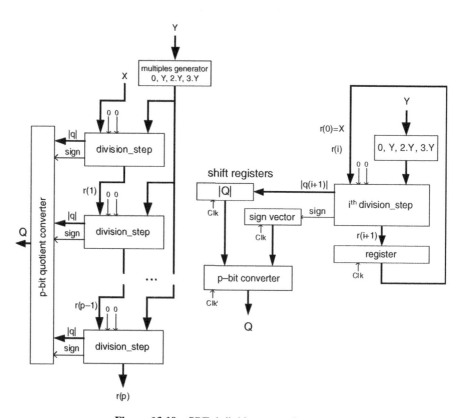

Figure 13.19   SRT-4 divider: general structure.

The cost and computation time of the SRT-4 basic cell of Figure 13.18 are given by

$$C_{\text{basic cell}} = 6.C_{\text{LUT}} + 6.C_{\text{mux}} + C_{\text{AND2}} + C_{\text{adder/subtractor}}(n+1), \qquad (13.28)$$

$$T_{\text{basic cell}} = T_{\text{LUT}} + 3.T_{\text{mux}} + T_{\text{adder/subtractor}}(n+1), \qquad (13.29)$$

where the multiplexer $4-1$ is assumed to be built up from 3 multiplexers $2-1$. The cost and computation time of the SRT-4 divider (Figure 13.19) are then

$$C(n, p) = C_{\text{mult generator}} + p.C_{\text{basic cell}} + C_{\text{converter}}(p), \qquad (13.30)$$

$$T(n, p) = T_{\text{mult generator}} + p.T_{\text{basic cell}} + T_{\text{converter}}(p). \qquad (13.31)$$

To simplify the multiples of Y computation, an alternative of the SRT-4 algorithm has been proposed with a quotient-digit set reduced to $\{-2, -1, 0, 1, 2\}$; so, the operations on Y are reduced to shifts only. Moreover, the range of the remainder can be restricted, typically $[-2.Y/3, 2.Y/3]$. The drawback of this method is the increased complexity of the quotient selection tables. Actually, more steps appear in the staircase borderlines of the $P-D$ plot; this means more bits are needed from both the divisor Y and the shifted remainder $4.r(i)$ to achieve a correct selection of q. The designer is faced with an increased hardware cost for look-up tables, to be balanced against some hardware savings for Y multiples generation. Finally, depending on the available design resources and options, the basic step complexity could slow down the overall process: multiples are computed only once while p cycles are needed to produce the full quotient.

Example 13.9 (Complete VHDL source code available.) Generate a generic n-bit base-4 SRT divider with 2's complement remainder. The division step of Figure 13.18 is:

```
entity srt_r4_step is
  port (
    r: in STD_LOGIC_VECTOR (N+2 downto 0);
    b: in STD_LOGIC_VECTOR (N-1 downto 0);
    Q: out STD_LOGIC_VECTOR (2 downto 0);
    r_n: out STD_LOGIC_VECTOR (N+2 downto 0)
  );
end srt_r4_step;

architecture behavior of srt_r4_step is
signal mult_m: std_logic_vector(N+1 downto 0);
signal Q_digit: std_logic_vector(2 downto 0);
begin
  selection: Qsel port map (rt=>r(N+2 downto N-1),
            d_2=>b(N-2), q=>Q_digit);
```

```
  Q<=Q_digit;
  mult_m<=b*Q_digit(1 downto 0);

  add_subtract: process(mult_m,Q_digit,r)
  begin
    if Q_digit(2)='1' then
      r_n<=r - mult_m;
    else r_n<=r+mult_m;
    end if;
  end process;
end behavior;
```

The divider structure is:

```
entity div_SRT_r4 is
  port (
    X: in STD_LOGIC_VECTOR (N-1 downto 0);
    Y: in STD_LOGIC_VECTOR (N-1 downto 0);
    Q: out STD_LOGIC_VECTOR (P-1 downto 0);
    R: out STD_LOGIC_VECTOR (N-1 downto 0)
  );
end div_SRT_r4;

architecture div_arch of div_SRT_r4 is

type connect is array (0 to P) of STD_LOGIC_VECTOR (N+2
downto 0); Signal wires: connect;
type Qmatrix is array (0 to P-2) of STD_LOGIC_VECTOR (2
downto 0); Signal Q_digit: Qmatrix;
Signal add,subst,the_adjust: STD_LOGIC_VECTOR (P-1 downto 0);
signal adjust: STD_LOGIC;
begin
  wires(0)<="0"&X&"00";
  divisor: for i in 0 to P-1 generate
  div_in: if i mod 2=0 generate
    in_step: srt_r4_step port map (r=>wires(i),
                  b=>Y, Q=>Q_digit(i), r_n=>wires(i+1));
      wires(i+2)<=wires(i+1)(N downto 0)&"00";--X 4
    end generate;
  end generate;
  adjust<=wires(P-1)(N+2);

  correction_step: process (adjust, wires(P))
  begin
    if adjust='0' then
      R<=wires(P-1)(N-1 downto 0);
    else R<=wires(P-1)(N-1 downto 0)+Y;
    end if;
  end process;
```

```
Quotient: process(Q_digit)
begin
  for i in 0 to P-1 loop
    if i mod 2=0 then
      if Q_digit(i)(2)='1' then--positive
        add(P-i-1 downto P-i-2)<=Q_digit(i)(1 downto 0);
        subst(P-i-1 downto P-i-2)<="00";
      else--negative
        add(P-i-1 downto P-i-2)<="00";
        subst(P-i-1 downto P-i-2)<=Q_digit(i)(1 downto 0);
      end if;
    end if;
  end loop;
end process;
  the_adjust(0)<=adjust; the_adjust(P-1 downto 1)<=(others=>
  '0');
  Q<=add-subst-the_adjust;
end div_arch;
```

13.2.5 Convergence Dividers

Two convergence dividers are presented in this section: the *Newton–Raphson* divider and the *Goldschmidt* divider. Basically, the complexity of the involved algorithms is proved to be better than the one of recurrence dividers (Chapter 6). Nevertheless, the overall performances will depend on the performance of the multiplication involved in each step. So the use of these dividers should be considered as part of a more complex system such as, for example, an arithmetic and logic unit.

13.2.5.1 Newton–Raphson Divider
The basic cell of the Newton–Raphson divider computes (formula (6.62))

$$x_{i+1} = x_i.(2 - d.x_i),$$

that is, two dependent multiplications and a subtraction. As quoted in Chapter 6, a B's complement operation may replace the subtraction. The basic cell and the general structure of an n-digit, precision p Newton–Raphson divider are shown in Figure 13.20, where it is assumed that $p \geq n$. The basic cell can iteratively use the same multiplier; this would require two clock cycles per step. Assuming $d \geq 0.1$, $t + 2$ digit LUT inputs ensure a t-bit precision (error $< B^{-t}$) in the first estimation. Then, the minimum precision p will be achieved after $k = \lceil \log_2 p/t \rceil$ steps. The final multiplication of the dividend by $x(k)$, the inverse of the divisor, is implicit in the general structure of Figure 13.20.

The cost and computation time of the Newton–Raphson basic cell, using a single multiplier within two clock cycles, are given by

$$C_{\text{basic cell}} = C_{\text{multiplier}}(p) + C_{\text{subtractor}}(p), \tag{13.32}$$

$$T_{\text{basic cell}} = 2.T_{\text{multiplier}}(p) + T_{\text{subtractor}}(p). \tag{13.33}$$

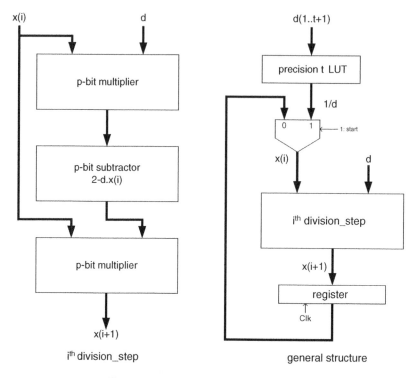

Figure 13.20   Newton–Raphson divider.

The cost and computation time of the Newton–Raphson divider as shown in Figure 13.20 are then

$$C(p) = C_{\text{LUT}}(2^{(t+2)} \times t) + C_{\text{mux}} + C_{\text{multiplier}}(p) + C_{\text{subtractor}}(p) + C_{\text{register}}(p),$$
(13.34)

$$T(k, p) = T_{\text{LUT}}(2^{(t+2)} \times t) + T_{\text{mux}} + (2.k + 1).T_{\text{multiplier}}(p) + k.T_{\text{subtractor}},$$
(13.35)

where (i) a k-step sequential implementation, (ii) a 2-cycle division step (one multiplier), and (iii) a shared (with basic cell) multiplier for the final multiplication have been assumed.

Example 13.10   (Complete VHDL source code available.) Generate a generic n-bit base-2 Newton–Raphson inverter. The division step of Figure 13.20 is:

```
entity newton_raphson_step is
generic(L: integer);
  port (
    r: in std_logic_vector (L downto 0);
```

```
     d: in std_logic_vector (N-1 downto 0);
     r_n: out std_logic_vector (2*L downto 0)
);
end newton_raphson_step;

architecture behavioral of newton_raphson_step is
signal r_x_d, r_x_d_neg: std_logic_vector (L+N downto 0);
signal r_n_long: std_logic_vector (2*L+N+1 downto 0);
begin
  r_x_d<=r*d;
  r_x_d_neg<=not(r_x_d)+1;
  r_n_long<=r*r_x_d_neg;
  r_n<=r_n_long(2*L+N downto N);
end behavioral;
```

The inverter structure is:

```
entity newton_raphson is
  port (
    X: in std_logic_vector (N-1 downto 0);
    Q: out std_logic_vector (P downto 0)
  );
end newton_raphson;

architecture behavioral of newton_raphson is
constant LOG_P: natural:=log_base_2(P);
constant LOG_F: natural:=log_base_2(F);
type rmd is array (0 to LOG_P-1) of std_logic_vector (P
downto 0);
signal r: rmd;
signal lut_val: std_logic_vector (F downto 0);

begin
  lut: LUT_Newton_Raphson port map (x=>X, l=>lut_val);
  r(LOG_F-1)(P downto P-F)<=lut_val;
  gen_p: for i in LOG_F to LOG_P-1 generate
    cell: newton_raphson_step generic map(2**i)
      port map (r=>r(i-1)(P downto P-2**i),
        d=>X, r_n=>r(i)(P downto P-2**(i+1)));
    end generate;
  Q<=r(LOG_P-1)(P downto 0);
end behavioral;
```

13.2.5.2 Goldschmidt Divider The basic cell of the Goldschmidt divider computes (formulas (6.72))

$$d(i) = d(i-1).(2 - d(i-1))$$

and

$$D(i) = D(i-1).(2 - d(i-1)).$$

$D(0)$ and $d(0)$ are initially set to the dividend and the divisor, respectively. The basic cell and the general structure of the Goldschmidt divider are presented in Figure 13.21. As stated in Chapter 6, a look-up table procedure can refine those initial values in order to speed up the convergence process. This alternative is not represented in the basic cell scheme. An important feature of the Goldschmidt basic cell is the independence of the multipliers; so the multiplications may be either performed in parallel or pipelined into a unique multiplier. The pipeline alternative is quite attractive for its lower cost in most design environments. A typical case is that of Xilinx FPGA, where flip-flops are readily available in every slice.

The cost and computation time of the Goldschmidt basic cell, using a single pipelined multiplier, are given by

$$C_{\text{basic cell}} = C_{\text{ppmultiplier}}(p) + C_{\text{subtractor}}(p), \tag{13.36}$$

$$T_{\text{basic cell}} = T_{\text{ppmultiplier}}(p) + T_{\text{subtractor}}(p). \tag{13.37}$$

The cost and computation time of the Goldschmidt divider as shown in Figure 13.21 are then

$$C(p) = 2.C_{\text{mux}} + C_{\text{ppmultiplier}}(p) + C_{\text{subtractor}}(p) + 2.C_{\text{register}}(p), \tag{13.38}$$

$$T(k, p) = T_{\text{mux}} + k.T_{\text{ppmultiplier}}(p) + k.T_{\text{subtractor}}. \tag{13.39}$$

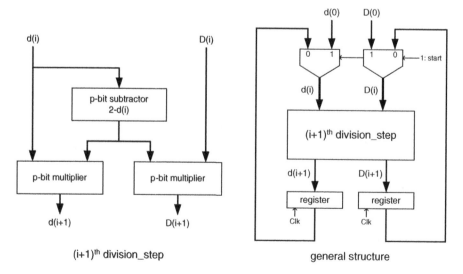

Figure 13.21 Goldschmidt divider: basic cell and general structure.

Example 13.11 (Complete VHDL source code available.) Generate a generic n-bit base-2 Goldschmidt divider. The division step of Figure 13.21 is:

```
entity goldschmidt_step is
generic(L: integer:=4);
port (
    r: in std_logic_vector (P-1 downto 0);
    d: in std_logic_vector (P-1 downto 0);
    r_n: out std_logic_vector (P-1 downto 0);
    d_n: out std_logic_vector (P-1 downto 0)
);
end goldschmidt_step;

architecture behavioral of goldschmidt_step is
signal d_neg: std_logic_vector (L downto 0);
signal d_neg_long: std_logic_vector (P downto 0);
signal r_n_long, d_n_long: std_logic_vector (P+L downto 0);
begin
  d_neg_long<=not('0'&d);
  d_neg<=d_neg_long(P downto P-L);
  r_n_long<=r*d_neg;
  r_n<=r_n_long(P+L-1 downto L);
  d_n_long<=d*d_neg;
  d_n<=d_n_long(P+L-1 downto L);
end behavioral;
```

The divider structure is:

```
entity goldschmidt is
port (
  X: in std_logic_vector (N-1 downto 0);
  Y: in std_logic_vector (N-1 downto 0);
  Q: out std_logic_vector (P-1 downto 0)
);
end goldschmidt;

architecture behavioral of goldschmidt is
type re is array (0 to LOGP+1) of std_logic_vector (P-1
downto 0);
signal r, d: re;
begin
  r(0)(P-1 downto P-N)<=X; r(0)(P-N-1 downto 0)<=
  (others=>'0');
  d(0)(P-1 downto P-N)<=Y; d(0)(P-N-1 downto 0)<=
  (others=>'0');
  gen_p: for i in 0 to LOG_P-1 generate
      cell: goldschmidt_step generic map(2**(i+1))
            port map (r(i),d(i),r(i+1), d(i+1));
```

```
    end generate;
    celln: goldschmidt_step generic map(P)
        port map (r(LOG_P), d(LOG_P), r(LOG_P+1), d(LOG_P+1));
    Q<=r(LOG_P+1)(P-1 downto 0);
end behavioral;
```

13.2.5.3 Comparative Data Between Newton–Raphson (NR) and Goldschmidt (G) Implementations From the algorithmic point of view the main facts are (Chapter 6):

- G algorithm converges toward the quotient while NR algorithm doesn't, it converges toward the inverse of the divisor and then multiplies by the dividend.
- G is not self-correcting, errors propagate; NR compensates errors.
- Multiplications are independent in G, not in NR.
- Neither algorithm can provide the exact remainder.
- Both algorithms essentially consist of two multiplications and one subtraction (or base complement).
- For both algorithms, convergence rate is quadratic.

At the implementation level the main consequences of the preceding are the following:

- An additional multiplication is needed by NR; this consumes time, increases cost or both but this can be negligible in the overall performance.
- The rounding operations are more critical for G.
- Multiplications can be made in parallel or pipelined in G; this can save time at a reasonable cost.

As a conclusion on convergence methods, one can state that those algorithms compete with each other. The technology at hand, the performance criteria, and a number of constraints are finally the key for a reasonable algorithm selection. Experimental designers feel that as well as sound theoretical options, smart designing techniques may significantly improve the overall quality of a particular algorithm implementation. Newton–Raphson dividers have been used, among others, by Intel and IBM ([INT1989], [MAR1990]); Goldschmidt dividers appear in some IBM and Advanced Micro Device processors ([AND1967], [OBE1999]).

13.3 BIBLIOGRAPHY

[AND1967] S. F. Anderson, J. G. Earle, R. E. Goldschmidt, and D. M. Powers, The IBM System/360 Model 91: Floating-point execution unit. *IBM J. Res. Dev.* **11**: 34–53 (1967).

[ERC1987] M. D. Ercegovac and T. Lang, On-the-fly conversion of redundant into conventional representations. *IEEE Trans. Comput.* **36**(7): 895–897 (1987).

[ERC1992] M. D. Ercegovac and T. Lang, On-the-fly rounding. *IEEE Trans. Comput.* **41**(12): 1497–1503 (1992).

[ERC1994] M. D. Ercegovac and T. Lang, *Division and Square-Root: Digit-Recurrence Algorithms and Implementations.* Kluwer Academic Publishers, New York, 1994.

[INT1989] *Intel i860 64-bit Microprocessor Programmer's Reference Manual*, 1989.

[MAR1990] P. W. Markstein, Computation of the elementary functions on the IBM RISC System/6000 Processor. *IBM J. Res. Dev.*, 111–119 (1990).

[MON1994] P. Montuschi, and L. Ciminiera, Over-redundant digit sets and the design of digit-by-digit division units. *IEEE Trans. Comput.*, **43**(3): 269–277 (1994).

[OBE1999] S. F. Oberman, Floating point division and square root algorithms and implementation in the AMD-K7 Microprocessor. *Proceedings of the 14th IEEE Symposium on Computer Arithmetic*, 106–115 (1999).

14

OTHER ARITHMETIC OPERATORS

This chapter is devoted to implementations of arithmetic functions reviewed in Chapter 7. As in the preceding chapters, several alternatives will be proposed for the algorithms previously described. As mentioned before, the ever-increasing availability of fast, low-cost memory blocks (ROM, RAM) motivates the development of affordable logical circuits as alternatives to micro-programmed implementations. On the other hand, reconfigurable devices such as the FPGA offer another approach for low-cost circuits implementing special functions normally not efficient within general-purpose arithmetic units. This chapter presents combinational and sequential circuits, FPGA implementation, and VHDL models for most of the algorithms developed in Chapter 7.

14.1 BASE CONVERSION

14.1.1 General Base Conversion

The general conversion algorithm described in Algorithm 7.1 for natural numbers has a mainly theoretical interest. In the context of general-purpose binary computers, a general circuit to convert n-digit base-B_1 numbers into m-digit base-B_2 ones doesn't seem to warrant practical interest. A block diagram is nevertheless presented in Figure 14.1 to illustrate a possible implementation of such a circuit, assuming a binary coding for the digits in both bases.

Synthesis of Arithmetic Circuits: FPGA, ASIC, and Embedded Systems
By Jean-Pierre Deschamps, Géry J. A. Bioul, and Gustavo D. Sutter
Copyright © 2006 John Wiley & Sons, Inc.

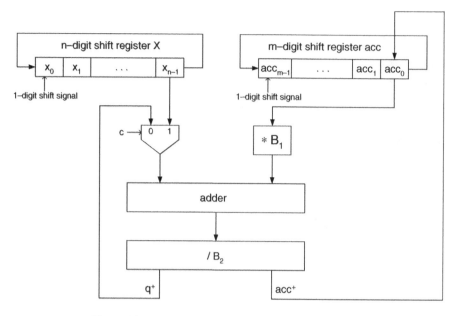

Figure 14.1   Block diagram of a general base converter.

The multiplier by B_1, the adder stage, and the divider by B_2 are binary operators defined as follows:

- The multiplier-by-B_1 has a $(1 + \lfloor \log_2 (B_2 - 1) \rfloor)$-bit input and a $(2 + \lfloor \log_2 (B_2 - 1) \rfloor + \lfloor \log_2 B_1 \rfloor)$-bit output.
- The adder stage inputs are the output of the multiplier, and a $(1 + \lfloor \log_2 (B_1 - 1) \rfloor)$-bit digit; the output is a $(1 + \lfloor \log_2 (B_2 . B_1 - 1) \rfloor)$-bit digit.
- The divider-by-B_2 input is the output of the adder while the outputs, namely, the quotient q^+ and the remainder acc^+, are, respectively, $(1 + \lfloor \log_2 (B_1 - 1) \rfloor)$-bit and $(1 + \lfloor \log_2 (B_2 - 1) \rfloor)$-bit digits.

At the start, the shift register X of Figure 14.1 is loaded with the n-digit base-B_1 number to be converted, while the result will be stored in the m-digit shift register acc initially set to zero. Control signals are then generated as follows, according to Algorithm 7.1:

- at the first step ($i = 0; j = 0$) the multiplexer control signal c is set to 1, then ($j \geq 1$) c is reset to 0 until i is incremented (j back to zero);
- each time j is incremented, a shift signal generates a circular 1-digit right-shift of register acc;

- each time j is reset to zero (from $m - 1$), i is incremented and shift signals generate circular 1-digit right-shifts of both registers X and acc; the multiplexer control signal c is set to 1, then reset to zero as soon as $j = 1$.

The cost $C(n, m)$ and computation time $T(n, m)$ of the implementation suggested by Figure 14.1 are

$$
\begin{aligned}
C(n, m) = {} & C_{\text{register}}(n, B_1) + C_{\text{register}}(m, B_2) + C_{\text{mux2}} \\
& (1 + \lfloor \log_2 (B_1 - 1) \rfloor) + C_{\text{multB1}} + C_{\text{adder}} + C_{\text{divB2}}, \\
T(n, m) = {} & n.m.(T_{\text{shift}} + T_{\text{multB1}} + T_{\text{adder}} + T_{\text{divB2}}),
\end{aligned} \tag{14.1}
$$

where m is such that

$$
B_1^n \le B_2^m; \tag{14.2}
$$

m is thus computed as

$$
m = \lceil n. \log_{B2} B_1 \rceil. \tag{14.3}
$$

14.1.2 BCD to Binary Converter

14.1.2.1 Nonrestoring 2^p Subtracting Implementation   To make sense, this implementation assumes that the successive powers of 2 are expressed in BCD and that a BCD subtractor is available. Figure 14.2 presents a block diagram of a possible implementation. The BCD-coded 2^p ($p = 0, 1, \ldots, n - 1$) may be read from a look-up table. The allowed range for a given BCD number X is

$$
X \le 2^n - 1, \tag{14.4}
$$

that is,

$$
n > \log_2 X.
$$

Figure 14.2 presents a combinational (part (a)) and a sequential (part (b)) implementation circuit. In both instances, the successive BCD powers of 2 have to be generated. The sequential circuit requires a multiplexer to initialize the process by loading the data X_{BCD}. The complemented successive sign-bits are the desired binary components of X.

The cost $C(n)$ and computation time $T(n)$ of the combinational implementation suggested by Figure 14.2a are (excluding registers)

$$
\begin{aligned}
C(n) &= n.C_{\text{adder-subtractorBCD}}, \\
T(n) &= n.T_{\text{adder-subtractorBCD}}.
\end{aligned} \tag{14.5}
$$

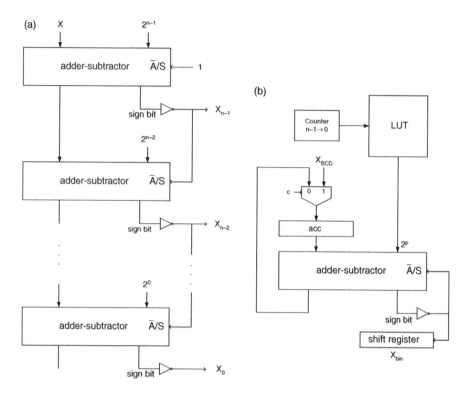

Figure 14.2   Nonrestoring BCD to binary converter.

For the sequential implementation, the computation time behavior is roughly the same while the cost $C(n)$ is reduced to

$$C(n) = C_{\text{adder-subtractor BCD}} + C_{\text{mux2}}(m) + C_{\text{LUT}}. \qquad (14.6)$$

14.1.2.2   Shift-and-Add BCD to Binary Converter   According to the first step of Algorithm 7.2, the leftmost BCD digit is multiplied by ten and added to the next right neighbor digit. The next iteration steps then consist of multiplying the successive results by ten and adding the next right neighbor digit. Multiplication by ten (binary coded 1010) is handled by a double shift as illustrated in the following.

$$X(i).1010 = (X(i).100 + X(i)).10, \qquad (14.7)$$

where $X(i)$ is a 4-bit binary number (BCD code). The current step i is then completed by adding $X(i - 1)$ to the result.

A customized multiplier by ten can be designed or, alternatively, operation (14.7) can be handled by a 2-bit shift, a sum, and a 1-bit shift. Figure 14.3 depicts a 2-adder

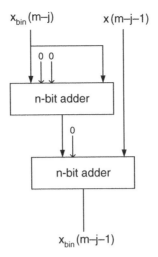

Figure 14.3 Shift-and-add BCD to binary converter: elementary step.

circuit to carry out the elementary step:

$$X_{\text{bin}}(m - j - 1) = (X_{\text{bin}}(m - j).100 + X_{\text{bin}}(m - j)).10 + X(m - j - 1). \quad (14.8)$$

Figure 14.4 presents a combinational (part (a)) and a sequential (part (b)) implementation circuit. The sequential circuit requires a multiplexer to initialize the process by loading the data $X_{\text{BCD}}(m - 1)$. The cost $C(m)$ and computation time $T(m)$ of the combinational implementation of Figure 14.4a are

$$C(m) = 2.(m - 1).C_{\text{adder}}(n),$$
$$T(m) = 2.(m - 1).T_{\text{adder}}(n). \quad (14.9)$$

For the sequential implementation the computation time behavior is roughly the same while the cost $C(n)$ is reduced to

$$C(n) = 2.C_{\text{adder}}(n) + C_{\text{mux2}}(n) + C_{\text{BCDreg}}(m - 1) + C_{\text{acc}}(n), \quad (14.10)$$

Formula (14.3) for $B_2 = 10$ and $B_1 = 2$ gives

$$m = \lceil n.\log_{10} 2 \rceil, \quad \text{that is, } m/n \approx 0.3. \quad (14.11)$$

Thus, comparing formulas (14.9) to (14.5), we find better cost and computation time behaviors for the shift-and-add implementation. Moreover, the binary adder is simpler than the BCD adder-subtractor needed in the nonrestoring implementation of Section 14.1.2.1.

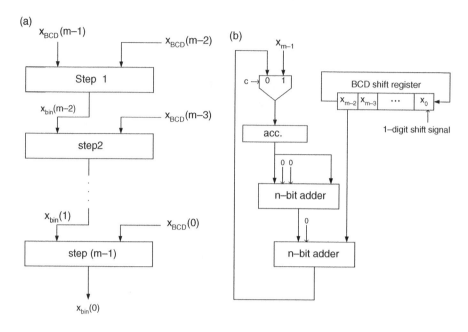

Figure 14.4   Shift-and-add BCD to binary converter.

14.1.3   Binary to BCD Converter

The binary-to-BCD conversion procedure described by Algorithm 7.3 basically consists in doubling the BCD partial result $bcd(i)$ and adding the next bit from the binary expression to be converted. Two basic procedures have to be defined:

1. Add a 1-bit number to a BCD number.
2. Multiplication by two of a BCD number.

Since the binary digit x also stands for the BCD number $000x$, the first procedure is just a straight BCD sum. The multiplication by two is less straightforward for BCD numbers than for binary numbers (shift). The procedure `BCDx2_step`, described in Chapter 7, may be carried out in parallel on each BCD digit. Assuming that each digit is 1-bit left shifted, a carry is computed and set to 0 whenever the shifted digit is not greater than 9; otherwise the carry is set to 1 and a correction of $(0110)_2$ is added modulo 16 to the shifted BCD digit. The computed carry can be used to feed the correction input of the mod 16 adder. Since the next left neighbor digit has also been 1-bit shifted, the carry will stand at the rightmost position of the next digit without generating carry propagation. The Boolean expression (Chapter 7)

$$y(i+1)_0 = x(i)_3 \vee (x(i)_2 . (x(i)_1 \vee x(i)_0)) \tag{14.12}$$

generates a carry $y(i+1)_0 = 0$ whenever the BCD digit $X(i)$ is not greater than 4, that is,

$$(x(i)_3 x(i)_2 x(i)_1 x(i)_0) \leq (0100),$$

$y(i+1)_0 = 1$ otherwise. Figure 14.5a shows the carry circuit. Figure 14.5b displays the BCDx2_step circuit.

The circuits of Figure 14.6 implement the full conversion process described in Algorithm 7.3. Observe that the first three steps are trivial, so one can initialize the computation scheme with the partial result $bcd(n-3) = (0\ x_{n-1}\ x_{n-2}\ x_{n-3})$; the next partial results $bcd(n-j)$ are then iteratively multiplied by two and added to $(0\ 0\ 0\ x_{n-j-1})$. The cost $C(n)$ and computation time $T(n)$ of the combinational

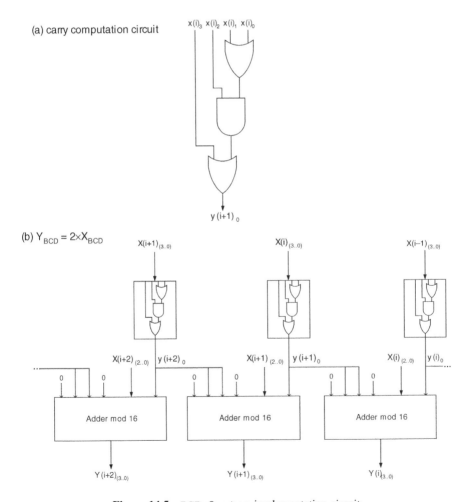

Figure 14.5 BCDx2_step implementation circuit.

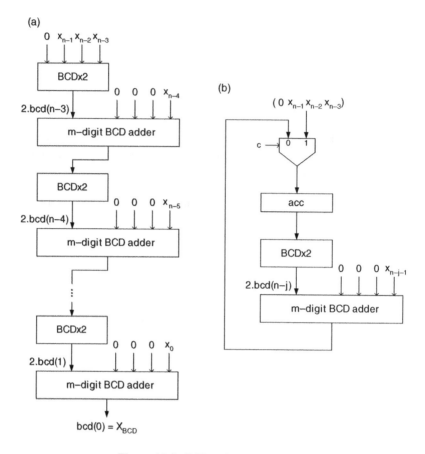

Figure 14.6   BCD to binary converter.

implementation of Figure 14.6a are

$$C(n) = (n-3).(C_{\text{BCD}x2}(m) + C_{\text{adderBCD}}(m)),$$
$$T(n) = (n-3).(T_{\text{BCD}x2}(m) + T_{\text{adderBCD}}(m)). \qquad (14.13)$$

For the sequential implementation (Figure 14.6b) the computation time behavior is roughly the same while the cost $C(n)$ is reduced to

$$C(n) = C_{\text{BCD}x2}(m) + C_{\text{adderBCD}}(m) + C_{\text{mux2}}(m) + C_{\text{acc}}(m). \qquad (14.14)$$

The relation (14.11) between m and n still holds.

14.1.4 Base-*B* to RNS Converter

Let

$$X = x_{n-1}, x_{n-2}, \ldots, x_1, x_0, \quad x_i \in \{0, 1, \ldots, B - 1\}, \tag{14.15}$$

be an *n*-digit *B*-ary number to be converted in a RNS system defined by the non redundant set of moduli $\{m_j\}, j = 1, 2, \ldots, s$. One assumes that

$$B^n - 1 \leq M = \Pi_{1 \leq j \leq s} m_j, \tag{14.16}$$

otherwise the conversion would be modulo *M*.

The most straightforward circuit to convert base-*B* to RNS consists of a set of dividers by the respective moduli m_j (Figure 14.7). The remainders R_j are the desired RNS components: results of the successive reductions modulo m_j. The design problem is thus that of the synthesis of integer dividers by m_j. Optimization techniques generally start from the selection of the set of moduli. As quoted in Chapters 8 and 15, specific moduli can lead to better algorithms and higher performance implementations.

The converter circuit presented in Figure 14.8 implements Algorithm 7.4. It is assumed that the x_i and m_j are binary coded, as well as the outputs: RNS components r_j, and R_j. The circuit first computes the RNS expressions of $x_i.B^i$ ($i = 0, 1, \ldots, n - 1$) then adds them. This operation corresponds to *s* multioperand additions mod m_j ($j = 0, 1, \ldots, s$).

The *n.s* values $r_j(i)$ of ($x_i.B^i$) mod m_j are read from *n* look-up tables LUT_i such as those displayed in Figure 7.1 to illustrate Example 7.5. Since the factor B^i is implicit, only $\lceil \log_2(B) \rceil$ binary inputs are needed to address any LUT_i and $\Sigma_j(\lceil \log_2(m_j) \rceil)$ binary outputs are required to code the RNS expression of ($x_i.B^i$). The size C_{LUT}

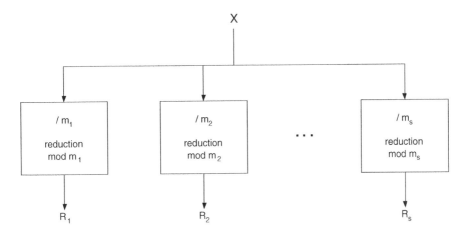

Figure 14.7 Base-*B* to RNS converter.

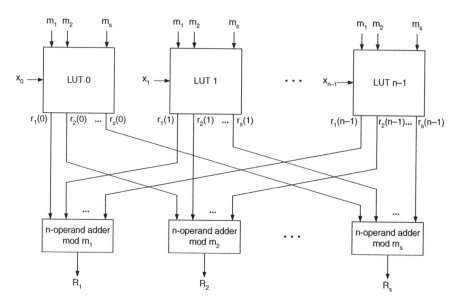

Figure 14.8 Base-*B* to RNS converter using LUTs.

of *LUT$_i$*, expressed as a number of binary entries, is given by

$$C_{\text{LUT}} = B.\Sigma_{1 \le j \le s} \log_2 m_j, \qquad \forall i. \tag{14.17}$$

The *n*-operand adder mod m_j can be designed as a tree of 2-operand adders. From the cost and computation time point of view, the choice of the modulus set is critical. The cost $C(n, s)$ and computation time $T(n, s)$ of the circuit presented in Figure 14.8 are

$$C(n, s) = n.C_{\text{LUT}} + \Sigma_j C_{n\text{-operandadder}}(m_j),$$
$$T(n, s) = T_{\text{LUT}} + \max_j (T_{n\text{-operandadder}}(m_j)). \tag{14.18}$$

14.1.5   CRT RNS to Base-*B* Converter

The implementation proposed in this section is based on the Chinese remainder theorem (Algorithm 7.6); it is assumed that $\{m_i^*\}$ and $\{(1/m_i^*) \bmod m_i\}$ are available from look-up tables. The circuit presented in Figure 14.9 is made up of *s* multipliers mod *M*, *s* multipliers mod m_j ($j = 1, 2, \ldots, s$), and 1 multioperand adder mod *M*.

If binary coding is assumed, $\log_2 m_i$ bits will be required for the inputs and outputs of the respective multipliers mod m_j, and $\log_2 M$ bits for the multipliers mod *M* and the multioperand adder mod *M*. The cost $C(M, s)$ and computation

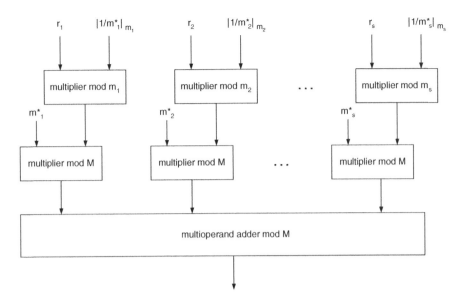

Figure 14.9 *CRT* RNS to base-*B* converter.

time $T(M, s)$ of the circuit presented in Figure 14.9 are

$$C(M, s) = 2.C_{\text{LUT}} + \Sigma_j C_{\text{mult}}(m_j) + s.C_{\text{mult}}(M) + C_{s\text{-operandadder}}(M),$$
$$T(M, s) = T_{\text{LUT}} + \max_j (T_{\text{mult}}(m_j)) + T_{\text{mult}}(M) + T_{s\text{-operandadder}}(M). \quad (14.19)$$

The size of the look-up tables is bounded by $s.\log_2 m$ bits (m is the greatest modulus).

Given the range M, a set of pairwise prime moduli $\{m_i\}$ may be selected as minimal according to some particular criterion ([GAR1959], [SZA1967]). So if a strategy is defined to select the set of moduli from a given range M, the cost and time complexity can actually be computed as a function of M only.

If a parameterized circuit can be synthesized to achieve mod m_i multiplication, a low-cost iterative sequential circuit can be designed (Figure 14.10).

The cost $C(M, s)$ and computation time $T(M, s)$ of the circuit presented in Figure 14.10 are roughly

$$C(M, s) = 2.C_{\text{LUT}}(m) + C_{\text{mult}}(m) + C_{\text{mult}}(M) + C_{2\text{-operandadder}}(M) + C_{\text{acc}}(M),$$
$$T(M, s) = s.(T_{\text{LUT}}(m) + T_{\text{mult}}(m) + T_{\text{mult}}(M) + T_{2\text{-operandadder}}(M)), \quad (14.20)$$

where m stands for the greatest modulus of the set $\{m_j\}$. It is assumed that

$$T_{\text{LUT}}(m_j) \cong T_{\text{LUT}}(m); \quad C_{\text{LUT}}(m_j) = C_{\text{LUT}}(m), \quad \forall j,$$
$$T_{\text{mult}}(m_j) \cong T_{\text{mult}}(m); \quad C_{\text{mult}}(m_j) = C_{\text{mult}}(m), \quad \forall j. \quad (14.21)$$

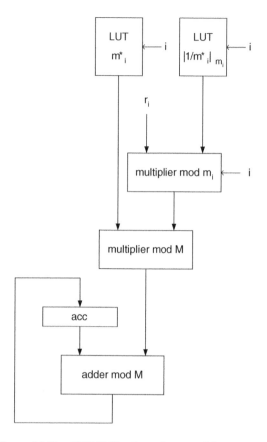

Figure 14.10   CRT RNS to base-B sequential converter.

The preceding examples emphasize the fact that RNS converter design mainly consists of finite field operator design (Chapters 8 and 15). Nevertheless, this result may be seen as a part of the more general problem coming out of the complete design of a RNS system; this involves the modulus set selection, the RNS arithmetic unit design, and the coding–decoding process implementation. Typically, modulus selection plays a prominent role in the complexity of RNS arithmetic operators, including coding–decoding units. Special moduli such as Mersenne numbers, $2^n - 1$ or more generally $B^n - c$, with $c \ll B^n$ (Chapter 8), are often considered; as quoted before carefully selected moduli can lead to specific algorithms for which implementations can be derived with better time/cost characteristics.

14.1.6   RNS to Mixed-Radix System Converter

Since the conversion from mixed-radix system to base-B is straightforward, the following implementation can be set to achieve the RNS to base-B conversion.

According to Chapters 3 and 7, let

$$X = B_{n-1}.x_{n-1} + \cdots + B_1.x_1 + B_0.x_0, \quad 0 \le x_i \le b_i - 1, \tag{14.22}$$

be a mixed-radix expression in a system defined by the set of radices $\{b_0, b_1, \ldots, b_{n-1}\}$ setting down the set of weights $\{B_0, B_1, \ldots, B_{n-1}\}$ such that $B_0 = 1$ and

$$B_i = \Pi_{0 \le j \le i-1} b_j, \quad i = 1, 2, \ldots, n - 1. \tag{14.23}$$

The mixed-radix `digit_extraction` algorithm 7.7 achieves the conversion from a source RNS system with a set of moduli $\{m_i\}$ to a target mixed-radix system with the same set $\{m_i\}$ as radix set. The m_i do not need to be ordered by size or whatever; that same order must be respected for both source and target systems. Nevertheless, ordering the residues r_i from left to right, by decreasing the size of the m_i, makes trivial the iterative RNS coding of the successive residues with respect to the left side moduli, as required by substep 2 of each iteration step. In the iterative circuit, presented in Figure 14.11, the initial RNS expression is denoted $R_1(s..1)$, assuming $m_i > m_j, \forall i, j$ such that $i > j$. The first mixed-radix digit x_0 (substep 1) is readily extracted as $R_1(1)$. The next stage (substep 2) achieves the RNS subtraction $R_1(s..1) - R_1(1)$; as the m_i are ordered by increasing size from right to left, $R_1(1) < m_{i>1}, \forall i$, the RNS components of $R_1(1)$ are repeatedly $R_1(1)$. The third stage (substep 3) divides its RNS input $R_2(s..2)$ by m_1: this operation is achieved through componentwise multiplication mod $m_{i>1}$ of $R_2(i)$ by $(1/m_1)$ mod m_i. From the result $R_2^*(s..2)$, the next mixed-radix digit x_1 is extracted as $R_2^*(2)$. Step k starts with input $R_k^*(s..k)$; x_{k-1} is first extracted as $R_k^*(k)$ then (substep 2) the subtractor computes $R_{k+1}(s..k + 1) = R_k^*(s..k) - R_k^*(k)$; the third substep finally multiplies $R_{k+1}(s..k + 1)$ by $1/m_k$ through componentwise multiplication mod $m_{i>k}$ of $R_{k+1}(i)$ by $(1/m_k)$ mod m_i.

Example 14.1   This example illustrates a converter for a RNS source system with the following set of 6 (s) moduli (ordered by size) defining a range of 23881935:

$$m_6 = 31, m_5 = 29, m_4 = 23, m_3 = 15, m_2 = 11, m_1 = 7 \tag{14.24}$$

The LUT design is first tackled by computing the inverses of m_i modulo $m_{j>i}$. Table 14.1 displays the contents of the LUTs related to the respective $(1/m_i)$.

According to Figure 14.11, the specific circuit can be built with 5 $(s - 1)$ sets (LUT + subtractor + multiplier) as shown in Figure 14.12, where a numeric example is worked out. The source RNS expression is given as $R_1 = (11, 10, 9, 8, 7, 6)$ for which the target mixed-radix expression is computed as $X = (12, 22, 3, 3, 8, 6)$.

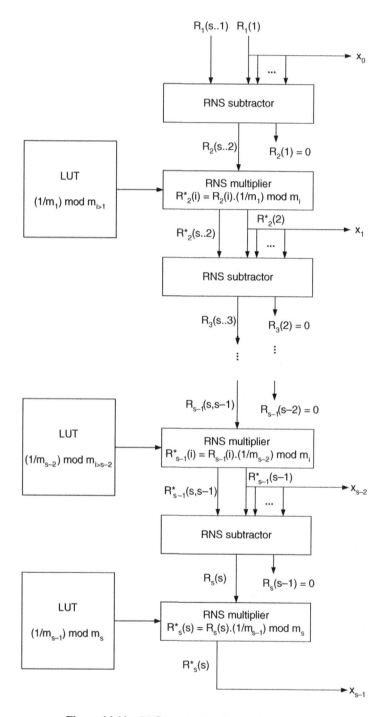

Figure 14.11   RNS to mixed-radix system converter.

TABLE 14.1 Inverse $(1/m_i)$ mod $m_{j>i}$

mod ↓	1/7	1/11	1/15	1/23	1/29
11	8				
15	13	11			
23	10	21	20		
29	25	8	2	24	
31	9	17	29	27	15

The mixed-radix system weights, computed from (14.24), are

$$B_0 = 1, B_1 = 7, B_2 = 77, B_3 = 1155, B_4 = 26565, B_5 = 770385,$$

from which the decimal value of X can be computed as 9832808.

If standard LUT and RNS operators are used, the cost $C(s)$ and computation time $T(s)$ of the circuit presented in Figure 14.11 are

$$C(s) = (s - 1).(C_{\text{LUT}} + C_{\text{mult}} + C_{\text{subt}}),$$
$$T(s) = (s - 1).(T_{\text{mult}} + T_{\text{subt}}); \tag{14.25}$$

otherwise, the sizes of the operators may decrease step by step as appears in figures 14.11 and 14.12. Observe that the sizes of the moduli are also involved in the overall complexity. The sequential implementation of the circuit depicted in Figure 14.11 is given in figure 14.13. In that case, the maximum size is required for the subtractor and multiplier units. Three multiplexers are used to initialize the process with the source RNS expression $R_1(s..1)$, and to extract the first mixed-radix digit x_0. As the size (number of residues) of $R_i^*(s..i)$ decreases step by step after each subtraction, a special mod shift operator is needed to adjust the result $R_{i+1}(s..i + 1)$ before the multiplier stage. In the same way, a connecting box sets the connections $R_i^*(i)$ to the subtractor. The cost $C(s, m)$ and computation time $T(s, m)$ of the circuit presented in Figure 14.13 are

$$C(s) = C_{\text{LUT}} + 2.C_{\text{mux}}(s, m) + C_{\text{mux}}(m) + C_{\text{mult}}(s, m)$$
$$+ C_{\text{subt}}(s, m) + C_{\text{conn}}(s, m) + C_{m.\text{shift}}(s, m) + C_{\text{acc}}(s, m) + C_{\text{reg}}(s, m),$$
$$T(s) = (s - 1).(T_{\text{mult}}(s, m) + T_{\text{subt}}(s, m) + T_{\text{mux}}(s, m) + T_{\text{conn}} + T_{\text{shift}}), \tag{14.26}$$

where m stands for the greatest modulus m_s.

For practical implementations, binary coding is used for residues and moduli; so standard digital cells may be used to synthesize all the elements of Figures 14.11 to 14.13. Observe that the circuits presented are conceptually independent of the coding selected; the available technology will lead the choice.

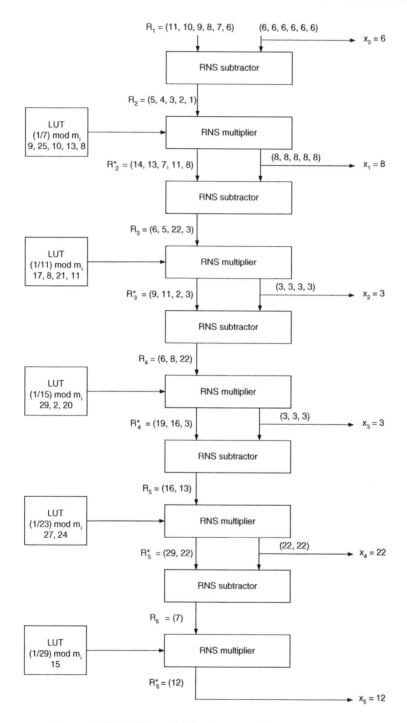

Figure 14.12   RNS to mixed-radix conversion (Example 14.1).

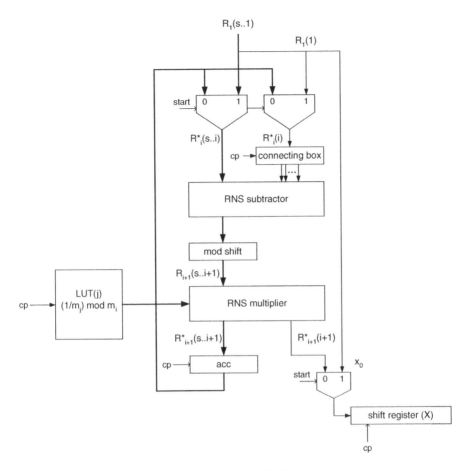

Figure 14.13   RNS to mixed-radix converter.

14.2 POLYNOMIAL COMPUTATION CIRCUITS

Polynomial approximation methods are often used to calculate special functions such as logarithmic, exponential, of trigonometric (Chapter 7). In Section 7.3.2 a recursive multilevel computation scheme was proposed as a generalization of the Hörner expansion technique to compute polynomials: the generalized Hörner expansion (GHE). Using formulas (7.30)–(7.32), the example suggested in Section 7.3.2 is implemented in this section. The example consists of a 3-level GHE implementing a degree-63 polynomial to be computed in 9 multiply-and-add steps. First, 16 degree-3 polynomials can be computed (3 steps); then four degree-15 polynomials are worked out using the degree-3 polynomials as primitives (3 steps); another 3 steps are finally needed to compute the degree-63 polynomial using the degree-16 ones as primitives. The respective cells to be implemented correspond to the following polynomials:

First stage cells (Figure 14.14a)

$$C_i^4(x) = c_{4i+3}.x^3 + c_{4i+2}.x^2 + c_{4i+1}.x + c_{4i}$$
$$= ((c_{4i+3}.x + c_{4i+2}).x + c_{4i+1}).x + c_{4i}, \quad i = 0, 1, \ldots, 15. \qquad (14.27)$$

Second stage cells (Figure 14.15a)

$$C_j^{16}(x) = ((C_{4j+3}^4(x).x^4 + C_{4j+2}^4(x)).x^4$$
$$+ C_{4j+1}^4(x)).x^4 + C_{4j}^4(x), \quad j = 0, 1, 2, 3. \qquad (14.28)$$

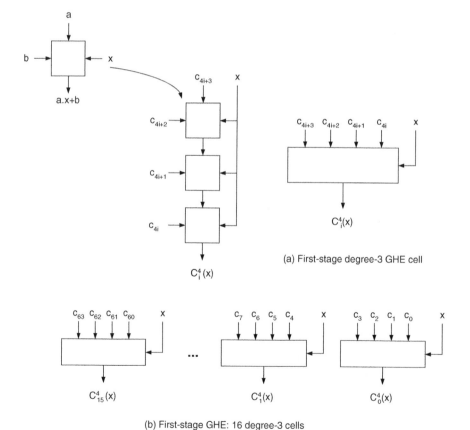

(a) First-stage degree-3 GHE cell

(b) First-stage GHE: 16 degree-3 cells

Figure 14.14   GHE degree-63 polynomial: first stage.

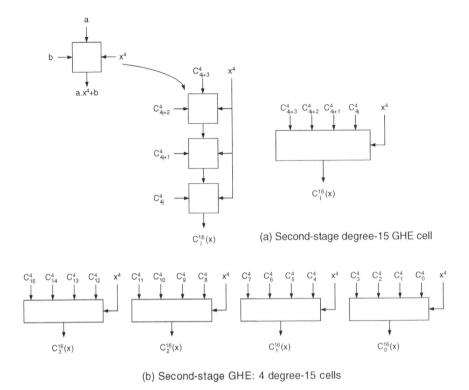

(a) Second-stage degree-15 GHE cell

(b) Second-stage GHE: 4 degree-15 cells

Figure 14.15   GHE degree-63 polynomial: second stage.

Third stage cell (Figure 14.16a)

$$C^{64}(x) = ((C_3^{16}(x).x^{16} + C_2^{16}(x)).x^{16} + C_1^{16}(x)).x^{16} + C_0^{16}(x). \qquad (14.29)$$

The 16 first-stage cells are represented in figure 14.14b; the 4 second-stage cells are represented in Figure 14.15b, and the full 3-stage circuit is shown in Figure 14.16b.

Each cell of Figure 14.16 implements recursively the function $a.x^k + b$: three times in this example. Observe that the sizes of the operands increase with the stage level. The practical implementation of the basic cells may depart from the direct application of the Hörner scheme. Nevertheless, the synthesis problem may become quickly unmanageable, for example, dealing with integer numbers with a significant precision; observe that the inputs of the output cell of Figure 14.16 are made up of four degree-15 polynomials and one power of x (x^{16}). Actually, as far as a sufficient precision is desired for the intermediate polynomial results, the number of binary variables to handle may become prohibitive for hardware implementation (LUT or circuits). Firmware approaches can be

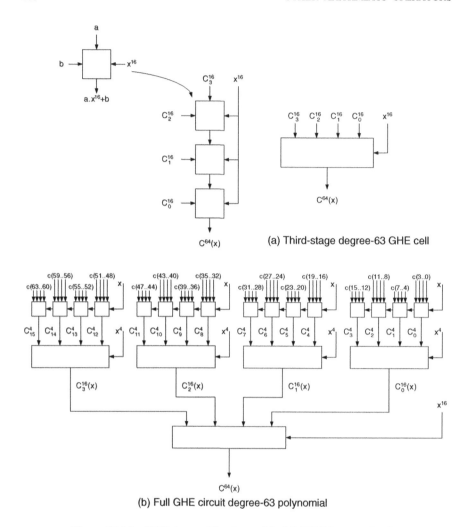

(a) Third-stage degree-63 GHE cell

(b) Full GHE circuit degree-63 polynomial

Figure 14.16   GHE degree-63 polynomial: full GHE 3-stage circuit.

suitable alternatives. Special applications of particular finite fields (e.g., GF(2)) look more realistic for a full hardware implementation.

The overall circuit complexity will depend on the cell cost and computation time. According to the degree of the polynomial to synthesize, several cell sizes can be foreseen. Whenever this size is selected, the value k (power of x) is set for the next stage level. Cell sizes and number of stages are optimization parameters to be considered by the designer according to the time/cost constraints. As an alternative to the synthesis of degree-63 polynomials, one could have considered a 6-level tree using cells $C^2(x)$, $C^4(x)$, $C^8(x)$, ..., $C^{64}(x)$.

The cost and computation time of an s-level tree using cells $C^t(x)$ $(t = r^{s-i};$ $0 \geq i \geq s - 1)$ to synthesize polynomials of degree $n = r^s - 1$, are given by

$$C(n) = \Sigma_{0 \leq i \leq s-1} r^i . C(r^{s-i}),$$

$$T(n) = \Sigma_{0 \leq i \leq s-1} T(r^{s-i}), \tag{14.30}$$

where $C(r^{s-i})$ and $T(r^{s-i})$ stand for the cost and computation time of computing cell $C^t(x)$, $t = r^{s-i}$.

14.3 LOGARITHM OPERATOR

This section presents an implementation for binary logarithms computation using multiplicative normalization. As shown in Section 7.3.3.1, the main (logarithm) sequence may be computed in another base different from the auxiliary sequence, built up in binary. The implementation, displayed in Figure 14.17, handles all data and results in binary. Algorithm 7.8 (logarithm computation by multiplicative normalization) assumes that the numerical values of $(1 + 2^{-i})$, $(1 - 2^{-i})$, $\ln (1 + 2^{-i})$, and $\ln (1 - 2^{-i})$ are available. In practical implementations, those values are read out from a look-up table to be preset by the designer. As the precision of the result is linear (1 bit-result per step), then, for p-bit precision, $2p$ logarithms $\ln (1 \pm 2^{-i})$ and $2p$ values (1 ± 2^{-i}) have to be precomputed and loaded. On the other hand, the precision of the stored values has to be defined too: if p is the required precision for the result, at least p bits are needed per LUT entry. Actually, to cope with the errors generated by rounding and error propagation, some more bits have to be included. Nevertheless, $4p^2$ is a fair order of magnitude of the LUT cost. The argument X is in $[1/2, 2]$.

A counter, not represented in the circuit displayed in figure 14.17, may be used to increment step number i. LUT outputs are thus updated while, at the same time, a combinational circuit computes $x_{-i}(i)$ and $x_{-i}(i).not\ x_{-i-1}(i)$. During the second phase of step i, the registers X and acc are loaded with $X(i + 1)$ and $acc(i + 1)$, respectively. The final result is stored in the register acc after step p:

$$\ln x = acc(p).$$

Observe that the stop condition test $(x(i) = 1\ ?)$, which is optional, is not represented in Figure 14.17.

The cost and computation time are given by

$$C(p) = 4.C_{\text{LUT}}(p^2) + 2.C_{\text{mux4}}(p) + C_{\text{mux2}}$$
$$+ C_{\text{multiplier}}(p) + C_{\text{subtractor}}(p)$$
$$+ C_{\text{comb.circ.}}(p) + C_{\text{acc}}(p) + C_{\text{reg}}(p),$$

$$T(p) = p.(\max (T_{\text{LUT}}(p^2), (T_{\text{comb.circ.}}(p) + T_{\text{mux2}}))$$
$$+ T_{\text{mux4}}(p) + T_{\text{multiplier}}(p)), \tag{14.31}$$

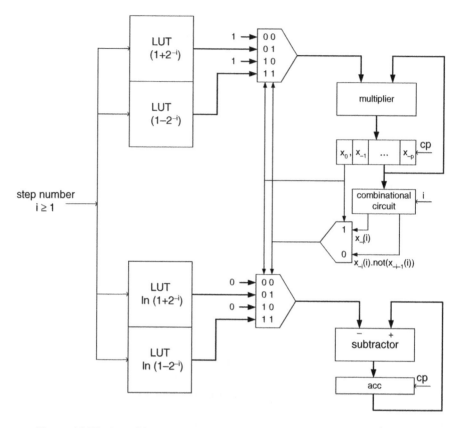

Figure 14.17 Logarithm computation circuit using multiplicative normalization.

where the delay $T_{\text{multiplier}}(p)$ of the multiplier is assumed greater than the one of the subtractor.

14.4 EXPONENTIAL OPERATOR

This section presents an implementation for binary exponential computation using additive normalization, as described in Section 7.3.3.2. The implementation of Algorithm 7.9, presented in Figure 14.18, is somewhat similar to the preceding one. It also handles binary coded data and results; the same look-up tables are required to read out the numerical values of $(1 + 2^{-i})$, $(1 - 2^{-i})$, $\ln (1 + 2^{-i})$, and $\ln (1 - 2^{-i})$. Nevertheless, the argument X is now in $[-1, 1[$, so to implement the auxiliary sequence computations, a signed-number subtractor is needed. As the main sequence (computing e^x) starts with 1 and always multiplies by positive numbers (1 ± 2^{-i}), the multiplier device can be simpler, dealing with natural numbers only. Figure 14.18 assumes a 2's complement coding for the argument X, so the first bit of $x(i)$ may be used to control the multiplexers selecting the LUT outputs.

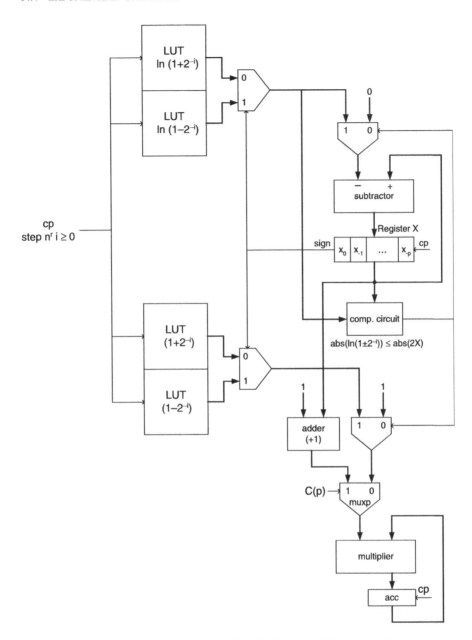

Figure 14.18 Exponential computation circuit using additive normalization.

The X register requires a length corresponding to the desired precision, plus some additional bits (not represented) to cope with rounding errors. As quoted in Chapter 7, the sign of $x(i)$ can be used to select the candidate values to subtract (auxiliary sequence) and to multiply (main sequence) but a further comparison is needed to proceed. This comparison ensures that $x(i)$ will either decrease or stay unchanged after each step. A specific comparison circuit (comp. circuit) can be designed to compare the absolute values of $2.x(i)$ and $\ln c(i)$; $c(i) = 1 + 2^{-i}$ if $x_0(i) \geq 0$, $c(i) = 1 - 2^{-i}$ otherwise. If $\ln c(i)$ is smaller or equal to $2.x(i)$, then $\ln c(i)$ is subtracted from $x(i)$ while $y(i)$ is multiplied by $c(i)$; otherwise the operations are neutralized: subtracting 0 from $x(i)$ and multiplying $y(i)$ by 1. For that purpose, the output of comp. circuit is used to control the multiplexers located at the respective inputs of the subtractor and multiplier. According to the assumptions of Example 7.10 (Section 7.3.3.2), $x(0) = X$ and $y(0) = 1$, the steps are numbered from 0 to $p - 1$, while $x(i + 1)$ and $y(i + 1)$ are computed at step i. After step $i = p - 1$, a final multiplication of $y(p)$ by $(1 + x(p))$ doubles the precision. This operation, appropriately timed, is materialized in figure 14.18 by a signed adder connected to the output of register X. The output of this adder is connected to the multiplier through a multiplexer (muxp) controlled by $C(p)$: $C(p) = 1$ at step p.

As in the preceding implementation circuit, a counter (not represented) may be used to generate step number i. During the second phase of clock pulse i, registers X and acc are loaded with $x(i + 1)$ and $y(i + 1)$, respectively. At step p, register acc only has to be loaded, while the control input of multiplexer muxp is set to $C(p) = 1$. The final result is stored in register acc after step p:

$$e^x = acc(p).$$

Observe that the stop condition test ($x(i) = 0$?), which is optional, is not represented in Figure 14.18.

The cost and computation time are given by

$$C(p) = 4.C_{\text{LUT}}(p^2) + 5.C_{\text{mux2}}(p) + C_{\text{multiplier}}(2p) + C_{\text{subtractor}}(p)$$
$$+ C_{\text{adder}}(p) + C_{\text{comp.circ.}}(p) + C_{\text{acc}}(p) + C_{\text{reg}}(p),$$

$$T(p) = p.(T_{\text{LUT}}(p^2) + 3.T_{\text{mux2}}(p) + T_{\text{comp.circ.}}(p)$$
$$+ T_{\text{multiplier}}(2.p)) + T_{\text{adder}}(p) + T_{\text{mux2}}(p) + T_{\text{multiplier}}(p), \qquad (14.32)$$

where the delay $T_{\text{multiplier}}(2.p)$ of the multiplier is assumed greater that the one of the subtractor.

14.5 SINE AND COSINE OPERATORS

This section presents an implementation for sine and cosine computation using CORDIC Algorithm 7.10 as described in Section 7.3.4. The circuit presented in Figure 14.19 is basically made up of three loop-circuits. The auxiliary sequence (upper loop) computes the successive values of the residual rotation angle a_i. A lookup table provides the successive values of $\tan^{-1} 2^{-i}$ to be added or subtracted from the current angle value a_i stored in register A: if a_i is negative, a_{i+1} is computed as

$$a_{i+1} = a_i + \tan^{-1} 2^{-i}. \qquad (14.33)$$

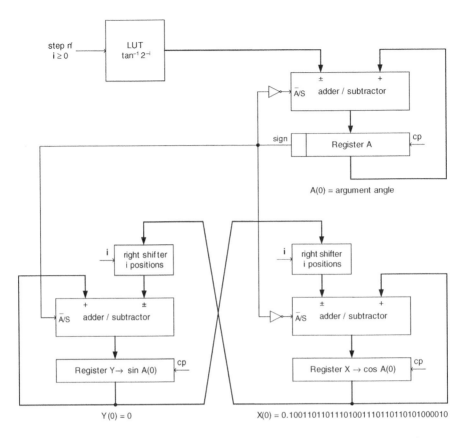

Figure 14.19 CORDIC algorithm implementation for sine and cosine computation.

Otherwise, if a_i is positive,

$$a_{i+1} = a_i - \tan^{-1} 2^{-i}. \qquad (14.34)$$

The initial value a_0 is the argument angle a.

The main sequences, computing sin a (register Y) and cos a (register X) are built up according to the following rules:

$$Y(0) = 0; \ X(0) = 0.1001101101110100111011011010101000010$$

(binary value of $1/k$).

If a_i is positive,

$$x_{i+1} = x_i - y_i.2^{-i}; \quad y_{i+1} = y_i + x_i.2^{-i}. \qquad (14.35)$$

Otherwise, if a_i is negative,

$$x_{i+1} = x_i + y_i.2^{-i}; \quad y_{i+1} = y_i - x_i.2^{-i}. \tag{14.36}$$

As quoted in Section 7.3.4, the argument a can be selected in $[-99.88°, +99.88°]$. In practice, the range can be restricted to $[-90°, +90°]$. 2's Complement notation is appropriate to represent data and (intermediate) results; operand length needs to cope with the required precision p, so LUT, adders/subtractors, shifters, and registers are accordingly designed. As before, a counter (not represented) may be used to generate the step number. Index i actualizes LUT addresses and shifter range control. After p steps, registers Y and X hold sin a and cos a, respectively. The cost and computation time are given by

$$C(p) = C_{LUT}(p^2) + 3.C_{adder/subtractor}(p)$$
$$+ 2.C_{shifter}(p) + 3.C_{Reg}(p) + 2.C_{Inv},$$
$$T(p) = p.(T_{LUT}(p^2) + T_{adder/subtractor}(p)). \tag{14.37}$$

14.6 SQUARE ROOTERS

This section presents implementations of binary square rooters based on restoring shift-and-subtract Algorithm 7.12, nonrestoring shift-and-subtract Algorithm 7.13, and the Newton–Raphson method of Section 7.4.4.

14.6.1 Restoring Shift-and-Subtract Square Rooter (Naturals)

The circuits presented in Figures 14.20 and 14.21, implementing Algorithm 7.12, are somewhat similar to the restoring divider presented in Chapter 13. Binary 2's complement notation is assumed. The restoring process is achieved by a multiplexer selecting the previous remainder in case of a negative result from the subtraction step. The key difference rests on the expression $P(i)$ to be subtracted from the successive remainder $R(i-1)$. The final result $Q(-1)$ is built up by concatenation of the complemented sign bits, from $q(n-1)$ to $q(0)$. The function $P(i)$ is computed as (formula (7.82) of Chapter 7)

$$P(i) = (4.Q(n-i) + 1).2^{2(n-i)}. \tag{14.38}$$

To achieve this function (14.38), pseudo-operators are displayed in Figure 14.20 as shifters: they stand for the rules to be respected to connect $Q(n-i)$ to the subtractor input $P(i)$; input $P(i)$ is made up of $Q(n-i)$, followed, from left to right, by the string `01` then by a string of $2.(n-i)$ zeros.

At step 1, registers are initialized as

$$R(0) = X; \quad P(1) = 2^{2(n-1)}; \quad Q(n-1) = 0. \tag{14.39}$$

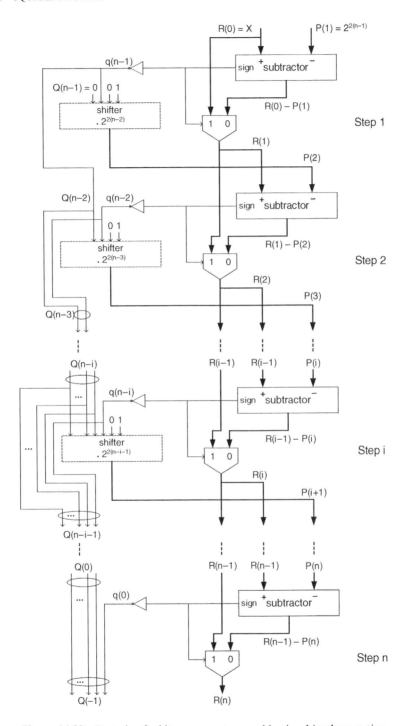

Figure 14.20 Restoring $2n$-bit square rooter, combinational implementation.

After n steps the integer square root of X and the remainder R are stored as

$$X^{1/2} = Q(-1); \quad R = R(n). \tag{14.40}$$

The cost and computation time of a combinational $2n$-bit square rooter (n-bit square root), as shown in Figure 14.20, are given by

$$C(p) = n.(C_{\text{subtractor}}(2.n) + C_{\text{mux2}}(2.n) + C_{\text{inv}}),$$
$$T(p) = n.(T_{\text{subtractor}}(2.n) + T_{\text{mux2}}(2.n)), \tag{14.41}$$

where T_{mux2} is assumed greater than the inverter delay.

The sequential implementation presented in Figure 14.21 needs a nontrivial *indexed shifter* device to connect $Q(n - i - 1)$ to the subtractor input, through register $P(i + 1)$. Synchronized registers ensure one digit per step. The cost and computation time of a sequential $2n$-bit square rooter (n-bit square root), as shown in Figure 14.21, are given by

$$C(p) = C_{\text{subtractor}}(2.n) + C_{\text{mux2}}(2.n) + C_{\text{indshifter}}(2.n) + C_{\text{inv}} + 6.C_{\text{reg}}(2.n),$$
$$T(p) = n.(T_{\text{subtractor}}(2.n) + \max{(T_{\text{mux2}}(2.n), T_{\text{inv}} + T_{\text{indshifter}}(2.n))}). \tag{14.42}$$

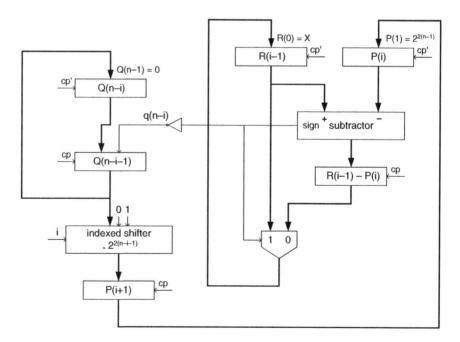

Figure 14.21   Restoring $2n$-bit square rooter, sequential implementation.

14.6.2 Nonrestoring Shift-and-Subtract Square Rooter (Naturals)

The circuits presented in Figures 14.22 and 14.23, implementing Algorithm 7.13, are somewhat similar to the nonrestoring divider presented in Chapter 13. Binary 2's complement notation is assumed. The nonrestoring feature brings on two main differences with respect to the circuits of Figures 14.20 and 14.21. One first observes that the arithmetic cell is now an adder/subtractor whose operation (Add/Subtract) is controlled by the sign of the preceding partial remainder. The other key difference rests on the expression $P(i)$ to be added/subtracted from the successive remainder $R(i-1)$. The final result $Q(-1)$ is still built up by concatenation of the complemented sign bits, from $q(n-1)$ to $q(0)$. Function $P(i)$ is now computed as (formulas (7.90) and (7.91) of Chapter 7)

$$P(i) = (4.Q(n-i) + \text{'01'}).2^{2(n-i)}, \quad \text{if } R(i-1) \geq 0 \quad (q(n-i+1) = 0) \quad (14.43)$$

or

$$P(i) = Pstar(i) = (4.Q(n-i) + \text{'11'}).2^{2(n-i)}, \atop \text{if } R(i-1) < 0 \quad (q(n-i+1) = 1). \qquad (14.44)$$

Formulas (14.43) and (14.44) may be merged as

$$P(i) = (4.Q(n-i) + 2.q(n-i+1) + 1).2^{2(n-i)}. \qquad (14.45)$$

This allows the use of a unique set of connecting rules materialized by the pseudo-operators (shifters) in Figure 14.22 or the indexed shifters in Figure 14.23.

At step 1, registers are initialized as

$$R(0) = X; \quad P(1) = 2^{2(n-1)}; \quad Q(n-1) = 0. \qquad (14.46)$$

As X is positive, the first arithmetic operation is a subtraction: $R(0) - P(1)$. After n steps the integer square root of X and the remainder R are stored as

$$X^{1/2} = Q(-1); \quad R = R(n). \qquad (14.47)$$

If $R(n)$ is negative, the final remainder needs to be adjusted to the last positive partial remainder. The cost and computation time of a combinational $2n$-bit square rooter (n-bit square root), as shown in Figure 14.22, are given by

$$C(p) = n.(C_{\text{adder/subtractor}}(2.n) + C_{\text{inv}}),$$
$$T(p) = n.(T_{\text{adder/subtractor}}(2.n) + T_{\text{inv}}). \qquad (14.48)$$

The sequential implementation presented in Figure 14.23 needs an indexed shifter device to connect $Q(n-i-1)$ to the subtractor input, through register $P(i+1)$.

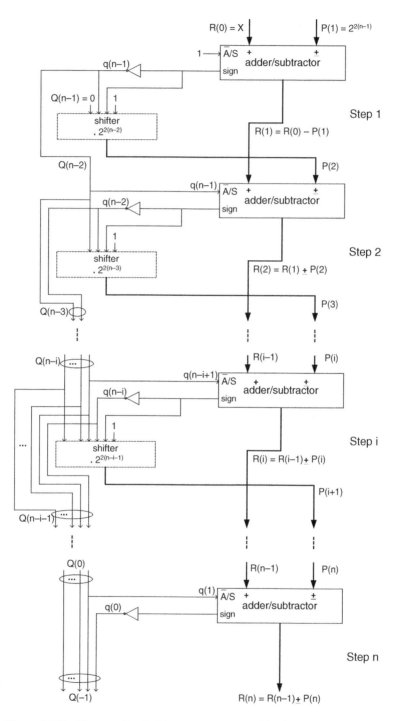

Figure 14.22 Nonrestoring $2n$-bit square rooter, combinational implementation.

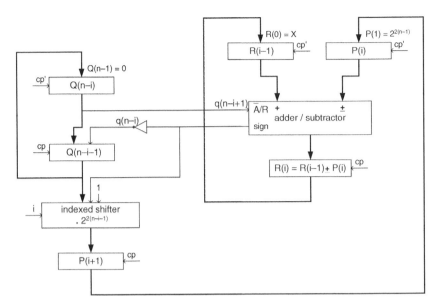

Figure 14.23   Nonrestoring $2n$-bit square rooter, sequential implementation.

Synchronized registers ensure one digit per step. The cost and computation time of a sequential $2n$-bit square rooter (n-bit square root), as shown in Figure 14.23, are given by

$$C(p) = C_{\text{adder/subtractor}}(2.n) + C_{\text{indshifter}}(2.n) + C_{\text{inv}} + 6.C_{\text{reg}}(2.n),$$
$$T(p) = n.(T_{\text{adder/subtractor}}(2.n) + T_{\text{inv}} + T_{\text{indshifter}}(2.n)). \tag{14.49}$$

14.6.3   Newton–Raphson Square Rooter (Naturals)

The iteration equation (7.96)

$$x_{i+1} = \tfrac{1}{2}x_i \cdot (3 - X \cdot x_i^2), \tag{14.50}$$

converges toward the inverse square root $1/X^{1/2}$. A dedicated implementation is depicted in Figure 14.24. The iteration step involves one squaring unit, two multipliers, and one dedicated (3's complement) subtractor. A pseudo-operator (shifter) stands for a right-shift operation, readily achieved by an appropriate loop-connection to the multiplexer represented in the general structure of Figure 14.24. The operators are designed according to the required precision p. A look-table (LUT), addressed by the truncated argument X_t, provides a first t-bit evaluation of $1/X^{1/2}$; its dimension is (roughly) $t.2^t$ bits.

The outputs of the operators are rounded up at the required precision p; several bits may be added to cope with rounding and error propagation. The argument length is assumed not greater than $2p$. Although the argument is a natural number, the

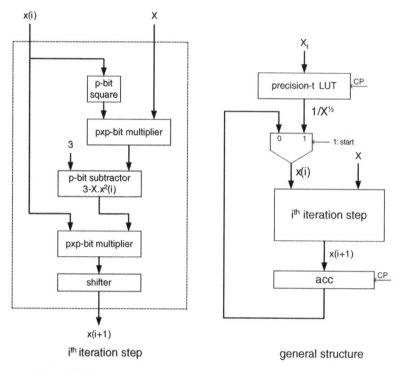

Figure 14.24 Newton–Raphson iteration circuit: $2p$-bit square rooter.

intermediate results, as well as the final one, are not. Floating-point notation is recommended to optimize the overall precision. Example 14.2 illustrates this point.

Example 14.2 Let

$$X = 01011011$$

whose 16-bit precision root $X^{\frac{1}{2}}$ is $1001.1000101000011 - \varepsilon < 2^{-13}$.

Assume that the LUT value $x(0)$ of $1/X^{1/2}$ is 11.2^{-5}, 2-bit approximation corresponding to the inverse of $X_t^{1/2}$ root of the truncated 2-bit value $X_t = 01100000$ (rounded up). The following table shows the first two steps of computation with rounding to 16 bits.

i	$x(i)^2$	$X.x(i)^2$	$3 - X.x(i)^2$	$x(i+1) = \frac{1}{2}x(i).(3 - X.x(i)^2)$
0	1001.2^{-10}	1100110011.2^{-10}	100011001101.2^{-10}	1101001100111.2^{-16}
1	1010111001000101.2^{-22}	1111011111001010.2^{-16}	1000001000001101.2^{-14}	1101011010011010.2^{-19}

After step 2 ($i = 1$), the approximation of $X^{1/2}$ is

$$X.x(2) = 1011011 \times 1101011010011010.2^{-19} = 1001.1000100100010111110$$

showing 10-bit accuracy.

The cost and computation time of the Newton–Raphson square rooter (p-bit square root), as shown in Figure 14.24, are given by

$$
\begin{aligned}
C(p) = {}& C_{\text{LUT}}(t \times 2^t) + C_{\text{square}}(p) + 2.C_{\text{multiplier}}(p) \\
& + C_{\text{subtractor}}(p) + C_{\text{mux2}}(p) + C_{\text{acc}}(p), \\
T(p) = {}& T_{\text{LUT}}(t \times 2^t) + k.(T_{\text{square}}(p) \\
& + 2.T_{\text{multiplier}}(p) + T_{\text{subtractor}}(p) + T_{\text{mux2}}(p)),
\end{aligned}
\tag{14.51}
$$

where $k = \log_2 (p/t)$.

Comment 14.1 An important question about algorithms using look-up tables, and convergence algorithms in particular, is the evaluation of the exact (i.e., minimum) amount of bits necessary for any prescaling or intermediate calculations to ensure a correct result within the desired precision. Theoretically, the Newton–Raphson algorithms provide a quadratic convergence rate, doubling the number of exact bits at each step. In practice, the accuracy of the look-up tables together with the rounding errors could slow down that rate, unless additional bits are provided to represent LUT data and intermediate results. Actually, the accurate calculus of rounding errors is not a straightforward matter. This mathematical problem has been treated extensively in the literature ([COR1999], [DAS1995], [TAN1991]). Using extra-bits is a safe and easy way to ensure correctness; nevertheless, a careful error computation can lead to significant savings.

14.7 BIBLIOGRAPHY

[COR1999] M. A. Cornea-Hasegan, R. A. Golliver, and P. Markstein, Correctness proofs outline for Newton–Raphson based floating-point divide and square root algorithms. *Proceedings of the 14th IEEE Symposium on Computer Arithmetic*, 1999, pp. 96–105.

[DAS1995] D. DasSarma and D. W. Matula, Faithfull bipartite ROM reciprocal tables. *Proceedings of the 12th IEEE Symposium on Computer Arithmetic*, 1995, pp. 17–28.

[GAR1959] H. L. Garner, The residue number system. *IRE Trans. Electron. Comput.*, **EC 8**: 140–147 (1959).

[SZA1967] N. S. Szabo and R. I. Tanaka, *Residue Arithmetic and Its Applications to Computer Technology*. McGraw-Hill, New York, 1967.

[TAN1991] P. K. Tang, Table look-up algorithms for elementary functions and their error analysis. *Proceedings of the 10th IEEE Symposium on Computer Arithmetic*, 1991, pp. 232–236.

15

CIRCUITS FOR FINITE FIELD OPERATIONS

This chapter deals with the synthesis of circuits implementing the main finite field operations: addition, subtraction, product, exponentiation, and inversion. The reason why these operations should be implemented in hardware, instead of just being programmed for some target microprocessor, is the reduction of the computation time. This is particularly true in the case of computer and communications systems including the execution of cryptographic algorithms for security purposes: they use very long operands so that their software-only execution time could become prohibitively long for some real-time applications. An efficient solution is the development of specific circuits (coprocessors) executing the most time-consuming operations.

15.1 OPERATIONS IN Z_m

15.1.1 Adders and Subtractors

The structure of a base-B modulo m adder is shown in Figure 15.1. It is based on Algorithm 8.2. Its cost is equal to

$$C_{\text{mod-adder}}(n) = 2.C_{\text{adder}}(n) + n.C_{\text{mux2-1}}. \tag{15.1}$$

Synthesis of Arithmetic Circuits: FPGA, ASIC, and Embedded Systems
By Jean-Pierre Deschamps, Géry J. A. Bioul, and Gustavo D. Sutter
Copyright © 2006 John Wiley & Sons, Inc.

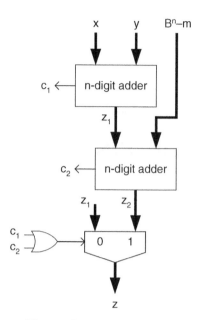

Figure 15.1   Modulo m adder.

If every adder is a ripple-carry adder made up of full-adder cells, then its computation time is equal to

$$T_{\text{mod-adder}}(n) \cong (n+1).T_{\text{FA}} + T_{\text{mux}}. \tag{15.2}$$

The structure of a modulo m subtractor is shown in Figure 15.2. It is based on Algorithm 8.4. Its cost and computation time are practically the same as in the case of the modulo m adder.

$$C_{\text{mod-subtractor}}(n) = 2.C_{\text{adder}}(n) + n.C_{\text{mux2}-1}. \tag{15.3}$$

If every n-digit adder is a ripple-carry adder made up of full-adder cells, then its computation time is equal to

$$T_{\text{mod-subtractor}}(n) \cong (n+1).T_{\text{FA}} + T_{\text{mux}}. \tag{15.4}$$

Example 15.1   (Complete VHDL source code available.) Generate VHDL models of binary ($B = 2$) modulo m adders and subtractors:

```
entity mod_adder is
port (
  x, y: in std_logic_vector(n-1 downto 0);
```

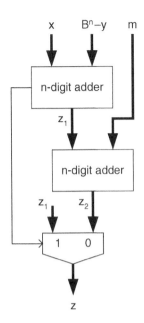

Figure 15.2 Modulo m subtractor.

```
  z: out std_logic_vector(n-1 downto 0)
);
end mod_adder;

architecture circuit of mod_adder is
  signal z1, z2: std_logic_vector(n-1 downto 0);
  signal c1, c2: std_logic;
  signal long_x, long_y, long_result1, long_z1, minus_m,
  long_result2: std_logic_vector(n downto 0);
begin
  long_x<='0'&x; long_y<='0'&y;
  long_result1<=long_x+long_y;
  c1<=long_result1(n);
  z1<=long_result1(n-1 downto 0);
  long_z1<='0'&z1;
  minus_m<=conv_std_logic_vector((2**n)-m, n+1);
  long_result2<=long_z1+minus_m;
  c2<=long_result2(n);
  z2<=long_result2(n-1 downto 0);
  z<=z1 when (c1 or c2)='0' else z2;
end circuit;

entity mod_subtractor is
port (
  x, y: in std_logic_vector(n-1 downto 0);
```

```
  z: out std_logic_vector(n-1 downto 0)
);
end mod_subtractor;

architecture circuit of mod_subtractor is
  signal z1, z2, inv_y: std_logic_vector(n-1 downto 0);
  signal c1: std_logic;
  signal long_x, long_inv_y, long_result1:
  std_logic_vector(n downto 0);
begin
  long_x<='0'&x;
  inversion: for i in 0 to n-1 generate
    inv_y(i)<=not(y(i));
  end generate;
  long_inv_y<='0'&inv_y;
  long_result1<=long_x+long_inv_y+'1';
  c1<=long_result1(n);
  z1<=long_result1(n-1 downto 0);
  z2<=z1+conv_std_logic_vector(m, n);
  z<=z1 when c1='1' else z2;
end circuit;
```

15.1.2 Multiplication

15.1.2.1 Multiply and Reduce A first multiplier structure is shown in Figure 15.3. It is based on Algorithm 8.5. As regards the division, observe that the divider is greater than the dividend ($p = x.y < m.m < m.B^n$). Furthermore, it can be assumed that $m \geq B^{m-1}$; in the contrary case all numbers could be

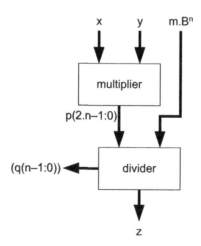

Figure 15.3 Multiply and reduce algorithm implementation.

represented with one digit less. The cost and computation time are equal to

$$C_{\text{multiply-reduce}}(n) = C_{\text{multiplier}}(n,\, n) + C_{\text{divider}}(2.n,\, n),$$
$$T_{\text{multiply-reduce}}(n) = T_{\text{multiplier}}(n,\, n) + T_{\text{divider}}(2.n,\, n).$$

As regards the computation time observed, if an SRT divider (see Chapter 13, Section 13.2.3) is used, the total computation time is a linear function of n.

15.1.2.2 Shift and Add Another multiplier structure can be deduced from Algorithm 8.6. It's an iterative circuit whose basic cell is shown in Figure 15.4. The total cost and computation time are equal to

$$C_{\text{shift-add}}(n) = n.(C_{\text{multiplier}}(n,\, 1) + C_{\text{adder}}(n+1) + C_{\text{divider}}(n+2,\, 2)),$$
$$T_{\text{shift-add}}(n) = n.(T_{\text{multiplier}}(n,1) + T_{\text{adder}}(n+1) + T_{\text{divider}}(n+2,\, 2)).$$

In base $B = 2$, Algorithm 8.9 can be used. The corresponding iterative circuit is shown in Figure 15.5. Observe that, in Figure 15.5b,

$p_1 = 2.p$, where $0 \le p < m$, such that $0 \le p_1 < 2.m$;

$p_2 = p_1 - w$, where $w = m$ or $w = m - y$, with $0 \le y < m$, so that $-m \le p_2 < 2.m$;

$p_3 = p_2 - m$ if $0 \le p_2 < 2.m$, so that $-m \le p_3 < m$; $p_3 = p_2 + m$ if $-m \le p_2 < 0$, so that $0 \le p_3 < m$; conclusion: $-m \le p_3 < m$.

Thus p_2 is an $(n+2)$-bit number and p_3 an $(n+1)$-bit number.

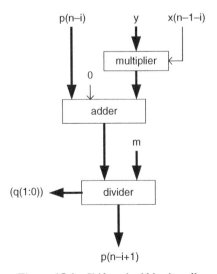

Figure 15.4   Shift-and-add basic cell.

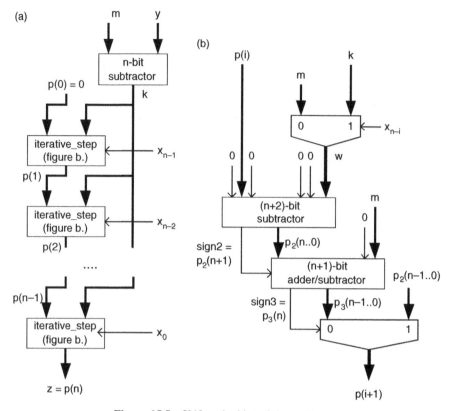

Figure 15.5   Shift-and-add modular multiplier.

Example 15.2   (Complete VHDL source code available.) Generate the VHDL model of a binary ($B = 2$) modulo m shift-and-add multiplier:

```
entity iterative_step is
port (
  p, k: in std_logic_vector(n-1 downto 0);
  x: in std_logic;
  next_p: out std_logic_vector(n-1 downto 0)
);
end iterative_step;

architecture circuit of iterative_step is
  signal w, module: std_logic_vector(n-1 downto 0);
  signal sign2, sign3: std_logic;
  signal p1, long_w, p2: std_logic_vector(n+1 downto 0);
  signal long_module, p3: std_logic_vector(n downto 0);
begin
  module<=conv_std_logic_vector(m, n);
```

```
  w<=module when x='0' else k;
  long_w<="00"&w;
  p1<='0'&p&'0'; p2<=p1-long_w; sign2<=p2(n+1);
  long_module<='0'&module;
  with sign2 select p3<=p2(n downto 0)+long_module when '1',
  p2(n downto 0)-long_module when others;
  sign3<=p3(n);
  next_p<=p2(n-1 downto 0) when sign3='1'
  else p3(n-1 downto 0);
end circuit;

entity mod_multiplier is
port (
  x, y: in std_logic_vector(n-1 downto 0);
  z: out std_logic_vector(n-1 downto 0)
);
end mod_multiplier;

architecture circuit of mod_multiplier is
  component iterative_step..end component;
  signal p: p_vector;
  signal k, module: std_logic_vector(n-1 downto 0);
begin
  module<=conv_std_logic_vector(m, n);
  k<=module-y;
  p(0)<=zero;
  iteration: for i in 0 to n-1 generate
  step: iterative_step port map (p(i), k, x(n-i-1), p(i+1));
  end generate;
  z<=p(n);
end circuit;
```

15.1.2.3 Montgomery Multiplication

The iterative circuit of Figure 15.6a implements the Montgomery multiplication (Algorithm 8.10). It can be used for computing the modular product (Algorithm 8.12) or the modular exponentiation (Algorithm 8.15).

The cost of the Montgomery cell of Figure 15.6b is equal to

$$C_{\text{Montgomery-cell}}(n) = 2.C_{\text{adder}}(n+2) + 2.n.C_{\text{AND}}$$

and its computation time, if ripple-carry adders are used, is

$$T_{\text{Montgomery-cell}}(n) = (n+3).T_{\text{FA}} + T_{\text{AND}}.$$

The total cost is equal to

$$C_{\text{Montgomery}}(n) = n.(2.C_{\text{adder}}(n+2) + 2.n.C_{\text{AND}}) + C_{\text{subtractor}}(n+2) + n.C_{\text{mux2-1}}.$$

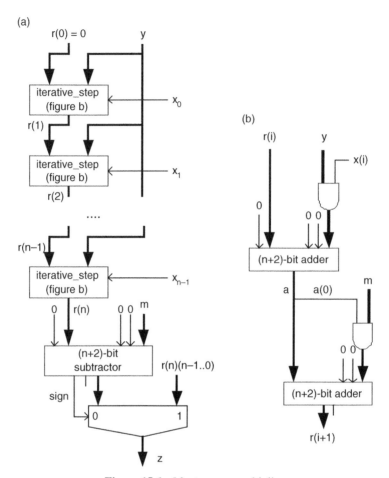

Figure 15.6 Montgomery multiplier.

If ripple adders are used then

$$C_{\text{Montgomery}}(n) \cong 2.n.(n+2).C_{\text{FA}}$$

and

$$T_{\text{Montgomery}}(n) \cong 4.n.T_{\text{FA}}. \tag{15.5}$$

Example 15.3 (Complete VHDL source code available.) Generate the VHDL model of a binary ($B = 2$) Montgomery multiplier:

```
entity Montgomery_step is
port (
  r: in std_logic_vector(n downto 0);
  y: in std_logic_vector(n-1 downto 0);
```

```
  x: in std_logic;
  next_r: out std_logic_vector(n downto 0)
);
end Montgomery_step;

architecture circuit of Montgomery_step is
  signal long_r, y_by_x, a, m_by_a, two_r:
  std_logic_vector(n+1 downto 0);
  signal module: std_logic_vector(n-1 downto 0);
begin
  long_r<='0'&r;
  and_gates1: for i in 0 to n-1 generate
    y_by_x(i)<=y(i) and x;
  end generate;
  y_by_x(n)<='0'; y_by_x(n+1)<='0';
  a<=long_r+y_by_x;
  module<=conv_std_logic_vector(m, n);
  and_gates2: for i in 0 to n-1 generate
    m_by_a(i)<=module(i) and a(0);
  end generate;
  m_by_a(n)<='0'; m_by_a(n+1)<='0';
  two_r<=a+m_by_a;
  next_r<=two_r(n+1 downto 1);
end circuit;

entity Montgomery_multiplier is
port (
  x, y: in std_logic_vector(n-1 downto 0);
  z: out std_logic_vector(n-1 downto 0)
);
end Montgomery_multiplier;

architecture circuit of Montgomery_multiplier is
  component Montgomery_step...end component;
  signal r: r_vector;
  signal module: std_logic_vector(n-1 downto 0);
  signal long_r_n, long_module, dif:
  std_logic_vector(n+1 downto 0);
begin
  module<=conv_std_logic_vector(m, n);
  r(0)<=zero;
  iteration: for i in 0 to n-1 generate
    step: Montgomery_step port map (r(i), y, x(i), r(i+1));
  end generate;
  long_r_n<='0'&r(n); long_module<="00"&module;
  dif<=long_r_n - long_module;
  with dif(n+1) select z<=dif(n-1 downto 0) when '0',
  r(n)(n-1 downto 0) when others;
end circuit;
```

15.1.2.4 *Modulo (B^k-c) Reduction*   In the case where $m = B^k - c$ for some small c the modulo m reduction can be performed with Algorithm 8.13. The corresponding cell is shown in Figure 15.7a: z is assumed to be a q-digit number and c a t-digit one. The value of new_z is smaller than the initial value z. After several steps, a number last_z is obtained with the following properties: last_z mod $m = x$ mod m, and last_z $< B^n$. If $m \geq B^{n-1}$, then x mod $m =$ last_z $- q.m$ for some $q < B$. The total cost and computation time depend on the number of steps, that is, on the particular values of x and m. An additional one-step divider is necessary in order to generate the final result (Figure 15.7b).

Example 15.4   (Complete VHDL source code available.) Generate the VHDL model of a circuit that computes x modulo $2^n - c$, x being a $2.n$-bit number (e.g., the result of multiplying two n-bit numbers). The structure of the data path is shown in Figure 15.8.

```
entity data_path is
port (
  x: in std_logic_vector(2*n-1 downto 0);
  clk, sel, enable: in std_logic;
  z: out std_logic_vector(n-1 downto 0);
  equal_zero: out std_logic
);
end data_path;

architecture circuit of data_path is
  signal next_x1, next_x0, x1, x0, y1, y0:
  std_logic_vector(n-1 downto 0);
  signal x1_by_c, long_x0, y: std_logic_vector(2*n-1 downto 0);
  signal long_y0, minus_m, y0_minus_m:
  std_logic_vector(n downto 0);
```

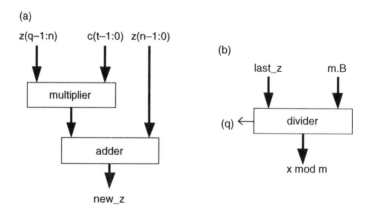

Figure 15.7   (a) Mod m reduction cell. (b) Divider.

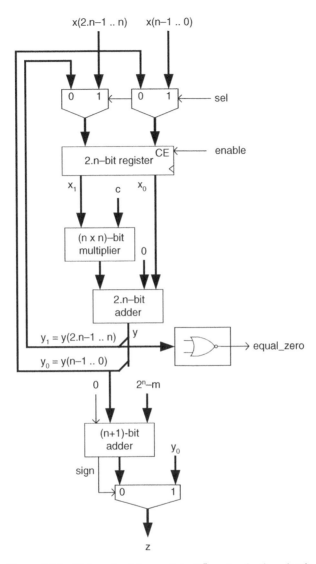

Figure 15.8 Data path of the modulo $(2^n - c)$ reduction circuit.

```
begin
  with sel select next_x1<=y1 when '0', x(2*n-1 downto n)
  when others;
  with sel select next_x0<=y0 when '0', x(n-1 downto 0)
  when others;
  process(clk)
  begin
    if clk'event and clk='1' then
      if enable='1' then x1<=next_x1; x0<=next_x0; end if;
    end if;
```

```
    end process;
    x1_by_c<=x1*conv_std_logic_vector(c, n);
    long_x0<=zero&x0;
    y<=x1_by_c+long_x0;
    y1<=y(2*n-1 downto n); y0<=y(n-1 downto 0);
    equal_zero<='1' when y1=zero else '0';
    long_y0<='0'& y0;
    minus_m<=conv_std_logic_vector(2**n - m, n+1);
    y0_minus_m<=long_y0+minus_m;
    with y0_minus_m(n) select z<=y0_minus_m(n-1 downto 0)
    when '1', y0 when others;
end circuit;

entity control_unit is
port (
  clk, reset, start, equal_zero: in std_logic;
  done, sel, enable: out std_logic
);
end control_unit;

architecture rtl of control_unit is
  subtype internal_state is natural range 0 to 3;
  signal state: internal_state;
begin
  process(clk, reset)
  begin
    case state is
      when 0=>sel<='1'; enable<='0'; done<='1';
      when 1=>sel<='1'; enable<='0'; done<='1';
      when 2=>sel<='1'; enable<='1'; done<='0';
      when 3=>sel<='0'; enable<='1'; done<='0';
    end case;
    if reset='1' then state<=0;
    elsif clk'event and clk='1' then
    case state is
      when 0=>if start='0' then state<=state+1; end if;
      when 1=>if start='1' then state<=state+1; end if;
      when 2=>state<=state+1;
      when 3=>if equal_zero='1' then state<=0;end if;
    end case;
    end if;
  end process;
end rtl;

entity mod_reduction is
port (
  x: in std_logic_vector(2*n-1 downto 0);
  clk, reset, start: in std_logic;
  z: out std_logic_vector(n-1 downto 0);
```

```
  done: out std_logic
);
end mod_reduction;

architecture circuit of mod_reduction is
  component data_path...end component;
  component control_unit...end component;
  signal sel, enable, equal_zero: std_logic;
begin
  component1: data_path port map(x, clk, sel, enable, z,
  equal_zero);
  component2: control_unit port map(clk, reset, start,
  equal_zero, done, sel, enable);
end circuit;
```

As was mentioned in Chapter 8, for some particular values of m (still) more specific algorithms can be used.

Example 15.5 (Complete VHDL source code available.) The circuit of Figure 15.9 implements a mod 239 reduction circuit based on the algorithm of Example 8.5.

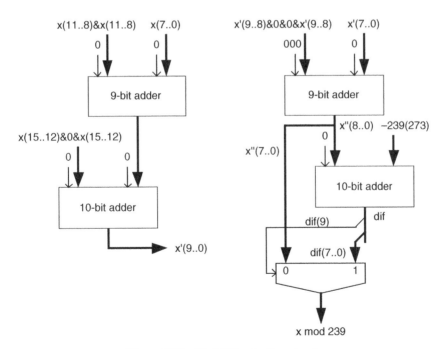

Figure 15.9 Mod 239 reduction circuit.

Observe that if $x = x(15)\, x(14) \cdots x(0)$ and $x' = x'(9)\, x'(8) \cdots x'(0)$, then

$$x_2 = x(15)x(14)x(13)x(12), \quad x_1 = x(11)x(10)x(9)x(8), \quad x_0 = x(7)x(6)\cdots x(0),$$
$$17.x_1 = 16.x_1 + x_1 = x(11)x(10)x(9)x(8)x(11)x(10)x(9)x(8),$$
$$33.x_2 = 32.x_2 + x_2 = x(15)x(14)x(13)x(12)0x(15)x(14)x(13)x(12),$$
$$x'_1 = x'(9)x'(8), x'_0 = x'(7)x'(6)\cdots x'(0),$$
$$17.x'_1 = 16.x'_1 + x'_1 = x'(9)x'(8)00x'(9)x'(8).$$

If x can be any 16-bit number, then

$$17.x_1 + x_0 \le 17.15 + 255 = 510 \quad \text{(a 9-bit number)},$$
$$x'_1 = 33.x_2 + 17.x_1 + x_0 \le 33.15 + 17.15 + 255 = 1005 \quad \text{(a 10-bit number)},$$
$$x'' = 17.x'_1 + x'_0 \le 17.15 + 255 = 510 \quad \text{(a 9-bit number)}.$$

15.1.2.5 Exponentiation The data path of Figure 15.10 allows us to execute Algorithm 8.15. Its cost and computation time are equal to

$$C_{\text{exponentiation}}(n) = C_{\text{Montgomery}}(n) + 2.n.C_{\text{FF}} + n.(2.C_{\text{mux2-1}} + C_{\text{mux4-1}}), \quad (15.6)$$
$$T_{\text{exponentiation}}(n) = T_{\text{Montgomery}}(n) + T_{\text{FF}} + T_{\text{mux4-1}} + T_{\text{mux2-1}}.$$

If ripple-carry adders are used within the Montgomery multiplier, then (15.5):

$$C_{\text{exponentiation}}(n) \cong 2.n.(n+2).C_{\text{FA}} + 2.n.C_{\text{FF}} + n.(2.C_{\text{mux2-1}} + C_{\text{mux4-1}}), \quad (15.7)$$
$$T_{\text{exponentiation}}(n) \cong 4.n.T_{\text{FA}} + T_{\text{FF}} + T_{\text{mux4-1}} + T_{\text{mux2-1}}.$$

For large values of n:

$$C_{\text{exponentiation}}(n) \cong 2.n^2.C_{\text{FA}}, \quad (15.8)$$
$$T_{\text{exponentiation}}(n) \cong 4.n.T_{\text{FA}}.$$

Example 15.6 (Complete VHDL source code available.) Generate the VHDL model of a circuit that computes y^x modulo m, where x and y are two n-bit numbers. The structure of the data path is shown in Figure 15.10.

```
entity exp_data_path is
port (
  y: in std_logic_vector(n-1 downto 0);
  sel: in std_logic_vector(1 downto 0);
  clk, enable_e, enable_y: in std_logic;
  z: out std_logic_vector(n-1 downto 0)
);
end exp_data_path;
```

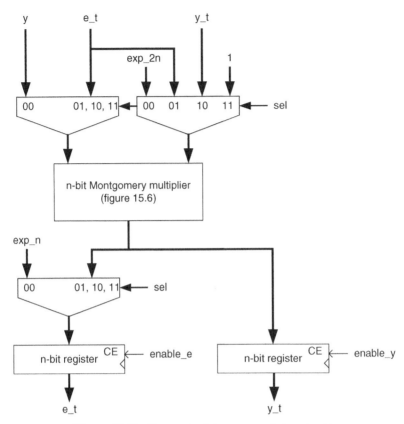

Figure 15.10   Data path of the exponentiation circuit.

```
architecture circuit of exp_data_path is
   signal mult_in1, mult_in2, mult_out, reg_e_in, e_t, y_t,
   exp_n, exp_2n: std_logic_vector(n-1 downto 0);
   component Montgomery_multiplier...end component;
begin
   exp_2n<=conv_std_logic_vector(int_exp_2n, n);
   with sel select mult_in1<=y when "00", e_t when others;
   with sel select mult_in2<=exp_2n when "00", e_t when "01",
   y_t when "10", one when others;
   multiplier: Montgomery_multiplier
   port map (mult_in1, mult_in2, mult_out);
   exp_n<=conv_std_logic_vector(int_exp_n, n);
   with sel select reg_e_in<=exp_n when "00", mult_out when
   others;
   process(clk)
   begin
      if clk'event and clk='1' then
      if enable_e='1' then e_t<=reg_e_in; end if;
      end if;
```

```vhdl
  end process;
  process(clk)
  begin
    if clk'event and clk='1' then
    if enable_y='1' then y_t<=mult_out; end if;
    end if;
  end process;
  z<=e_t;
end circuit;

entity exp_control_unit is
port (
  clk, reset, start: in std_logic;
  x: in std_logic_vector(n-1 downto 0);
  done, enable_e, enable_y: out std_logic;
  sel: out std_logic_vector(1 downto 0)
);
end exp_control_unit;

architecture rtl of exp_control_unit is
  subtype internal_state is natural range 0 to 7;
  signal state: internal_state;
  subtype count is integer range 0 to n-1;
  signal counter: count;
begin
  process(clk, reset)
  begin
    case state is
    when 0=>sel<="00"; enable_e<='0'; enable_y<='1';
    done<='1';
    when 1=>sel<="00"; enable_e<='0'; enable_y<='1';
    done<='1';
    when 2=>sel<="00"; enable_e<='1'; enable_y<='0';
    done<='0';
    when 3=>sel<="00"; enable_e<='0'; enable_y<='1';
    done<='0';
    when 4=>sel<="01"; enable_e<='1'; enable_y<='0';
    done<='0';
    when 5=>sel<="10"; enable_e<='1'; enable_y<='0';
    done<='0';
    when 6=>sel<="10"; enable_e<='0'; enable_y<='0';
    done<='0';
    when 7=>sel<="11"; enable_e<='1'; enable_y<='0';
    done<='0';
    end case;
    if reset='1' then state<=0; counter<=n-1;
    elsif clk'event and clk='1' then
      case state is
        when 0=>if start='0' then state<=state+1; end if;
```

```
      when 1=>if start='1' then state<=state+1; end if;
      when 2=>state<=state+1;
      when 3=>state<=state+1; counter<=n-1;
      when 4=>if x(counter)='1' then state<=state+1;
      else state<=6; end if; counter<=counter - 1;
      when 5=>state<=state+1;
      when 6=>if counter<0 then state<=state+1;
      else state<=4; end if;
      when 7=>state<=0;
    end case;
    end if;
  end process;
end rtl;

entity exponentiate is
port (
  x, y: in std_logic_vector(n-1 downto 0);
  clk, reset, start: in std_logic;
  z: out std_logic_vector(n-1 downto 0);
  done: out std_logic
);
end exponentiate;

architecture circuit of exponentiate is
  component exp_data_path...end component;
  component exp_control_unit...end component;
  signal sel: std_logic_vector(1 downto 0);
  signal enable_e, enable_y: std_logic;
begin
  first_component: exp_data_path port map (y, sel, clk,
  enable_e, enable_y, z);
  second_component: exp_control_unit port map (clk, reset,
  start, x, done, enable_e, enable_y, sel);
end circuit;
```

15.2   INVERSION IN *GF(p)*

For relatively small p, the value of x^{-1} for all $x \in \{1, \ldots, p-1\}$ can be computed in advance and stored within a look-up table. For larger values of p, Algorithm 8.16 can be used. As was already mentioned in Section 8.2, the value of $c(i)$ belongs to the interval $-p/2 < c(\mathrm{i}) < p/2$, so that

$$-B^{n-1} < c(i) < B^{n-1}$$

and all $c(i)$ can be represented with n digits in reduced B's complement form. The value of $r(i)$ belongs to the interval $p \geq r(i) \geq 1$ so that all $r(i)$ are n-digit base B numbers. The data path of the corresponding circuit is shown in Figure 15.11.

Example 15.7 (Complete VHDL source code available.) Generate the VHDL model of a circuit that computes $z = x^{-1}$ modulo p, where x and p are two n-bit numbers, with $x < p$. The structure of the data path is shown in Figure 15.11.

```
entity inv_data_path is
port (
  x: in std_logic_vector(n-1 downto 0);
  clk, first_step, enable: in std_logic;
  gt_one: out std_logic;
  z: out std_logic_vector(n-1 downto 0)
);
end inv_data_path;
```

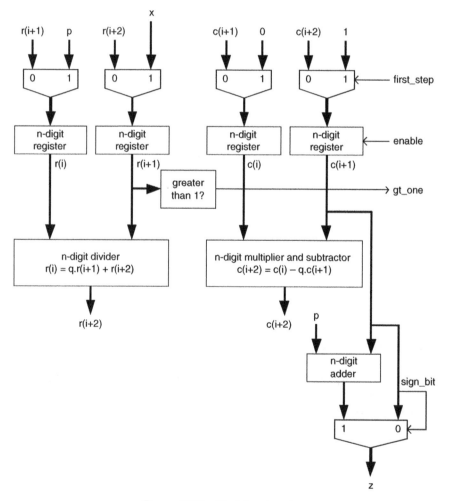

Figure 15.11 Modulo p inverter.

```
architecture circuit of inv_data_path is
  signal r_i, r_iplus1, r_iplus2, c_i, c_iplus1, c_iplus2,
  next_r_i, next_r_iplus1, next_c_i, next_c_iplus1, zero,
  one, module, q: std_logic_vector(n-1 downto 0);
  component functional_divider...end component;
  component functional_multiplier...end component;
begin
  zero<=conv_std_logic_vector(0, n);
  one<=conv_std_logic_vector(1, n);
  module<=conv_std_logic_vector(p, n);
  with first_step select next_r_i<=r_iplus1 when '0', module
  when others;
  with first_step select next_r_iplus1<=r_iplus2 when '0', x
  when others;
  with first_step select next_c_i<=c_iplus1 when '0', zero
  when others;
  with first_step select next_c_iplus1<=c_iplus2 when '0', one
  when others;
  divider: functional_divider port map (r_i, r_iplus1, q,
  r_iplus2);
  multiplier: functional_multiplier port map (c_i, c_iplus1,
  q, c_iplus2);
  process(clk)
  begin
    if clk'event and clk='1' then
    r_i<=next_r_i; r_iplus1<=next_r_iplus1; c_i<=next_c_i;
    end if;
  end process;
  process(clk)
  begin
    if clk'event and clk='1' then
      if enable='1' then c_iplus1<=next_c_iplus1; end if;
    end if;
  end process;
  gt_one<='1' when r_iplus1>one else '0';
  with c_iplus1(n-1) select z<=c_iplus1 when '0', c_iplus1+
  module when others;
end circuit;

entity inv_control_unit is
port (
  clk, reset, start, gt_one: in std_logic;
  first_step, enable, done: out std_logic
);
end inv_control_unit;

architecture rtl of inv_control_unit is
  subtype internal_state is natural range 0 to 3;
  signal state: internal_state;
begin
```

```
process(clk, reset)
begin
  case state is
    when 0=>first_step<='1'; enable<='0'; done<='1';
    when 1=>first_step<='1'; enable<='0';
    done<='1';
    when 2=>first_step<='1'; enable<='1'; done<='0';
    when 3=>first_step<='0';
    if gt_one='1' then enable<='1'; else enable<='0';
    end if; done<='0';
  end case;
  if reset='1' then state<=0;
  elsif clk'event and clk='1' then
    case state is
      when 0=>if start='0' then state<=state+1; end if;
      when 1=>if start='1' then state<=state+1; end if;
      when 2=>state<=state+1;
      when 3=>if gt_one='0' then state<=0; end if;
    end case;
  end if;
end process;
end rtl;

entity field_inverter is
port (
  x: in std_logic_vector(n-1 downto 0);
  clk, reset, start: in std_logic;
  z: out std_logic_vector(n-1 downto 0);
  done: out std_logic
);
end field_inverter;

architecture circuit of field_inverter is
  component inv_data_path...end component;
  component inv_control_unit...end component;
  signal first_step, enable, gt_one: std_logic;
begin
  first_component: inv_data_path port map (x, clk, first_step,
  enable, gt_one, z);
    second_component: inv_control_unit port map (clk, reset,
start,
  gt_one, first_step, enable, done);
end circuit;
```

15.3   OPERATIONS IN $Z_p[x]/f(x)$

An adder or a subtractor in $Z_p[x]/f(x)$ is just a set of n modulo p adders or subtractors working in parallel (Algorithms 8.17 and 8.18).The same occurs with the

multiplication of a polynomial by an element of Z_p: the corresponding circuit is a set of modulo p multipliers working in parallel (Section 8.3.2, procedure `by_coefficient`). In order to multiply two polynomials, Algorithm 8.22 can be used. The corresponding data path is shown in Figure 15.12. Initially, $a(x)$ must be stored in a (nonrepresented) p-ary shift register, which implements the `right_rotate` function.

The circuit of Figure 15.12 includes $2.n$ multipliers and n adders. For relatively large values of p, the corresponding cost could be excessive. Another solution is a completely sequential implementation (see next example).

Example 15.8 (Complete VHDL source code available.) Generate a sequential multiplier based on Algorithm 8.21. A possible data path is shown in Figure 15.13. The VHDL descriptions of the data path and of the control unit are the following:

```
entity poly_data_path is
port (
  a, b: in polynomial;
  clk, clear_z, write_enable: std_logic;
  addr_i, addr_j: in address;
  z: inout polynomial
);
end poly_data_path;
```

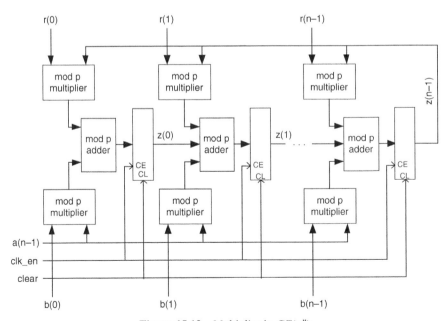

Figure 15.12 Multiplier in $GF(p^n)$.

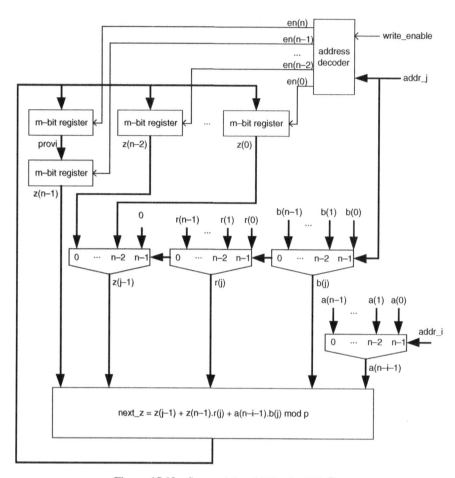

Figure 15.13   Sequential multiplier in $GF(p^n)$.

```
architecture circuit of poly_data_path is
   component main_operation ... end component;
   signal z_jminus1, z_nminus1, r_j, a_nminus, b_j, next_z,
   provi: std_logic_vector(m-1 downto 0);
   signal en: std_logic_vector(n downto 0);
   signal r: polynomial;
begin
   r<=poly_module;
   --multiplexers
   z_jminus1<=z(n-addr_j-2) when addr_j<n-1 else zero;
   r_j<=r(n-addr_j-1) when addr_j<n else zero;
   b_j<=b(n-addr_j-1) when addr_j<n else zero;
   a_nminus<=a(n-addr_i-1);
   --
```

```
z_nminus1<=z(n-1);
main_component: main_operation port map(z_jminus1,
z_nminus1, r_j, a_nminus, b_j, next_z);
--address decoder:
process(addr_j, write_enable)
begin
  for i in 0 to n-2 loop
  if addr_j=n-i-1 and write_enable='1' then en(i)<='1';
  else en(i)<='0'; end if;
  end loop;
  if addr_j=n and write_enable='1' then en(n-1)<='1'; else
  en(n-1)<='0'; end if;
  if addr_j=0 and write_enable='1' then en(n)<='1'; else
  en(n)<='0'; end if;
end process;
--
registers: for i in 0 to n-2 generate
  process(clk)
  begin
    if clk'event and clk='1' then
      if clear_z='1' then z(i)<=zero;
      elsif en(i)='1' then z(i)<=next_z; end if;
    end if;
  end process;
end generate;
process(clk)
begin
  if clk'event and clk='1' then
    if clear_z='1' then z(n-1)<=zero;
    elsif en(n-1)='1'then z(n-1)<=provi; end if;
  end if;
end process;
process(clk)
begin
  if clk'event and clk='1' then
    if en(n)='1' then provi<=next_z; end if;
  end if;
end process;
end circuit;

library ieee; use ieee.std_logic_1164.all;
use work.mypackage.all;
entity poly_control_unit is
port (
  clk, reset, start: in std_logic;
  addr_i, addr_j: inout natural;
  clear_z, write_enable, done: out std_logic
);
end poly_control_unit;
```

```
architecture rtl of poly_control_unit is
  subtype internal_state is natural range 0 to 5;
  signal state: internal_state;
begin
  process(clk, reset)
  begin
    case state is
    when 0=>clear_z<='0'; write_enable<='0'; done<='1';
    when 1=>clear_z<='0'; write_enable<='0'; done<='1';
    when 2=>clear_z<='1'; write_enable<='0'; done<='0';
    when 3=>clear_z<='0'; write_enable<='0'; done<='0';
    when 4=>clear_z<='0'; write_enable<='1'; done<='0';
    when 5=>clear_z<='0'; write_enable<='0'; done<='0';
    end case;
    if reset='1' then state<=0;
    elsif clk'event and clk='1' then
    case state is
    when 0=>if start='0' then state<=state+1; end if;
    when 1=>if start='1' then state<=state+1; end if;
    when 2=>addr_i<=0; state<=state+1;
    when 3=>addr_j<=0; state<=state+1;
    when 4=>if addr_j=n then state<=state+1; else
    addr_j<=addr_j+1; end if;
    when 5=>if addr_i=n-1 then state<=0;
      else addr_i<=addr_i+1; state<=3; end if;
    end case;
    end if;
  end process;
end rtl;
```

15.4   INVERSION IN $GF(p^n)$

In order to execute Algorithm 8.24, the computation resources that correspond to the procedures should be synthesized. Most of them (invert, by_coefficient, sub) have been studied in the preceding sections. The shift procedure can be implemented with a barrel shifter ([ULL1984]). The implementation of the degree procedure could be based on the following iterative algorithm.

Algorithm 15.1   Degree Computation

```
state(n):=0;
for i in 1..n-1 loop
  if state (n-i+1)=0 and a(n-i)=0 then state (n-i):=0;
  else state (n-i):=1; end if;
end loop;
degree_a:=count (state);
```

where the `count` function returns the number of 1's in `state`. The corresponding iterative circuit includes $n - 1$ cells. Each of them computes state(i) as a function of $a(i)$ and `state(i+1)`: if `state(i+1)` $= 0$ and $a(i) = 0$ then `state(i)` $= 0$; in all other cases `state(i)` $= 1$. The circuit that generates the output `degree` is an $(n - 1)$-to-$\log_2(n)$ binary counter.

The data path corresponding to Algorithm 8.24 is shown in Figure 15.14. It is made up of the following computation resources:

degree: implements the degree procedure,

shifter: implements the shift procedure,

coefficient_inverter: implements the invert procedure,

subtractor: implements the sub procedure,

coefficient_multiplier: implements the by_coefficient procedure.

Furthermore, a lot of memory (registers) and connection (multiplexers) resources are necessary. The computation time is proportional to the number of executions of the main iteration (Algorithm 8.24, **while** t>0 **loop**...**end** **loop**). As the degree of r is reduced at every step, the maximum number of iteration steps is n.

Example 15.9 (Complete VHDL source code available.) Generate the VHDL model of a circuit that computes $(a(x))^{-1}$ modulo $f(x)$:

```
entity polynomial_inverter is
port (
  a: in polynomial;
  result: out polynomial;
  start, clk, reset: in std_logic;
  done: out std_logic
);
end polynomial_inverter;

architecture circuit of polynomial_inverter is
  component degree ... end component;
  component selector ... end component;
  component shifter ... end component;
  component coefficient_inverter ... end component;
  component subtractor ... end component;
  component coefficient_multiplier ... end component;
  signal u, next_u, v, next_v, c, next_c, e, next_e, k_by_v,
  k_by_v_shifted, r, k_by_e, k_by_e_shifted, cc, r_a:
  polynomial;
  signal m, next_m, t, next_t, j, deg_v: index;
  signal uu_m, next_u_m, u_m, v_t, v_t_inverted, k, k_v:
  coefficient; signal load, sign, t_equal_zero: std_logic;
  signal mux_control: std_logic_vector (1 downto 0);
  subtype state_type is natural range 0 to 8;
```

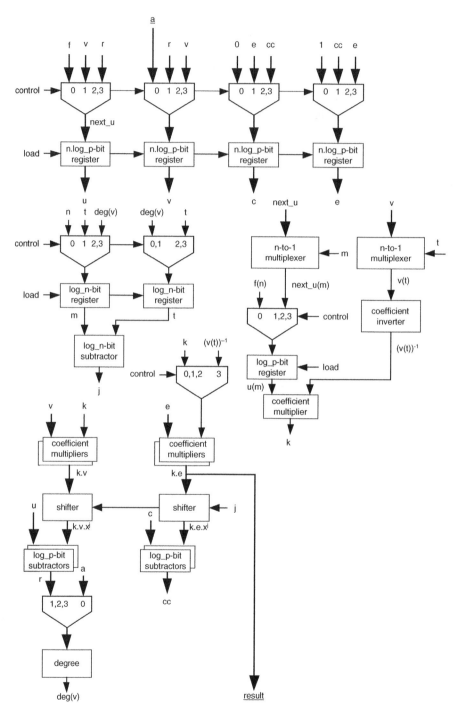

Figure 15.14 Inverter in $GF(p^n)$.

```
  signal state: state_type;
begin
  --data_path
  f<=irreducible; f_n<=irreducible_n;
  process (clk)
  begin
    if clk'event and clk='1' then
      if load='1' then u<=next_u; v<=next_v; c<=next_c;
      e<=next_e; m<=next_m; t<=next_t; u_m<=next_u_m;
      end if;
    end if;
end process;
process (mux_control, f_n, f, v, r, a, e, cc, t, deg_v, k,
v_t_inverted, uu_m)
begin
    case mux_control is
      when "00"=>next_u<=f; next_v<=a;
        next_c<=zero_polynomial; next_e<=one_polynomial;
        next_m<=conv_std_logic_vector(n, logn);
        next_t<=deg_v; next_u_m<=f_n; r_a<=a; k_v<=k;
      when "01";=>next_u<=v; next_v<=r; next_c<=e;
        next_e<=cc; next_m<=t; next_t<=deg_v;
        next_u_m<=uu_m; r_a<=r; k_v<=k;
      when "10"=>next_u<=r; next_v<=v; next_c<=cc;
        next_e<=e; next_m<=deg_v; next_t<=t;
        next_u_m<=uu_m; r_a<=r; k_v<=k;
      when others=>next_u<=r; next_v<=v; next_c<=cc;
        next_e<=e; next_m<=deg_v; next_t<=t;
        next_u_m<=uu_m; r_a<=r; k_v<=v_t_inverted;
   end case;
end process;
j<=m - t;
selector1: selector port map (next_u, m, uu_m);
selector2: selector port map (v, t, v_t);
inverter: coefficient_inverter port map (v_t, v_t_inverted);
multiplier1: coefficient_multiplier
port map (u_m, v_t_inverted, k);
multipliers1:
for i in 0 to n-1 generate
  multiplier2: coefficient_multiplier
  port map (v(i), k, k_by_v(i));
end generate;
shifter1: shifter port map (k_by_v, j, k_by_v_shifted);
subtractors1: for i in 0 to n-1 generate
  subtractor1: subtractor port map
  (u(i), k_by_v_shifted(i), r(i));
end generate;
multipliers2:
for i in 0 to n-1 generate
```

```
  multiplier3: coefficient_multiplier
   port map (e(i), k_v, k_by_e(i));
end generate;
shifter2: shifter port map (k_by_e, j, k_by_e_shifted);
subtractors2: for i in 0 to n-1 generate
   subtractor2: subtractor port map
   (c(i), k_by_e_shifted(i), cc(i));
end generate;
degree1: degree port map (r_a, deg_v);
sign<='0' when conv_integer(t)>=conv_integer(deg_v)
else '1';
t_equal_zero<='1' when conv_integer(t)=0 else '0';
result<=k_by_e;
--control unit:
process (clk, reset, state)
begin
   case state is
      when 0=>load<='0'; mux_control<="11"; done<='1';
      when 1=>load<='0'; mux_control<="11"; done<='1';
      when 2=>load<='1'; mux_control<="00"; done<='0';
      when 3=>load<='0'; mux_control<="00"; done<='0';
      when 4=>load<='1'; mux_control<="01"; done<='0';
      when 5=>load<='1'; mux_control<="10"; done<='0';
      when 6=>load<='0'; mux_control<="01"; done<='0';
      when 7=>load<='0'; mux_control<="10"; done<='0';
      when 8=>load<='0'; mux_control<="11"; done<='0';
   end case;
   if reset='1' then state<=0;
   elsif clk'event and clk='1' then
```

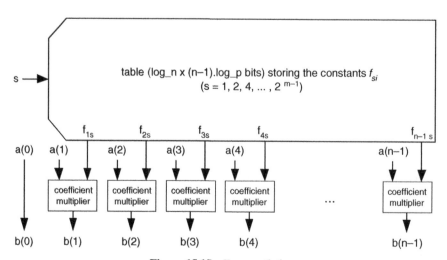

Figure 15.15 Exponentiation.

```
case state is
  when 0=>if start='0' then state<=1; end if;
  when 1=>if start='1' then state<=2; end if;
  when 2=>state<=3;
  when 3=>if t_equal_zero='1' then state<=8;
    elsif sign='0' then state<=4;
    else state<=5; end if;
```

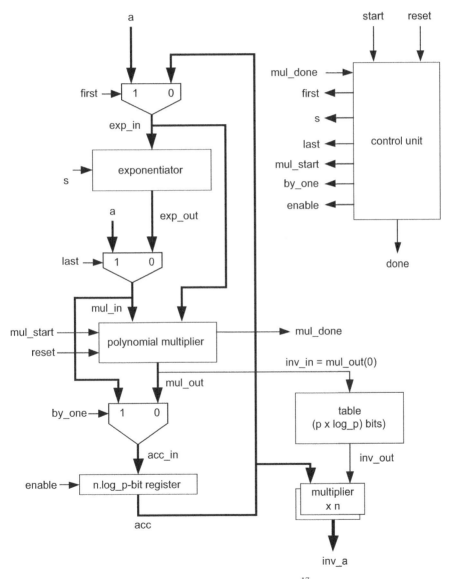

Figure 15.16 Inverter in $GF(239^{17})$.

```
    when 4=>state<=6;
    when 5=>state<=7;
    when 6=>if t_equal_zero='1' then state<=8;
       elsif sign='0' then state<=4;
       else state<=5; end if;
    when 7=>if t_equal_zero='1' then state<=8;
       elsif sign='0' then state<=4;
       else state<=5; end if;
    when 8=>state<=0;
  end case;
  end if;
 end process;
end circuit;
```

As mentioned earlier (Chapter 8), a different method of inversion can be used if $f(x)$ is a binomial. In particular, if

$$f(x) = x^n - c, \quad p \bmod n = 1, n = 2^m + 1,$$

then Algorithm 8.28 can be used. The computation resources corresponding to the procedures must be synthesized. Most of them (`multiply`, `invert`, `by_ coefficient`) have been studied in the preceding sections. The *exponentiation* procedure can be implemented by a table storing the coefficients f_{si} (Appendix 8.1) for $s = 1, 2, 2^2, \ldots, 2^{m-1}$, and $n - 1$ coefficient multipliers (Figure 15.15).

Example 15.10 (Complete VHDL source code available.) The circuit of Figure 15.16 implements the inversion in $GF(p^n)$ with

$$p = 239 \quad \text{and} \quad f(x) = x^{17} - 2.$$

As p is small, the inversion in $GF(239)$ is implemented by a table storing $x^{-1} \bmod 239$ for all x in $\{1, 2, \ldots, 238\}$.

15.5 BIBLIOGRAPHY

[AGN1991] G. B. Agnew, R. C. Mullin and S. A. Vanstone, An implementation of elliptic curve cryptosystems over F_2^{155}. *Cryptology*, **3**: 63–79 (1991).

[DES2002] J.-P. Deschamps and G. Sutter, FPGA implementation of modular multipliers. In: *Proceedings of the XVII Design of Circuits and Integrated Systems Conference*, Santander, Spain, November 2002, pp. 107–112.

[DES2003] J.-P. Deschamps and G. Sutter, Multiplication in a finite extension ring, In: *Proceedings of the XVIII Design of Circuits and Integrated Systems Conference*, Ciudad Real, Spain, November 2003, pp. 181–185.

[KIM2002] C. H. Kim, S. Oh and J. Lim, A new hardware architecture for operations in $GF(2^n)$. *IEEE Trans. Comput.* **51**(1): 90–92 (2002).

[PAA1996] C. Paar, A new architecture for a parallel finite field multiplier with low complexity on composite fields. *IEEE Trans. Comput.*, **45**(7):856–861 (1996).

[ULL1984] J. Ullman, *Computational Aspects of VLSI*, Computer Science Press, Maryland, 1984.

[WOO2000] A. D. Woodbury, D. V. Bailey, and C. Paar, Elliptic curve cryptography on smart cards without coprocessors. IFIP CARDIS 71–92 (2000).

16

FLOATING-POINT UNIT

There are many data processing applications (e.g., image and voice processing), which use a large range of values and need a relatively high precision. In such cases, instead of encoding the information in the form of integers or fixed-point numbers, an alternative solution is a floating-point representation (Chapter 3). In the first section of this chapter, a method is proposed for defining a particular floating-point representation system as a function of the application specification. The next section is devoted to the algorithms for executing the basic arithmetic operations. The two following sections define the main rounding methods and introduce the concept of guard digit. Finally, the last few sections propose basic implementations of the arithmetic operations, namely, addition and subtraction, multiplication, division, and square root.

16.1 FLOATING-POINT SYSTEM DEFINITION

Assume that a set of real numbers x belonging to the interval

$$-x_{max} \le x \le x_{max}$$

is represented in such a way that the following specifications are satisfied:

d_1 is the maximum distance between small exactly-represented non zero numbers;

Synthesis of Arithmetic Circuits: FPGA, ASIC, and Embedded Systems
By Jean-Pierre Deschamps, Géry J. A. Bioul, and Gustavo D. Sutter
Copyright © 2006 John Wiley & Sons, Inc.

d_2 is the maximum distance between large exactly-represented numbers;

x_{min} is the maximum distance between 0 and the smallest exactly-represented numbers:

where the adjectives *small* and *large* refer to the absolute value of the corresponding numbers.

Every number x will be represented in the form $\pm s.b^e$, with $b \geq 2$, s being the significand and e the exponent.

In order to make the implementation of the arithmetic operations easier (Section 16.2), the two following conditions must be satisfied:

1. The significand s is represented in base $B = b$.
2. The significand belongs to the interval

$$1 \leq s \leq B - ulp. \tag{16.1}$$

Thus x is expressed in the form

$$(1.s_{-1}s_{-2}..s_{-p}).B, \quad \text{where } e_{min} \leq e \leq e_{max}.$$

The values of p, e_{min}, and e_{max} are chosen in such a way that

$$B^{e_{min}} \leq x_{min}, \quad \text{that is, } e_{min} \leq \log_B (x_{min}), \tag{16.2}$$

$$B^{-p}.B^{e_{min}} \leq d_1, \quad \text{that is, } e_{min} - p \leq \log_B (d_1), \tag{16.3}$$

$$B^{-p}.B^{e_{max}} \leq d_2, \quad \text{that is, } e_{max} - p \leq \log_B (d_2), \tag{16.4}$$

$$2.B^{e_{max}} \geq x_{max} \quad \text{that is, } e_{max} \geq \log_B (x_{max}/2). \tag{16.5}$$

Example 16.1   Define a floating-point representation system where

$$x_{max} = 2^{30}, x_{min} = 2^{-30}, d_1 = 2^{-50}, d_2 = 2^{10}.$$

Choose $B = 2$. A straightforward solution of the system (16.2)–(16.5) is

$$e_{max} = \log_2 (x_{max}/2) = 29.$$

$$p = e_{max} - \log_2 (d_2) = 30 - 10 = 20,$$

$$e_{min} = \min \{p + \log_2 (d_1), \log_2 (x_{min})\} = \min \{20 - 50, -30\} = -30.$$

The smallest nonzero exactly-represented positive number is 2^{-30}; the distance between small exactly-represented numbers is

$$2^{-20}.2^{-30} = 2^{-50};$$

the largest exactly-represented positive number is

$$(1.11 \ldots 11).2^{29} \cong 2^{30};$$

the distance between large exactly-represented numbers is

$$(0.00 \ldots 01).2^{29} = 2^{-20}.2^{29} = 2^9 < 2^{10}.$$

16.2 ARITHMETIC OPERATIONS

First analyze the main arithmetic operations and generate the corresponding computation algorithms.

16.2.1 Addition of Positive Numbers

Given two positive floating-point numbers $s_1.B^{e1}$ and $s_2.B^{e2}$ their sum $s.B^e$ is computed as follows.

Assume that e_1 is greater than or equal to e_2; then (*alignment*) the sum of $s_1.B^{e1}$ and $s_2.B^{e2}$ can be expressed in the form $s.B^e$, where

$$s = s_1 + s_2/(B^{e1-e2}) \text{ and } e = e_1. \tag{16.6}$$

The value of s belongs to the interval

$$1 \le s \le 2.B - 2.ulp \tag{16.7}$$

so that s could be greater than or equal to B. If it is the case, that is, if

$$B \le s \le 2.B - 2.ulp, \tag{16.8}$$

then (*normalization*) substitute s by s/B, and e by $e + 1$, so that the value of $s.B^e$ is the same as before, and the new value of s satisfies

$$1 \le s \le 2 - (2/B).ulp \le B - ulp, B \ge 2. \tag{16.9}$$

The significands s_1 and s_2 of the operands are multiples of *ulp*. If e_1 is greater than e_2, the value of s could no longer be a multiple of *ulp* and some rounding function should be applied to s. Assume that

$$s' < s < s'' = s' + ulp,$$

s' and s'' being two successive multiples of *ulp*. Then the *rounding* function associates to s either s' or s'', according to some rounding strategy. According to (16.9) and to the fact that 1 and $B - ulp$ are multiples of *ulp*, it is obvious that

$$1 \le s' < s'' \le B - ulp.$$

Nevertheless, if condition (16.8) does not hold, that is, if

$$1 \le s < B, \tag{16.10}$$

s could belong to the interval

$$B - ulp < s < B, \tag{16.11}$$

so that *rounding*(*s*) could be equal to *B*. A new *normalization step* would be necessary, that is, substitution of $s = B$ by $s = 1$ and e by $e + 1$.

Algorithm 16.1   Sum of Positive Numbers

```
if e1>=e2 then e:=e1; s:=s1+(s2/B*(e1-e2));
else e:=e2; s:=(s1/B*(e2-e1))+s2; end if;
if s>=B then e:=e+1; s:=s/B; end if;
s:=round(s);
if s>=B then e:=e+1; s:=s/B; end if;
```

Examples 16.2   Assume that $B = 10$ and $ulp = 10^{-4}$, so that the numbers are represented in the form $s.10^e$ where $1 \le s \le 9.9999$.

1. Compute $z = (3.4375 \times 10^3) + (2.5491 \times 10^{-1})$:

    alignment: $z = (3.4375 + 0.00025491) \times 10^3 = 3.43775491 \times 10^3$,

    $3.43775491 < 10$,

    rounding: $s \cong 3.4378$,

    $3.4378 < 10$,

    $z = 3.4378 \times 10^3$

2. Compute $z = (9.4375 \times 10^3) + (8.6247 \times 10^2)$:

    alignment: $z = (9.4375 + 0.86247) \times 10^3 = 10.29997 \times 10^3$,

    normalization: $s = 1.029997, e = 4$,

    rounding: $s \cong 1.0300$,

    $1.0300 < 10$,

    $z = 1.0300 \times 10^4$.

3. Compute $z = (9.4375 \times 10^3) + (5.6247 \times 10^2)$:

    alignment: $z = (9.4375 + 0.56247) \times 10^3 = 9.99997 \times 10^3$,

    $9.99997 < 10$,

    rounding: $s \cong 10.0000$,

    normalization: $s = 1.0000, e = 4$,

    $z = 1.0000 \times 10^4$.

Comment 16.1   The addition of two positive numbers could produce an *overflow*, as the final value of e could be greater than e_{max}.

16.2.2 Difference of Positive Numbers

Given two positive floating-point numbers $s_1.B^{e1}$ and $s_2.B^{e2}$ their difference $s.B^e$ is computed as follows:

Assume that e_1 is greater than or equal to e_2; then (*alignment*) the difference between $s_1.B^{e1}$ and $s_2.B^{e2}$ can be expressed in the form $s.B^e$, where

$$s = s_1 - s_2/(B^{e1-e2}) \text{ and } e = e_1. \tag{16.12}$$

The value of s belongs to the interval

$$-(B - ulp) \le s \le B - ulp. \tag{16.13}$$

If s is negative, then it is substituted by $-s$ and the sign of the final result will be modified accordingly. If s is equal to 0, an exception `equal_zero` could be raised. It remains to consider the case where

$$0 < s \le B - ulp.$$

The value of s could be smaller than 1. In order to normalize the significand, a procedure

procedure `leading_zeroes(s:` **in** fixed_point; `k:` **out** natural`)`

must be executed: it counts the number of initial 0's of the representation of s. In other words, it looks for the minimum exponent k such that $s.B^k \ge 1$. Then s is substituted by $s.B^k$ and e by $e - k$. Thus, the relation (16.10) holds, that is,

$$1 \le s < B.$$

It remains to round (up or down) the significand and to normalize it if necessary.

Algorithm 16.2   Difference of Positive Numbers

```
if e1>=e2 then e:=e1; s:=s1-(s2/B**(e1-e2));
else e:=e2; s:=(s1/B**(e2-e1))-s2; end if;
if s<0 then s:=-s; sign:=1; end if;
leading_zeroes(s, k);
s:=s*(B**k); e:=e-k;
s:=round(s);
if s>=B then e:=e+1; s:=s/B; end if;
```

Examples 16.3   Assume again that $B = 10$ and $ulp = 10^{-4}$, so that the numbers are represented in the form $s.10^e$ where $1 \le s \le 9.9999$. For computing the difference, the 10's complement system is used.

1. Compute $z = (3.4518 \times 10^{-1}) - (7.2471 \times 10^3)$:

   alignment: $z = (0.00034518 - 7.2471) \times 10^3$

   $\qquad\qquad = (00.00034518 + 92.75289999 + 1) \times 10^3$

   $\qquad\qquad = 92.75324518 \times 10^3$,

   change the sign: $-s = 07.24675481 + 1 = 7.24675482$,

   $7.24675482 \geq 1$,

   rounding: $-s = 7.2468$,

   $7.2468 < 10$

   $z = -7.2468 \times 10^3$.

2. Compute $z = (1.0014 \times 10^3) - (9.9491 \times 10^2)$:

   alignment: $z = (1.0014 - 0.99491) \times 10^3$

   $\qquad\qquad = (01.0014 + 99.00508 + 1) \times 10^3 = 00.00649 \times 10^3$,

   $00.00649 < 0$,

   leading zeroes: $s = 6.4900, e = 0$,

   rounding: $s = 6.4900$,

   $6.4900 < 10$,

   $z = 6.4900 \times 10^0$.

3. Compute $z = (1.0714 \times 10^4) - (7.1403 \times 10^2)$:

   alignment: $z = (1.0714 - 0.071403) \times 10^4$

   $\qquad\qquad = (01.0714 + 99.928596 + 1) \times 10^4 = 00.999997 \times 10^4$,

   $00.999997 > 0$,

   leading zeroes: $s = 9.99997, \quad e = 3$,

   rounding: $s = 10.0000$,

   normalization: $s = 1.0000, \quad e = 4$,

   $z = 1.0000 \times 10^4$.

Comment 16.2   The difference of two positive numbers could produce an *underflow*, as the final value of e could be smaller than e_{min}.

16.2.3   Addition and Subtraction

Given two floating-point numbers $(-1)^{sign1}.s_1.B^{e1}$ and $(-1)^{sign2}.s_2.B^{e2}$, and a control variable `operation`, an algorithm is defined for computing

$$z = (-1)^{sign}.s.B^e = (-1)^{sign1}.s_1.B^{e1} + (-1)^{sign2}.s_2.B^{e2}, \text{ if operation} = 0,$$

$$z = (-1)^{sign}.s.B^e = (-1)^{sign1}.s_1.B^{e1} - (-1)^{sign2}.s_2.B^{e2}, \text{ if operation} = 1.$$

TABLE 16.1

operation	$sign_1$	$sign_2$	actual operation
0	0	0	$s_1 + s_2$
0	0	1	$s_1 - s_2$
0	1	0	$-(s_1 - s_2)$
0	1	1	$-(s_1 + s_2)$
1	0	0	$s_1 - s_2$
1	0	1	$s_1 + s_2$
1	1	0	$-(s_1 + s_2)$
1	1	1	$-(s_1 - s_2)$

Once the significands have been aligned, the actual operation (addition or subtraction of the significands) depends on the values of operation, $sign_1$, and $sign_2$ (Table 16.1).

The following algorithm, based on Algorithms 16.1 and 16.2 as well as Table 16.1, computes z.

Algorithm 16.3   Addition and Subtraction

```
if e1>=e2 then e:=e1; s2:=s2/B**(e1-e2);
else e:=e2; s1:=s1/B**(e2-e1); end if;
sign:=sign1;
if operation xor sign1 xor sign2=0 then
  s:=s1+s2;
  if s>=B then e:=e+1; s:=s/B; end if;
  s:=round(s);
  if s>=B then e:=e+1; s:=s/B; end if;
else
  s:=s1-s2;
  if s<0 then s:=-s; sign:=1-sign; end if;
  leading_zeroes(s, k);
  s:=s*(B**k); e:=e-k;
  s:=round(s);
  if s>=B then e:=e+1; s:=s/B; end if;
end if;
```

As regards the hardware implementation, the following equivalent algorithm is better.

Algorithm 16.4   Addition and Subtraction, Second Version

```
if operation=1 then sign2:=1-sign2; end if;
if e1<e2 then swap(sign1, sign2); swap(s1, s2); swap (e1, e2);
end if;
e:=e1; s2:=s2/B**(e1-e2); sign:=sign1;
if sign xor sign2=0 then
```

```
    s:=s1+s2;
    if s>=B then e:=e+1; s:=s/B; end if;
else
    if (e1=e2) and (s1<s2) then swap(s1, s2); sign:=1-sign;
    end if;
    s:=s1-s2;
    leading_zeroes(s, k);
    s:=s*(B**k); e:=e-k;
end if;
s:=round(s);
if s>=B then e:=e+1; s:=s/B; end if;
```

16.2.4 Multiplication

Given two floating-point numbers $(-1)^{sign1}.s_1.B^{e1}$ and $(-1)^{sign2}.s_2.B^{e2}$, their product $(-1)^{sign}.s.B^e$ is computed as follows:

$$sign = sign_1 \ xor \ sign_2, \quad s = s_1.s_2, \quad e = e_1 + e_2. \tag{16.14}$$

The value of s belongs to the interval

$$1 \le s \le (B - ulp)^2, \tag{16.15}$$

and could be greater than or equal to B. If it is the case, that is, if

$$B \le s \le (B - ulp)^2, \tag{16.16}$$

then (*normalization*) substitute s by s/B, and e by $e + 1$. The new value of s satisfies

$$1 \le s \le (B - ulp)^2/B = B - 2.ulp + (ulp)^2/B < B - ulp \tag{16.17}$$

($ulp < B$ so that $2 - ulp/B > 1$).

It remains to round the significand and to normalize if necessary.

Algorithm 16.5 Multiplication

```
sign:=sign1 xor sign2; s:=s1*s2; e:=e1+e2;
if s>=B then e:=e+1; s:=s/B; end if;
s:=round(s);
if s>=B then e:=e+1; s:=s/B; end if;
```

Examples 16.4 Assume that $B = 10$ and $ulp = 10^{-4}$, so that the numbers are represented in the form $s.10^e$, where $1 \le s \le 9.9999$.

1. Compute $z = (3.4382 \times 10^3) \times (2.5471 \times 10^{-1})$:

   $z = 8.75743922 \times 10^2$,

   $8.75743922 < 10$,

   rounding: $s \cong 8.7574$,

   $8.7574 < 10$,

   $z = 8.7574 \times 10^{-2}$.

2. Compute $z = (9.4300 \times 10^3) \times (8.6200 \times 10^2)$:

   $z = 81.2866 \times 10^5$,

   normalization: $s = 8.12866$, $e = 6$,

   rounding: $s \cong 8.1287$,

   $8.1287 < 10$,

   $z = 8.1287 \times 10^6$.

3. Compute $z = (4.7619 \times 10^2) \times (2.1000 \times 10^3)$:

   $z = 9.99999 \times 10^5$,

   $9.99999 < 10$,

   rounding: $s \cong 10.00$,

   normalization: $s = 1, e = 6$,

   $z = 1.0000 \times 10^6$.

Comment 16.3   The product of two real numbers could produce an *overflow* as the final value of e could be greater than e_{max}.

16.2.5 Division

Given two floating-point numbers $(-1)^{sign1}.s_1.B^{e1}$ and $(-1)^{sign2}.s_2.B^{e2}$ their quotient $(-1)^{sign}.s.B^e$ is computed as follows:

$$sign = sign_1 \text{ xor } sign_2, \quad s = s_1/s_2, \quad e = e_1 - e_2 \qquad (16.18)$$

The value of s belongs to the interval

$$1/B < s \leq B - ulp, \qquad (16.19)$$

and could be smaller than 1. If that is the case, that is if $s = s_1/s_2 < 1$, then

$$s_1 < s_2, \quad s_1 \leq s_2 - ulp, \quad s_1/s_2 \leq 1 - ulp/s_2 < 1 - ulp/B,$$

and

$$1/B < s < 1 - ulp/B. \qquad (16.20)$$

Then (*normalization*) substitute s by $s.B$, and e by $e - 1$. The new value of s satisfies

$$1 < s < B - ulp \tag{16.21}$$

It remains to round the significand.

Algorithm 16.6   Division

```
sign:=sign1 xor sign2; s:=s1/s2; e:=e1 - e2;
if s<1 then e:=e-1; s:=s*B; end if;
s:=round(s);
```

Examples 16.5   Assume that $B = 10$ and $ulp = 10^{-4}$, so that the numbers are represented in the form $s.10^e$, where $1 \le s \le 9.9999$.

1. Compute $z = (3.4375 \times 10^3)/(2.5491 \times 10^{-1})$:

   $z = 1.3485152 \times 10^2$,

   $1.3485152 \ge 1$,

   rounding: $s \cong 1.3485$,

   $z = 1.3485 \times 10^2$.

2. Compute $z = (2.5491 \times 10^{-1})/(3.4375 \times 10^3)$:

   $z = 0.74155564 \times 10^{-4}$,

   normalization: $s = 7.4155564, e = -5$,

   rounding: $s \cong 7.4156$,

   $z = 7.4156 \times 10^{-5}$.

Comment 16.4   The quotient of two real numbers could produce an *underflow*, as the final value of e could be smaller than e_{min}.

16.2.6   Square Root

Given a positive floating-point number $s_1.B^{e1}$, its square root $s.B^e$ is computed as follows:

$$\text{if } e_1 \text{ is even,} \quad s = (s_1)^{1/2}, e = e_1/2; \tag{16.22}$$

$$\text{if } e_1 \text{ is odd,} \quad s = (s_1/B)^{1/2}, e = (e_1 + 1)/2. \tag{16.23}$$

In the first case (16.22),

$$1 \le s \le (B - ulp)^{1/2} < B - ulp. \tag{16.24}$$

In the second case (16.23),

$$(1/B)^{1/2} \leq s \leq 1, \tag{16.25}$$

and (*normalization*) s must be substituted by $s.B$ and e by $e - 1$, so that

$$1 \leq s < B.$$

It remains to round the significand and to normalize if necessary.

Algorithm 16.7 Square Root

```
if (e1 mod 2)=1 then s1:=s1/B; e1:=e1+1; end if;
s:=square_root(s1); e:=e1/2;
if s<1 then e:=e-1; s:=s*B; end if;
s:=round(s);
if s>=B then e:=e+1; s:=s/B; end if;
```

Examples 16.6 Assume that $B = 10$ and $ulp = 10^{-4}$, so that the numbers are represented in the form $s.10^e$, where $1 \leq s \leq 9.9999$.

1. Compute $z = (9.9491 \times 10^2)^{1/2}$:

   2 even ,

   $z = 3.1542194 \times 10^1$

   rounding: $s \cong 3.1542$,

   $z = 3.1542 \times 10^1$

2. Compute $z = (3.4518 \times 10^{-1})^{1/2}$:

   -1 odd,

   $s = 0.34518, \quad e = 0,$

   $z = 0.5875202 \times 10^0,$

   normalization: $s = 5.875202, \quad e = -1,$

   rounding: $s \cong 5.8752,$

   $z = 5.8752 \times 10^{-1}.$

3. Compute $z = (9.9999 \times 10^3)^{1/2}$:

   3 odd ,

   $s = 0.99999, \quad e = 4,$

   $z = 0.9999949 \times 10^2,$

   normalization: $s = 9.999949, \quad e = 1,$

   some rounding schemes (e.g., toward infinite) generate $s \cong 10.0000,$

   normalization: $s = 1.0000, \quad e = 2,$

   $z = 1.0000 \times 10^2.$

Comments 16.5   The square rooting of a real number could produce an *underflow*, as the final value of e could be smaller than e_{min}.

16.3   ROUNDING SCHEMES

Given a real number x and a floating-point representation system, the following situations could occur:

1. $|x| < s_{min}.B^{e_{min}}$, that is, an *underflow* situation.
2. $|x| > s_{max}.B^{e_{max}}$, that is an *overflow* situation.
3. $|x| = s.B^e$, where $e_{min} \leq e \leq e_{max}$ and $s_{min} \leq s \leq s_{max}$.

In the third case, either s is a multiple of *ulp*, in which case a rounding operation is not necessary, or it is included between two multiples s' and s'' of *ulp*:

$$s' < s < s''.$$

The *rounding* operation associates to s either s' or s'', according to some rounding strategy. The most common are the following ones.

Definitions 16.1

1. The *truncation* (*round toward 0*, *chopping*) method is accomplished by dropping the extra digits, that is,

    $round(s) = s'$ if s is positive, $round(s) = s''$ if s is negative.

2. The *round toward plus infinity* is defined by

    $$round(s) = s'', \quad \text{whatever the sign of } x,$$

    and the *round toward minus infinity* by

    $$round(s) = s'.$$

3. The *round to nearest* method associates s with the closest value, that is,

    if $s < s' + ulp/2, round(s) = s'$,   and if $s > s' + ulp/2, round(s) = s''$.

If the distances to s' and s'' are the same, that is, if $s = s' + ulp/2$, there are several options. For instance:

$round(s) = s'$;
$round(s) = s''$;
$round(s) = s'$ if s' is an even multiple of *ulp*, $round(s) = s''$ if s'' is an even multiple of *ulp*;
$round(s) = s'$ if s' is an odd multiple of *ulp*, $round(s) = s''$ if s'' is an odd multiple of *ulp*.

The preceding schemes (*round to the nearest*) produce the smallest absolute error, and the two last ones (*tie to even, tie to odd*) also produce the smallest average absolute error (*unbiased*, 0-*bias* representation systems).

Assume now that the exact result of an operation, after normalization, is

$$s = 1.s_{-1}s_{-2}s_{-3} \ldots s_{-p}|s_{-(p+1)}s_{-(p+2)}s_{-(p+3)}..$$

where *ulp* is equal to B^{-p} (the | symbol indicates the separation between the digit which corresponds to the *ulp* and the following ones). Whatever the chosen rounding scheme (Definitions 16.1), it is not necessary to have previously computed all the digits $s_{-(p+1)}$ $s_{-(p+2)}$ ...; it is sufficient to know whether all the digits $s_{-(p+1)}$ $s_{-(p+2)}$ ... are equal to 0, or not. For example, the following algorithm computes *round(s)* if the *round to the nearest, tie to even* scheme is used:

Algorithm 16.8   Round to the Nearest, Tie to Even

```
s1:=1.s(-1) s(-2)...s(-p);
s2:=s-s1-s(-(p+1)).ulp/B;  --s2=0.00..0|0 s_(p+2) s_(p+3)..
if s(-(p+1))<B/2 then round:=s1;
elsif s(-(p+1))>B/2 then round:=s1+ulp;
elsif s(-(p+1))=B/2 and s2>0 then round:=s1+ulp;
elsif s(-(p+1))=B/2 and s2=0 and (s(-p) mod 2)=0 then
round:=s1;
elsif s(-(p+1))=B/2 and s2=0 and (s(-p) mod 2)=1 then
round:=s1+ulp;
end if;
```

In order to execute the preceding algorithm it is sufficient to know

the value of $s_1 = 1.s_{-1} \, s_{-2} \, s_{-3} \ldots s_{-p}$,
the value of $s_{-(p+1)}$,
whether $s_2 = 0.00 \ldots 0 \mid 0 \, s_{-(p+2)} \, s_{-(p+3)} \ldots$ is equal to 0, or not.

16.4   GUARD DIGITS

Consider the exact result r of an operation, before normalization. According to (16.7), (16.13), (16.15), (16.19), (16.24) and (16.25),

$$r < B^2, \quad \text{that is }, r = r_1 r_0.r_{-1}r_{-2}r_{-3} \ldots r_{-p}|r_{-(p+1)}r_{-(p+2)}r_{-(p+3)} \ldots.$$

The normalization operation (if necessary) is accomplished by

dividing the result by B (sum of positive numbers, multiplication),
multiplying the result by B (division, square root),
multiplying the result by B^k (difference of positive numbers).

Furthermore, if the operation is a difference of positive numbers (Algorithm 16.2), consider two cases:

- if $e_1 - e_2 \geq 2$, then $r = s_1 - s_2/(B^{e1-e2}) > 1 - B/B^2 = 1 - 1/B \geq 1/B$ (as $B \geq 2$), so that the number k of leading zeroes is equal to 0 or 1, and the normalization operation (if necessary, i.e., $k = 1$) is accomplished by multiplying the result by B;
- if $e_1 - e_2 \leq 1$, then the result before normalization is either

$$r_0.r_{-1}r_{-2}r_{-3} \ldots r_{-p}|r_{-(p+1)}00 \ldots \quad (e_1 - e_2 = 1)$$

or

$$r_0.r_{-1}r_{-2}r_{-3} \ldots r_{-p}|000 \ldots \quad (e_1 - e_2 = 0)$$

A consequence of the preceding analysis is that the result after normalization can be either

$$r_0.r_{-1}r_{-2}r_{-3} \ldots r_{-p}|r_{-(p+1)}r_{-(p+2)}r_{-(p+3)} \cdots$$

$$\text{(no normalization operation),} \tag{16.26}$$

or

$$r_1.r_0r_{-1}r_{-2} \ldots r_{-p+1}|r_{-p}r_{-(p+1)}r_{-(p+2)} \cdots \quad \text{(divide by B),} \tag{16.27}$$

or

$$r_{-1}.r_{-2}r_{-3}r_{-4} \ldots r_{-(p+1)}|r_{-(p+2)}r_{-(p+3)}r_{-(p+4)} \cdots \quad \text{(multiply by B),} \tag{16.28}$$

or

$$r_{-k}.r_{-(k+1)}r_{-(k+2)} \ldots r_{-p}r_{-(p+1)}0 \ldots 0|00 \ldots$$

$$\text{(multiply by} B^k \text{ where } k > 1). \tag{16.29}$$

For executing a rounding operation, the worst case is (16.28). In particular, for executing Algorithm 16.8, it is necessary to know

the value of $s_1 = r_{-1}.r_{-2} \, r_{-3} \, r_{-4} \ldots r_{-(p + 1)}$,
the value of $r_{-(p + 2)}$,
whether $s_2 = 0.00 \ldots 0 \mid 0 \; r_{-(p + 3)} \, r_{-(p + 4)} \ldots$ is equal to 0, or not.

The conclusion is that the result r of an operation, before normalization, must be computed in the form

$$r \cong r_1 r_0.r_{-1}r_{-2}r_{-3}\ldots r_{-p}|r_{-(p+1)}r_{-(p+2)}T,$$

that is with two *guard digits* $r_{-(p+1)}$ and $r_{-(p+2)}$, and an additional *sticky digit T* equal to 0 if all the other digits ($r_{-(p+3)}$, $r_{-(p+4)}$, ...) are equal to 0, and equal to any positive value otherwise.

After normalization, the significand will be obtained in the following general form:

$$s \cong 1.s_{-1}s_{-2}s_{-3}\ldots s_{-p}|s_{-(p+1)}s_{-(p+2)}s_{-(p+3)}.$$

The new version of Algorithm 16.8 is the following:

Algorithm 16.9 Round to the Nearest, Tie to Even, Second Version

```
s1:=1.s(-1) s(-2)...s(-p);
if s(-(p+1))<B/2 then round:=s1;
elsif s(-(p+1))>B/2 then round:=s1+ulp;
elsif s(p+2)>0 or s(p+3)>0 then round:=s1+ulp;
elsif s(-p) mod 2=0 then round:=s1;
else round:=s1+ulp;
end if;
```

16.5 ADDER-SUBTRACTOR

An adder-subtractor based on Algorithm 16.4 will now be synthesized. It is made up of four parts, namely, *alignment, addition, normalization*, and *rounding*.

16.5.1 Alignment

The *alignment* circuit implements the three first lines of the algorithm, that is,

```
if operation=1 then sign2:=1 - sign2; end if;
if e1<e2 then swap(sign1, sign2); swap(s1, s2); swap (e1, e2);
end if;
e:=e1; s2:=s2/B**(e1-e2); sign:=sign1;
```

An example of the implementation is shown in Figure 16.1. The principal component is a shifter.

Given a $(2.p + 4)$-component vector

$$[a(0)a(-1)a(-2)\ldots a(-(2.p+3))],$$

the shifter generates a $(2.p + 4)$-component output vector

$$[00\ldots 0a(0)a(-1)\ldots a(d-(2.p+3))], \quad \text{where } 0 \leq d \leq p+3.$$

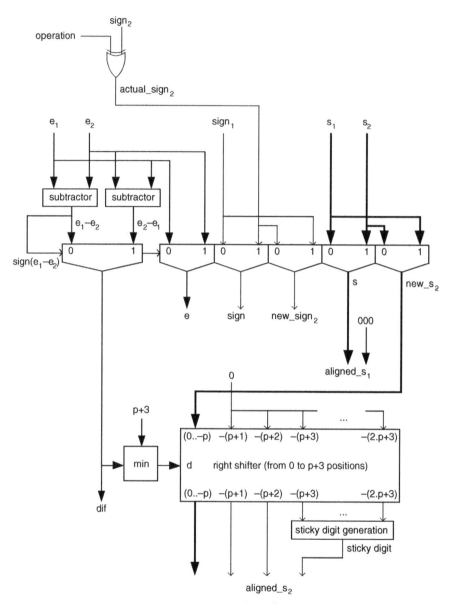

Figure 16.1 Alignment circuit.

The sticky-digit circuit generates an output value 1 if at least one of its inputs is positive. If $B = 2$, the sticky-digit circuit is an OR circuit. Observe that if $e_1 - e_2$ is equal to $p + 3$, then the shifter output is equal to

```
[0 0..0 new_s₂(0) new_s₂(-1)..new_s₂(-p)].
```

Taking into account that `new_s₂` is either s_1 or s_2, i.e. a normalized significand, `new_s₂(0)` is positive. Thus the sticky digit is equal to 1 and the value of `aligned_s₂` is

```
[0 0..0 0 1].
```

If $e_1 - e_2$ were greater than $p + 3$, the value of `aligned_s₂` should be the same, so that it is not necessary to shift `new_s₂` more than $p + 3$ positions.

16.5.2 Additions

Depending on the respective signs of the aligned operands, one of the following operations must be executed:

- if they have the same sign, the sum `aligned_s₁ + aligned_s₂` must be computed;
- if they have different signs, the difference `aligned_s₁ - aligned_s₂` is computed, and if the difference is negative, the alternative difference `aligned_s₂ - aligned_s₁` must be computed.

In the circuit of Figure 16.2 two additions are performed in parallel:

```
result=aligned_s₁±aligned_s₂,
```

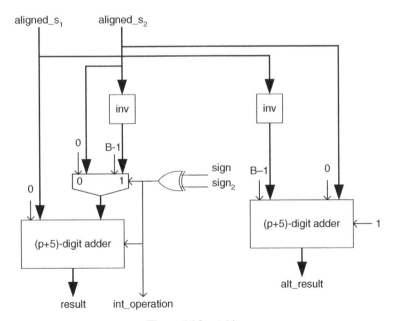

Figure 16.2 Adders.

where the actual operation is selected with the signs of the operands, and

```
alt_result=aligned_s₂-aligned_s₁.
```

16.5.3   Normalization

The *normalization* circuit executes the following part of Algorithm 16.4:

```
if sign xor sign=0 then
  s:=s1+s2;
  if s>=B then e:=e+1; s:=s/B; end if;
else
  if (e1=e2) and (s1<s2) then swap(s1, s2); sign:=1-sign;
  end if;
  s:=s1-s2;
  leading_zeroes(s, k);
  s:=s*(B**k); e:=e-k;
end if;
```

If the number of leading zeroes is greater than $p + 3$, that is, $s_1 - s_2 < B^{-(p+2)}$, then $s_2 > s_1 - B^{-(p+2)}$. If e_1 were greater than e_2, then $s_2 \leq (B - ulp)/B = 1 - B^{-(p+1)}$ so that $1 - B^{-(p+1)} \geq s_2 > s_1 - B^{-(p+2)} \geq 1 - B^{-(p+2)}$, that is, $B^{-(p+1)} < B^{-(p+2)}$: impossible! Thus the only case where the number of leading zeroes can be greater than $p + 3$ is when $e_1 = e_2$ and $s_1 = s_2$. If more than $p + 3$ leading 0's are detected in the circuit of Figure 16.3, a zero_flag is raised.

As the arithmetic operations have already been performed (*addition* circuit, Figure 16.2), it remains to execute the following algorithm where operation is the internal operation computed in Figure 16.2:

```
if operation=0 then
  s:=result;
  if s>=B then e:=e+1; s:=s/B; end if;
else
  if (e1=e2) and (s1<s2) then s:=alt_result; sign:=1-sign;
  else s:=result; end if;
  leading_zeroes(s, k);
  s:=s*(B**k); e:=e-k;
end if;
```

A possible implementation is shown in Figure 16.3.

16.5.4   Rounding

An example of the *rounding* circuit implementation is shown in Figure 16.4. If the *round to the nearest, tie to even* method is used (Algorithm 16.8), the block named *rounding decision* computes the following Boolean function *decision*:

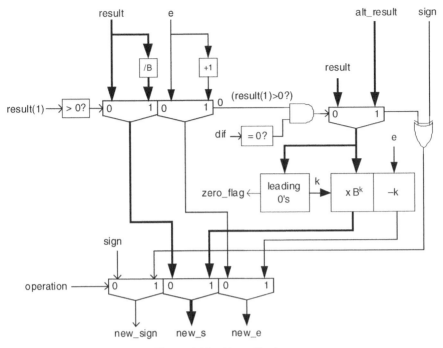

Figure 16.3 Normalization.

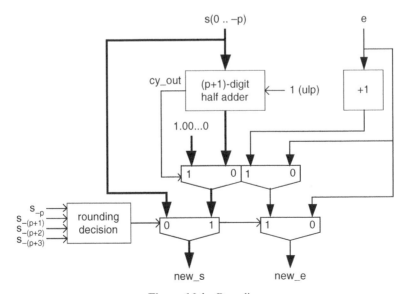

Figure 16.4 Rounding.

```
if s(-(p+1))<B/2 then decision:=0;
elsif s(-(p+1))>B/2 then decision:=1;
elsif (s(-(p+1))=B/2) and (s(-(p+2))>0) and (s(-(p+3))>0)
then decision:=1;
elsif (s(-(p+1))=B/2) and (s(-(p+2))=0) and (s(-(p+3))=0) and
((s(-p) mod 2)=0) then decision:=0;
elsif (s(-(p+1))=B/2) and (s(-(p+2))=0) and (s(-(p+3))=0) and
((s(-p) mod 2)=1) then decision:=1;
end if;
```

Example 16.7 (Complete VHDL code available.) Generate the VHDL model of a
generic floating-point adder-subtractor. It is made up of four blocks:

1. Alignment (Figure 16.1):

```
entity alignment is
port (
  sign1, sign2, operation: in std_logic;
  e1, e2: in integer;
  s1, s2: in digit_vector(0 downto -p);
  dif: inout natural;
  sign, new_sign2: out std_logic;
  e: out natural;
  aligned_s1, aligned_s2: out digit_vector(0 downto -(p+3))
);
end alignment;

architecture behavior of alignment is
  signal actual_sign2: std_logic;
  signal s, new_s2: digit_vector(0 downto -p);
  signal shift_length: natural;
  signal sticky: digit;
begin
  actual_sign2<=operation xor sign2;
  swap: process(sign1, actual_sign2, e1, e2, s1, s2, s)
  begin
    if e1<e2 then
      dif<=e2-e1; e<=e2; sign<=actual_sign2;
      new_sign2<=sign1; s<=s2; new_s2<=s1;
    else
      dif<=e1 - e2; e<=e1; sign<=sign1;
      new_sign2<=actual_sign2; s<=s1; new_s2<=s2;
    end if;
    aligned_s1(-(p+1))<=0; aligned_s1(-(p+2))<=0;
    aligned_s1(-(p+3))<=0;
    for i in 0 downto -p loop aligned_s1(i)<=s(i); end loop;
```

```
  end process swap;
  barrel_shifter: process(dif, shift_length, new_s2, sticky)
    variable a: digit_vector(0 downto -(2*p+3));
    variable acc_or: digit;
  begin
    for i in -(p+1) downto -(2*p+3) loop a(i):=0; end loop;
    for i in 0 downto -p loop a(i):=new_s2(i); end loop;
    if dif<p+3 then shift_length<=dif;
    else shift_length<=p+3; end if;
    if shift_length>0 then
      for j in 1 to shift_length loop
        for i in -(2*p+3) to -1 loop a(i):=a(i+1); end loop;
        a(0):=0;
      end loop;
    end if;
    acc_or:=0;
    for i in -(p+3) downto -(2*p+2) loop
      if (a(i)>0) or (acc_or>0) then acc_or:=1; end if;
    end loop;
    sticky<=acc_or;
    aligned_s2<=a(0 downto -(p+2))&sticky;
  end process barrel_shifter;
end behavior;
```

2. Addition (Figure 16.2):

```
entity addition is
port (
  sign, sign2: in std_logic;
  aligned_s1, aligned_s2: in digit_vector(0 downto -(p+3));
  int_operation: inout std_logic;
  result, alt_result: out digit_vector(1 downto -(p+3))
);
end addition;

architecture rtl of addition is
  signal long_s, long_s2: digit_vector(1 downto -(p+3));
  signal inv_s, inv_s2: digit_vector(1 downto -(p+3));
  signal carry1: mybit_vector(1 downto -(p+3));
  signal carry2: mybit_vector(1 downto -(p+3));
begin
  int_operation<=sign xor sign2;
  long_s<=0&aligned_s1; long_s2<=0&aligned_s2;
  inverters1: for i in -(p+3) to 1 generate
    inv_s2(i)<=B-1-long_s2(i) when int_operation='1'
    else long_s2(i);
  end generate;
```

```vhdl
  inverters2: for i in -(p+3) to 1 generate
  inv_s(i)<=B-1-long_s(i);
  end generate;
  carry1(-(p+3))<=int_operation;
  first_adder: for i in -(p+3) to 0 generate
    carry1(i+1)<='1' when long_s(i)+inv_s2(i) +
    conv_integer(carry1(i))>B-1 else '0';
    result(i)<=(long_s(i)+inv_s2(i) +
    conv_integer(carry1(i))) mod B;
  end generate;
  result(1)<=(long_s(1)+inv_s2(1)+conv_integer(carry1(1)))
  mod B;
  carry2(-(p+3))<='1';
  second_adder: for i in -(p+3) to 0 generate
    carry2(i+1)<='1' when inv_s(i)+long_s2(i) +
    conv_integer(carry2(i))>B-1 else '0';
    alt_result(i)<=(inv_s(i)+long_s2(i) +
    conv_integer(carry2(i))) mod B;
  end generate;
  alt_result(1)<=(inv_s(1)+long_s2(1)+
  conv_integer(carry2(1))) mod B;
end rtl;
```

3. Normalization (Figure 16.3):

```vhdl
entity normalization is
port (
  sign, operation: in std_logic;
  e, dif: in natural;
  result, alt_result: in digit_vector(1 downto -(p+3));
  new_sign, zero_flag: out std_logic;
  new_s: out digit_vector(0 downto -(p+3));
  new_e: out natural
);
end normalization;

architecture behavior of normalization is
  signal result_div_B, s1, s, s2: digit_vector(0 downto -
  (p+3));
  signal exp1, k, exp2: natural;
  signal sign1, sign2: std_logic;
begin
  divide_by_B: for i in -(p+3) to 0 generate
    result_div_B(i)<=result(i+1);
  end generate;
  s1<=result(0 downto -(p+3)) when result(1)=0 else
  result_div_B;
  exp1<=e when result(1)=0 else e+1; sign1<=sign;
  s<=alt_result(0 downto -(p+3)) when (dif=0) and
```

```
  (result(1)>0) else result(0 downto -(p+3));
  leading_zeroes: process(s)
    variable var_k: natural;
  begin
    var_k:=0;
    for i in 0 downto -(p+3) loop
      if s(i)>0 then exit; end if;
      var_k:=var_k+1;
    end loop;
    if var_k=p+4 then zero_flag<='1'; else zero_flag<='0';
    end if;
    k<=var_k;
  end process leading_zeroes;
  shift_k: process (s, k)
    variable a: digit_vector(0 downto -(p+3));
  begin
    a:=s;
    if k>0 then
      for i in 1 to k loop
        for i in 0 downto -(p+2) loop a(i):=a(i-1); end loop;
      a(-(p+3)):=0;
      end loop;
    end if;
    s2<=a;
  end process shift_k;
  exp2<=e-k;
  sign2<=not(sign) when (dif=0) and (result(1)>0) else sign;
  new_s<=s1 when operation='0' else s2;
  new_e<=exp1 when operation='0' else exp2;
  new_sign<=sign1 when operation='0' else sign2;
end behavior;
```

  4. Rounding (Figure 16.4):

```
entity rounding is
port (
  s: in digit_vector(0 downto -(p+3));
  e: in natural;
  new_s: out digit_vector(0 downto -p);
  new_e: out natural
);
end rounding;

architecture behavior of rounding is
begin
    process(s)
      variable carry: digit_vector(1 downto -p);
      variable sum: digit_vector(0 downto -p);
```

```
   begin
     if s(-(p+1))<B_div_2 then new_s<=s(0 downto -p);
       new_e<=e;
     elsif (s(-(p+1))>B_div_2) or (s(-(p+2))>0) or
       (s(-(p+3))>0) or (s(-p) mod 2=1) then
       --plus ulp
       carry(-p):=1;
       for i in -p to 0 loop
         if s(i)+carry(i)>B-1 then carry(i+1):=1;
         else carry(i+1):=0; end if;
         sum(i):=(carry(i)+s(i)) mod B;
       end loop;
       ------
       if carry(1)=1 then
         new_s(0)<=1;
         for i in -1 downto -p loop new_s(i)<=0; end loop;
         new_e<=e+1;
       else new_s<=sum; new_e<=e; end if;
     else new_s<=s(0 downto -p); new_e<=e;
     end if;
   end process;
end behavior;
```

It remains to assemble the four blocks (Figure 16.5):

```
entity adder_subtractor is
port (
  sign1, sign2, operation: in std_logic;
  e1, e2: in integer;
  s1, s2: in digit_vector(0 downto -p);
  sign, zero_flag: out std_logic;
  e: out natural;
  s: out digit_vector(0 downto -p)
);
end adder_subtractor;

architecture circuit of adder_subtractor is
  component alignment...end component;
  component addition...end component;
  component normalization...end component;
  component rounding...end component;
  signal sign_a, sign2_a, int_operation: std_logic;
  signal e_a, dif, e_n: natural;
  signal aligned_s1, aligned_s2, s_n:
    digit_vector(0 downto -(p+3));
  signal result, alt_result:
    digit_vector(1 downto -(p+3));
```

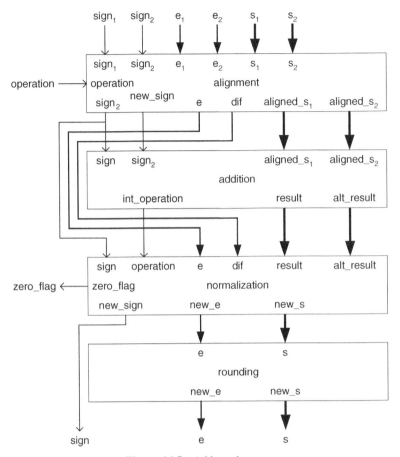

Figure 16.5 Adder-subtractor.

```
begin
  alignment_component: alignment port map (sign1, sign2,
    operation, e1, e2, s1, s2, dif, sign_a, sign2_a, e_a,
    aligned_s1, aligned_s2);
  addition_component: addition port map (sign_a, sign2_a,
    aligned_s1, aligned_s2, int_operation, result, alt_result);
  normalization_component: normalization port map (sign_a,
    int_operation, e_a, dif, result, alt_result,
    sign, zero_flag, s_n, e_n);
  rounding_component: rounding port map (s_n, e_n, s, e);
end circuit;
```

16.6 MULTIPLIER

A basic multiplier deduced from Algorithm 16.5 is shown in Figure 16.6. The rounding circuit is the same as in the case of the adder-subtractor (Figure 16.4).

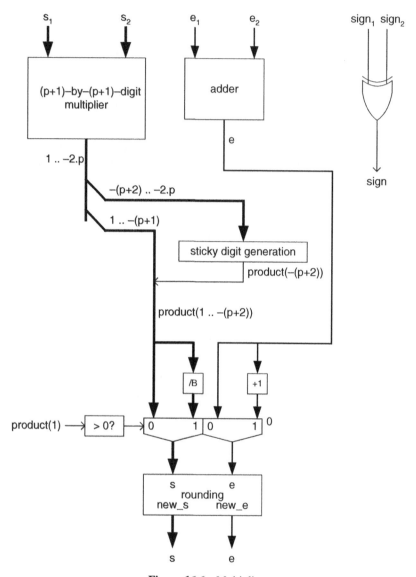

Figure 16.6   Multiplier.

Example 16.8   (Complete VHDL code available.) Generate the VHDL model of a generic floating-point multiplier. It is made up of four blocks:

1. *Multiplication.* The multiplication circuit corresponds to a $(p + 1)$-by-$(p + 1)$ multiplier, an adder and a XOR gate—Figure 16.6—and generates the exact value of the product. Any type of multiplier can be used (Chapter 12). In this model, a simple parallel multiplier has been used:

```
entity multiplication is
port (
  s1, s2: in digit_vector(0 downto -p);
  sign1, sign2: in std_logic;
  e1, e2: in integer;
  s: out digit_vector(1 downto -2*p);
  sign: out std_logic;
  e: out integer
);
end multiplication;

architecture circuit of multiplication is
  component basic_base_B_mult...end component;
  ...
end circuit;
```

2. *Generation of the guard digits*. This block computes the sticky digit and concatenates its value with positions 1 down to $-(p + 1)$ of the exact product:

```
entity guard_digits is
port (
  s: in digit_vector(1 downto -2*p);
  product: out digit_vector(1 downto -(p+2))
);
end guard_digits;

architecture behavior of guard_digits is
begin
  process(s)
    variable acc_or: digit;
  begin
    acc_or:=0;
    for i in -(p+2) downto -2*p loop
      if (s(i)>0) or (acc_or>0) then acc_or:=1; end if;
    end loop;
    product<=s(1 downto -(p+1))&acc_or;
  end process;
end behavior;
```

3. *Normalization*. This block updates the significand as well as the exponent if the value of product (Figure 16.6) is greater than or equal to B:

```
entity normalization is
port (
  e: in natural;
  product: in digit_vector(1 downto -(p+2));
  new_s: out digit_vector(0 downto -(p+3));
  new_e: out natural
);
end normalization;
```

```
architecture rtl of normalization is
  signal product_div_B: digit_vector(0 downto -(p+3));
begin
  divide_by_B: for i in -(p+3) to 0 generate product_div_
  B(i)<=product(i+1); end generate;
  new_s<=product(0 downto -(p+2))&0 when product(1)=0 else
  product_div_B;
  new_e<=e when product(1)=0 else e+1;
end rtl;
```

4. The rounding block is the same as before (Figure 16.4):

```
entity rounding is
port (
  s: in digit_vector(0 downto -(p+3));
  e: in natural;
  new_s: out digit_vector(0 downto -p);
  new_e: out natural
);
end rounding;
```

It remains to assemble the four blocks:

```
entity fp_multiplier is
port (
  sign1, sign2: in std_logic;
  e1, e2: in integer;
  s1, s2: in digit_vector(0 downto -p);
  sign: out std_logic;
  e: out natural;
  s: out digit_vector(0 downto -p)
);
end fp_multiplier;

architecture circuit of fp_multiplier is
  component multiplication...end component;
  component guard_digits...end component;
  component normalization...end component;
  component rounding...end component;
  signal e_m, e_n: natural;
  signal s_m: digit_vector(1 downto -2*p);
  signal s_g: digit_vector(1 downto -(p+2));
  signal s_n: digit_vector(0 downto -(p+3));
begin
  multiplication_component: multiplication
    port map (s1, s2, sign1, sign2, e1, e2, s_m, sign, e_m);
  guard_digits_component: guard_digits port map (s_m, s_g);
  normalization_component: normalization
```

```
    port map (e_m, s_g, s_n, e_n);
  rounding_component: rounding port map (s_n, e_n, s, e);
end circuit;
```

If a carry-save multiplier is used, the part of the circuit that generates the value of product can be modified (Figure 16.7). The multiplier generates two $(2.p + 2)$-digit numbers u and v (stored-carry encoding of the product). Then it remains to generate the carry cy corresponding to the position number $-(p + 1)$ as well as the sticky_digit.

The computation of cy can be performed with any one of the methods described in Chapter 11. The sticky_digit can be generated directly from the stored-carry representation (Chapter 8 of [ERC2004]). For that purpose, observe that the equality

$$u(-(p+2), -(p+3), \ldots, -2.p) + v(-(p+2), -(p+3), \ldots, -2.p)$$

$$= 0 \bmod B^{p-1}$$

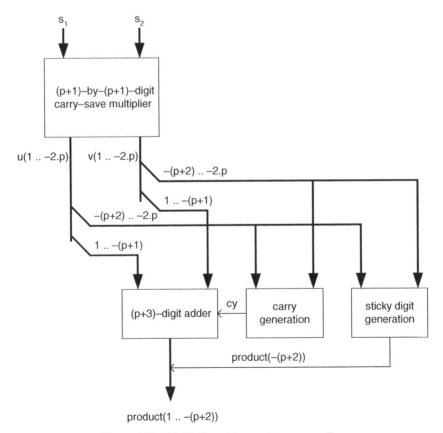

Figure 16.7 Multiplier with stored-carry encoding.

is equivalent to

$$u(-(p+2),\ -(p+3),\ldots,\ -2.p) + v(-(p+2),\ -(p+3),\ldots,\ -2.p)$$
$$+ (B^{p-1} - 1) = (B^{p-1} - 1) \bmod B^{p-1}, \qquad (16.30)$$

where

$$B^{p-1} - 1 = (B-1, B-1, \ldots, B-1).$$

First encode the result of (16.30) in stored-carry form, that is,

$$u(-(p+2),\ -(p+3),\ldots,\ -2.p) + v(-(p+2),\ -(p+3),\ldots,\ -2.p)$$
$$+ (B^{p-1} - 1) = s(-(p+2),\ -(p+3),\ldots,\ -2.p)$$
$$+ c(-(p+2),\ -(p+3),\ldots,\ -2.p).$$

Relation (16.30) is equivalent to

$$s(-(p+2),\ -(p+3),\ldots,\ -2.p) + c(-(p+2),\ -(p+3),\ldots,\ -2.p)$$
$$= (B-1, B-1, \ldots, B-1),$$

and the preceding relation only holds if, for every position i, the sum $s(i) + c(i)$ is equal to $B-1$. Thus the sticky digit is equal to 0 if, and only if,

$$s(i) + c(i) = B - 1, \quad \forall i \text{ in } \{-(p+2),\ -(p+3),\ldots,\ -2.p\}.$$

The corresponding circuit is shown in Figure 16.8. The comp block works as follows:

$$\text{if } a + b = B - 1 \text{ then } comp(a, b) = 1; \text{ else } comp(a, b) = 0.$$

If $B = 2$ the comp circuit is a 2-input XOR gate.

16.7 DIVIDER

A basic divider deduced from algorithm 16.6 is shown in Figure 16.9. The inputs of the $(p+1)$-digit divider are s_1/B and s_2 (Comment 6.1), so that the dividend is smaller than the divisor. The precision is chosen equal to $p + 3$ digits. Thus (see Section 6.1) the outputs quotient and remainder satisfy the relation

$$(s_1/B).B^{p+3} = s_2.q + r, \quad \text{where } r < s_2,$$

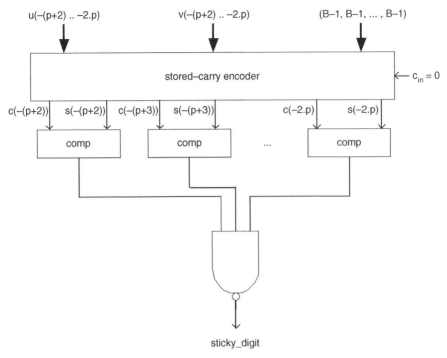

Figure 16.8 Sticky digit generation.

that is,

$$s_1/s_2 = q.B^{-(p+2)} + (r/s_2).B^{-(p+2)} \quad \text{where } (r/s_2).B^{-(p+2)} < B^{-(p+2)}.$$

The sticky digit is equal to 1 if $r > 0$ and to 0 if $r = 0$. The final approximation of the exact result is

$$\text{quotient} = q.B^{-(p+2)} + \text{sticky\_digit}. B^{-(p+3)}.$$

Example 16.9 (Complete VHDL code available.) Generate the VHDL model of a generic floating-point divider. It is made up of three blocks:

1. *Division.* This block includes the $(p+2)$-digit divider, the subtractor, the xor gate, and the sticky digit generation circuit. Any type of divider can be used (Chapter 13). In this model a modified (dividend $= s_1/B$) restoring divider has been used:

```
entity division is
port (
   s1, s2: in digit_vector(0 downto -p);
   sign1, sign2: in std_logic;
   e1, e2: in integer;
```

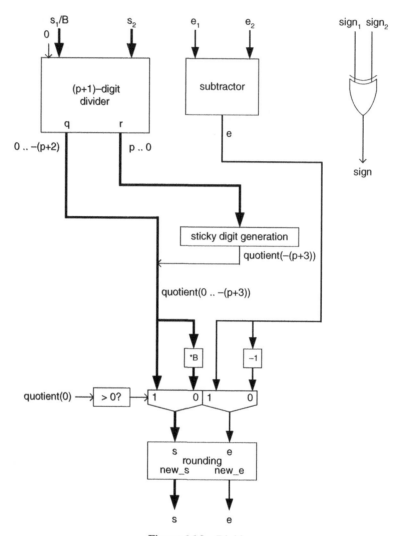

Figure 16.9   Divider.

```
   s: out digit_vector(0 downto -(p+3));
   sign: out std_logic;
   e: out integer
);

end division;

architecture circuit of division is
   component modif_div_rest_baseB...end component;
   ...
end circuit;
```

2. *Normalization.* This circuit multiplies the quotient by *B*, and decreases the exponent accordingly, if the quotient is smaller than 1:

```
entity normalization is
port (
  e: in natural;
  s: in digit_vector(0 downto -(p+3));
  new_s: out digit_vector(0 downto -(p+3));
  new_e: out natural
);
end normalization;
architecture rtl of normalization is
  signal quotient_by_B: digit_vector(0 downto -(p+3));
begin
  multiply_by_B: for i in -(p+2) to 0 generate
    quotient_by_B(i)<=s(i-1);
  end generate;
  quotient_by_B(-(p+3))<=0;
  new_s<=quotient_by_B when s(0)=0 else s;
  new_e<=e-1 when s(0)=0 else e;
end rtl;
```

3. *Rounding.* The rounding circuit is the same as before, or even simpler (it is not necessary to normalize after rounding):

```
entity rounding is
port (
  s: in digit_vector(0 downto -(p+3));
  e: in natural;
  new_s: out digit_vector(0 downto -p);
  new_e: out natural
);
end rounding;
```

It remains to assemble the three parts:

```
entity fp_divider is
port (
  sign1, sign2: in std_logic;
  e1, e2: in integer;
  s1, s2: in digit_vector(0 downto -p);
  sign: out std_logic;
  e: out natural;
  s: out digit_vector(0 downto -p)
);
end fp_divider;
```

```
architecture circuit of fp_divider is
  component division...end component;
  component normalization...end component;
  component rounding...end component;
  signal e_d, e_n: natural;
  signal s_d, s_n: digit_vector(0 downto -(p+3));
begin
  divider_component: division
  port map (s1, s2, sign1, sign2, e1, e2, s_d, sign, e_d);
  normalization_component: normalization port map (e_d, s_d,
  s_n, e_n);
  rounding_component: rounding port map (s_n, e_n, s, e);
end circuit;
```

16.8 SQUARE ROOT

A basic square-rooter deduced from Algorithm 16.7 is shown in Figure 16.10. If e_1 is even, the square-rooter input is

$$s' = (s_1.B^p).B^{p+4} = s_1.B^{2.(p+2)}, \qquad (16.31)$$

and if e_1 is odd

$$s' = (s_1.B^p).B^{p+3} = (s_1/B).B^{2.(p+2)}. \qquad (16.32)$$

In both cases s' can be represented as a $(2.p + 5)$-digit integer. The square-root algorithms (Chapter 7) generate Q and R such that

$$s' = Q^2 + R, \quad \text{where } R \le 2.Q, \qquad (16.33)$$

so that $Q^2 \le s'$ and $(Q + 1)^2 > s'$. Observe that Q is a $(p + 3)$-digit integer and R a $(p + 4)$-digit integer; then, according to (16.33),

$$s'.B^{-2.(p+2)} = (Q.B^{-(p+2)})^2 + R.B^{-2.(p+2)},$$

so that $(Q.B^{-(p+2)})^2 \le s'.B^{-2.(p+2)}$ and $(Q.B^{-(p+2)} + B^{-(p+2)})^2 > s'.B^{-2.(p+2)}$.

Thus $Q.B^{-(p+2)}$ is the square root of either s_1 (16.31) or s_1/B (16.32), with a precision of $p + 2$ fractional digits. The sticky digit is equal to 1 if $R > 0$ and to 0 if $R = 0$. The final approximation of the exact result is

$$\text{root} = Q.B^{-(p+2)} + \text{sticky\_digit}. \, B^{-(p+3)}.$$

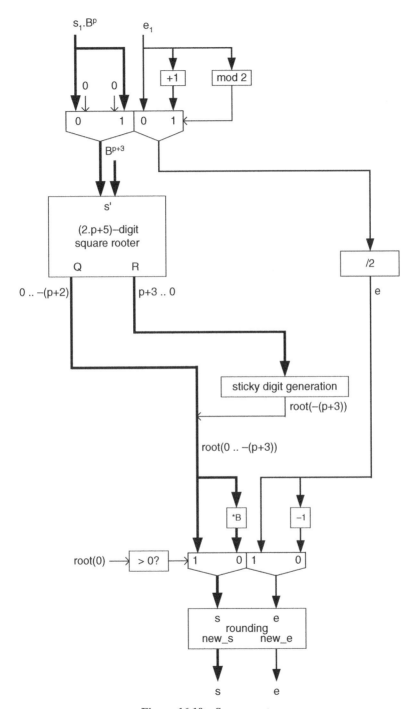

Figure 16.10 Square-rooter.

16.9 COMMENTS

1. Unless the result r of an arithmetic operation is known to be included between either $-x_{max}$ and $-x_{min}$, or x_{min} and x_{max}, an *underflow* ($|r| < x_{min}$) or *overflow* ($|r| > x_{max}$) could exist. The corresponding circuits should include additional components for detecting this type of situation. For example, the adder-subtractor of Figure 16.5 generates the signal `zero_flag` if the result is equal to 0.

2. A similar comment can be done if some special values (e.g., ∞, $-\infty$) must be represented.

3. The proposed circuits are straightforward implementations of the basic algorithms described in Section 16.2. A lot of improvements have been proposed in order to reduce the latency of the corresponding circuits. See, for example, [ERC2004] and [PAR2000].

4. The exponents e are integers included in some predefined interval $e_{min} \leq e \leq e_{max}$. Any type of representation can be used. For example (already mentioned in Chapter 3), in the *ANSI/IEEE single-precision floating-point system* the exponents are excess-127 integers.

16.10 BIBLIOGRAPHY

[ERC2004] M. Ercegovac and T. Lang, *Digital Arithmetic*. Morgan Kaufmann, San Francisco, 2004.

[OBE1997] S. F. Oberman and M. J. Flynn, Design issues in division and other floating-point operations. *IEEE Trans. Comput.*, **46**(2): 154–161 (1997).

[PAR2000] B. Parhami, *Computer Arithmetic: Algorithms and Hardware Design*. Oxford University Press, New York, 2000.

INDEX

Synthesis of Arithmetic Circuits: FPGA, ASIC, and Embedded Systems
By Jean-Pierre Deschamps, Géry J. A. Bioul, and Gustavo D. Sutter
Copyright © 2006 John Wiley & Sons, Inc.

Printed and bound by CPI Group (UK) Ltd, Croydon, CR0 4YY

27/10/2024

14580331-0003